Analysis, Synthesis, and Design of Chemical Processes

ISBN 0-13-570565-7

9 780135 705650

90000

Analysis, Synthesis, and Design of Chemical Processes

Richard Turton
West Virginia University

Richard C. Bailie
West Virginia University

Wallace B. Whiting
University of Nevada-Reno

Joseph A. Shaeiwitz
West Virginia University

To join a Prentice Hall PTR Internet mailing list, point to:
http://www.prenhall.com/mail_lists/

Prentice Hall, PTR
Upper Saddle River, New Jersey 07458

Library of Congress Cataloging-in-Publication Data

Analysis, synthesis, and design of chemical processes / Richard Turton
    . . . [et al.].
        p.   cm.
    Includes bibliographical references and index.
    ISBN 0-13-570565-7
    1. Chemical processes.   I.   Turton, Richard, 1955-   .
TP155.7.A53   1998
660'.2812—dc21   97-34025
                                                       CIP

Acquisitions editor: Bernard M. Goodwin
Cover design director: Jerry Votta
Manufacturing manager: Julia Meehan
Marketing manager: Betsy Carey
Compositor/Production services: Pine Tree Composition, Inc.

© 1998 by Prentice Hall PTR
Prentice-Hall, Inc.
A Simon & Schuster Company
Upper Saddle River, New Jersey 07458

Prentice Hall books are widely used by corporations and government agencies for training, marketing, and resale.

The publisher offers discounts on this book when ordered in bulk quantities. For more information contact:

Corporate Sales Department
Phone: 800-382-3419
Fax: 201-236-7141
E-mail: corpsales@prenhall.com

Or write:

Prentice Hall PTR
Corp. Sales Dept.
One Lake Street
Upper Saddle River, New Jersey 07458

Printed in the United States of America
10 9 8 7 6 5 4 3 2 1

ISBN: 0-13-570565-7

Prentice-Hall International (UK) Limited, *London*
Prentice-Hall of Australia Pty. Limited, *Sydney*
Prentice-Hall Canada Inc., *Toronto*
Prentice-Hall Hispanoamericana, S.A., *Mexico*
Prentice-Hall of India Private Limited, *New Delhi*
Prentice-Hall of Japan, Inc., *Tokyo*
Simon & Schuster Asia Pte. Ltd., *Singapore*
Editora Prentice-Hall do Brasil, Ltda., *Rio de Janeiro*

Contents

Contents ix

APPENDICES

Preface

This book represents the culmination of many years of teaching experience in the design program at West Virginia University. Although this program has evolved over more than thirty years and is still evolving, it is fair to say that the current program has gelled over the last ten years through the concerted effort of the authors to integrate design throughout the undergraduate curriculum in chemical engineering.

We view design as the focal point of chemical engineering practice. Far more than the development of a set of specifications for a new chemical plant, design is that creative activity through which engineers continuously improve the operation of facilities to create products that enhance the quality of life. Whether developing the grass roots plant, proposing and guiding process modifications, or troubleshooting and implementing operational strategies for existing equipment, engineering design requires a broad spectrum of knowledge and intellectual skills to be able to analyze the big picture and the minute details and, most importantly, to know when to concentrate on each.

Our vehicle for helping students develop and hone their design skills is process design rather than plant design, covering synthesis of the entire chemical process through topics relating to the preliminary sizing of equipment, flowsheet optimization, economic evaluation of projects, and the operation of chemical processes. The purpose of this text is to assist chemical engineering students in making the transition from solving well-posed problems in a specific subject to integrating all the knowledge that they have gained in their undergraduate education and applying it to solving open-ended process problems. Many of the "nuts and bolts" issues regarding plant design (for example, what schedule pipe to use for a given stream or what corrosion allowance to use for a vessel in a certain service) are not covered. Although such issues are clearly important to the practicing engineer, several excellent handbooks and textbooks are available to address such problems, and these are cited in the text where they apply.

As a result of our integrated approach to design, we have divided this book into six sections. Section 0, the first chapter in the book, covers the principal diagrams used by chemical engineers. In particular, details of the most important diagram for the analysis of chemical processes are given, namely the Process Flow Diagram (PFD).

Section 1 covers the engineering economic aspects of a process, including the material needed for the Fundamentals of Engineering (FE or EIT) exam required as the first step toward professional registration.

Section 2 looks at the common features of all processes and explains how and why we choose the operating conditions in a process. This section also includes some guidelines for preliminary process design.

Section 3 focuses on the performance of existing processes and equipment. This material is substantially different from that found in most textbooks. We consider equipment that is already built and operating and show how to analyze, evaluate, and modify the performance of the system, including process troubleshooting to determine the cause of a process upset.

Section 4 looks at the synthesis of a chemical process. The minimum information required to simulate a process is covered as are the basics of using a process simulator. This section also covers process optimization and heat integration techniques.

Section 5 addresses the role of the professional engineer in society. Separate chapters on ethics and professionalism; health, safety, and the environment; and oral and written communication cover topics crucial to an engineer's success but sometimes overlooked in design courses. An entire chapter is devoted to addressing some of the common mistakes that students make in written reports.

Finally, three appendices are included. Appendix A gives a series of cost charts for equipment. This cost information is also included in the CAPCOST© program for evaluating fixed capital investment introduced in Chapter 2. Appendix B gives the preliminary design information for four chemical processes: dimethyl ether, acrylic acid, acetone, and heptenes production. This information is used in many of the end-of-chapter problems in the book. These processes can also be used as the starting point for more detailed analyses, for example, optimization studies. Appendix C gives six case study problems suitable for individual or group design projects.

For a one-term design course, we recommend including the following core:

- Chapter 1
- Section 2 (Chapters 6–9)
- Chapter 17
- Chapter 18
- Section 5 (Chapters 20–23)

For programs in which engineering economics is not a prerequisite to the design course, Chapters 2 through 5 should also be included. If students have previously covered engineering economics, Chapter 19 (Optimization) could be substituted.

For the second term of a two-term sequence, we recommend Chapters 10 through 14 (and Chapter 19 if not included in the first design course) plus design projects. If time permits, we strongly recommend Chapter 15 (Regulating Process Conditions) and Chapter 16 (Process Troubleshooting), as these tend to solidify as well as to extend the concepts of Chapters 10 through 14. Section 3 (Chapters 10–16) addresses the analysis of existing processes and mirrors the type of work that an entry-level process engineer will encounter in the first few years of employment at a chemical process facility.

The chapters, however, can be covered in many different sequences, depending on the background of the students entering the design course. At West Virginia University, for example, we cover Chapters 1, 10–16, 2–5, 19, 21, and 20 (in that order) because the students have covered the material of Chapters 6–9, 17, 18, much of 19, 22, and 23 in prerequisite courses. The second semester is devoted almost entirely to a large-group design project. In addition, during the two-semester sequence, we give our students a sequence of individual design projects. Some examples of these projects are given in Appendix C. Additional projects are available from the authors. Projects C.1, C.3, and C.5 cover the analysis of existing processes and should not be assigned without some coverage of Section 3. The other projects (C.2, C.4, and C.6) are open-ended design projects for new processes. These can be given as individual or small-group projects (3–4 students).

We have found that the most effective way both to enhance and to examine student progress is through oral presentations in addition to the submission of written reports. During these oral presentations, individual students or a student group defend their results to a faculty panel.

As design is at its essence a creative, dynamic, challenging, and iterative activity, we welcome feedback on and encourage experimentation with this design textbook. We hope that students and faculty will find the excitement in teaching and learning engineering design that has sustained us over the years.

Finally, we would like to thank those people who have been instrumental to the successful completion of this book. First, thanks are given to all the undergraduate chemical engineering students at West Virginia University over the years, particularly during the period 1987–1997. Their feedback and criticism have been a constant source of ideas and stimulation. Second, we would like to thank those people who have read, criticized, and used parts of this text in the course of its preparation. In particular, we would like to recognize Dr. Mark Stadtherr of the University of Notre Dame and Dr. Susan Montgomery of the University of Michigan for their helpful criticism and support. Finally, on a personal note we (RT, RCB, and WBW) would like to thank our long suffering wives (Becky, Judy, and Patricia) for their continued support, love, and patience throughout this prolonged endeavor.

List of Nomenclature

Symbol	Definition	SI Units
A	Equipment Cost Attribute	
A	Area	m^2
A	Absorption Factor	
A	Annuity Value	$/time
$A/F,i,n$	Sinking Fund Factor	
$A/P,i,n$	Capital Recovery Factor	
BV	Book Value	$
C	Equipment Cost	$
C or c	Molar Concentration	$kmol/m^3$
C_{BM}	Bare Module Cost	$
COM	Cost of Manufacture	$/time
C_p	Heat Capacity	kJ/kg°C or kJ/kmol°C
CCP	Cumulative Cash Position	$
CCR	Cumulative Cash Ratio	
D	Diameter	m
D	Amount Allowed for Depreciation	$
d	Yearly Depreciation Allowance	$/yr
$DCFROR$	Discounted Cash Flow Rate of Return	
DMC	Direct Manufacturing Cost	$/time
$DPBP$	Discounted Payback Period	years
E	Money Earned	$
E_{act} or E	Activation Energy	kJ/kmol
$EAOC$	Equivalent Annual Operating Cost	$/yr
ECC	Equivalent Capitalized Cost	$

f_q	Quantity Factors for Trays	
F	Future Value	$
F	Molar Flowrate	kmol/s
F	Equipment Module Cost Factor	
F	Correction for Multipass Heat Exchangers	
F	Future Value	$
F_d	Drag Force	N/m^2 or kPa
f	Friction Factor	
f	Rate of Inflation	
$F/A,i,n$	Uniform Series Compound Amount Factor	
FCI	Fixed Capital Investment	$
$F/P,i,n$	Single Payment Compound Amount Factor	
FMC	Fixed Manufacturing Costs	$/time
F_{Lang}	Lang Factor	
G	Gas Flowrate	kg/s, kmol/s
GE	General Expenses	$/time
h	Individual Heat Transfer Coefficient	W/m^{2}K
H	Enthalpy or Specific Enthalpy	kJ or kJ/kg
I	Cost Index	
i	Compound Interest	
i'	Effective Interest Rate Including Inflation	
INPV	Incremental Net Present Value	$
IPBP	Incremental Payback Period	years
k	Thermal Conductivity	W/m K
k_o	Pre-exponential Factor for Reaction Rate Constant	depends on molecularity of reaction
K_p	Equilibrium Constant	depends on reaction stoichiometry
k_{reac} or k_i	Reaction Rate Constant	depends of molecularity of reaction
L	Liquid Flowrate	kg/s or kmol/s
$\dot{m}$	Flowrate	kg/s
m	Partition Coefficient (y/x)	
n	Life of Equipment	years
n	Years of Investment	years
N	Number of Trays or Stages	
N	Molar Flowrate	kmol/s
NPSH	Net Positive Suction Head	m of liquid
NPV	Net Present Value	$
N_{toG}	Number of Transfer Units	
OBJ	Objective Function	usually $ or $/time
P	Pressure	bar or kPa
P	Present Value	$

P^*	Vapor Pressure	bar or kPa
$P/A,i,n$	Uniform Series Present Worth Factor	
PBP	Payback Period	year
PC	Project Cost	$
$P/F,i,n$	Single Payment Present Worth Factor	
PVR	Present Value Ratio	
Q or q	Rate of Heat Transfer	W or MJ/h
r	Reaction Rate	$kmol/m^3$ or $kmol/kg$ cat s
R	Gas Constant	kJ/kmol K
R	Residual Funds Needed	$
R	Reflux Ratio	
Re	Reynolds Number	
$ROROI$	Rate of Return on Investment	
$ROROII$	Rate of Return on Incremental Investment	
S	Salvage Value	$
SF	Stream Factor	
t	Time	s, min, h, yr
T	Temperature	K, R, °C, or °F
u	Flow Velocity	m/s
U	Overall Heat Transfer Coefficient	W/m^2K
V	Volume	m^3
$\dot{v}$	Volumetric Flowrate	m^3/s
W or W_S	Work	kJ/kg
WC	Working Capital	$
X	Conversion	
X	Base Case Ratio	
x	Mole or Mass Fraction	
y	Mole or Mass Fraction	
YOC	Yearly Operating Cost	$/yr
YS	Yearly Cash Flow (Savings)	$/yr
z	Distance	m

Greek Symbols

α	Multiplication Cost Factor	
α	Relative Volatility	
ε	Void Fraction	
ε	Pump Efficiency	
η	Selectivity	
λ	Heat of Vaporization	kJ/kg
μ	Viscosity	kg/m s
ξ	Selectivity	

ρ	Density	kg/m^3
τ	Space Time	s

Subscripts

1	Base time
2	Desired time
a	Required attribute
ACT	Actual
Aux	Auxiliary Buildings
b	Base Attribute
BM	Bare Module
Cont	Contingency
d	Without Depreciation
E	Contractor Engineering Expenses
eff	Effective Interest
eq	Equivalent
Fee	Contractor Fee
FTT	Transportation, etc.
GR	Grass Roots
k	Year
L	Installation Labor
L	Without Land Cost
m	Number of Years
M	Materials for Installation
M	Material Cost Factor
max	Maximum
min	Minimum
nom	Nominal Interest
O	Construction Overhead
Off	Offsites and Utilities
OL	Operating Labor
P	Equipment at Manufacturer's Site (Purchased)
P	Pressure Cost Factor
RM	Raw Materials
rxn	Reaction
s	Simple Interest
Site	Site Development
TM	Total Module
UT	Utilities
WT	Waste Treatment

Superscripts

DB	Double Declining Balance Depreciation
o	Cost Includes Pressure and Material Factors

SL	Straight Line Depreciation
$SOYD$	Sum of the Years Depreciation
´	Includes Effect of Inflation on Interest

Additional Nomenclature

Table 1.2	Convention for Specifying Process Equipment
Table 1.3	Convention for Specifying Process Streams
Table 1.7	Abbreviations for Equipment and Materials of Construction
Table 1.10	Convention for Specifying Instrumentation and Control Systems

0

Chemical Process Diagrams

The first section of this book consists of a single chapter: Chapter 1 covers the important diagrams routinely used by chemical engineers to help design and understand chemical processes. We start the book with this section and chapter because nearly all the technical information that we present in the remainder of the book is, in some way, related to the three principal diagrams that are presented in Chapter 1. These three diagrams are the block flow diagram (BFD), process flow diagram (PFD), and the piping and instrument diagram (P&ID).

Design is an evolutionary process that can be represented by the sequence of process diagrams describing it. This evolution is illustrated in Figure 0.1. A chemical engineer might begin the design process by sketching out a very crude block flow diagram in which only the feed and product streams are identified. This diagram is sometimes known as an *input-output* diagram. From this simple starting point, the engineer might then break down the process into its basic functional elements such as the reaction and separation sections. The engineer would also identify recycle streams and consider additional unit operations in order to obtain the desired temperature and pressure conditions required by the reactor and separator sections. By identifying these basic elements, a *generic process block flow diagram* can be drawn. After material balance calculations have been performed, a *preliminary block flow diagram* (BFD) can be drawn, and estimates of process flows can be made. As more complete material and energy balances are performed and preliminary equipment specifications are calculated, the diagram becomes more sophisticated, and a *process flow diagram* (PFD) comes into being. Finally, as the mechanical and instrumentation details of the process are considered, this information is recorded in the *piping and instrumentation diagram* (P&ID). The existence of the latter three levels of diagrams (BFD, PFD, and P&ID)

1

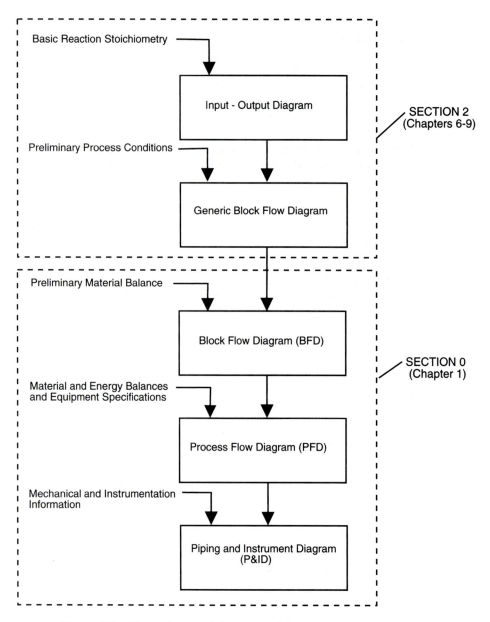

Figure 0.1   The evolution of the principal flow diagrams in a chemical process.

is universal among the chemical process industry, and it is extremely important that the chemical engineering student be familiar with them and be able to use them without fear.

The input-output diagram, the generic process flow diagram that deals with the structure (in terms of the topology of the process), and the conditions chosen for a given piece of equipment or unit operation, are not covered in Chapter 1. This information is addressed in Section 2. Only the information contained in the three principal flow diagrams and the general conventions used in constructing these diagrams are considered in this section. However, to obtain a full understanding of the evolutionary process of design that results in these diagrams, it is recommended that Section 2 also be consulted.

Chapter 1: Essential Flow Diagrams for Understanding Processes

The technical diagrams commonly used by chemical engineers are presented. These diagrams include the block flow diagram (BFD), the process flow diagram (PFD), and the piping and instrumentation diagram (P&ID). A standard method for presenting a PFD is given and illustrated using a process to produce benzene via the catalytic hydrodealkylation of toluene.

1

Essential Flow Diagrams for Understanding Processes

The chemical process industry (CPI) is involved in the production of a wide variety of products that improve the quality of our lives and generate income for the companies and their stockholders. In general, chemical processes are complex and chemical engineers in industry encounter a variety of chemical process flow diagrams. These processes often involve substances of high chemical reactivity, high toxicity, and high corrosivity operating at high pressures and temperatures. These characteristics can lead to a variety of potentially serious consequences including explosions, environmental damage, and threats to people's health. It is essential that errors or omissions resulting from missed communication between persons and/or groups involved in the design and operation do not occur when dealing with chemical processes. Visual information is the clearest way to present material and is least likely to be misinterpreted. For these reasons, it is essential that chemical engineers be able to formulate appropriate process diagrams and be skilled in analyzing and interpreting diagrams prepared by others.

> **The most effective way of communicating information about a process is through the use of flow diagrams.**

This chapter presents and discusses the more common flow diagrams encountered in the chemical process industry. These diagrams evolve from the time a process is conceived in the laboratory through the design, construction, and the many years of plant operation. The most important of these diagrams are described and discussed in this chapter.

The following narrative is taken from Kauffman [1 adapted by permission of the American Institute of Chemical Engineers, AIChE copyright © 1986, all rights reserved] and describes a representative case history related to the development of a new chemical process. It shows how teams of engineers work together to provide a plant design and introduces the types of diagrams that will be explored in this chapter.

> *The research and development group at ABC Chemicals Company worked out a way to produce alpha-beta souptol (ABS). Process engineers assigned to work with the development group have pieced together a continuous process for making ABS in commercial quantities and have tested key parts of it. This work involved hundreds of* **block flow diagrams,** *some more complex than others.* **Based on information derived from these block flow diagrams, a decision was made to proceed with this process.**
>
> *A process engineering team from ABC's central office carries out the detailed process calculations, material and energy balances, equipment sizing, etc. Working with their drafting department, they produced a series of* **PFDs (Process Flow Diagrams)** *for the process. As problems arise and are solved, the team may revise and redraw the PFDs. Often the work requires several rounds of drawing, checking, and revising.*
>
> *Specialists in distillation, process control, kinetics, and heat transfer are brought in to help the process team in key areas. Some are company employees and others are consultants.*
>
> *Since ABC is only a moderate-sized company, it does not have sufficient staff to prepare the 120* **P&IDs (Piping and Instrumentation Diagrams)** *needed for the new ABS plant. ABC hires a well-known engineering and construction firm (**E&C Company),** DEFCo, to do this work for them. The company assigns two of the ABC process teams to work at DEFCo to coordinate the job. DEFCo's process engineers, specialists, and drafting department prepare the P&IDs. They do much of the detailed engineering (pipe sizes, valve specifications, etc.) as well as the actual drawing. The job may take two to six months. Every drawing is reviewed by DEFCo's project team and by ABC's team. If there are disagreements, the engineers and specialists from the companies must resolve them.*
>
> *Finally, all the PFDs and the P&IDs are completed and approved. ABC can now go ahead with the construction. They may extend their contract with DEFCo to include this phase, or they may go out for construction bids from a number of sources.*

This narrative describes a typical sequence of events taking a project from its initial stages through plant construction. If DEFCo had carried out the construction, ABC could go ahead and take over the plant or DEFCo could be contracted to carry out the start-up and to commission the plant. Once satisfactory performance specifications have been met, ABC would take over the operation of the plant and commercial production would begin.

From conception of the process to the time the plant starts up, two or more years will have elapsed and millions of dollars will have been spent with no revenue from the plant. The plant must operate successfully for many years to produce sufficient income to pay for all plant operations and to repay the costs asso-

ciated with designing and building the plant. During this operating period, many unforeseen changes are likely to take place. The quality of the raw materials used by the plant may change, product specifications may be raised, production rates may need to be increased, the equipment performance will decrease because of wear, the development of new and better catalysts, the changing costs of utilities, the introduction of new environmental regulations, or the appearance of improved equipment on the market.

As a result of these unplanned changes, plant operations must be modified. While the operating information on the original process diagrams remains informative, the actual performance taken from the operating plant will be different. The current operating conditions will appear on updated versions of the various process diagrams, which will act as a primary basis for understanding the changes taking place in the plant. These process diagrams are essential to an engineer who has been asked to diagnose operating problems, to solve problems in operations, to debottleneck systems for increased capacity, and to predict the effects of making changes in operating conditions. All of these activities are essential in order to maintain profitable plant operation.

In this chapter, we will concentrate on three diagrams that are important to chemical engineers: Block Flow, Process Flow, and Piping and Instrumentation Diagrams. Of these three diagrams, we will find that the most useful to chemical engineers is the Process Flow Diagram (PFD). The understanding of the PFD represents a central goal of this textbook.

1.1   BLOCK FLOW DIAGRAMS (BFD)

The block flow diagram is introduced early in the education of chemical engineers. In the first courses in material and energy balances, often the initial step was to convert a word problem into a simple visual block flow diagram. This diagram was a series of blocks connected with input and output flow streams. It included operating conditions (temperature and pressure) and other important information such as conversion and recovery, given in the problem statement. It did not provide details regarding what was involved within the blocks, but concentrated on the main flow of streams through the process.

The block flow diagram can take one of two forms. First, a block flow diagram may be drawn for a single process. Alternatively, a block flow diagram may be drawn for a complete chemical complex involving many different chemical processes. We differentiate between these two types of diagram by calling the first a block flow process diagram and the second a block flow plant diagram.

1.1.1   Block Flow Process Diagram

An example of a block flow process diagram is shown in Figure 1.1, and the process illustrated is described below.

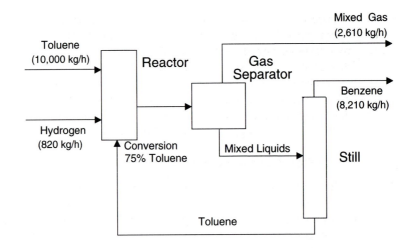

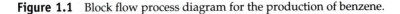

Reaction : $C_7H_8 + H_2 = C_6H_6 + CH_4$

Figure 1.1   Block flow process diagram for the production of benzene.

Toluene and hydrogen are converted in a reactor to produce benzene and methane. The reaction does not go to completion, and excess toluene is required. The noncondensable gases are separated and discharged. The benzene product and the unreacted toluene are then separated by distillation. The toluene is then recycled back to the reactor and the benzene removed in the product stream.

This block flow diagram gives a clear overview of the production of benzene, unobstructed by the many details related to the process. Each block in the

Table 1.1   Conventions and Format Recommended for Laying out a Block Flow Process Diagram

1. Operations shown by blocks
2. Major flow lines shown with arrows giving direction of flow
3. Flow goes from left to right whenever possible
4. Light stream (gases) toward top with heavy stream (liquids and solids) toward bottom
5. Critical information unique to process supplied
6. If lines cross, then the horizontal line is continuous and the vertical line is broken. (hierarchy for all drawings in this book)
7. Simplified material balance provided

diagram represents a process function and may, in reality, consist of several pieces of equipment. The general format and conventions used in preparing block flow process diagrams are presented in Table 1.1.

Although much information is missing from Figure 1.1, it is clear that such a diagram is very useful for "getting a feel" for the process. Block flow process diagrams often form the starting point for developing a PFD. They are also very helpful in conceptualizing new processes and explaining the main features of the process without getting bogged down in the details.

1.1.2   Block Flow Plant Diagrams

An example of a block flow plant diagram for a complete chemical complex is illustrated in Figure 1.2. This block flow plant diagram is for a coal to higher alcohol fuels plant. Clearly, this is a complicated process in which there are a number of alcohol fuel products produced from a feedstock of coal. Each block in this diagram represents a complete chemical process (compressors and turbines are also shown as trapezoids), and we could, if we wished, draw a block flow process diagram for each block in Figure 1.2. The advantage of a diagram such as Figure 1.2 is that it allows us to get a complete picture of what this plant does and how all the different processes interact. On the other hand, in order to keep the diagram relatively uncluttered, only limited information is available about each process unit. The conventions for drawing block flow plant diagrams are essentially the same as given in Table 1.1.

Both types of block flow diagrams are useful for explaining the overall operation of chemical plants. For example, consider that you have just joined a large chemical manufacturing company that produces a wide range of chemical products from the site to which you have been assigned. You would most likely be given a *block flow plant diagram* to orient you to the products and important areas of operation. Once assigned to one of these areas, you would again likely be provided with a *block flow process diagram* describing the operations in your particular area.

In addition to the orientation function described above, block flow diagrams are used to sketch out and screen potential process alternatives. Thus, they are used to convey information necessary to make early comparisons and eliminate competing alternatives without having to make detailed and costly comparisons.

1.2   Process Flow Diagram (PFD)

The process flow diagram (PFD) represents a quantum step up from the block flow diagram in terms of the amount of information that it contains. The PFD contains the bulk of the chemical engineering data necessary for the design of a chemical process. For all of the diagrams discussed in this chapter, there are no universally accepted standards. The PFD from one company will probably con-

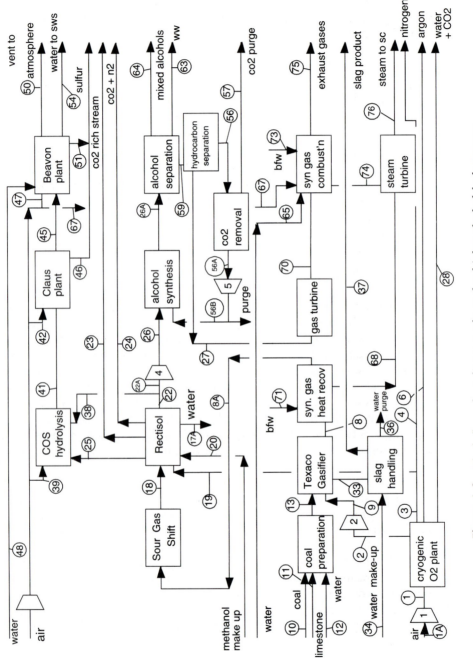

Figure 1.2 Block flow plant diagram of a coal to higher alcohol fuels process.

10

tain slightly different information than the PFD for the same process from another company. Having made this point, it is fair to say that most PFDs convey very similar information. A typical commercial PFD will contain the following information:

1. All the major pieces of equipment in the process will be represented on the diagram along with a description of the equipment. Each piece of equipment will have assigned a unique equipment number and a descriptive name.

2. All process flow streams will be shown and identified by a number. A description of the process conditions and chemical composition of each stream will be included. These data will be displayed either directly on the PFD or included in an accompanying flow summary table.

3. All utility streams supplied to major equipment that provides a process function will be shown.

4. Basic control loops, illustrating the control strategy used to operate the process during normal operations, will be shown.

It is clear that the PFD is a complex diagram that requires a substantial effort to prepare. It is essential that to avoid errors in presentation and interpretation it should remain uncluttered and be easy to follow. Often PFDs are drawn on large sheets of paper (Size D: 24″ × 36″), and several connected sheets may be required for a complex process. Because of the page size limitations associated with this text, complete PFDs cannot be presented here. Consequently, certain liberties have been taken in the presentation of the PFDs in this text. Specifically, certain information will be presented in accompanying tables and only the essential process information will be included on the PFD. The resulting PFDs will retain clarity of presentation, but the reader must refer to the flow summary and equipment summary tables in order to extract all the required information about the process.

Before we discuss the various aspects of the PFD, it should be noted that the PFD and the process that we describe in this chapter will be used throughout the book. The process is the hydrodealkylation of toluene to produce benzene. This is a well-studied and well-understood commercial process that is still used today. The PFD that we present in this chapter for this process is technically feasible but is in no way optimized. In fact, there are many improvements to the process technology and economic performance that can be made. Many of these improvements will become evident when the appropriate material is presented. This allows the techniques provided throughout this text to be applied to identify both technical and economic problems in the process and to make the necessary process improvements. Therefore, as we proceed through the text, we will identify weak spots in the design, make improvements, and eventually obtain an optimized process flow diagram.

The basic information provided by a PFD can be categorized into one of the following:

1. Process Topology
2. Stream Information
3. Equipment Information

We will look at each aspect of the PFD separately. After we have addressed each of the three topics above, we will bring all the information together and present the PFD for the benzene process.

1.2.1   Process Topology

Figure 1.3 is a skeleton process flow diagram for the production of benzene (see also the block flow process diagram in Figure 1.1). This skeleton diagram illustrates the location of the major pieces of equipment and the connections that the process streams make between equipment. The location of and interaction between equipment and process streams is referred to as the process topology.

Equipment is represented symbolically by "icons" that identify specific unit operations. Although the American Society of Mechanical Engineers (ASME) [2] publishes a set of symbols to use in preparing flow sheets, it is not uncommon for companies to use in-house symbols. A comprehensive set of symbols is also given by Austin [3]. Whatever set of symbols is used, there is seldom a problem in identifying the operation represented by each icon. Figure 1.4 contains a list of the symbols used in process diagrams presented in this text. This list covers over 90% of those needed in fluid (gas or liquid) processes.

Figure 1.3 shows that each major piece of process equipment is identified by a number on the diagram. A list of the equipment numbers along with a brief descriptive name for the equipment is printed along the top of the diagram. The location of these equipment numbers and names roughly corresponds to the horizontal location of the corresponding piece of equipment. The convention for formatting and identifying the process equipment is given in Table 1.2.

Table 1.2 provides the information necessary for the identification of the process equipment icons shown in a PFD. As an example of how to use this information, consider the unit operation P-101A/B and what each number or letter means.

> **P**-101A/B identifies the equipment as a pump
>
> P-**1**01A/B indicates that the pump is located in area 100 of the plant
>
> P-1**01**A/B indicates that this specific pump is number 01 in unit 100.
>
> P-101**A/B** indicates that a back-up pump is installed. Thus, there are two identical pumps P-101A and P-101B. One pump will be operating while the other is idle.

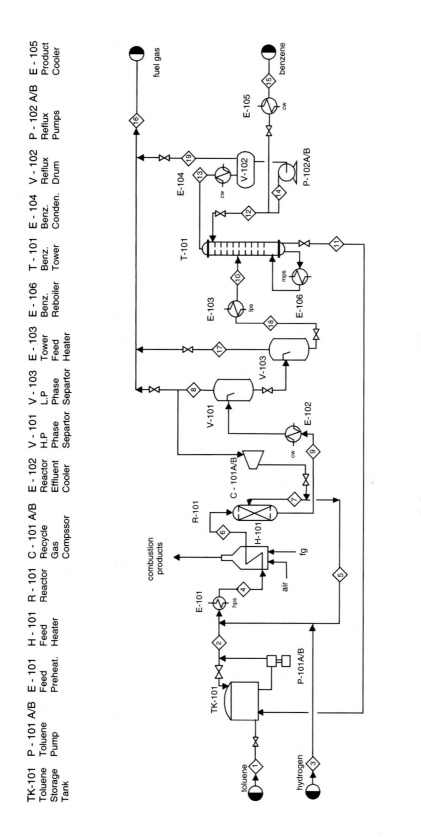

Figure 1.3 Skeleton Process Flow Diagram (PFD) for the production of benzene via the hydrodealkylation of toluene.

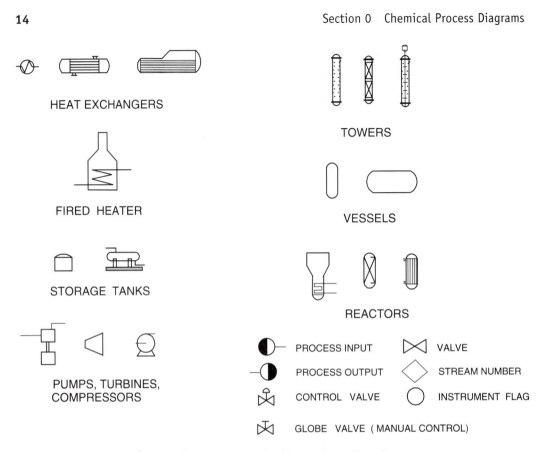

HEAT EXCHANGERS

TOWERS

FIRED HEATER

VESSELS

STORAGE TANKS

REACTORS

PUMPS, TURBINES,
COMPRESSORS

PROCESS INPUT   VALVE

PROCESS OUTPUT   STREAM NUMBER

CONTROL VALVE   INSTRUMENT FLAG

GLOBE VALVE (MANUAL CONTROL)

Figure 1.4   Symbols for building process flow diagrams.

The 100 area designation will be used for the benzene process throughout this text. Other processes presented in the text will carry other area designations. Along the top of the PFD each piece of process equipment is assigned a descriptive name. From Figure 1.3 it can be seen that Pump P-101 is called the "toluene pump." This name will be commonly used in discussions about the process and is synonymous with P-101.

During the life of the plant, many modifications will be made to the process; often it will be necessary to replace or eliminate process equipment. When a piece of equipment wears out and is replaced by a new unit that provides essentially the same process function as the old unit, then it is not uncommon for the new piece of equipment to inherit the old equipment's name and number (often an additional letter suffix will be used, e.g., H-101 might become H-101A). On the other hand, if a significant process modification takes place, then it is usual to use new equipment numbers and names. The following example, taken from Figure 1.3, illustrates this concept.

Table 1.2   Conventions Used for Identifying Process Equipment

Process Equipment	General Format XX-YZZ A/B
	XX are the identification letters for the equipment classification
	C - Compressor or Turbine
	E - Heat Exchanger
	H - Fired Heater
	P - Pump
	R - Reactor
	T - Tower
	TK - Storage Tank
	V - Vessel
	Y designates an area within the plant
	ZZ are the number designation for each item in an equipment class
	A/B identifies parallel units or backup units not shown on a PFD
Supplemental Information	Additional description of equipment given on top of PFD

Example 1.1

Operators report frequent problems with E-102, which are to be investigated. The PFD for the plant's 100 area is reviewed, and E-102 is identified as the "Reactor Effluent Cooler." The process stream entering the cooler is a mixture of condensable and non-condensable gases at 600°C that are partially condensed to form a two-phase mixture. The coolant is water at 30°C. These conditions characterize a complex heat transfer problem. In addition, operators have noticed that the pressure drop across E-102 fluctuates wildly at certain times, making control of the process difficult. Because of the frequent problems with this exchanger, it is recommended that E-102 be replaced by two separate heat exchangers. The first exchanger cools the effluent gas and generates steam needed in the plant. The second exchanger uses cooling water to reach the desired exit temperature of 38°C. These exchangers are to be designated as E-107 (reactor effluent boiler) and E-108 (reactor effluent condenser).

The E-102 designation is retired and not reassigned to the new equipment. There can be no mistake that E-107 and E-108 are new units in this process and that E-102 no longer exists.

Referring back to Figure 1.3, it can be seen that each of the process streams is identified by a number in a diamond box located on the stream. The direction of the stream is identified by one or more arrowheads. The process stream numbers are used to identify streams on the PFD and the type of information that is typically given for each stream will be discussed in the next section.

Also identified on Figure 1.3 are utility streams. Utilities are needed services that are available at the plant. Chemical plants are provided with a range of central utilities that include electricity, compressed air, cooling water, refrigerated water, steam, condensate return, inert gas for blanketing, chemical sewer, waste water treatment, and flares. A list of the common services is given in Table 1.3, which also provides a guide for the identification of process streams.

Each utility is identified by the initials provided in Table 1.3. As an example, let us locate E-102 in Figure 1.3. The notation, cw, associated with the non-process stream flowing into E-102 indicates that cooling water is used as a coolant.

Electricity used to power motors and generators is an additional utility that is not identified directly on the PFD or in Table 1.3 but is treated separately. Most of the utilities shown are related to equipment that add or remove heat within the

Table 1.3   Conventions for Identifying Process and Utility Streams

Process Streams
All conventions shown in Table 1.1 apply.
Diamond symbol located in flow lines.
Numerical identification (unique for that stream) inserted in diamond.
Flow direction shown by arrows on flow lines.

Utility Streams	
lps	Low Pressure Steam: 3–5 barg (sat) [‡]
mps	Medium Pressure Steam: 10–15 barg (sat) [‡]
hps	High Pressure Steam: 40–50 barg (sat) [‡]
htm	Heat Transfer Media (Organic): to 400°C
cw	Cooling Water: From cooling tower 30°C returned at less than 45°C[†]
wr	River Water: From river 25°C returned at less than 35°C
rw	Refrigerated Water: In at 5°C returned at less than 15°C
rb	Refrigerated Brine: In at −45°C returned at less than 0°C
cs	Chemical Waste Water with high COD
ss	Sanitary Waste Water with high BOD, etc.
el	Electric Heat (specify 220, 440, 660V service)
ng	Natural Gas
fg	Fuel Gas
fo	Fuel Oil
fw	Fire Water

[‡]These pressures are set during the preliminary design stages and typical values vary within the ranges shown.
[†]Above 45°C, significant scaling occurs.

process in order to control temperatures. This is common for most chemical processes.

1.2.2 Stream Information

From the process flow diagram, Figure 1.3, the identification of the process streams is clear. For small diagrams containing only a few operations, the characteristics of the streams such as temperatures, pressures, compositions, and flowrates can be shown directly on the figure, adjacent to the stream. This is not practical for a more complex diagram. In this case, only the stream number is provided on the diagram. This indexes the stream to information on a flow summary or stream table, which is often provided below the process flow diagram. In this text the flow summary table is provided as a separate attachment to the PFD.

The stream information that is normally given in a flow summary table is given in Table 1.4. It is divided into two groups—required information and optional information—that may be important to specific processes. The flow sum-

Table 1.4   Information Provided in a Flow Summary

Essential Information
Stream Number
Temperature (°C)
Pressure (bar)
Vapor Fraction
Total Mass Flow Rate (kg/h)
Total Mole Flow Rate (kmol/h)
Individual Component Flow Rates (kmol/h)
Optional Information
Component Mole Fractions
Component Mass Fractions
Individual Component Flow Rates (kg/h)
Volumetric Flow Rates (m³/h)

Significant Physical Properties
  Density
  Viscosity
  Other

Thermodynamic Data
  Heat Capacity
  Stream Enthalpy
  K-values

Stream Name

Table 1.5   Flow Summary Table for the Benzene Process Shown in Figure 1.3 (and Figure 1.5)

Stream Number	1	2	3	4	5	6	7	8
Temperature (°C)	25	59	25	225	41	600	41	38
Pressure (bar)	1.90	25.8	25.5	25.2	25.5	25.0	25.5	23.9
Vapor Fraction	0.0	0.0	1.00	1.0	1.0	1.0	1.0	1.0
Mass Flow (tonne/h)	10.0	13.3	0.82	20.5	6.41	20.5	0.36	9.2
Mole Flow (kmol/h)	108.7	144.2	301.0	1204.4	758.8	1204.4	42.6	1100.8
Component Mole Flow (kmol/h)								
Hydrogen	0.0	0.0	286.0	735.4	449.4	735.4	25.2	651.9
Methane	0.0	0.0	15.0	317.3	302.2	317.3	16.95	438.3
Benzene	0.0	1.0	0.0	7.6	6.6	7.6	0.37	9.55
Toluene	108.7	143.2	0.0	144.0	0.7	144.0	0.04	1.05

mary table for the benzene process, Figure 1.3, is given in Table 1.5 and contains all the required information listed in Table 1.4.

With information from the PFD (Figure 1.3) and the flow summary table (Table 1.5), problems regarding material balances and other problems are easily analyzed. To start gaining experience in working with information from the PFD, the following examples are provided.

Example 1.2

Check the overall material balance for the benzene process shown in Figure 1.3. From the figure, we identify the input streams as Stream 1 (toluene feed) and Stream 2 (hydrogen feed) and the output streams as Stream 15 (product benzene) and Stream 16 (fuel gas). From the flow summary table, these flows are listed as (units are in $(10^3 \text{ kg})/\text{h}$):

Input:	Output:
Stream 3— 0.82	Stream 15— 8.21
Stream 1— 10.00	Stream 16— 2.61
Total 10.82×10³ kg/h	Total 10.82×10³ kg/h

Balance is achieved since Output = Input.

Example 1.3

Determine the conversion per pass of toluene to benzene in R-101 in Figure 1.3. Conversion is defined as

$$\varepsilon = (\text{benzene produced})/(\text{total toluene introduced})$$

9	10	11	12	13	14	15	16	17	18	19
654	90	147	112	112	112	38	38	38	38	112
24.0	2.6	2.8	3.3	2.5	3.3	2.3	2.5	2.8	2.9	2.5
1.0	0.0	0.0	0.0	0.0	0.0	0.0	1.0	1.0	0.0	1.0
20.9	11.6	3.27	14.0	22.7	22.7	8.21	2.61	0.07	11.5	0.01
1247.0	142.2	35.7	185.2	290.7	290.7	105.6	304.6	4.06	142.2	0.90
652.6	0.02	0.0	0.0	0.02	0.0	0.0	178.0	0.67	0.02	0.02
442.3	0.88	0.0	0.0	0.88	0.0	0.0	123.2	3.10	0.88	0.88
116.0	106.3	1.1	184.3	289.46	289.46	105.2	2.85	0.26	106.3	0.0
36.0	35.0	34.6	0.88	1.22	1.22	0.4	0.31	0.03	35.0	0.0

From the PFD, the input streams to R-101 are shown as Stream 6 (reactor feed) and Stream 7 (recycle gas quench), and the output stream is Stream 9 (reactor effluent stream). From the information in Table 1.5 (units are kmol/h):

toluene introduced = 144 (Stream 6) + 0.04 (Stream 7) = 144.04 kmol/h

benzene produced = 116 (Stream 9) − 7.6 (Stream 6) − 0.37 (Stream 7) = 108.03 kmol/h

$$\varepsilon = 108.03/144.04 = 0.75$$

Alternatively, we can write

moles of benzene produced = toluene in − toluene out = 144.04 − 36.00 = 108.04 kmol/h

$$\varepsilon = 108.04/144.04 = 0.75$$

1.2.3   Equipment Information

The final element of the PFD is the equipment summary. This summary provides the information necessary to estimate the costs of equipment and furnish the basis for the detailed design of equipment. Table 1.6 provides the information needed for the equipment summary for most of the equipment encountered in fluid processes.

The information presented in Table 1.6 is used in preparing the equipment summary portion of the PFD for the benzene process. The equipment summary for the benzene process is presented in Table 1.7, and details of how we estimate and choose the various equipment parameters will be discussed in Chapter 9.

Table 1.6   Equipment Descriptions for PFD and PIDs

Equipment Type
Description of Equipment
Towers
Size (height and diameter), Pressure, Temperature Number and Type of Trays Height and Type of Packing Materials of Construction
Heat Exchangers
Type: Gas-Gas, Gas-Liquid, Liquid-Liquid, Condenser, Vaporizer Process: Duty, Area, Temperature, and Pressure for both streams. No. of Shell and Tube Passes Materials of Construction: Tubes and Shell
Tanks
See vessels
Vessels
Height, Diameter, Orientation, Pressure, Temperature, Materials of Construction
Pumps
Flow, Discharge Pressure, Temperature, ΔP, Driver Type, Shaft Power, Materials of Construction
Compressors
Actual Inlet Flow Rate, Temperature, Pressure, Driver Type, Shaft Power, Materials of Construction
Heaters (fired)
Type, Tube Pressure, Tube Temperature, Duty, Fuel, Material of Construction
Others
Provide Critical Information

Table 1.7   Equipment Summary for Toluene Hydrodealkylation PFD

Heat Exchangers	E-101	E-102	E-103	E-104	E-105	E-106
Type	Fl.H.	Fl.H.	MDP	Fl.H.	MDP	Fl.H.
Area (m^2)	36	763	11	35	12	80
Duty (MJ/h)	15190	46660	1055	8335	1085	9045
Shell						
Temp. (°C)	225	654	160	112	112	185
Pres. (bar)	26	24	6	3	3	11
Phase	Vap.	Par. Cond.	Cond.	Cond.	l	Cond.
MOC	316SS	316SS	CS	CS	CS	CS
Tube						
Temp. (°C)	258	40	90	40	40	147
Pres. (bar)	42	3	3	3	3	3
Phase	Cond.	l	l	l	l	Vap.
MOC	316SS	316SS	CS	CS	CS	CS

Vessels/Tower/ Reactors	V-101	V-102	V-103	TK-101	T-101	R-101
Temperature (°C)	38	112	38	55	147	660
Pressure (bar)	24	2.5	3.0	2.0	3.0	25
Orientation	Vertical	Horizn'l	Vertical	Horizn'l	Vertical	Vertical
MOC	CS	CS	CS	CS	CS	316SS
Size						
Height/Length (m)	3.5	3.9	3.5	5.9	29	14.2
Diameter (m)	1.1	1.3	1.1	1.9	1.5	2.3
Internals	s.p.		s.p.		42 sieve trays 316SS	catalyst packed bed-10m

Pumps/Compressors	P-101 (A/B)	P-102 (A/B)	C-101 (A/B)	Heater		H-101
Flow (kg/h)	13000	22700	6,770	Type		Fired
Fluid Density (kg/m^3)	870	880	8.02	MOC		316SS
Power (shaft) (kW)	14.2	3.2	49.1	Duty (MJ/h)		27040

(continued)

Table 1.7   Equipment Summary for Toluene Hydrodealkylation PFD (*continued*)

Pumps/Compressors	P-101 (A/B)	P-102 (A/B)	C-101 (A/B)	Heater	H-101
Type/Drive	Recip./ Electric	Centrf./ Electric	Centrf./ Electric	Radiant Area (m²)	106.8
Efficiency (Fluid Power/Shaft Power)	0.75	0.50	0.75	Convective Area (m²)	320.2
MOC	CS	CS	CS	Tube P (bar)	26.0
Temp. (in) (°C)	55	112	38		
Pres. (in) (bar)	1.2	2.2	23.9		
Pres. (out) (bar)	27.0	4.4	25.5		

Key:

MOC	Materials of construction	Par	Partial
316SS	Stainless steel type 316	F.H.	Fixed head
CS	Carbon steel	Fl.H.	Floating head
Vap	Stream being vaporized	Rbl	Reboiler
Cond	Stream being condensed	s.p.	Splash plate
Recipr.	Reciprocating	l	Liquid
Centrf.	Centrifugal	MDP	Multiple double pipe

1.2.4   Combining Topology, Stream Data, and Control Strategy to Give a PFD

Up to this point, we have kept the amount of process information displayed on the PFD to a minimum. A more representative example of a PFD for the benzene process is shown in Figure 1.5. This diagram includes all of the elements found in Figure 1.3, some of the information found in Table 1.5, plus additional information on the major control loops used in the process.

Stream information is added to the diagram by attaching "information flags." The shape of the flags indicates the specific information provided on the flag. Figure 1.6 illustrates all the flags used in this text. These information flags play a dual role. They provide information needed in the plant design leading to plant construction and in the analysis of operating problems during the life of the plant. Flags are mounted on a staff connected to the appropriate process stream. More than one flag may be mounted on a staff. An example illustrating the different information displayed on the PFD is given below.

Example 1.4

We locate Stream 1 in Figure 1.5 and note that immediately following the stream identification diamond a staff is affixed. This staff carries three flags containing the following stream data:

1. Temperature of 25 °C
2. Pressure of 1.9 bar
3. Mass flow rate of 10.0×10^3 kg/h

The units for each process variable are indicated in the key provided at the left-hand side of Figure 1.5.

With the addition of the process control loops and the information flags, the PFD starts to become cluttered. Therefore, in order to preserve clarity, it is necessary to limit what data are presented with these information flags. Fortunately, flags on a PFD are easy to add, remove, and change, and even temporary flags may be provided from time to time.

The information provided on the flags is also included in the flow summary table. However, often it is far more convenient when analyzing the PFD to have certain data directly on the diagram.

Not all process information is of equal importance. General guidelines for what data should be included in information flags on the PFD are difficult to define. However, as a minimum, information critical to the safety and operation of the plant should be given. This includes temperatures and pressures associated with the reactor, flowrates of feed and product streams, and stream pressures and temperatures that are substantially higher than the rest of the process. Additional needs are process specific. Some examples of where and why information should be included directly on a PFD are given below.

Example 1.5

Acrylic acid is temperature sensitive and polymerizes at 90°C when present in high concentration. It is separated by distillation and leaves from the bottom of the tower. In this case, a temperature and pressure flag would be provided for the stream leaving the reboiler.

Example 1.6

In the benzene process, the feed to the reactor is substantially hotter than the rest of the process and is crucial to the operation of the process. In addition, the reaction is exothermic, and the reactor effluent temperature must be carefully monitored. For this reason Stream 6 (entering) and Stream 9 (leaving) have temperature flags.

Example 1.7

The pressures of the streams to and from R-101 in the benzene process are also important. The difference in pressure between the two streams gives the pressure drop across the reactor. This, in turn, gives an indication of any maldistribution of gas through the catalyst beds. For this reason, pressure flags are also included on Streams 6 and 9.

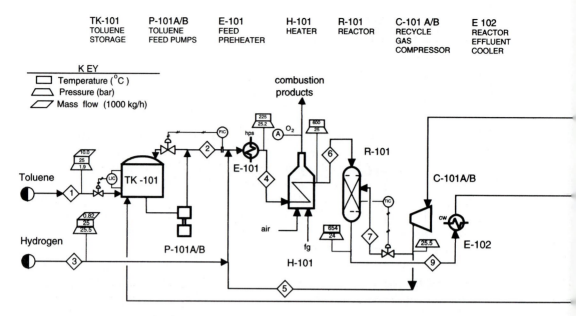

Figure 1.5 Benzene process flow diagram (PFD) for the production of benzene via the hydrodealkylation of toluene.

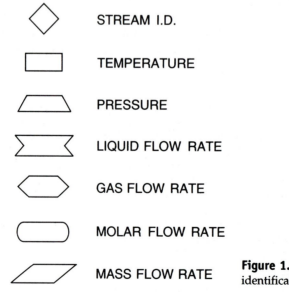

Figure 1.6 Symbols for stream identification.

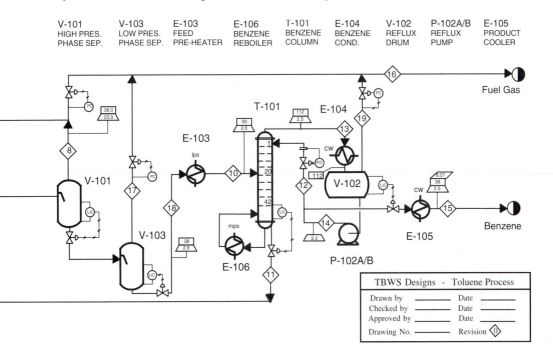

Of secondary importance is the fact that flags are useful in reducing the size of the flow summary table. For pumps, compressors, and heat exchangers, the mass flows are the same for the input and output streams, and complete entries in the stream table are not necessary. If the input (or output) stream is included in the stream table, and a flag is added to provide the temperature (in the case of a heat exchanger) or the pressure (in the case of a pump) for the other stream, then there is no need to present this stream in the flow summary table.

Example 1.8

Follow Stream 13 leaving the top of the benzene column in the benzene PFD given in Figure 1.5 and in Table 1.5. This stream passes through the benzene condenser, E-104, into the reflux drum, V-102. The majority of this stream then flows into the reflux pump, P-102, and leaves as Stream 14, while the remaining non-condensables leave the reflux drum in Stream 19. The mass flowrate and component flowrates of all these streams are given in Table 1.5. The stream leaving E-104 is not included in the stream table. Instead, a flag giving the temperature (112°C) was provided on the diagram (indicating condensation without sub-cooling). An additional flag, showing the pressure following the pump, is also shown. In this case the entry for Stream 14 could be omitted from the stream table, since it is simply the sum of Streams 12 and 15, and no information would be lost.

More information could be included in Figure 1.5 had space for the diagram not been limited by text format. It is most important that the PFD remains uncluttered and easy to follow in order to avoid errors and misunderstandings. Adding additional material to Figure 1.5 risks sacrificing clarity.

The flow table presented in Table 1.5, the equipment summary presented in Table 1.7, and Figure 1.5 taken together constitute all the information contained on a commercially produced PFD.

The PFD is the first comprehensive diagram drawn for any new plant or process. It provides all of the information needed to understand the chemical process. In addition, sufficient information is given on the equipment, energy, and material balances to establish process control protocol and to prepare cost estimates to determine the economic viability of the process.

Many additional drawings are needed to build the plant. All the process information required can be taken from this PFD. As described in the narrative at the beginning of this chapter, the development of the PFD is most often carried out by the operating company. Subsequent activities in the design of the plant are often contracted out.

The value of the PFD does not end with the construction of the plant. It remains the document that best describes the process and it is used in the training of operators and new engineers. It is consulted regularly to diagnose operating problems that arise and to predict the effects of changes on the process.

1.3   PIPING AND INSTRUMENTATION DIAGRAM (P&ID)

The piping and instrumentation diagram (P&ID) or mechanical flow diagram (MFD) provides information needed by engineers to begin planning for the construction of the plant. The P&ID includes every mechanical aspect of the plant except the information given in Table 1.8. The general conventions used in drawing P&IDs are given in Table 1.9.

Each PFD will require many P&IDs to provide the necessary data. Figure 1.7 is a representative P&ID for the distillation section of the benzene process shown in Figure 1.5. The P&ID presented in Figure 1.7 provides information on the piping, and this is included as part of the diagram. As an alternative, each pipe can be numbered, and the specifics of every line can be provided in a sepa-

Table 1.8   Exclusions from Piping and Instrumentation Diagram

1. Operating conditions   T, P
2. Stream flows
3. Equipment locations
4. Pipe routing
   a. Pipe lengths
   b. Pipe fittings
5. Supports, structures, and foundations

Table 1.9   Conventions in Constructing Piping and Instrumentation Diagrams

For Equipment—Show Every Piece Including
Spare units
Parallel units
Summary details of each unit

For Piping—Include All Lines Including Drains, Sample Connections and Specify
Size (use standard sizes)
Schedule (thickness)
Materials of construction
Insulation (thickness and type)

For Instruments—Identify
Indicators
Recorders
Controllers
Show instrument lines

For Utilities—Identify
Entrance utilities
Exit utilities
Exit to waste treatment facilities

rate table accompanying this diagram. When possible, the physical size of the larger-sized unit operations is reflected by the size of the symbol in the diagram.

Utility connections are identified by a numbered box in the P&ID. The number within the box identifies the specific utility. The key identifying the utility connections is shown in a table on the P&ID.

All process information that can be measured in the plant is shown on the P&ID by circular flags. This includes the information to be recorded and used in process control loops. The circular flags on the diagram indicate where the information is obtained in the process and identifies the measurements taken and how the information is dealt with. Table 1.10 summarizes the conventions used to identify information related to instrumentation and control. The following example illustrates the interpretation of instrumentation and control symbols.

Example 1.9

Consider the benzene product line leaving the right-hand side of the P&ID in Figure 1.7. The flowrate of this stream is controlled by a control valve that receives a signal from a level measuring element placed on V-102. The sequence of instrumentation is as follows:

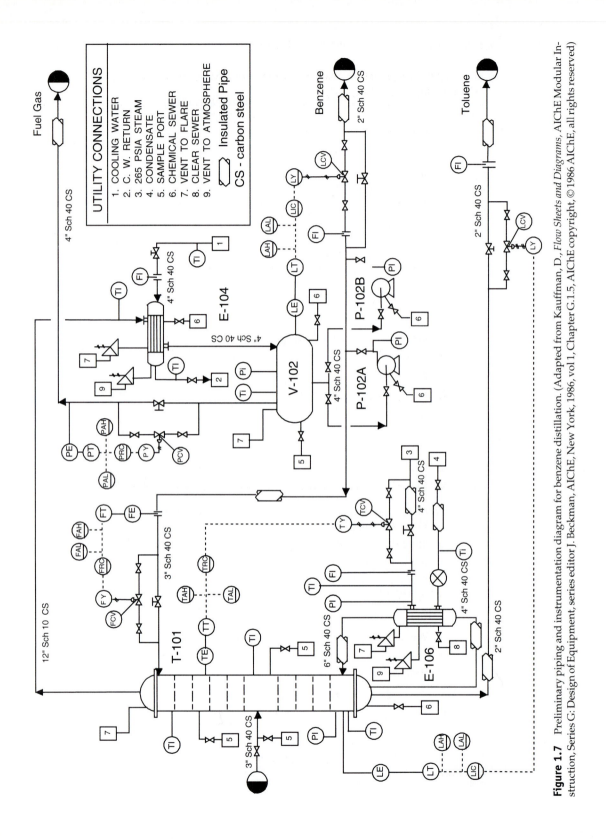

Figure 1.7 Preliminary piping and instrumentation diagram for benzene distillation. (Adapted from Kauffman, D., *Flow Sheets and Diagrams*, AIChE Modular Instruction, Series G: Design of Equipment, series editor J. Beckman, AIChE, New York, 1986, vol 1, Chapter G.1.5, AIChE copyright, ©1986 AIChE, all rights reserved)

Table 1.10   Conventions Used for Identifying Instrumentation on P&IDs (ISA standard ISA-S5-1, [4])

Location of Instrumentation

◯	Instrument located in plant
⊖	Instrument located on front of panel in control room
⊙	Instrument located on back of panel in control room

Meanings of Identification Letters (XYY)	
First Letter (X)	**Second or Third Letter (Y)**
A Analysis	Alarm
B Burner flame	
C Conductivity	Control
D Density or specific gravity	
E Voltage	Element
F Flowrate	
H Hand (manually initiated)	High
I Current	Indicate
J Power	
K Time or time schedule	Control station
L Level	Light or low
M Moisture or humidity	Middle or intermediate
O	Orifice
P Pressure or vacuum	Point
Q Quantity or event	
R Radioactivity or ratio	Record or print
S Speed or frequency	Switch
T Temperature	Transmit
V Viscosity	Valve, damper, or louver
W Weight	Well
Y	Relay or compute
Z Position	Drive

Identification of Instrument Connections	
———————	Capillary
—⫻———	Pneumatic
····················	Electrical

A level sensing element (LE) is located on the reflux drum V-102. A level transmitter (LT) also located on V-102 sends an electrical signal (designated by a dashed line) to a level indicator and controller (LIC). This LIC is located in the control room on the control panel or console (as indicated by the horizontal line under LIC) and can be observed by the operators. From the LIC, an electrical signal is sent to an instrument (LY) that computes the correct valve position and in turn sends a pneumatic signal (designated by a solid line with cross hatching) to activate the control valve (LCV). In order to warn operators of potential problems, two alarms are placed in the control room. These are a high-level alarm (LAH) and a low-level alarm (LAL), and they receive the same signal from the level transmitter as does the controller.

This control loop is also indicated on the PFD of Figure 1.5. However, the details of all the instrumentation are condensed into a single symbol (LIC), which adequately describes the essential process control function being performed. The control action that takes place is not described explicitly in either drawing. However, it is a simple matter to infer that if there is an increase in the level of liquid in V-101, the control valve will open slightly and the flow of benzene product will increase, tending to lower the level in V-101. For a decrease in the level of liquid, the valve will close slightly.

The details of the other control loops in Figures 1.5 and 1.7 are left to problems at the end of this chapter. It is worth mentioning that in virtually all cases of process control in chemical processes, the final control element is a valve. Thus, all control logic is based on the effect that a change in a given flowrate has on a given variable. The key to understanding the control logic is to identify which flowrate is being manipulated to control which variable. Once this has been done, it is a relatively simple matter to see in which direction the valve should change in order to make the desired change in the control variable. The response time of the system and type of control action used—for example, proportional, integral, or differential—is left to the instrument engineers and is not covered in this text.

The final control element in nearly all chemical process control loops is a valve.

The P&ID is the last stage of process design and serves as a guide by those who will be responsible for the final design and construction. Based on this diagram

1. Mechanical engineers and civil engineers will design and install pieces of equipment.
2. Instrument engineers will specify, install, and check control systems.
3. Piping engineers will develop plant layout and elevation drawings.
4. Project engineers will develop plant and construction schedules.

Before final acceptance, the P&IDs serve as a checklist against which each item in the plant is checked.

The P&ID is also used to train operators. Once the plant is built and is operational, there are limits to what operators can do. About all that can be done to correct or alter performance of the plant is to open, close, or change the position of a valve. Part of the training would pose situations and require the operators to be able to describe what specific valve should be changed, how it should be changed, and what to observe in order to monitor the effects of the change. Plant simulators (similar to flight simulators) are sometimes involved in operator training. These programs are sophisticated, real-time process simulators that show a trainee operator how quickly changes in controlled variables propagate through the process. It is also possible for such programs to display scenarios of process upsets so that operators can get training in recognizing and correcting such situations. These types of programs are very useful and cost effective in initial operator training. However, the use of P&IDs is still very important in this regard.

The P&ID is particularly important for the development of start-up procedures where the plant is not under the influence of the installed process control systems.

Example 1.10

Consider the start-up of the distillation column shown in Figure 1.7. What sequence would be followed? The procedure is beyond the scope of this text but it would be developed from a series of questions such as

    a. What valve should be opened first?
    b. What should be done when the temperature of… reaches …?
    c. To what value should the controller be set?
    d. When can the system be put on automatic control?

These last three sections have followed the development of a process from a simple block flow diagram through the PFD and finally to the P&ID. Each step showed additional information. This can be seen by following the progress of the distillation unit as it moves through the three diagrams described.

1. Block Flow Diagram (BFD) (see Figure 1.1): The column was shown as a part of one of the three process blocks.
2. Process Flow Diagram (PFD) (see Figure 1.5): The column was shown as the following set of individual equipment: a tower, condenser, reflux drum, reboiler, reflux pumps, and associated process controls.
3. Piping and Instrumentation Diagram (P&ID) (see Figure 1.7): The column was shown as a comprehensive diagram that includes additional details

such as pipe sizes, utility streams, sample taps, numerous indicators, and so on. It is the only unit operation on the diagram.

The value of these diagrams does not end with the start-up of the plant. The design values on the diagram are changed to represent the actual values determined under normal operating conditions. These conditions form a "base case" and are used to compare operations throughout the life of the plant.

1.4   ADDITIONAL DIAGRAMS

During the planning and construction phases of a new project, many additional diagrams are needed. Although these diagrams are essential to the successful completion of the project, they do not possess additional process information. Consequently, we give only a brief overview below.

A **Utility Flowsheet** may be provided that shows all the headers for utility inputs and outputs available along with the connections needed to the process. It provides information on the flows and characteristics of the utilities used by the plant.

Additional drawings are necessary to locate all of the equipment in the plant. **Plot Plans and Elevation Diagrams** are provided that locate the elevation and placement of all of the major pieces of equipment such as towers, vessels, pumps, heat exchangers, and so on. When constructing these drawings, it is necessary to consider and to provide for access for repairing equipment, removing tube bundles from heat exchangers, replacement of units, and so on. What remains to be shown is the addition of the structural support and piping.

Piping Isometrics are drawn for every piece of pipe required in the plant. These drawings are 3-D sketches of the pipe run indicating the elevations and orientation of each section of pipe. For comprehensive plants, it was also common to build a **Scale Model** so the system could be viewed in three dimensions and modified to remove any potential problems. **Computer Aided Design (CAD)** programs are now common tools for dealing with this problem. They provide an opportunity to view the local equipment topology from any angle at any location inside the plant. You can actually "walk through" the plant and preview what will be seen when the plant is built. The ability to "view" the plant before construction will be made even more realistic with the help of **Virtual Reality** software. With this new tool it will be possible not only to "walk through" the plant but actually to touch the equipment, turn valves, and climb to the top of distillation columns, or "perform" other tasks.

Vessel Sketches, Logic Ladder Diagrams, Wiring Diagrams, Site Plans, Structural Support Diagrams and many other drawings are routinely used but add little to our understanding of the basic chemical processes that take place.

Computers are being used more and more to do the tedious work associated with all of these drawing details. The creative work comes in the development of the concepts provided in the block flow diagram and the process development required to produce the PFD. The computer can help with the drawings but cannot create a new process. Computers are valuable in many aspects of the design process where the size of equipment to do a specific task is to be determined. Computers may also be used when considering performance problems that deal with the operation of existing equipment. However, they are severely limited in dealing with diagnostic problems that are required throughout the life of the plant.

The diagrams presented here are largely in SI units, although there are occasions where British units appear. The most noticeable exception is in the sizing of piping, where pipes are specified in inches and pipe schedule. This remains the way they are produced and purchased in the United States. A process engineer today must be comfortable with SI, conventional metric, and British units.

1.5 SUMMARY

In this chapter, you have learned that the three principal types of diagrams used to describe the flow of chemical streams through a process are the block flow diagram (BFD), the process flow diagram (PFD), and the piping and instrumentation diagram (P&ID). These diagrams describe a process in increasing detail.

Each diagram serves a different purpose. The block flow diagram is useful in conceptualizing a process or a number of processes in a large complex. Little stream information is given, but a clear overview of the process is presented. The process flow diagram contains all the necessary information to complete material and energy balances on the process. In addition, important information such as stream pressures, equipment sizes, and major control loops are included. Finally, the piping and instrumentation diagram contains all the process information necessary for the construction of the plant. These data include pipe sizes and the location of all instrumentation for both the process and utility streams.

In addition to the three diagrams listed above, there are a number of other diagrams used in the construction and engineering phase of a project. However, these diagrams contain little additional information about the process.

The **PFD** is the single most important diagram for the chemical/process engineer and will form the basis of much of the discussion covered in this book.

REFERENCES

1. Kauffman, D., "Flow Sheets and Diagrams," *AIChE Modular Instruction, Series G: Design of Equipment*, series editor J. Beckman, American Institute of Chemical Engineers, New York, 1986, vol 1, Chapter. G1.5. Reproduced by permis-

sion of the American Institute of Chemical Engineers, AIChE copyright ©
1986, all rights reserved.

2. *Graphical Symbols for Process Flow Diagrams*, ASA Y32.11, American Society of
   Mechanical Engineers, New York, 1961.

3. D. G. Austin, *Chemical Engineering Drawing Symbols*, George Godwin, London, 1979.

4. *Instrument Symbols and Identification*, Instrument Society of America, Standard ISA-S5-1, Research Triangle Park, NC, 1975.

PROBLEMS

Note: Problems 1–9, are from Kauffman [1] and are reproduced by permission of
the American Institute of Chemical Engineers, AIChE copyright © 1986, all rights
reserved.

1. What are the three principal types of flowsheets used in the chemical process
   industries?

2. Which of the three principal types of flowsheets would one use to:
   a. give a group of visiting chemical engineering students an overview of a
      plant's process?
   b. make a preliminary capital cost estimate?
   c. trace down a fault in a control loop?

3. In what type of flowsheet could one expect to find pipe diameters and materials of construction?

4. To what extent are instruments and controls indicated in each of the three
   types of flowsheets?

5. On which of the three principal types of flowsheets would one expect to find:
   a. relief valves?
   b. which pipe lines need insulation?
   c. which control loops are needed for normal operation?
   d. rectangles shown, rather than symbols, that resemble pieces of equip-
      ment?
   e. whether a controller is to be located in the control room or in the plant?

6. Would you expect the process design to include more PFDs or P&IDs?

7. Prepare the simplest principal type of flowsheet for the following process,
   and indicate the flowrates of the principal chemical components:

   > A refinery stream containing paraffins and a mixture of aromatics (benzene,
   > toluene, xylene, and heavier aromatics) is extracted with a liquid solvent to re-
   > cover the aromatics. The solvent and aromatics are separated by distillation, with
   > the solvent recycled to the extraction column. The aromatics are separated in

three columns, recovering benzene, toluene, and mixed xylenes, in that order.
The feed stream consists of the following:

paraffins	*300,000 kg/h*
benzene	*100,000 kg/h*
toluene	*180,000 kg/h*
xylene	*70,000 kg/h*
heavy aromatics	*40,000 kg/h*

A 3-to-1 weight ratio of solvent to aromatics is used.

8. Liquid is pumped from an elevated vessel through the tube side of a water cooled heat exchanger. The fluid flow is controlled by a flow-rate controller in the control room. The pump has a spare. Sketch a portion of the most detailed principal type of flow diagram that would be used to illustrate this process.

9. Figure P1.9 is a portion of a P&ID. Find at least six errors in it. All errors are in items actually shown on the drawing. Do not cite "errors of omission" ("such and such not shown"), since this is only a portion of the P&ID.

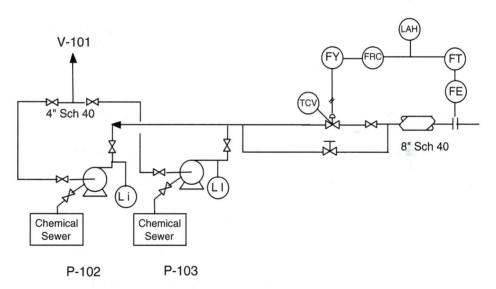

Figure P1.9   A portion of a P&ID containing several errors.

10. In a process to separate and purify propane from a mixture of propane and heavier straight-chain saturated hydrocarbons (e.g., n-butane, n-pentane, etc.), the feed stream is fed to the 18th tray of a 24-tray distillation column. The overhead vapor stream from the column is totally condensed in a water cooled heat exchanger prior to being fed to an overhead reflux drum. The

liquid product from the drum is sent to the reflux pump (which has a spare), and the discharge from the pump is split into two streams. One of these streams is the overhead reflux to the column and is fed back to the column on Tray 1. The second liquid stream from the pump discharge is the overhead product and is sent to storage.

The bottom of the distillation column is used to store the liquid leaving the bottom plate. From the bottom of the column a liquid stream leaves and is immediately split into two. One stream is the bottom product, which is sent for further processing in Unit 400. The other stream is sent to a thermosyphon reboiler where a portion of the stream is vaporized by condensing low pressure steam on the other side of the exchanger. The partially vaporized stream from the reboiler is returned to the column just below the 24th tray. The two phase mixture separates, with the vapor portion passing upward through the bottom plate to provide the vapor flow in the column. The liquid portion returns to the liquid accumulated at the bottom of the column.

For the process described above, draw a PFD. You may assume that the process is Unit 200, and you should identify and number all the equipment appropriately.

11. For the process described in Problem 10 above, the following control scheme has been suggested for the overhead portion of the column:

    The flow of overhead product going to storage is controlled by a signal from the liquid level indicator on the reflux drum, which is used to control the position of a pneumatic control valve in the product line (pipe). The flow of reflux back to the column is also regulated by a pneumatic control valve, which uses the signal (electrical) from a flow indicator on the overhead product line to adjust the valve such that the flow of reflux is always in a certain proportion to the product flow.

    On the PFD developed in Problem 10, add the controls to give the desired control action described above.

    Bonus Points: Can you describe how the control scheme should operate, i.e., what valve opens or closes, etc., when the level of liquid rises above or falls below its set-point value?

12. For the process described in Problem 10, the flow of bottom product sent to Unit 400 is controlled by a pneumatic valve that receives a signal from a liquid level indicator that senses the level of accumulated liquid in the bottom of the column. Add this control loop to the PFD developed in Problem 10.

13. For the process described in Problem 10, it is desired to control the purity of the top product. If we assume that the pressure of the column does not vary (not necessarily a good assumption), we may infer the product purity from the temperature of the top tray. Devise a control scheme to implement a feedback loop to control the top product purity. Draw this control loop on the PFD for Problem 10.

14. Drying oil (DO) is produced by thermally decomposing acetylated castor oil (ACO) according to the following reaction:

$$ACO(l) \overset{340°C}{\underset{heat}{\rightarrow}} DO(l) + CH_3COOH(g)$$

The process to produce DO is fairly straightforward and is described below:

ACO is fed from storage (off site) to a small horizontal storage vessel, V-101. From V-101, ACO liquid at 30°C is fed to a feed pump (P-101 A/B) where it is pressurized to 2 barg. The flow of ACO is controlled by a flow control valve situated on the discharge side of the pump. The ACO is fed to a reactor feed furnace (H-101) where the temperature is increased to 340°C, and the stream leaving the furnace is sent directly to a reactor (R-101), containing inert ceramic packing, where the decomposition reaction takes place. The single-pass conversion of ACO to DO in the reactor is 40%. The stream leaving the reactor is then fed to a gas-liquid separator (V-102) where the acetic acid flashes off and leaves in the overhead vapor stream. The heavy DO and ACO liquids have very low vapor pressures and consequently do not vaporize appreciably and leave the vessel as hot liquid product. This hot liquid stream, at 310°C, leaves V-102 and is then fed to a waste heat boiler (E-101) where the hot oil is cooled to 160°C by exchanging heat with boiler feed water to produce medium pressure steam at 10 barg. The temperature of the cooled oil stream is controlled by adjusting the set point on the level controller on E-101. This level controller in turn regulates the level of water in E-101 by adjusting the flowrate of boiler feed water. The cooled oil stream, at a pressure of 1.3 barg, is sent to Unit 200 for further processing.

For the process described above, draw a PFD showing the following details:

Equipment numbers and description
Basic control loops
Temperature and pressure flags

1

Engineering Economic Analysis of Chemical Processes

In this section, we concentrate on the evaluation of the economics of a chemical process. In order for a chemical engineer or cost engineer to evaluate the economic impact of a new (or existing) chemical process, certain technical information must be available. Although this information is gleaned from a variety of sources, it is generally presented in the form of the technical diagrams discussed in Chapter 1.

In this section, methods to evaluate the economics of a chemical process are covered. The term economics refers to the evaluation of capital costs and operating costs associated with the construction and operation of a chemical process. The methods by which the one-time costs associated with the construction of the plant and the continuing costs associated with the daily operation of the process are combined into meaningful economic criteria are provided.

This material is treated in the following chapters:

Chapter 2: Estimation of Capital Cost
The common types of estimates are presented along with the basic relationships for scaling costs with equipment size. The concept of cost inflation is presented, and some common cost indexes are presented. The concept of total fixed capital investment to construct a new process is discussed, and the cost module approach to estimating is given. Finally, a software program (CAPCOST©) to evaluate fixed capital costs is described.

Chapter 3: Estimation of Manufacturing Costs
The basic components of the manufacturing cost of a process are presented. A method is given to relate the total cost of manufacturing (COM) to five el-

ements: fixed capital investment, cost of operating labor, cost of raw materials, cost of utilities, and cost of waste treatment. A modular method to estimate the cost of operating labor is presented. Also provided is a list of costs for common chemicals (used throughout the text) and common utilities used in chemical processes.

Chapter 4: Engineering Economic Analysis

The concept of the time value of money is discussed. The following topics are presented: simple and compound interest, effective and nominal interest rates, annuities, cash flow diagrams, and discount factors. In addition, the concepts of depreciation, inflation, and taxation are covered.

Chapter 5: Profitability Analysis

The ideas discussed in Chapter 4 are extended to evaluate the profitability of chemical processes. Profitability criteria using nondiscounted and discounted bases are presented and include: net present value (NPV), discounted cash flow rate of return (DCFROR), and payback period (PBP). A discussion of evaluating equipment alternatives using equivalent annualized operating costs (EAOC) and other methods is presented. Finally, the idea of incremental investment is presented and discussed.

CHAPTER

2

Estimation
of Capital Costs

In the previous chapter, we learned about the information provided on a process flow diagram, including a stream table and an equipment summary table. In the next four chapters, we will use this information as a basis for estimating:

1. How much money (capital cost) it takes to build a new chemical plant.
2. How much money (operating cost) it takes to operate a chemical plant.
3. How to combine Items 1 and 2 to provide several distinct types of composite values reflecting process profitability.
4. How to select a "best process" from competing alternatives.
5. How to estimate the economic value of making process changes and modifications to an existing processes.

In this chapter, we concentrate on the estimation of capital costs. Capital cost pertains to the costs associated with construction of a new plant or modifications to an existing chemical manufacturing plant.

2.1 CLASSIFICATIONS OF CAPITAL COST ESTIMATES

There are five generally accepted classifications of capital cost estimates that you are most likely to encounter in the process industries [1,2,3]:

1. Order of magnitude estimate
2. Study estimate
3. Preliminary estimate
4. Definitive estimate
5. Detailed estimate

The information required to perform each of these estimates is provided in Table 2.1. The accuracy range and the approximate cost for performing each type of estimate is provided in Table 2.2 and Figure 2.1.

Figure 2.1 illustrates the accuracy range associated with each estimate classification along with the costs associated with carrying out the estimation. Each of the estimating classifications is represented by a block in Figure 2.1. The top and bottom edges of the block give the high and low estimates of the error range, respectively. In all classifications, there is a greater range above the estimate than below, meaning that underestimates of capital costs are more common than overestimates.

Example 2.1

The estimated capital cost for a chemical plant using the study estimate method was calculated to be $2,000,000. If the plant were to be built, over what range would you expect the actual capital estimate to vary?

Highest Expected Cost = ($2.0×10^6)(1.3) = $2.6×10^6
Lowest Expected Cost = ($2.0×10^6)(0.8) = $1.6×10^6
or from $0.6×10^6 over to $0.4×10^6 under the estimate.

The primary reason that capital costs are underestimated stems from the failure to include all of the operating units needed in the process. Typically, as a design progresses, the need for additional process units is uncovered.

Typical costs for performing a cost estimate are provided in Figure 2.1. The cost is bracketed between the value shown on the right- and left-hand edges of the block representing the estimating technique. These costs correspond to those presented in Table 2.2.

Example 2.2

Compare the capital costs of a plant estimated at 2.0×10^6 using the order-of-magnitude estimate with that obtained by using the study method used in Example 1. Compare the costs of using these two methods.

For the order-of-magnitude estimate:

Highest Expected Value: ($2.0×10^6)(1.4) = $2.8×10^6
Lowest Expected Value: ($2.0×10^6)(0.8) = $1.6×10^6

The upper limit increased from $2.6×10^6 to $2.8×10^6 (or by $0.2×10^6) and the lower bound remained the same.

The cost of the estimation for the order-of-magnitude estimate is given in Table 2.2 as $3,000 compared to $20,000 for the study estimate.

Table 2.1   Summary of Capital Cost Estimating Classifications (references [1], [2], and [3])

Order of Magnitude (also Known as Ratio or Feasibility) Estimate:

Data: This type of estimate typically relies on cost information for a complete process taken from previously built plants. This cost information is then adjusted using appropriate scaling factors, for capacity, and for inflation, to provide the estimated capital cost.

Diagrams: Normally requires only a block flow diagram.

*Accuracy:** Range +40% to −20%

Study (also known as Major Equipment or Factored) Estimate:

Data: This type of estimate utilizes a list of the major equipment found in the process. This includes all pumps, compressors and turbines, columns and vessels, fired heaters and exchangers. Each piece of equipment is roughly sized and the approximate cost determined. The total cost of equipment is then factored to give the estimated capital cost.

Diagrams: Based on PFD as described in Chapter 1. Costs from generalized charts.

*Accuracy:** Range +30% to −20%

Preliminary Design (also known as Scope) Estimate:

Data: This type of estimate requires more accurate sizing of equipment than used in the study estimate. In addition, approximate layout of equipment is made along with estimates of piping, instrumentation, and electrical requirements. Utilities are estimated.

Diagrams: Based on PFD as described in Chapter 1. Includes vessel sketches for major equipment, preliminary plot plan, and elevation diagram.

*Accuracy:** Range +25% to −15%

Definitive (also known as Project Control) Estimate:

Data: This type of estimate requires preliminary specifications for all the equipment, utilities, instrumentation, electrical, and off-sites.

Diagrams: Final PFD, vessel sketches, plot plan, and elevation diagrams, utility balances, and a preliminary P&ID.

*Accuracy:** Range +15% to −7%

Detailed (also known as Firm or Contractor's) Estimate:

Data: This type of estimate requires complete engineering of the process and all related off-sites and utilities. Vendor quotes for all expensive items will have been obtained. At the end of a detailed estimate the plant is ready to go to the construction stage.

Diagrams: Final PFD and P&ID, vessel sketches, utility balances, plot plan and elevation diagrams, piping isometrics. All diagrams are required to complete the construction of the plant if it is built.

*Accuracy:** Range +6% to −4%

---

*The accuracy range indicates that the capital cost estimate can be expected to cost between the limits shown. For example, the order of magnitude estimate shows an accuracy range of between +40% to −20%. The actual cost of building a chemical plant can be expected to cost between 40% higher and 20% lower than the cost predicted.

Table 2.2   Typical Costs for Making Cost Estimates (1996) (Adapted from references [1] and [2] by special permission from *Chemical Engineering* (October 10, 1977). Copyright © 1977 by McGraw Hill Inc., NY 10020 and The McGraw-Hill Companies (1991).)

Cost of Project	Less Than $2 Million	$2 Million to $10 Million	$10 Million to $100 Million
Classification of estimate	Estimate cost	Estimate cost	Estimate cost
Order of magnitude	$ 3,200	$ 6,400	$ 13,900
Study	$ 21,500	$ 42,900	$ 64,400
Preliminary	$ 53,700	$ 85,800	$ 139,500
Definitive	$ 85,800	$171,700	$ 343,400
Detailed	$214,600	$558,000	$1,073,000

Capital cost estimates are essentially "paper-and-pencil" studies. The cost of making an estimate indicates the personnel hours required in order to complete the estimate. We observe from Figure 2.1 the trend between the accuracy of an estimate and the cost of the estimate. If we require greater accuracy in our capital cost estimate, we must expend more time and money in conducting the estimate. This is the direct result of the greater detail required for the more accurate estimating techniques. Table 2.2 provides additional detail on the costs and relates the cost of the estimate to the cost of the plant being estimated.

What cost estimation technique is appropriate? At the beginning of Chapter 1 a short narrative was given that introduced the evolution of a chemical process leading to a final design and construction of a chemical plant. Cost estimates are performed at each stage of this evolution.

There are hundreds of process systems examined at the block flow diagram level for each process that makes it to the construction stage. Most of the processes initially considered are screened out before any detailed cost estimates are made. Two major areas dominate this screening process. To continue process development the process must be both technically sound and economically attractive.

A typical series of cost estimates that would be carried out in the narrative presented in Chapter 1 are:

- Preliminary feasibility estimates (order-of-magnitude or study estimates) are made to compare several process alternatives.
- More accurate estimates (preliminary or definitive estimates) are made for the most profitable processes identified in the feasibility study.
- Detailed estimates are then made for the more promising alternatives that remain after the preliminary estimates.
- Based on the results from the detailed estimate, a final decision is made whether to go ahead with the construction of a plant.

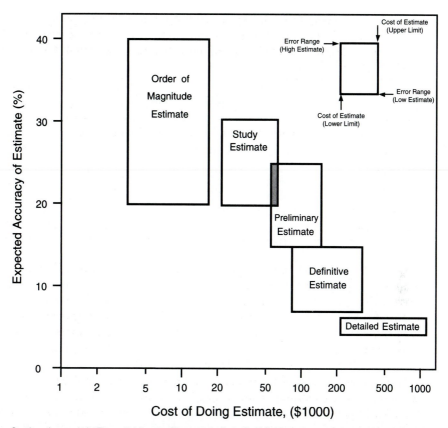

Caution: the x-axis in Figure 2.1 is a nonlinear scale (logarithmic). Most of us perceive graphical information to be linear and special attention must be given to interpreting the significance of a nonlinear plot such as the one in Figure 2.1. Figure 2.1 shows the position of the highest cost detailed estimate to be about three times the distance from the y-axis as the highest cost for an order-of-magnitude estimate. This could easily be interpreted as meaning that the cost of a detailed estimate is about three times that of the order-of-magnitude estimate. This represents a fatal error in interpretation. Using values from Figure 2.1 (1000 and 15) provides this ratio of the maximum cost of a detailed estimate to an order-of-magnitude to be about 80/1 rather than 3/1.

Figure 2.1   Costs and accuracies of different capital cost estimating classifications. (Adapted from references [1] and [2])

This text focuses on the preliminary and study estimation classifications based on a PFD (as presented in Chapter 1). This approach will provide estimates accurate in the range of +30% to –20%.

In this section, it is assumed that all processes considered are technically sound and attention is focused on the economic estimation of capital costs. The technical aspects of processes will be considered in later chapters.

2.2   ESTIMATING PURCHASED EQUIPMENT COSTS

To obtain an estimate of the capital cost of a chemical plant, the costs associated with major plant equipment must be known. For the presentation in this chapter, it is assumed that a PFD for the process is available. This PFD is similar to the one discussed in detail in Chapter 1 that included material and energy balances with each major piece of equipment identified, materials of construction selected, and the size/capacity roughly estimated from conditions on the PFD. Additional PFDs and equipment summary tables are given for four processes in Appendix B.

The most accurate estimate of the purchased cost of a piece of major equipment is provided by a current price quote from a suitable vendor (a seller of equipment). The next best alternative is to use cost data on previously purchased equipment of the same type. Another technique, sufficiently accurate for study and preliminary cost estimates, utilizes summary graphs available for various types of common equipment. This last technique is used for study estimates emphasized in this text and is discussed in detail in Section 2.3. Any cost data must be adjusted for any difference in unit capacity (see Section 2.2.1) and also for any elapsed time since the cost material was generated (see Section 2.2.2).

2.2.1   Effect of Capacity on Purchased Equipment Cost

The relationship between the purchased cost and an attribute of the equipment related to units of capacity is given by

$$\frac{C_a}{C_b} = \left(\frac{A_a}{A_b}\right)^n \tag{2.1}$$

where: A = Equipment cost attribute
C = Purchased cost
n = Cost exponent

Subscripts: a—refers to equipment with the required attribute
b—refers to equipment with the base attribute

The equipment cost attribute is the equipment parameter that is used to correlate capital costs. The equipment cost attribute is most often related to the unit capacity, and the term capacity is commonly used to describe and identify this attribute. Some typical values of cost exponents and unit capacities are given in Table 2.3. From Table 2.3 it can be seen that the following information is given:

1. A description of the type of equipment
2. The units in which the capacity is measured

Table 2.3   Typical Values of Cost Exponents for a Selection of Process Equipment (All data from Table 25–14, reference [2]. Reproduced by permission of McGraw-Hill Companies, 1991)

Equipment Type	Range of Correlation	Units of Capacity	Cost Exponent n
Reciprocating compressor with motor drive	220 to 3000	kW	0.70
Heat exchanger shell and tube carbon steel	5 to 50	m^2	0.44
Vertical tank carbon steel	1 to 40	m^3	0.52
Single-stage blower	0.5 to 4	m^3/s	0.64
Jacketed kettle glass lined	3 to 10	m^3	0.65

3. The range of capacity over which the correlation is valid
4. The cost exponent (values shown for n vary between 0.44 and 0.7)

Eq. (2.1) can be rearranged to give:

$$C_a = K(A_a)^n \qquad (2.2)$$

where: $K = C_b/(A_b)^n$

Eq. (2.2) is a straight line with a slope of n when the log of C_a is plotted versus the log of A_a. To illustrate this relationship we have plotted, in Figure 2.2, the typical cost of a single-stage blower versus the capacity of the blower given as the volumetric flow-rate. The value for the cost exponent, n, from this curve is 0.64.

The value of the cost exponent, n, used in Eqs. (2.1) and (2.2) varies depending upon the class of equipment being represented (see Table 2.3). The value of n for different equipment is often around 0.6. Replacing n in Eq. (2.1) and/or Eq. (2.2) by 0.6 provides the relationship referred to as the *six-tenths-rule*.

Example 2.3

Use the six-tenths-rule to estimate the % increase in purchased cost when the capacity of a piece of equipment is doubled.

Using Eq. (2.1) with $n = 0.6$: $C_a/C_b = (2/1)^{0.6} = 1.52$
% increase = $((1.52-1.00)/1.00)(100) = 52\%$

This simple example illustrates a concept referred to as the *economy of scale*. While the equipment capacity was doubled, the purchased cost of the equipment increased by only 52%. This leads to the generalization shown as the following highlight.

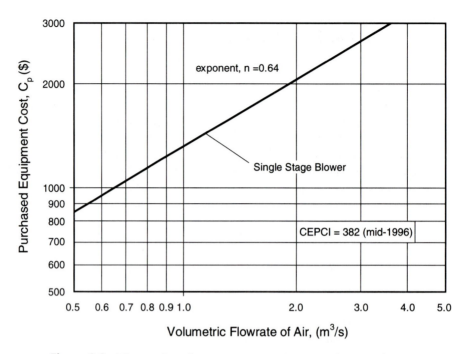

Figure 2.2   The purchased cost versus capacity curve for a single stage air blower.

The larger the equipment the lower the cost of equipment per unit of capacity.

Special care must be taken in using the six-tenths-rule for a single piece of equipment. The cost exponent may vary considerably from 0.6. The use of this rule for a total chemical process is more reliable and is discussed in Section 2.3.

Example 2.4

Compare the error for the scale-up of a heat exchanger by a factor of 5 using the six-tenth-rule in place of the cost exponent given in Table 2.3.

Using Eq. (2.1):

Cost ratio using six-tenths-rule (i.e., $n = 0.60$) $= 5.0^{0.60} = 2.63$
Cost ratio using ($n = 0.44$) from Table 2.3 $= 5.0^{0.44} = 2.03$

% Error $= (2.63 - 2.03)/(2.03)(100) = 29.5$ %

In the last two examples, the relative costs of equipment of differing size were calculated. It is necessary to have cost information on the equipment at some "base case" in order to be able to determine the cost of other similar equipment. This base-case information must allow for the constant, K, in Eq. (2.2), to be evaluated. This base-case cost information may be obtained from a current bid provided by a manufacturer for the needed equipment or from company records of prices paid for similar equipment.

Example 2.5

The purchased cost of a recently acquired heat exchanger with an area of 100 square meters was \$10,000.
   Determine:

a. the constant K in Eq. (2.2)
b. the cost of a new heat exchanger of 180 m^2.

a. From Table 2.3: $n = 0.44$: For Eq. (2.2):

$$K = C_b/(A_b)^n = 10,000/(100)^{0.44} = 1320 \ \{\$/(m^2)^{0.44}\}$$

b. $C_a = (1320)(180^{0.44}) = \$13,000$

There are additional techniques that allow for the price of equipment to be estimated that do not require information from either of the sources given above. One of these techniques is discussed in Section 2.3.

2.2.2   Effect of Time on Purchased Equipment Cost

When you depend upon past records or published correlations for price information, you must be able to update these costs to take into account changing economic conditions (inflation). You can do this by using

$$C_2 = C_1\left(\frac{I_2}{I_1}\right) \tag{2.3}$$

where: $C =$ Purchased Cost
            $I =$ Cost Index
Subscripts: 1—refers to base time when cost is known
               2—refers to time when cost is desired

There are several cost indices used by the chemical industry to adjust for the effects of inflation. Several of these cost indices are plotted in Figure 2.3.

All indices in Figure 2.3 show similar inflationary trends with time. The indices most generally accepted in the chemical industry and reported in the back page of every issue of *Chemical Engineering* are:

1. The Marshall and Swift Equipment Cost Index
2. The Chemical Engineering Plant Cost Index

Table 2.4 provides values for both the Marshall and Swift Equipment Cost Index and the Chemical Engineering Plant Cost Index from 1978 to the present.

Unless otherwise stated, the Chemical Engineering Plant Cost Index (CEPCI) will be used in this text to account for inflation. This is a composite index, and the items that are included in the index are described in Table 2.5.

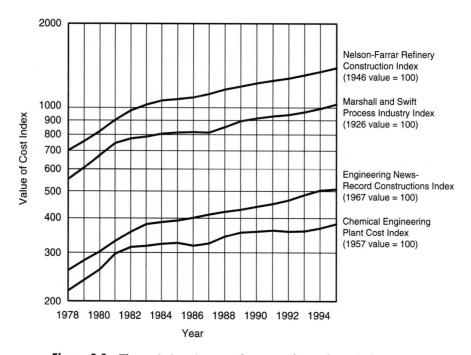

Figure 2.3   The variations in several commonly used cost indexes over the seventeen-year period 1978 to 1995.

Table 2.4   Values for the Chemical Engineering Plant Cost Index and the Marshall and Swift Equipment Cost Index from 1978 to 1996

Year	Marshall & Swift Equipment Cost Index	Chemical Engineering Plant Cost Index
1978	552	219
1979	607	239
1980	675	261
1981	745	297
1982	774	314
1983	786	317
1984	806	323
1985	813	325
1986	817	318
1987	814	324
1988	852	343
1989	895	355
1990	915	358
1991	931	361
1992	943	358
1993	964	359
1994	993	368
1995	1028	381
1996 (mid year)	1037	382

Example 2.6

The purchased cost of a heat exchanger of 500 m^2 area in mid-1978 was $25,000.

a. Estimate the cost of the same heat exchanger in mid-1996 using the two indices introduced above.

b. Compare the results.

From Table 2.4:	Year	1978	mid-1996
Marshal and Swift Index		552	1037
Chemical Engineering Plant Cost Index		219	382

Table 2.5   The Basis for the Chemical Engineering Plant Cost Index

Components of Index	Weighting of Component (%)
Equipment, Machinery, and Supports:	
(a) Fabricated equipment	37
(b) Process machinery	14
(c) Pipe, valves, and fittings	20
(d) Process instruments and controls	7
(e) Pumps and compressors	7
(f) Electrical equipment and materials	5
(g) Structural supports, insulation, and paint	10
	<u>100</u> 61% of total
Erection and installation labor	22
Buildings, materials, and labor	7
Engineering and supervision	10
Total	100

a. Marshal and Swift: Cost = ($25,000)(1037/552) = $46,970
 Chemical Engineering: Cost = ($25,000)(382/219) = $43,610
b. Average Difference: (($46,970 − 43,610)/($43,970+43,610)/2)(100) = 7.4%

2.3   ESTIMATING THE TOTAL CAPITAL COST OF A PLANT

Consider the costs associated with building a new home:

> *The purchased cost of all the materials that are needed to build a home does not represent the cost of the home. The final cost reflects the cost of property, the cost for delivering materials, the cost of construction, the cost of a driveway, the cost for hooking up utilities, and so on.*

In a parallel process, the capital cost for a chemical plant must take into consideration many costs other than the purchased cost of the equipment. In Table 2.6, we present a summary of the costs that you must consider in the evaluation of the total capital cost of a chemical plant.

The estimating procedures to obtain the full capital cost of the plant are described in this section. If you need an estimate of the capital cost for a process plant, where you have access to a previous estimate for a similar plant with a different capacity, you may apply the principles already introduced for the scaling of purchased costs of equipment, namely:

Table 2.6   Factors Affecting the Costs Associated with Evaluation of Capital Cost of Chemical Plants (From references [2] and [4])

Factor Associated with the Installation of Equipment	Symbol	Comments
(1) Direct Project Expenses		
(a) Equipment f.o.b. cost	C_p	Purchased cost of equipment at manufacturer's site. (f.o.b. = free on board)
(b) Materials required for installation	C_M	Includes all piping, insulation and fire proofing, foundations and structural supports, instrumentation and electrical, and painting associated with the equipment.
(c) Labor to install equipment and material	C_L	Includes all labor associated with installing the equipment and materials mentioned in (a) and (b).
(2) Indirect Project Expenses		
(a) Freight, insurance, and taxes	C_{FIT}	Includes all transportation costs for shipping equipment and materials to the plant site; all insurance on the items shipped; and any purchase taxes that may be applicable.
(b) Construction overhead	C_O	Includes all fringe benefits such as vacation, sick leave retirement benefits, etc.; labor burden such as social security and unemployment insurance, etc.; and salaries and overhead for supervisory personnel.
(c) Contractor engineering expenses	C_E	Includes salaries and overhead for the engineering, drafting, and project management personnel on the project.
(3) Contingency and Fee		
(a) Contingency	C_{Cont}	A factor included to cover unforeseen circumstances. These may include loss of time due to storms and strikes, small changes in the design, and unpredicted price increases.
(b) Contractor fee	C_{Fee}	This fee varies depending on the type of plant and variety of other factors.
(4) Auxiliary Facilities		
(a) Site development	C_{Site}	Includes the purchase of land; grading and excavation of the site; installation and hookup of electrical, water, and sewer systems; and construction of all internal roads, walkways, and parking lots.

(continued)

Table 2.6   Factors Affecting the Costs Associated with Evaluation of Capital Cost of Chemical Plants (From references [2] and [4]) (*continued*)

Factor Associated with the Installation of Equipment	Symbol	Comments
(b) Auxiliary Buildings	C_{Aux}	Includes administration offices, maintenance shop and control rooms, warehouses, and service buildings (e.g., cafeteria, dressing rooms, and medical facility).
(c) Offsites and Utilities	C_{Off}	Include raw material and final product storage, raw material and final product loading and unloading facilities, all equipment necessary to supply required process utilities (e.g., cooling water, steam generation, fuel distribution systems, etc.), central environmental control facilities (e.g., waste water treatment, incinerators, flares, etc.), and fire protection systems.

1. The six-tenths-rule (Eq. (2.1) with n set to 0.6) may be used to scale up/down to a new capacity.
2. The Chemical Engineering Plant Cost Index should be used to update the capital costs (Eq. (2.3)).

The six-tenths-rule is more accurate in this application than it is for estimating the cost of a single piece of equipment. The increased accuracy results from the fact that multiple units are required in a processing plant. Some of the process units will have cost coefficients, n, less than 0.6. For this equipment the six-tenths-rule overestimates the costs of these units. In a similar way, costs for process units having coefficients greater than 0.6 are underestimated. When the sum of the costs are determined, these differences tend to cancel each other out.

The Chemical Engineering Plant Cost Index (CEPCI) can be used to account for changes that result from inflation. The CEPCI values provided in Table 2.4 are composite values that reflect the inflation of a mix of goods and services associated with the Chemical Process Industries (CPI).

> *You may be familiar with the more common consumer price index issued by the government. This represents a composite cost index that reflects the effect of inflation on the cost of living. This index considers the changing cost of a "basket" of goods comprising of items used by the "average" person. For example, the price of housing, cost of basic foods, cost of clothes and transportation, and so on, are included and weighted appropriately to give a single number reflecting the average cost of these goods. By comparing this number over time, we get an indication of the rate of inflation as it affects the "average" person.*

In a similar manner, the Chemical Engineering Plant Cost Index represents a "basket" of items directly related to the costs associated with the construction of chemical plants. A breakdown of the items included in this index is given in Table 2.5. The index is directly related to the effect of inflation on the cost of an "average" chemical plant.

Example 2.7

The capital cost of a 30,000 metric ton/year isopropanol plant in 1980 was estimated to be $5,000,000. Estimate the capital cost of a new plant with a production rate of 50,000 metric tons/year in mid-1996.

$$\text{Cost in 1996} = (\text{Cost in 1980})(\text{Capacity Correction})(\text{Inflation Correction})$$
$$= (\$5,000,000)(50,000/30,000)^{0.6} (382/261)$$
$$= (\$5,000,000)(1.36)(1.46) = \$9,940,000$$

In most situations, you will not have cost information available for the same process configuration; therefore, other estimating techniques must be used.

A simple technique to estimate the capital cost of a chemical plant is the Lang Factor method [5,6,7]. The cost determined from the Lang Factor represents the cost to build a major expansion to an existing chemical plant. The total cost is determined by multiplying the total purchased cost for all the major items of equipment by a constant. The major items of equipment are those shown in the process flow diagram. The constant multiplier is called the Lang Factor. Values for Lang Factors, F_{Lang}, are given in Table 2.7.

The capital cost calculation is determined using Eq. (2.4).

$$C_{TM} = F_{Lang} \sum_{i=1}^{n} C_{p,i} \tag{2.4}$$

where: C_{TM} = the capital cost of the plant,
$\quad\quad\quad\quad C_{p,i}$ = the purchased cost for the major equipment units
$\quad\quad\quad\quad n$ = the total number of individual units
$\quad\quad\quad\quad F_{Lang}$ = the Lang Factor (from Table 2.7)

Table 2.7   Lang Factors for the Estimation of Capital Cost for Chemical Plants (From references [5,6,7])

Type of Chemical Plant	Lang Factor = F_{Lang}
Fluid Processing Plant	4.74
Solid-Fluid Processing Plant	3.63
Solid Processing Plant	3.10
Capital Cost = (Lang Factor)(Sum of Purchased Costs of all Major Equipment)	

Plants processing only fluids have the largest Lang Factor, 4.74, and plants processing only solids have a factor of 3.10. Combination fluid-solid systems fall between these two values. The greater the Lang Factor, the lower the contribution of the purchased costs to the plant cost. For all cases, the purchased cost of the equipment is less than a third of the capital cost of the plant.

Example 2.8

Determine the capital cost for a major expansion to a fluid processing plant that has a total purchased equipment cost of $6,800,000.

$$\text{Capital costs} = (\$6,800,000)(4.74) = \$32,232,000$$

This estimating technique is insensitive to changes in process configuration, especially between processes in the same broad categories shown in Table 2.7. It cannot accurately account for the common problems of special materials of construction and high operating pressures. A number of alternative techniques are available. All require more detailed calculations using specific price information for the individual units/equipment.

The equipment module costing technique is one of these techniques. It is generally accepted as the best for making preliminary cost estimates and is used extensively in this text. This approach, introduced by Guthrie [8,9] in the late 1960s and early 1970s, forms the basis of many of the equipment module techniques in use today. This costing technique relates all costs back to the purchased cost of equipment evaluated for some base conditions. Deviations from these base conditions are handled by using a multiplying factor. This factor will depend on the following:

1. the specific equipment type
2. the specific materials of construction
3. the specific system pressure

The material provided in the next section is based upon information provided in *A Guide to Chemical Engineering Process Design and Economics*, by Ulrich [4]. The reader is strongly encouraged to review this excellent reference for more specific material.

Eq. (2.5) is used to calculate the bare module cost for each piece of equipment. The bare module cost is the sum of the direct and indirect costs shown in Table 2.6.

$$C_{BM}^o = C_p F_{BM}^o \tag{2.5}$$

where: C_{BM}^o = Bare module equipment cost: direct and indirect costs for each unit.

F_{BM}^o = Bare module cost factor: multiplication factor to account for the items in Table 2.6 plus the specific materials of construction and operating pressure.

C_p = Purchased cost for base conditions: equipment made of the most common material, usually carbon steel and operating at near ambient pressures.

Because of the importance of this cost estimating technique, it is described below in detail.

2.3.1   Bare Module Cost for Equipment at Base Conditions

The bare module equipment cost represents the sum of direct and indirect costs shown in Table 2.6. The conditions specified for the base case are

1. unit fabricated from most common material, usually carbon steel (CS)
2. unit operated at near ambient pressure

Eq. (2.5) is used to obtain the bare module cost for the base conditions. For these base conditions, the superscripts on the bare module cost factor and the bare module equipment cost are dropped. Thus C_{BM} and F_{BM} refer to the base conditions.

Table 2.8 represents a supplement to Table 2.6. It provides relationships and equations for the direct, indirect, contingency, and fee costs based on the purchased cost of the equipment. These equations are used to evaluate the bare module factor. The entries in Table 2.8 are described below.

Column 1: Lists the factors given in Table 2.6

Column 2: Provides equations used to evaluate each of the costs. These equations introduce multiplication cost factors, α_i. Each cost item, other than the purchased equipment cost, introduces a separate factor.

Column 3: For each factor, the cost is related to the purchased cost C_p by an equation of the form.

$$C_{XX} = C_p f(\alpha_i) \tag{2.6}$$

The function, $f(\alpha_i)$, for each factor is given in column 3.

From Table 2.8 and Eqs. (2.5) and (2.6), it can be seen that the bare module factor is given by

$$F_{BM} = [1 + \alpha_L + \alpha_{FIT} + \alpha_O \alpha_L + \alpha_E][1 + \alpha_M] \tag{2.7}$$

Table 2.8   Equations for Evaluating Direct, Indirect, Contingency, and Fee Costs

Factor	Basic Equation	Multiplying Factor to Be Used with Purchased Cost, C_p
1. Direct		
(a) Equipment	$C_p = C_p$	1.0
(b) Materials	$C_M = \alpha_M C_p$	α_M
(c) Labor	$C_L = \alpha_L(C_p + C_M)$	$(1.0 + \alpha_M)\alpha_L$
Total Direct	$C_{DE} = C_p + C_M + C_L$	$(1.0 + \alpha_M)(1.0 + \alpha_L)$
2. Indirect		
(a) Freight, etc.	$C_{FIT} = \alpha_{FIT}(C_p + C_M)$	$(1.0 + \alpha_M)\alpha_{FIT}$
(b) Over-head	$C_O = \alpha_O C_L$	$(1.0 + \alpha_M)\alpha_L\alpha_O$
(c) Engineering	$C_E = \alpha_E(C_p + C_M)$	$(1.0 + \alpha_M)\alpha_E$
Total Indirect	$C_{IDE} = C_{FIT} + C_O + C_E$	$(1.0 + \alpha_M)(\alpha_{FIT} + \alpha_L\alpha_O + \alpha_E)$
Bare Module	$C_{BM} = C_{IDE} + C_{DE}$	$(1.0 + a_M)(1.0 + a_L + a_{FIT} + a_L a_O + a_E)$
3. Contingency & Fee		
(a) Cont.	$C_{Cont} = \alpha_{Cont} C_{BM}$	$(1.0 + \alpha_M)(1.0 + \alpha_L + \alpha_{FIT} + \alpha_L\alpha_O + \alpha_E)\alpha_{Cont}$
(b) Fee	$C_{Fee} = \alpha_{Fee} C_{BM}$	$(1.0 + \alpha_M)(1.0 + \alpha_L + \alpha_{FIT} + \alpha_L\alpha_O + \alpha_E)\alpha_{Fee}$
Total Module	$C_{TM} = C_{BM} + C_{Cont} + C_{Fee}$	$(1.0 + \alpha_M)(1.0 + \alpha_L + \alpha_{FIT} + \alpha_L\alpha_O + \alpha_E) \times$ $(1.0 + \alpha_{Cont} + \alpha_{Fee})$

The values for the cost multiplying factors vary between equipment modules.

Example 2.9

The purchased cost for a carbon steel heat exchanger operating at ambient pressure is $10,000. For a heat exchanger module, Ulrich [4] provides the following cost multiplying factors.

Cost Multiplier	Value
α_M	0.71
α_L	0.37
α_{FIT}	0.08
α_O	0.70
α_E	0.15

Using the information given above determine the following:

a. Bare module cost factor, F_{BM}
b. Bare module cost, C_{BM}
c. Materials and labor costs to install the heat exchanger

a. Using Eq. (2.7):

$$F_{BM} = (1 + 0.37 + 0.08 + (0.7)(0.37) + 0.15)(1 + 0.71) = 3.18$$

b. From Eq. (2.5):

$$C_{BM} = (3.18)(\$10{,}000) = \$31{,}800$$

c. Installation Costs

Materials: From Table 2.8

$$C_M = \alpha_M\, C_p = (0.71)(\$10{,}000) = \$7{,}100$$

Labor: From Table 2.8

$$C_L = \alpha_L\,(1 + \alpha_M)\, C_p = (0.37)(1 + 0.71)(\$10{,}000) = \$6{,}330$$

Cost of Labor and Materials = $7,100 + $6,330 = $13,430

Fortunately, we do not have to repeat the procedure illustrated in the above example in order to estimate F_{BM} for every piece of equipment. This has already been done for a large number of equipment modules and the results are provided in Appendix A. In addition, other equipment may be found in the references cited in this appendix.

Different authors provide the same information in a number of different formats. We use a graphical presentation based on the information in Ulrich [4]. Copies of the graphs and tables for shell and tube heat exchangers given in Appendix A (Figures A.1, A.2, A.3, and Table A.2) are reproduced here as Figures 2.4, 2.5, and 2.6, and Table 2.9.

Figure 2.4 provides the information necessary to estimate the purchased price of the base exchanger, C_p, for a specified heat transfer area. To obtain the bare module cost, C_{BM}, using Eq. (2.5), the value for the bare module factor, F_{BM} must be known. This is obtained from Figure 2.6. To obtain this value both F_M and F_p must be known (x-axis is shown as the product $F_M F_p$), where

F_M = Materials factor to account for materials of construction
   (for carbon steel, $F_M = 1$)
F_p = Pressure factor to account for high pressure from Figure 2.5
   (for ambient pressure, $F_p = 1$)

In the following example, we use this information to determine the bare module cost for a heat exchanger. It should be noted that the pressure corrections for bare module cost presented in this book are all in terms of gauge pressure. Thus 1.0 bar is equal to 0.0 bar gauge or 0.0 barg.

Example 2.10

Find the mid-1996 bare module cost of a floating head shell and tube heat exchanger with a heat transfer area of 100 m². The operating pressure of the equipment is 1.0 bar with both

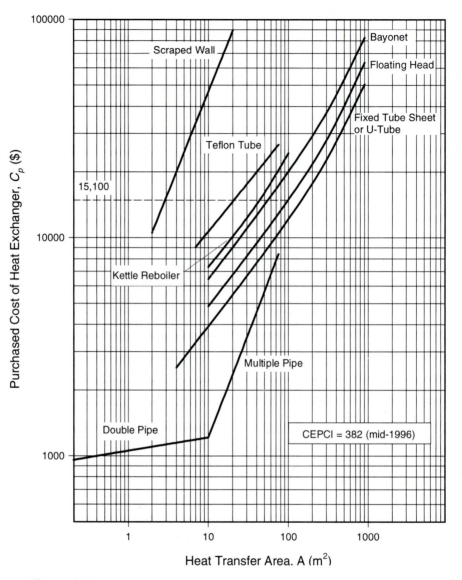

Figure 2.4 Purchased equipment costs for shell and tube heat exchangers (from Figure A.1).

shell and tube sides constructed of carbon steel. For this material and pressure the values of F_p and F_M are equal to 1.0.

From Figure 2.4: $C_p(1996) = \$ 15{,}100$ (evaluation path is shown on Figure 2.4)
$F_p = 1.0$ ($F_p = 1.0$ for $P < 10$ barg)

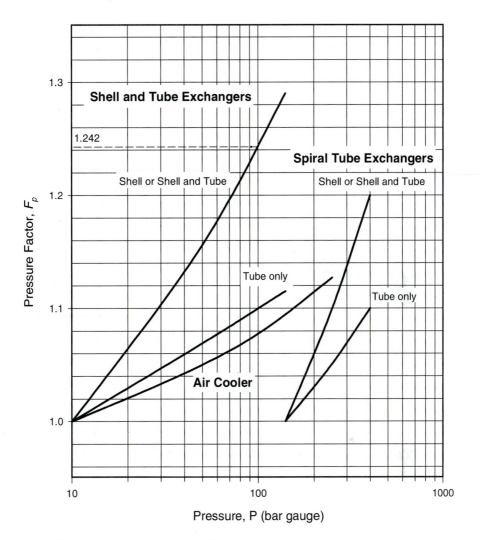

Figure 2.5   Pressure factors (F_p) for heat exchangers (from Figure A.2).

From Table 2.9: $F_M = 1.0$ (CS)
From Figure 2.6: $F_{BM} = 3.2$ where $F_p F_M = (1)(1) = 1$ (path shown on Figure)
Using Eq. (2.5): $C_{BM}(1996) = (\$15,100)(3.2) = \$48,300$

The value for the bare module cost factor of 3.2 in Example 2.10 is the same (within the accuracy of reading the graph) as the value of 3.18 evaluated using the individual values for α given in Example 2.9.

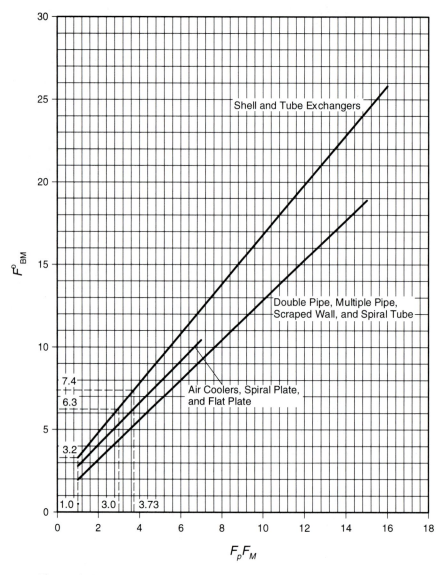

Figure 2.6 Bare module factors (F_{BM}^o) for heat exchangers (from Figure A.3).

2.3.2 Bare Module Cost for Non-Base Case Conditions

For equipment made from other materials of construction and/or operating at nonambient pressure, the values for F_M and F_p are greater than 1.0. In the equipment module technique, these additional costs are incorporated into the bare

Table 2.9 Material Factors Floating Head Heat Exchangers (From Table A.2 and references [4] and [7])

Shell Material	Tube Material	Material Factor, F_M
Carbon steel (CS)	Carbon steel (CS)	1.00
Carbon steel (CS)	Copper (Cu)	1.25
Copper (Cu)	Copper (Cu)	1.60
Carbon steel (CS)	Stainless steel (SS)	1.70
Stainless steel (SS)	Stainless steel (SS)	3.00
Carbon steel (CS)	Nickel alloy (Ni)	2.80
Nickel alloy (Ni)	Nickel alloy (Ni)	3.80
Carbon steel (CS)	Titanium (Ti)	7.20
Titanium (Ti)	Titanium (Ti)	12.00

module cost factor, F_{BM}. The bare module factor used for the base case, F_{BM}, is replaced with an actual bare module cost factor, F_{BM}^o in Eq. (2.5). The information needed to determine this actual bare module factor is provided in Appendix A.

Example 2.11

Repeat Example 2.10 except that the exchanger is made with stainless steel shell and tubes.

From Example 2.10: $C_p(1996) = \$15,100$, $F_p = 1$
From Table 2.9: $F_M = 3.0$
From Figure 2.6: $F_{BM}^o = 6.3$ using $F_p F_M = (1)(3.0) = 3.0$
Using Eq. (2.5): $C_{BM}^o(1996) = (\$15,100)(6.3) = \$95,100$

The purchased cost of the equipment for the stainless steel exchanger in Example 2.10 would be $C_p(1996) = (\$15,100)(3) = \$45,300$. If this equipment cost were multiplied by the bare module factor for the base case, the cost would become $C_{BM}^o = (\$45,300)(3.2) = \$145,000$. This is 52% greater than the \$95,100 calculated in Example 2.11. The difference between these two costs resulted from assuming, in the latter case, that all costs increased in direct proportion to the increase in material cost. This is far from the truth. Some costs such as freight, insulation, and foundations show little or no change with the cost of materials while other costs such as installation materials and labor costs are impacted to a varying extent. The method of equipment module costing accounts for these variations in the bare module factor.

To obtain the bare module cost factor, it is necessary to know the materials of construction. Table 2.10 provides information on the types of materials of construction to be used in different types of service. Table 2.10 may be used to select

Table 2.10   Table for the Selection of Materials of Construction (From Ulrich [4])

	Construction Material				
	Steels		Other Metals and Alloys		
	Carbon Steel (cs) and Alloy Steel (as)	Stainless Steels (ss)	Aluminum-Based (Al)	Copper-Based (Cu)	Nickel-Based (Ni)
Temperature (°C): Exposure	0 200 400 600	−200 0 200 400 600 800	−200 0 200	−200 0 200	−200 0 200 400 600
Aqueous Solutions					
Acetates	XXXX	AAAA	AAAA	AAAA	BBBB
Ammonium salts	BDDE	AAAA	BBCC	DEXX	AABC
Carbonates	BBBB	AAAA	DDEE	AABC	BBCC
Chlorides	DDEX	CCCD	DDEE	AACC	CCCC
Nitrates and nitrites	AABB	AAAA	BBDD	BBCD	AABC
Sulfates and Sulfites	BBBB	AAAA	DDDD	BBCC	AAAA
Acid Solutions and Wet Acid Vapors					
HBr	XXXX	XXXX	XXXX	XXXX	CCCC
HCl	XXXX	XXXX	XXXX	CCXX	CCCC
HF	XXXX	XXXX	XXXX	CCCC	CCCC
HNO_3	XXXX	AAAA	XXXX	XXXX	AAAA
H_2SO_4	XXXX	BBDX	DDXX	XXXX	ABCX
H_2PO_4	XXXX	AABC	XXXX	CCXX	AABC
Organic acids	XXXX	AABC	AABX	AABX	AAAA
Basic Solutions and Wet Vapors					
Ca (OH)$_2$	BBDD	AAAA	XXXX	AAAA	AAAA
NaOH	BBBB	AAAA	XXXX	XXXX	AAAA
NH_3OH	XXXX	AAAA	BBDD	XXXX	AABC
Food Intermediates					
Dairy products	XXXX	AAAA	DDXX	BBDD	AAAA
Fruit juices	XXXX	AAAA	ABDX	BBBB	BBBB
Sugar syrups	BBBB	AAAA	AAAA	BBBB	AAAA
Vegetable oils	BBBBBBBB	AAAAAAAA	AAAAA	BBDDDD	AAAAAAAA
Vinegar	XXXX	AAAA	BBBB	BBBB	AAAA
Gases (moist) and Cryogenic Liquids					
Air	AAAAAAABBDDDD	AAAAAAAAAAAABBBBBDDE	AAAAAAAAD	AAAAAAABDD	AAAAAAAAAAAAAABBD
Br$_2$	XXXXXXXXXXXX	XXXXXXXXXXXXXXX	XXXXX	CCCCCC	AAAAAAAAAAAA
CO$_2$	XXXXBBBBBBBBB	DDDDAAAAAAAAAAAA	BBBBAA	AABEEBBB	AAAAAAAAAAAA
Cl$_2$	XXXXXXXXXXXX	ABXXXXXXXXXXXXXX	XXXXXXX	XXXXXXX	AAAAAAAAAAAA
F$_2$	XXXXXXXXXXXX	AAABBXXXXXXXXXXXXXXX	XXXXXXXXX	XXXXXXXXX	AAAAAAAAAAAADDDDD
Flue gases	XXXBBBBBBBBB	XXXBBBBBBBBBBBBB	XXXBB	CCCCCC	AAAAAAAAAAAA
HBr	XXXXXXXXXXXX	BBDXXXXXXXXXXXXXX	XXXXXXA	XXXXXXXX	AABBBBBBBBBBBBBB
HCl	XXXXXXXXXXXX	BBDXXXXXXXXXXXXXX	XXXXXXA	XXXXXXXX	AABBBBBBBBBBBBBB
HF	XXXXXXXXXXXX	BBDXXXXXXXXXXXXXX	XXXXXBB	XXXXXXXX	AABBBBCCCCEEEEE
H$_2$	DDDXXXXXXXXX	AAAAAAAAAAAAAAAAA	AAAAAAAAA	AAAAAAAAA	AAAAAAAAAAAAAAA
H$_2$S	ABDDBBBBBBBBB	AAAAAAAAAAAAAAAA	AABDDBB	XXXXXXXX	AAABBBBBBBBBBBBB
Halogenated hydrocarbons	BBBDEXXXXXXXX	BBBBBBBBBBBBBBBBB	XXXXXX	CCCCCCC	AAAAAAAAAAAAAA
Hydrocarbons	BBBBBBBBBBBB	BBBBBBBBBBBBBBBBB	AAAAAAAAA	CCCCCCCCCC	AAAAAAAAAAAAAAA
NH$_3$	BBBBBBBBBBBB	BBBBBBBBBBBBBBBBB	BBBBBBB	XXXXXXXX	AAAAAAAAAAAAAAA
N$_2$	AAAAAAAAAAAA	AAAAAAAAAAAAAAAA	AAAAAAAAA	AAAAAAAAA	AAAAAAAAAAAAAAA
O$_2$	AAADXXXXXXXX	AAAAAAAAAADDDDEEXXXX	AAAAAAAAA	AAAAAAABDD	AAAAAAAAAAAAABBB
SO$_2$	BBXXBBBBBBBB	BBBXXBBBBBBBBBBBB	BBXXXBB	CCCCCCCC	AAAAAAAAAAAA
SO$_3$	BXXBBBBBBBBB	BXXBBBBBBBBBBBB	XXXBB	CCCCCC	AAAAAAAAAAAA
Steam	AAAAAAAAAAAA	AAAAAAAAAAAAAAAA	BBBBB	BBBCCC	AAAAAAAAAAAA
Liquids and Solvents					
Acetone	AAAA	AAAA	AAAA	AAAA	AAAA
Alcohols	BBBB	AAAA	AAAA	AAAA	AAAA
Dowtherm	AAAABBBBBB	AAAAAAAAA	AAABX	AAABBXX	AAAAAAAAA
Ethers	BBDD	AAAA	AAAA	AAAA	AAAA
Freon	XXXX	AAAAA	AAAAA	AAAAA	AAAAA
Glycols	AAAA	AAAA	AAAA	AAAA	AAAAA
Halogenated hydrocarbons	BBBD	BBBB	XXXX	CCDE	AAAA
Hydrocarbons	BBBB	BBBB	BBBB	CCCC	AAAA
Mercury	AAAAAAAAAAAA	CCCCCCCCCCCCCCCC	XXXXXX	XXXXXXX	CCCDDDDDDDEEEX
Molten alkali metals	AAAAAAAAAA	AAAAAAAAAAAAA	XXXXX	AAAAA	AAAAAAAAAAAA
Molten salts					
Halides	AAA	AAAAAA	X	X	AAAAA
Nitrates	AAAAAAAA	AAAAAAAA	XX	XX	AAAAAAAA
Sulfates	AAAAAAAAAA	AAAAAAAAAAAAAA	XXXX	XXXX	AAAAAAAAAAAA
Water					
Boiler feed	AAA	AAA	CCC	AAA	AAA
Brackish	XXX	BBC	XXX	ABD	AAA
Cooling tower	AAA	AAA	CCC	AAA	AAA
Fresh	XXX	AAA	XXX	ABD	AAA
Sea	XXX	CCD	XXX	CCC	AAA
Temperature (°C):	0 200 400 600	−200 0 200 400 600 800	−200 0 200	−200 0 200	−200 0 200 400 600

KEY

A   excellent or no limitations; B   modest limitations; C   special materials available at higher cost to minimize problems; D   limited in this regard; E   severely limited in this regard; X   unacceptable

Construction Material

Titanium-Based (Ti)	Plastics — Conventional (cp) and Fiberglass-Reinforced (frp)	Plastics — Fluorocarbon Plastics (fp)	Linings — Rubber-Lined (rl)	Linings — Glass-Lined (gl)	Refractories — Ceramics (c)	Refractories — Graphite (g)
-200 0 200 400 600	0 200	-200 0 200	-200 0 200	0 300	0 1000 2000	0 1000 2000
AAAA	AABCC	AAAAAAD	AABCCCD	AAA	A	A
AAAA	AAACC	AAAAAAD	AAACCCD	AAA	A	A
AAAA	AAACC	AAAAAAD	AAACCCD	AAA	A	A
AAAA	AAACC	AAAAAAD	AAACCCD	AAA	A	A
AAAA	AAACC	AAAAAAD	AAACCCD	AAA	A	A
AAAA	AAACC	AAAAAAD	AAACCCD	AAA	A	A
BBBB	AABCC	AAAAAAD	AABCCCD	AAA	A	A
CCCC	AABCC	AAAAAAD	AABCCCD	AAA	A	A
XXXX	ABCCC	AAAAAAD	ABCCCCD	XXX	C	X
BBBB	CCCEE	AAAAAAD	CCCCCCD	AAA	A	X
CCCX	CCCEE	AAAAAAD	CCCCCCD	AAA	A	X
AAAB	AABCC	AAAAAAD	AABCCCD	AAA	A	E
AAAA	AABCC	AAAAAAD	AABCCCD	AAA	A	B
AAAA	AAACC	AAAAAAD	AAACCCD	AAD	A	A
DDDD	AAACC	AAAAAAD	AAACCCD	AAD	C	A
AAAA	AAACC	AAAAAAD	AAACCCD	AAD	A	A
AAAA	AAACC	AAAAAAD	AAACCCD	AAA	A	A
AAAA	AAACC	AAAAAAD	AAACCCD	AAA	A	A
AAAA	AAACC	AAAAAAD	AAACCCD	AAA	A	A
AAAAAAAA	ABCCC	AAAAAAD	ABCCCCD	AAAA	A	A
AAAA	AABCC	AAAAAAD	AABCCCD	AAA	A	A
AAAAAAAAAAAABBBBD	AAACC	AAAAAAAAAD	CCCCAAACCCD	AAAA	AAAAAAAAA	AADXXXXXXX
XXXXXXXXXXXX	CCCCC	AAAAAD	CCCCCD	AAAA	AAAAAAAAA	BDXXXXXXXX
AAAAAAAAAAAAAA	AAACC	AAAAAAAD	CCAAACCD	AAAA	AAAAAAAAA	AADXXXXXXX
XXXXXXXXXXXXXXX	CCCCC	AAAAAAAD	CCCCCCCD	AAAA	AAAAAAAAA	AAAAAAAAAA
XXXXXXXXXXXXXXXXXX	DDDEE	DDDDDDDDEEE	DDDDDDDDEEE	DDDD	DDDDDCCCCC	DXXXXXXXXX
BXXBBBBBBBBBB	AAACC	AAAAAD	AAACCD	AAA	AAAAAAAAA	AADXXXXXXX
BBBBBBBBBBBBBBBB	AABCC	AAAAAAD	CCAABCCD	AAAA	AAAAAAAAA	AAAAAAAAAA
XXXXXXXXXXXXXXX	AABCC	AAAAAAD	CCAABCCD	AAAA	AAAAAAAAA	AAAAAAAAAA
XXXXXXXXXXXXXXX	DDDEE	DDDDDDDD	DDDDDDDD	XXXX	CCCCCCCCC	DXXXXXXXXX
AAAAAAAAAAAAAA	AABCC	AAAAAAD	CCAABCCD	AAAA	AAAAAAAAA	AAAAAAAAAA
AAAAAAAAAAAAAA	AABCC	AAAAAAAAAD	CCCCCAABCCD	AAAA	AAAAAAAAA	AAAAAAAAAA
AAAAAAAAAAAAAA	DDDDD	DDDDDDD	DDDDDDD	AAAA	AAAAAAAAA	AAABBDEXXX
AAAAAAAAAAAAAA	CCCCC	AAAAAAAAAD	CCCCCCCCCD	AAAA	AAAAAAAAA	AAABBDEAAA
BBBBXXXXXXXXXXXXX	AACCC	AAAAAAD	CCAACCCD	AAAA	AAAAAAAAA	AAAAAAAAAA
AAAAAAAAAAAAAA	AAACC	AAAAAAAAAD	CCCCCAAACCD	AAAA	AAAAAAAAA	AAAAAAAAAA
AAAAAAAAAAAABBBDD	AAACC	AAAAAAAAAD	CCCCCAAACCD	AAAA	AAAAAAAAA	AADXXXXXXX
BBBXXBBBBBBBBBB	ABCCC	AAAAAAD	CCABCCCD	AAAA	AAAAAAAAA	AADXXXXXXX
BXXBBBBBBBBBB	CCCEE	AAAAAD	CCCCCD	AAA	AAAAAAAAA	AADXXXXXXX
AAAAAAAAAAAAAA	AAACC	AAAAAD	AAACCD	AAA	AAAAAAAAA	AADXXXXXXX
AAAA	AACCC	AAAAAAD	AACCCCD	AAA	A	A
AAAAAAAA	AABCC	AAAAAAD	AABCCCD	AAA	A	A
AAAAAAAAA	AAACC	AAAAAAD	AAACCCD	AAAA	AAA	AAA
AAAA	CCCEE	AAAAAAD	CCCCCCD	AAA	A	A
AAAAA	AAAACC	AAAAAAAD	AAAACCCD	AAA	A	A
AAAA	AAACC	AAAAAAD	AAACCCD	AAA	A	A
AAAA	DDDDD	DDDDDDD	DDDDDDD	AAA	A	A
AAAA	CCCCC	AAAAAAD	CCCCCCCD	AAA	A	A
EEXXXXXXXXXXXX	AAACC	AAAAAAD	AAACCCD	AAAA	AACCCCCCCC	AAAAAAAAAA
AAAAAAABDEXX	DDD	DDDEE	DDDEE	AA	AACCCCCCCC	AAAAADDDDD
XXXX					AAAAAAAAA	AAAAAAAAAA
BBBBBBBB		AD	CD	A	AAAAAAAAA	AAAAAAAAAA
BBBBBBBBBB	XX	AAAD	CCCD	AA	AAAAAAAAA	ABDXXXXXXX
AAA	AACC	AAAAAD	AACCCD	AAA	A	A
AAA	AACC	AAAAAD	AACCCD	AAA	A	A
AAA	AACC	AAAAAD	AACCCD	AAA	A	A
AAA	AACC	AAAAAD	AACCCD	AAA	A	A
AAA	AACC	AAAAAD	AACCCD	AAA	A	A
-200 0 200 400 600	0 200	-200 0 200	-200 0 200	0 300	0 1000 2000	0 1000 2000

Reproduced from Ulrich, G. D., *A Guide to Chemical Engineering Process Design and Economics*, Wiley, New York 1984, Copyright © 1984, by John Wiley & Sons, Inc., reprinted by permission of John Wiley & Sons, Inc.

the appropriate materials of construction for a given piece of equipment from data provided on the PFD. However, the compatibility of the materials of construction with the process streams must be fully investigated before the final design is completed.

The following five-step algorithm is used to estimate actual bare module costs for equipment from the figures in Appendix A.

1. Using the suitable figure in Appendix A, obtain C_p for the desired piece of equipment. This is the purchased equipment cost for the base case (carbon steel construction and near ambient pressure).
2. Find the material of construction factor, F_M, and the pressure factor, F_p, from appropriate figures and tables in Appendix A.
3. Find the bare module factor, F_{BM}^o, taking into account the factors in Step 2.
4. Calculate the bare module cost of equipment, C_{BM}^o, from Eq. (2.5).
5. Update the cost from 1996 (CEPCI—382) to the present by using Eq. (2.3).

The next two example problems use this algorithm to estimate bare module costs.

Example 2.12

Find the bare module cost of a floating-head shell and tube heat exchanger with a heat transfer area of 100 m². The operating pressure of the equipment is 100 bar on both shell and tube sides and the construction of the shell and tubes is of stainless steel.

a. From Figure 2.4: $C_p(1996) = \$ 15,100$
b. From Figure 2.5: $F_p = 1.242$ (for $P = 100$ bar)
   From Table 2.9: $F_M = 3.0$ (SS/SS)
c. From Figure 2.6: $F_{BM}^o = 7.4$ (for $F_p F_M = (1.242)(3.0) = 3.73$)
d. Using Eq. (2.5): $C_{BM}^o(1996) = (7.4)(15,100) = \$ 111,700$

The last three problems all considered the same size heat exchanger made with different materials of construction and operating pressure. The results are summarized below.

Example	Pressure	Materials	F_{BM}^o	Cost
2.10	ambient	CS tubes/shell	3.2	$ 48,300
2.11	ambient	SS tubes/shell	6.3	$ 95,100
2.12	100 bar	SS tubes/shell	7.4	$111,700

These results reemphasize the point that the cost of the equipment depends upon the materials of construction and the pressure of operation.

Occasionally, the required estimation technique does not follow the five-step algorithm given above. An example is the costing of trays for a distillation column, an illustration of which is given below.

Example 2.13

Find the bare module cost (in 1996) of a stainless steel tower 3 m in diameter and 30 m tall. The tower has 40 stainless steel sieve trays and operates at 20 bar.

The costs of the tower and trays are calculated separately and then added together to obtain the total cost.

For the tower:

a. From Figure A.4: $C_p(1996) = \$\,132{,}400$
b. From Figure A.6: $F_p = 2.05$ (for $P = 20$ bar)
   From Table A.4: $F_M = 4.0$ (SS)
c. From Figure A.6: $F^o_{BM} = 16.5$, $F_p F_M = (2.05)(4.0) = 8.2$
d. Using Eq. (2.5): $C^o_{BM,tower}(1996) = C_p(1996)\,F^o_{BM} = (\$\,132{,}400)(16.5) = \$\,2{,}185{,}000$

For the trays:

From Figure A.7: $C_p(1996) = \$970$
$$C^o_{BM} = (C_p)(F_{BM})(N_{ACT})(f_q)$$
$f_q = 1.0$ (since number of trays > 20, Table A.5)
$F_{BM} = 2.0$ (SS, Table A.5)
$C^o_{BM,trays}(1996) = (970)(2.0)(40)(1.0) = \$\,77{,}600$

Tower plus trays:

$C^o_{BM,tower+trays}(1996) = \$\,2{,}185{,}000 + \$\,77{,}600 = \$\,2{,}263{,}000$

The evaluation of the modular cost for the distillation column in Example 2.13 did not follow the general format represented by Eq. (2.5). To avoid any problems resulting from not following the standard procedure, you should carefully read the graphs and tables provided in Appendix A when you seek needed data. You can best see this by reviewing the evaluation of the distillation column in Example 2.13.

2.3.3   Grass Roots and Total Module Costs

The term *grass roots* refers to a completely new facility in which we start the construction on essentially undeveloped land, a grass field. The term *total module cost* refers to the cost of making small-to-moderate expansions or alterations to an existing facility.

It is necessary to account for other costs in addition to the direct and indirect costs to estimate these costs. These additional costs are presented in Table 2.6 and can be divided into two groups.

> *Group 1: Contingency and Fee Costs: The contingency cost varies depending upon the reliability of the cost data and completeness of the process flowsheet you have available. This factor is included in the evaluation of the cost as a protection against oversights and faulty information. Unless otherwise stated, we will assume values of 15% and 3% of the bare module cost for contingency costs and fees, respectively. These are appropriate for systems that are well understood. Adding these costs to the bare module cost provides the* total module cost.
>
> > *Group 2: Auxiliary Facilities Costs: These include costs for site development, auxiliary buildings, and off-sites and utilities. These terms are generally unaffected by the materials of construction or the operating pressure of the process. Unless otherwise stated we assume these costs to be equal to a percentage (35 %) of the bare module costs for the base case conditions. Adding these costs to the total module cost provides the* grass roots cost.

The total module cost can be evaluated from

$$C_{TM} = \sum_{i=1}^{n} C_{TM, i}^{o} = 1.18 \sum_{i=1}^{n} C_{BM, i}^{o} \qquad (2.8)$$

and the grass roots cost can be evaluated from

$$C_{GR} = C_{TM} + 0.35 \sum_{i=1}^{n} C_{BM, i} \qquad (2.9)$$

where n represents the total number of pieces of equipment.

We illustrate the use of these equations by means of the following example.

Example 2.14

A small expansion to an existing chemical facility is being investigated and a preliminary PFD of the process is shown in Figure E2.14.

The expansion involves the installation of a new distillation column with a reboiler, condenser, pumps, and other associated equipment. A list of the equipment, sizes, materials of construction, and operating pressures is given in Table E2.14A. Using the charts in Appendix A, calculate the total module cost for this expansion in 1996.

The same algorithm presented above is used to estimate bare module costs for all equipment. This information is listed in Table E2.14B along with purchased equipment cost, pressure factors, material factors, and bare module factors.

The substitutions from Table 2.14B are made into Eqs. (2.8) and (2.9) to determine the total module cost and the grass roots cost.

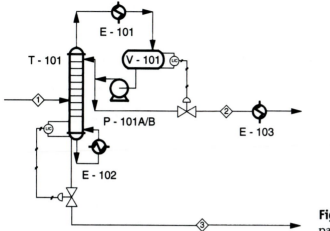

Figure E2.14   PFD for process expansion described in Example 2.14.

Table E2.14A   Information on Equipment Required for the Plant Expansion Described in Example 2.14

Equipment No.	Capacity/Size	Material of Construction*	Operating Pressure (barg[†])
E-101 Overhead condenser	Area = 170m^2 shell and tube (floating head)	Tube-CS Shell-CS	Tube = 5.0 Shell = 5.0
E-102 Reboiler	Area = 205m^2 shell and tube (floating head)	Tube-SS Shell-CS	Tube = 18.0 Shell = 6.0
E-103 Product cooler	Area = 10m^2 (double pipe)	All CS construction	Inner = 5.0 Outer = 5.0
P-101A/B Reflux pumps	Power$_{shaft}$ = 5kW centrifugal	Cast steel	Suction = 5.0
T-101 Aromatics column	Diameter = 2.1m Height = 23 m	Vessel-CS	Column = 5.0
	32 Sieve Trays	Trays-SS	
V-101 Reflux drum	Diameter = 1.8m Length = 6m Horizontal	Vessel-CS	Vessel = 5.0

*CS = Carbon Steel and SS = Stainless Steel

[†]barg = bar gauge, thus 0.0 barg = 1.0 bar

Table E2.14B   Results of Capital Cost Estimate for Example 2.14

Equipment	F_p	F_M	F_{BM}	C_p ($\times 10^3$)	C_{BM} ($\times 10^3$)	C_{BM}^o ($\times 10^3$)
E-101	1.0	1.0	3.3	20.0	66.0	66.0
E-102	1.02	1.7	4.3	22.0	72.6	94.6
E-103	1.0	1.0	1.95	1.2	2.3	2.3
P-101A/B	1.0	1.8	4.52	(2)(6.6)	(2)(21.8)	(2)(29.8)
T-101	1.1	1.0	4.4	81.0	340.2	356.4
32 Trays		2.0	2.0	19.2	23.0	38.4
V-101	1.1	1.0	3.2	14.0	43.4	44.8
Totals				170.6	591.1	662.1
CEPCI = 382						

$$\text{total module cost } (C_{TM}) = 1.18 \sum_{i=1}^{n} C_{BM,i}^o = 1.18(662.1) = \$781 \times 10^3$$

$$\text{grass roots cost } (C_{GR}) = C_{TM} + 0.35 \sum_{i=1}^{n} C_{BM,i} = 781 + 0.35(591.1) = \$988 \times 10^3$$

(Although the grass roots cost is not appropriate here since we only have a small expansion to an existing facility, it is shown above for completeness.)

2.3.4   A Computer Program [CAPCOST©] for Capital Cost Estimation Using the Equipment Module Approach

For processes involving only a few pieces of equipment, estimating the capital cost of the plant by hand is relatively easy. For complex processes with many pieces of equipment, these calculations become tedious. To avoid this we provide a computer program that allows the user to enter data interactively and obtain cost estimates in a fraction of the time required by hand calculations with less chance for error. The program [CAPCOST.BAS] is programmed in Microsoft VisualBasic®, and an executable copy of the program [CAPCOST.EXE] is supplied on the CD at the back of the book.

The program is written to be used in the Microsoft Windows™ programming environment. Instructions about the installation of the program are given in the documentation included in the README.DOC file on the CD. The program requires the user to input information about the equipment—for example, the capacity, operating pressure, and materials of construction. The cost data can be adjusted for inflation by inputing the current value of the CEPCI. Other information such as output file names and the number of the unit (100, 200, etc.) are also required.

The equipment options available to the user are given below:

HEAT EXCHANGERST
TOWERS (WITH SIEVE TRAYS AND PACKING)
VESSELS (HORIZONTAL AND VERTICAL)
PUMPS (WITH ELECTRIC DRIVES)
COMPRESSORS, BLOWERS, AND FANS
DRIVES FOR COMPRESSORS AND BLOWERS
TURBINES
FIRED HEATERS (REACTIVE AND NON-REACTIVE)
EVAPORATORS AND VAPORIZERS
USER ADDED EQUIPMENT

Enter the type of equipment required by using the mouse-activated buttons provided on the screen. You will be required to identify or enter the same information as you would need to do the calculations by hand, that is, operating pressure, materials of construction, and the size of the equipment. The same information as contained in the cost charts and tables in Appendix A are embedded in the program, and the program should give the same results as hand calculations using these charts.

When you have entered all the data for all the equipment you should press the FINISHED button, and a series of output options (e.g., printing, down loading to a file, etc.) will be made available. The following problem can be used to verify the correct use of the software.

Example 2.15

a. Repeat Example Problem 2.14 using the program CAPCOST provided in the back of the book and print out the results.

b. Compare results to the information found in Example 2.14.

a. The results obtained from CAPCOST are summarized in Table E2.15.

b. The results presented for the two cases are comparable. You should not conclude that one answer is better than the other. The number of significant figures presented by computer output is misleading. If you followed the hand calculation, you would recognize that the graphs cannot be read accurately (sometimes hard to assume two

Table E2.15   Summary of Results from CAPCOST Output for Problem 2.15

Equipment	$C_p(\$\times10^3)$	$C_{BM}^o(\$\times10^3)$
E-101	19.6	64.8
E-102	22.1	97.5
E-103	1.2	2.4
P-101A/B	(2)(6.5)	(2)(29.7)
T-101	81.6	364.0
32 Trays	19.4	38.8
V-101	13.2	43.4
Totals	170.1	370.3

$C_{TM} = \$791.0 \times 10^3$, $C_{GR} = \$998.0 \times 10^3$

significant figures). Even if you could read graphs precisely, the original information has an accuracy range of no better than ± 10 to 20%.

You are strongly advised to verify these results for yourself prior to using the program to solve problems in the back of this chapter.

2.4   SUMMARY

In this chapter, we reviewed some of the different types of capital cost estimating techniques that are available. We found that the accuracy of an estimate increases significantly with the time involved and the amount of data required. We reviewed the information required to make an equipment module estimate based on data from the major process equipment. We paid particular attention to the effect of operating pressure and materials of construction. The use of cost indices to adjust for the effects of inflation on equipment costs were considered, and the Chemical Engineering Plant Cost Index (CEPCI) was adopted for all inflation adjustments. The concepts of grass roots and total module costs were introduced in order to make estimates of the total capital required to build a brand new plant or make an expansion to an existing facility. To ease the calculation of the various costs, a computer program for cost estimation was introduced. You will find that mastery of this material is assumed in the remaining chapters in this text.

REFERENCES

1. Pikulik, A., and H. E. Diaz, "Cost Estimating for Major Process Equipment," *Chem. Eng.*, **84**(21), 106, 1977.

2. Peters, M. S., and K. D. Timmerhaus, *Plant Design and Economics for Chemical Engineers*, 4th Ed., McGraw-Hill, New York, 1991.

3. Perry, R. H., and C. H. Chilton, Eds., *Chemical Engineers Handbook*, 5th Ed., McGraw-Hill, New York, 1973.

4. Ulrich, G. D., *A Guide to Chemical Engineering Process Design and Economics*, Wiley, New York, 1984.

5. Lang, H. J., "Engineering Approach to Preliminary Cost Estimates," *Chem. Eng.*, **54**(9), 130, 1947.

6. Lang, H. J., "Cost Relationships in Preliminary Cost Estimates," *Chem. Eng.*, **54**(10), 117, 1947.

7. Lang, H. J., "Simplified Approach to Preliminary Cost Estimates," *Chem. Eng.*, **55**(6), 112, 1948.

8. Guthrie, K. M., "Capital Cost Estimating," *Chem. Eng.*, **76**(3), 114, 1969.

9. Guthrie, K. M., *Process Plant Estimating, Evaluation and Control*, Craftsman, Solana Beach, CA, 1974.

PROBLEMS

1. In 1986, our company bought a heat exchanger (heat transfer area = 50 m^2) for a cooling water application at moderate temperatures and pressures for $6,500. In 1990, we bought an exchanger (heat transfer area = 120 m^2) for a similar application for $11,000. What is your best estimate of the cost of a similar heat exchanger with a heat transfer area of 90 m^2 for use today (assume mid-1996 prices apply)?

2. What is meant by the economy of scale? When does this concept not apply to a process or piece of equipment? Give an example of when it does not apply.

3. Compare the results of Problem 1 above using the Marshall and Swift Index to account for the inflation (assuming Problem 1 was done using the CEPCI). What is the percentage difference between the two results?

4. The light gas separations unit of a certain refinery consists of two columns in series that are fed a mixture of propane, butanes, and higher hydrocarbons. The first column, the depropanizer, separates the propane (and small amounts of propylene) from the heavier material. The second column, the debutanizer, separates the butanes from the remaining hydrocarbons. A PFD and equipment summary table are given in Figure P2.4 and Table P2.4.

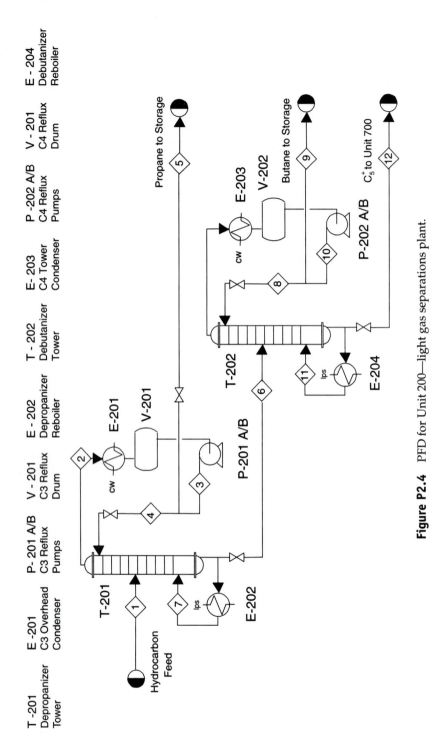

Figure P2.4 PFD for Unit 200—light gas separations plant.

Table P2.4 Equipment Summaries for Problem 4

Equipment	E-201	E-202	E-203	E-204	T-201	T-202	P-201A/B	P-202A/B	V-201	V-202
Type	Floating head	Floating head	Floating head	Floating head	Tower	Tower	Centrifugal pump	Centrifugal pump	Horizontal vessel	Horizontal vessel
Area	155 m²	45 m²	85 m²	20 m²	—	—	—	—	—	—
Shell Pressure	15 barg†	4 barg	5 barg	4 barg	—	—	—	—	—	—
Tube Pressure	4 barg	15 barg	4 barg	5 barg	—	—	—	—	—	—
MOC	Carbon Steel	Carbon steel	Carbon steel	Carbon steel	Carbon steel	Carbon steel	Cast steel	Cast steel	Carbon steel	Carbon steel
Diameter	—	—	—	—	0.95 m	1.00 m	—	—	1.25 m	1.25 m
Length or Height	—	—	—	—	19.0 m	21.0 m	—	—	3.75 m	3.75 m
Design Pressure	—	—	—	—	15 barg	5 barg	—	—	15 barg	5 barg
Internals	—	—	—	—	36 sieve trays	40 sieve trays	—	—	—	—
MOC	—	—	—	—	Stainless steel	Stainless steel	—	—	—	—
Power	—	—	—	—	—	—	1.3 kW	1.2 kW	—	—
Suction Pressure	—	—	—	—	—	—	16 barg	6 barg	—	—

†All pressures are entered as bar gauge, 0.0 barg = 1.0 bar

Using the CAPCOST program, estimate the total module cost and grass roots cost of this process unit.

5. Compare your answer from Problem 4 using the appropriate Lang Factor from Table 2.7. You should base your purchased equipment costs on those given in CAPCOST, which are for base purchased costs for carbon steel at atmospheric pressure. Comment on the accuracy of the Lang Factor method for this particular case.

6. Estimate the cost of the plant in Problem 4 if all the equipment were made of stainless steel.

7. Estimate the capital cost of a grass roots dimethyl ether (DME) facility using the information given in Appendix B.

8. Estimate the capital cost of a grass roots acetone production facility using the information given in Appendix B.

9. Estimate the capital cost of a grass roots acrylic acid production facility using the information given in Appendix B.

10. Estimate the capital cost of a grass roots heptene isomerization facility using the information given in Appendix B.

11. Estimate the capital cost of a grass roots toluene hydrodealkylation facility using the information given in Figure 1.5 and Table 1.7.

CHAPTER 3

Estimation
of Manufacturing Costs

The costs associated with the day-to-day operation of a chemical plant must be estimated before the economic feasibility of a proposed process can be assessed. This chapter introduces the important cost factors affecting the manufacturing cost and provides methods to estimate each factor. In order to estimate the manufacturing cost, we need process information provided on the PFD, an estimate of the fixed capital investment, and an estimate of the number of operators required to operate the plant. The fixed capital investment is the same as either the total module cost or the grass roots cost defined in Chapter 2. Manufacturing costs are expressed in units of $ per unit time in contrast to the capital costs that are expressed in $. How we treat these two costs, expressed in different units, to judge the economic merit of a process is covered in Chapters 4 and 5.

3.1 FACTORS AFFECTING THE COST OF MANUFACTURING A CHEMICAL PRODUCT

Many elements influence the cost of manufacturing chemicals. A list of the important costs involved, including a brief explanation of each cost, is given in Table 3.1.

The cost information provided in Table 3.1 is divided into three categories:

1. **Direct Manufacturing Costs:** These costs represent operating expenses that vary with production rate. When product demand drops, production rate is reduced below the design capacity. At this lower rate we would expect a reduction in the factors making up the direct manufacturing costs. These re-

Table 3.1   Factors Affecting the Cost of Manufacturing (COM) for a Chemical Product (from references 1, 2, 3)

Factor	Description of Factor
1. Direct Costs	***Factors that vary with the rate of production.***
A. Raw Materials	Costs of chemical feed stocks required by the process. Flow rates obtained from the PFD.
B. Waste Treatment	Costs of waste treatment to protect environment.
C. Utilities	Costs of utility streams required by process. Includes but not limited to:

	a. Fuel gas, oil, and/or coal
	  b. Electric power
	  c. Steam (all pressures)
	  d. Cooling water
	  e. Process water
	  f. Boiler feed water
	  g. Instrument air
	  h. Inert gas (nitrogen), etc.
	  i. Refrigeration

	Flowrates for utilities found on the PFD/PIDs.
D. Operating Labor	Costs of personnel required for plant operations.
E. Direct Supervisory and Clerical Labor	Cost of administrative/engineering and support personnel.
F. Maintenance and Repairs	Costs of labor and materials associated with equipment maintenance.
G. Operating Supplies	Costs of miscellaneous supplies that support daily operation not considered to be raw materials. Examples include: chart paper, lubricants, miscellaneous chemicals, filters, respirators and protective clothing for operators, etc.
H. Laboratory Charges	Costs of routine and special laboratory tests required for product quality control and troubleshooting.
I. Patents and Royalties	Cost of using patented or licensed technology.
2. Fixed Costs	***Factors not affected by the level of production.***
A. Depreciation	Costs associated with the physical plant (buildings, equipment, etc.). Legal operating expenses for tax purposes.
B. Local Taxes and Insurance	Costs associated with property taxes and liability insurance. Based on plant location and severity of the process.

(*continued*)

Table 3.1   Factors Affecting the Cost of Manufacturing (COM) for a Chemical Product (from references 1, 2, 3) (*continued*)

Factor	Description of Factor
2. Fixed Costs	***Factors not affected by the level of production.***
C. Plant Overhead Costs (sometimes referred to as factory expenses)	Catch-all costs associated with the operation of auxiliary facilities supporting the manufacturing process. Costs involve: payroll and accounting services, fire protection and safety services, medical services, cafeteria and any recreation facilities, payroll overhead and employee benefits, general engineering, etc.
3. General Expenses	***Costs associated with management level and administrative activities not directly related to the manufacturing process.***
A. Administration Costs	Costs for administration. Includes salaries, other administration, buildings, and other related activities.
B. Distribution and Selling Costs	Costs of sales and marketing required to sell chemical products. Includes salaries and other miscellaneous costs.
C. Research and Development	Costs of research activities related to the process and product. Includes salaries and funds for research related equipment and supplies, etc.

ductions may be directly proportional to the production rate (e.g., raw materials), or might be reduced slightly (e.g., maintenance costs or operating labor).

2. **Fixed Manufacturing Costs:** These costs are independent of changes in production rate. They include property taxes, insurance, and depreciation that are charged at constant rates even when the plant is not in operation.

3. **General Expenses:** These costs represent an overhead burden that is necessary to carry out business functions. They include management, sales, financing, and research functions. General expenses seldom vary with production level. However, items such as research and development and distribution and selling costs may decrease if extended periods of low production levels occur.

The equation used to evaluate the cost of manufacture using these costs becomes:

$$\text{Cost of Manufacture } (COM) = \text{Direct Manufacturing Costs } (DMC)$$

$$+ \text{ Fixed Manufacturing Costs } (FMC) + \text{General Expenses } (GE)$$

The approach we provide in this chapter is similar to that presented in other chemical engineering design texts [1,2,3].

The cost of manufacturing (COM) can be determined when the following costs are known or can be estimated:

1. Fixed Capital Investment (FCI): (C_{TM} or C_{GR})
2. Cost of Operating Labor (C_{OL}):
3. Cost of Utilities (C_{UT}):
4. Cost of Waste Treatment (C_{WT}):
5. Cost of Raw Materials (C_{RM}):

Table 3.2 provides information that allows us to obtain an estimate of the individual cost items identified in Table 3.1 (both tables carry the same identification of individual cost terms). With the exception of the cost of raw materials, waste treatment, utilities, and operating labor (all parts of the direct manufacturing costs), Table 3.2 presents equations that we will use to estimate each individual item. With each equation is presented a typical range for the constants (multiplication factors) to estimate an individual cost item. If no other information is available, we can use the midpoint value for each of these ranges to estimate the costs involved. It should be noted that you should always use the best information that is available in order to establish these constants. The method presented below should only be used when no other information on these costs is available.

By using the midpoint values given in Table 3.2 column 2, the resulting equations for the individual items are calculated in column 3. The cost items for each of the three categories are added together to provide the total cost for each category. The equations for estimating the costs for each of the categories are:

$$DMC = C_{RM} + C_{WT} + C_{UT} + 1.33C_{OL} + 0.069FCI + 0.03COM$$

$$FMC = 0.708C_{OL} + 0.168FCI$$

$$GE = 0.177C_{OL} + 0.009FCI + 0.16COM$$

We can obtain the total manufacturing cost obtained by adding these three cost categories together and solving for the total manufacturing cost (COM). The result is

$$COM = 0.304FCI + 2.73C_{OL} + 1.23(C_{UT} + C_{WT} + C_{RM}) \tag{3.1}$$

The cost of manufacture without depreciation (COM_d) is

$$COM_d = 0.180FCI + 2.73C_{OL} + 1.23(C_{UT} + C_{WT} + C_{RM}) \tag{3.2}$$

Table 3.2   Multiplication Factors Estimating Manufacturing Cost** (see also Table 3.1) (from references 1,2,3)

Cost Item from Table 3.1	Typical Range of Multiplying Factors	Value Used in Text
1. Direct Manufacturing Costs		
A. Raw Materials	C_{RM}*	
B. Waste Treatment	C_{WT}*	
C. Utilities	C_{UT}*	
D. Operating Labor	C_{OL}	C_{OL}
E. Direct Supervisory and Clerical Labor	$(0.1 - 0.25)C_{OL}$	$0.18C_{OL}$
F. Maintenance and Repairs	$(0.02 - 0.1)FCI$	$0.06FCI$
G. Operating Supplies	$(0.1 - 0.2)(\text{Line 1F})$	$0.009FCI$
H. Laboratory Charges	$(0.1 - 0.2)C_{OL}$	$0.15C_{OL}$
I. Patents and Royalties	$(0 - 0.06)COM$	$0.03COM$
Total Direct Manufacturing Costs	$C_{RM} + C_{WT} + C_{UT} + 1.33C_{OL}$ $+ 0.03COM + 0.069FCI$	
2. Fixed Manufacturing Costs		
A. Depreciation	$0.1FCI$***	$0.1FCI$***
B. Local Taxes and Insurance	$(0.014 - 0.05)FCI$	$0.032FCI$
C. Plant Overhead Costs	$(0.50 - 0.7)(\text{Line 1D}$ $+ \text{Line 1E} + \text{Line 1F})$	$0.708C_{OL} + 0.036FCI$
Total Fixed Manufacturing Costs	$0.708C_{OL} + 0.168FCI$	
3. General Manufacturing Expenses		
A. Administration Costs	$0.15(\text{Line 1D} + \text{Line 1E}$ $+ \text{Line 1F})$	$0.177C_{OL} + 0.009FCI$
B. Distribution and Selling Costs	$(0.02 - 0.2)COM$	$0.11COM$
C. Research and Development	$0.05COM$	$0.05COM$
Total General Manufacturing Costs	$0.177C_{OL} + 0.009FCI$ $+ 0.16COM$	
TOTAL COSTS	$C_{RM} + C_{WT} + C_{UT} + 2.215C_{OL}$ $+ 0.190COM + 0.246FCI$	

*Costs are evaluated from information given on the PFD and the unit cost.

**Costs are given in $ per unit time (usually per year).

***Depreciation costs will be covered separately in Chapter 4. The use of 10% of FCI is a crude approximation at best.

Example 3.1

The following cost information was obtained from a design for a 92,000 tonne/year nitric acid plant.

Fixed Capital Investment:	$ 11,000,000
Raw Material Cost	$ 7,950,000/yr
Waste Treatment Cost	$ 1,000,000/yr
Utilities	$ 356,000/yr
Direct Labor Cost	$ 300,000/yr

Determine:

   a. The manufacturing cost in $/yr and $/tonne of nitric acid.
   b. The percentage of manufacturing costs resulting from each cost category given in Table 3.1 and 3.2.

   a. Using Eq. (3.1):

   $COM = (0.304)(\$11,000,000) + (2.73)(\$300,000)$
   $+ (1.23)(\$356,000 + \$1,000,000 + \$7,950,000) = \$15,610,000/yr$
   $(\$15,610,000/yr)/(92,000 \text{ tonne}/yr) = \$170/tonne$

   b. From the relationships given in Table 3.2:

   Direct Costs $= \$7,950,000 + \$1,000,000 + \$356,000 + (1.33)(\$300,000) + (0.069)$
   $(\$11,000,000) + (0.03)(\$15,610,000) = \$10,932,000$
   Percentage of manufacturing cost $= (100)(10.932)/15.61 = 70\%$
   Fixed Costs $= (0.708)(\$300,000) + (0.168)(\$11,000,000) = \$2,060,000$
   Percentage of manufacturing cost $= (100)(2.060)/15.61 = 13\%$
   General Expenses $= (0.177)(\$300,000) + (0.009)(\$11,000,000) + (0.16)(\$15,610,000)$
   $= \$2,650,000$
   Percentage of manufacturing cost $= (100)(2.648)/15.61 = 17\%$

In Example 3.1, the direct costs were shown to dominate the manufacturing costs, accounting for about 70% of the manufacturing costs. Of these direct costs, the raw materials cost, the waste treatment cost, and the cost of utilities accounted for over $9,000,000 of the $10,900,000 direct costs. These three cost contributions are not dependent on any of the estimating factors provided in Table 3.2. In such a case, which is a common situation, the manufacturing cost is insensitive to the estimating factors provided in Table 3.2. The use of the mid-range values is acceptable for this situation.

3.2 COST OF OPERATING LABOR

The technique used to estimate operating labor requirements is based on the approach given by Ulrich [1] and is consistent with the module approach used in evaluating the capital cost in Chapter 2. Table 3.3 provides a list of the labor required for a variety of equipment modules.

Table 3.3   Operator Requirements for Various Process Equipment (from Ulrich, G. D., *A Guide to Chemical Engineering Process Design and Economics,* Wiley , New York, 1984, Copyright © 1984 John Wiley & Sons, Inc., reprinted by permission of John Wiley & Sons Inc.)

Equipment Type	Operators per Equipment per Shift
Auxiliary Facilities	
Air Plants	1.0
Boilers	1.0
Chimneys and Stacks	0.0
Cooling Towers	1.0
Water Demineralizers	0.5
Electric Generating Plants	0.5
Portable Electric Generating Plants	3.0
Electric Substations	0.0
Incinerators	2.0
Mechanical Refrigeration Units	0.5
Waste Water Treatment Plants	2.0
Water Treatment Plants	2.0
Process Equipment	
Evaporators	0.3
Vaporizers	0.05
Furnaces	0.5
Fans*	0.05
Blowers and Compressors*	0.15
Heat Exchangers	0.1
Towers	0.35
Vessels	0.0
Pumps*	0.0
Reactors	0.5

*For equipment with spares such as compressors and pumps, just count equipment plus spare as one item.

Table 3.3 includes only the process equipment involved in fluid processing plants and the auxiliary facilities. For units processing solids the reader is referred to the original reference [1].

The information in Table 3.3 provides the number of operators required per equipment unit per shift. A single operator works on the average 49 weeks

(3 weeks time off for vacation and sick leave) a year, five 8-hour shifts a week. This amounts to (49 weeks/year × 5 shifts/week) 245 shifts per operator per year. A chemical plant normally operates 24 hours/day. This requires (365 days/year × 3 shifts/day) 1095 operating shifts per year. The number of operators needed to be employed to provide this number of shifts is [(1095 shifts/yr)/(245 shifts/operator/yr)] or a little less than 4.5 operators. Four and one-half operators are hired for each operator needed in the plant at any time. This provides the needed operating labor but does not include any support or supervisory staff.

To estimate the cost of operating labor, we require the average hourly wage of an operator. Chemical plant operators are relatively highly paid, and a typical value of $42,000/yr ($21.00 per hour) in 1993 is assumed here [2], which is equivalent to $46,800/yr in mid-1996. The cost of labor depends considerably on the location of the plant and significant variations from the above figure may be expected. Historically, wage levels for chemical plant operators have grown slightly faster than the other cost indexes for process plant equipment given in Chapter 2. The *Oil and Gas Journal* and *Engineering News Record* provide appropriate indices to correct labor costs for inflation. Typically, labor costs are a small factor having little impact on the overall economics. Therefore, it is assumed that inflation of labor costs using the Chemical Engineering Plant Cost Index is sufficiently accurate for preliminary estimates and will be used throughout the text.

Example 3.2

Estimate the operating labor requirement and costs for the toluene hydrodealkylation facility shown in Figures 1.3 and 1.5.

From the PFD in Figure 1.5, the number and type of equipment is determined.

From Table 3.3, an estimate of the number of operators required per shift is made. This information is shown in Table E3.2.

Table E3.2   Results for the Estimation of Operating Labor Requirements for the Toluene Hydrodealkylation Process using the Equipment Module Approach (Example 3.2)

Equipment Type	Number of Equipment	Operators per Shift per Equipment	Operators per Shift
Compressors	1	0.15	0.15
Exchangers	7	0.10	0.7
Heaters/Furnaces	1	0.50	0.5
Pumps	2	0.00	0.0
Reactors	1	0.50	0.5
Towers	1	0.35	0.35
Vessels	4*	0.00	0.0
		TOTAL	2.2

*Include TK-101 as a vessel.

The number of operators required per shift (see column 4 Table E3.2) is calculated to be = 2.2.

Operating Labor = (4.5)(2.2) = 9.9   Rounding up to the nearest integer yields 10 operators.

Labor Costs(1996) = (10) × ($46,800) = $468,000/yr

3.3 UTILITY COSTS

The costs of utilities are directly influenced by the cost of fuel. Specific difficulties emerge when estimating the cost of fuel, and utilities such as electricity, steam, and thermal fluids are directly impacted. Figure 3.1 shows the general trends for fossil fuel costs over the past 15 years. The costs presented represent average values and are not site specific. These costs do not reflect the wide variability of cost and availability of various fuels throughout the country.

As seen from Figure 3.1, coal represents the lowest cost fossil fuel on an energy basis. Most coal is consumed near the "mine mouth" in large power plants to produce electricity. The electricity is transported by power lines to the consumer. At locations remote from mines, both the availability and cost of transportation reduce and/or eliminate much of the cost advantage of coal. Coal suffers further from its negative environmental impact.

The next lowest cost fuel source shown in Figure 3.1 is natural gas. Natural gas fuel is the least damaging fossil fuel energy supply with respect to the environment. It is transported by pipe lines throughout much of the country. The cost

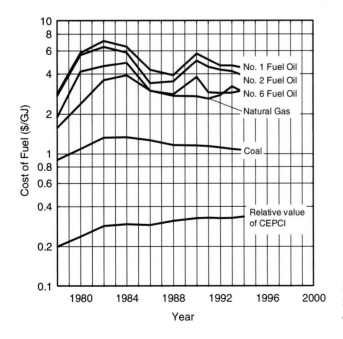

Figure 3.1 Changes in Fuel Prices from 1978 to 1994 (from reference 5).

is more uniform than coal throughout different regions of the country. However, regions in the country remain that are not yet serviced by the natural gas distribution system. In these regions, the use of natural gas is not an option that can be considered. Although natural gas is a mixture of several light hydrocarbons, it consists predominantly of methane. For the calculations used in this text we will assume that methane and natural gas are equivalent.

Petroleum is the final fossil fuel that is used in the chemical industry as an energy source. As shown in Figure 3.1, it is generally the highest cost fossil fuel source. It is most readily available near coastal regions where oil enters the country and refining takes place. The United States has become more dependent upon imported oil, which may be subject to large upsets in cost and domestic availability. Uncertainties in the availability of supplies, high cost, and large fluctuations in cost make this source of energy least attractive in most situations.

Figure 3.1 shows that fuel costs have increased somewhat more rapidly and in a much more chaotic fashion than the cost index (CEPCI) that we have used previously to correct costs for inflation. As a result of the regional variations in the availability and costs of fossil fuels along with the inability of the cost index to represent energy costs, we take the position that site-specific cost and availability information must be provided for a valid estimation of energy costs. We assume, in this text, that natural gas is the fuel of choice unless otherwise stated.

The PFD for the toluene hydrodealkylation process (Figure 1.5) represents the "battery-limits" plant. It does not provide the equipment necessary to produce the various "service or utility streams" that are shown and that are necessary for the plant to operate. Shown on the PFD are cooling water streams, steam streams for heating, electric power streams, and so on. These streams, termed utilities, were provided to enable the control of process stream temperatures as required by the process. These utilities can be supplied in a number of ways.

1. Purchasing from a public or private utility: In this situation no capital cost is involved and the utility rates charged are based upon consumption. In addition, the utility is delivered to the battery limits at known conditions.
2. Supplied by the company: A comprehensive "off-site" facility provides the utility needs for many processes at a common location. In this case, the rates charged to a process unit reflect the fixed capital and the operating costs required to produce the utility.
3. Self-generated and used by a single process unit: In this situation the capital cost for purchase and installation becomes part of the fixed capital cost of the process unit. Likewise, the related operating costs for producing that particular utility are directly charged to the process unit.

Table 3.4 identifies utilities that would likely be provided in a comprehensive chemical plant complex.

Table 3.4   Utilities Provided by Off-Sites for a Plant with Multiple Process Units (costs represent charges for utilities delivered to the battery limit of a process)

Utility	Description	Cost $/GJ	Cost $/ Common Unit
Air Supply	Pressurized and dried air		
	a. Process		$2.3/100 m^3
	b. Instrument		$4.7/100 m^3
Steam from Boilers	Process steam: Latent heat only		
	a. Low pressure (5 barg, 160°C)	3.17	$6.62/1000 kg
	b. Medium pressure (10barg, 184°C)	3.66	$7.31/1000 kg
	c. High pressure (41 barg, 254°C)	5.09	$8.65/1000 kg
Cooling Tower Water	Processes cooling water: 30°C to 40°C or 45°C	0.16[1]	$6.7/1000 m^3
Other Water	High purity water for		
	a. Process use		$0.04/1000 kg
	b. Boiler feed water[2]		$2.54/1000 kg
	c. Potable (drinking)		$0.26/1000 kg
	d. Deionized water		$1.00/1000 kg
Electrical Substation	Electric Distribution	16.8	$0.06/kWh
	a. 110 V		
	b. 220 V		
	c. 440 V		
Fuels	a. Fuel Oil (no. 2)	4.0	$170/m^3
	b. Natural Gas	2.5[3]	$0.085/std. m^3
	c. Coal (FOB mine mouth)	1.2	$31/tonne
Refrigeration	a. Moderately low T: 5°C	20	Based on Process
	b. Low T: −20°C	32	Cooling Duty
	c. Very low T: −50°C	60	
Thermal Systems	a. Moderately high T: to 330°C	4.9	Based on Process
	b. High T: to 400°C	5.2	Heating Duty
	c. Very high T: to 600°C	5.9	
Waste Disposal (solid and liquid)	(a) Non-Hazardous		$36/tonne
	(b) Hazardous		$145/tonne
Waste Water Treatment	(a) Primary (filtration)		$39/1000 m^3
	(b) Secondary (filtration + activated sludge)		$41/1000 m^3
	(c) Tertiary (filtration, activated sludge, and chemical processing)		$53/1000 m^3

[1]Based on $\Delta T_{cooling\ water}$ = 10°C. Cooling water return temperatures should not exceed 45°C due to excess scaling at higher temperatures.

[2]Approximately equal credit is given for condensate returned from exchangers using steam.

[3]Based on Lower Heating Value of Natural Gas.

In Option 2 above, the determination of the value of the utility stream is complex when new processes come on line, old processes are shut down, off-sites need replacement because of insufficient capacity, and so on. How these changes are reflected in the charges to various processes is complex. This is in addition to the complex assignment of "fair cost" when any facility produces a range of co-products. Consider the most common example, the production of process steam at various pressures. Each stream of different pressure steam is sent to a separate steam header for distribution to the various processes in the plant. A block flow diagram for such a facility is shown in Figure 3.2. From the figure it can be seen that all the steam is produced in a single boiler at high pressure. The needs for medium-pressure and low-pressure steam are met by dropping the pressure through expansion turbines that generate electricity used in the plant. The output from the block diagram in Figure 3.2 shows high-pressure steam, medium-pressure steam, low-pressure steam, and electricity as co-products.

The total fixed cost and operating costs for the steam plant are known. The problem is to assign a fair cost to each product so that each user (process unit) pays its "fair share" for each utility. There is no single answer to this problem. The final assignment of financial burden for the costs of the off-sites rests with the strategy used by your company. The costs are periodically changed to meet the costs involved in providing these off-site services.

The strategy used in this text is to treat the company-owned utilities as outside utilities. Thus the utilities are provided to the processes on the basis of a consumption or utilization charge. For this text the charges to be used will be the

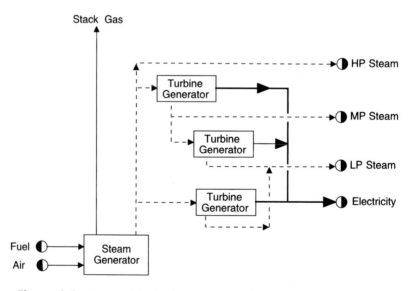

Figure 3.2 Process Block Flow Diagram for Multisteam/Electricity System.

costs shown in Table 3.4, unless otherwise stated in the problem. These charges represent the costs that any process unit would pay for each unit of utility.

For off-sites that produce a stream that will be used by a process and then recycled, the condition of the return stream must be specified. As an example, you will note that the cooling water stream carries a specification of 30°C to 40°C. This signifies that the process is provided a stream of cooling water at 30°C and is to return the coolant stream at 40°C. If the return temperature exceeds the 40°C specification, the process unit may be charged a penalty. The cooling tower at the off-site operates most effectively at these temperatures and the failure of a process to return the coolant stream increases the cost of this off-site operation. A penalty may be charged for cooling water, refrigeration, thermal systems, and steam utilities that do not meet the specifications for the return stream.

Example 3.3

Compare the hourly utility cost for two heat exchangers to remove energy from the same process stream (condensing stream at 80°C). The amount of heat to be removed is 11.5 GJ/h.

    a. Using cooling water entering at 30°C and leaving at 40°C.

    b. Using water entering at 30°C and leaving at 50°C (a penalty of 1% per degree over the 40°C is to be charged).

    c. Determine the relative flow of cooling water to the exchangers.

    d. Establish the effect on the cost of the heat exchangers

Assume that (1) the heat exchangers are to be designed for the different cooling water return streams, and (2) the heat transfer coefficients in the exchangers are the same.

    a. From Table 3.4: Utility cost is 0.16 \$/GJ

        Cost of Coolant = (11.5)(0.16) = \$1.84/h

    b. Returning the coolant at 10°C over the 40°C limit results in a 10% penalty.

        Cost of Coolant = (\$1.84)(1.1) = \$2.02/h

    c. From the heat balance $q = \dot{m}C_p\Delta T$

$$q = \dot{m}C_p\Delta T_{part\ a} = q = \dot{m}C_p\Delta T_{part\ b}$$
$$\dot{m}_{part\ a} / \dot{m}_{part\ b} = \Delta T_{part\ a}/\Delta T_{part\ b} = (40-30)/(50-30) = 0.5\ (50\%)$$

    d. Using the heat transfer design equation $q = UA\Delta T_{lm}$

$$q = (UA\Delta T_{lm})_{part\ a} = (UA\Delta T_{lm})_{part\ b}\text{ gives}$$
$$A_{part\ b}/A_{part\ a} = (\Delta T_{lm})_{part\ a}/(\Delta T_{lm})_{part\ b}$$

$$= \frac{\{(80-30)-(80-40)/\ln[(80-30)/(80-40)]\}}{\{(80-30)-(80-50)/\ln[(80-30)/(80-50)]\}} = \frac{44.8}{39.2} = 1.15$$

As the size (area) of the heat exchanger increases, the capital cost would be expected to increase. The optimum difference between the supply and return temperatures for cooling water in a given plant depends on many factors. For the above example, we can see that in general as the ΔT_{cw} increases the following observations can be made:

1. The size of the process heat exchangers will increase.
2. The flow of cooling water will decrease.
3. The fraction of the circulating water that must be evaporated in order to cool the water back to the supply temperature will increase.

Each of the above factors affects the optimum ΔT_{cw}. In general, for a chemical plant with many heat exchangers, using a cooling water temperature increase of between 10°C and 15°C will be close to the optimum. Another reason for limiting the temperature rise of cooling water is that over temperatures of about 45°C, excessive fouling of heat exchanger surfaces occurs.

A process unit may produce a stream that is normally provided from the central off-site facility. The most common situation occurs when the process involves a highly exothermic reaction, and the heat is removed to produce steam. This steam may be sent to the steam header and/or used in the process. If the process unit is an addition to an existing facility, this reduces the amount of steam that must be produced by the off-site system. What dollar value should the process unit be credited for this steam? Our policy here is to credit the unit with the "avoided cost" at the off-site. The off-site facility producing the steam is already built and there will be no reduction in the number of operators, equipment, insurance, and so on at the off-site facility. Therefore, these costs cannot be avoided. The savings come from a reduction in fuel consumed at the off-site. This reduction represents the "avoided cost." This number varies with the cost of fuel used at the off-site.

Example 3.4

Determine the credit that could be expected by installing a waste heat boiler on our reactor discharge. A total of 44 GJ/h could be recovered as low pressure steam (5 barg) by installing a waste heat boiler. Currently the central off-site uses natural gas as the fuel source. The thermal efficiency for this central boiler is reported to be 91%.

a. Determine the credit that would be received for this steam.
b. Compare this to the cost paid when this amount of (low pressure) steam is taken from the header.

a. From Table 3.4 cost of natural gas = $2.5/GJ
   Fuel saved: 44 GJ/(0.91) = 48.4 GJ/h
   Credit Received (Avoided Cost): (48.4)(2.5) = $121/h
b. From Table 3.4 cost of low pressure steam = $3.17/GJ
   Charge for low pressure steam = (44.0)(3.17) = $139.5/h

We note that the credit is somewhat smaller than the charge for the low pressure steam. This is to be expected since only the avoided cost is being saved.

3.4   RAW MATERIAL COSTS

The cost of raw materials can be estimated by using the current price listed in such publications as the *Chemical Marketing Reporter* [4]. A list of common chemicals and their selling prices at the middle of 1996 are given in Table 3.5. Current raw material and product chemical prices should be obtained from the current issue of the *Chemical Marketing Reporter* [4]. To locate costs for individual items, it is not sufficient to look solely at the current issue, as all chemicals are not listed in

Table 3.5   Costs of Some Common Chemicals (from *Chemical Marketing Reporter*, November 1993)

Chemical	Cost ($/kg)	Typical Shipping Capacity or Basis for Price
Acetaldehyde	1.00	Railroad Tank Cars
Acetic Acid	0.73	Railroad Tank Cars
Acetone	0.75	Railroad Tank Cars
Acrylic Acid	1.59	Railroad Tank Cars
Allyl Alcohol	2.20	F.O.B. Gulf Coast
Allyl Chloride	1.72	F.O.B. Gulf Coast
Benzene	0.27	Barge, Gulf Coast
Chlorine	0.25	Railroad Tank Car
Di-Methyl Ether	0.95	Railroad Tank Car
Ethanol (190 Proof)	0.62	Railroad Tank Car
Hydrogen Chloride (anhydrous)	0.33	Railroad Tank Car, Works
Iso-Propanol (99%)	0.79	Railroad Tank Car
(91%)	0.70	Railroad Tank Car
Methanol	0.16	F.O.B. Gulf Coast
Methyl Ethyl Ketone	0.88	Railroad Tank Car
Propylene		
(Polymer Grade)	0.30	F.O.B. Gulf Coast
(Chemical Grade)	0.28	F.O.B. Gulf Coast
Sulfur (Crude)	0.053	Railroad Car
Sulfuric Acid	0.075	Railroad Tank Car, Gulf Coast
Toluene	0.24	Barge, Gulf Coast
Mixed Xylenes	0.25	Barge, Gulf Coast
Ortho-Xylene	0.93	Railroad Tank Cars
Para-Xylene	0.35	Railroad Tank Cars
Meta-Xylene	0.46	Railroad Tank Cars

each issue. It is necessary to explore several of the most recent issues. In addition, for certain chemicals large seasonal price fluctuations may exist, and it may be advisable to look at the average price over a period of several months.

3.5   YEARLY COSTS AND STREAM FACTORS

Manufacturing and associated costs are most often reported in terms of $/yr. Information on a PFD is most often reported in terms of kg or kmol per hour or per second. In order to calculate the yearly cost of raw materials or utilities the fraction of time that the plant is operating in a year must be known. This fraction is known as the Stream Factor (*SF*), where:

$$\text{Stream Factor } (SF) = \frac{\text{Number of Days Plant Operates per Year}}{365} \quad (3.3)$$

Typical values of the stream factor are in the range of 0.96 to 0.90. Even the most reliable and well managed plants will typically shut down for two weeks a year for scheduled maintenance giving an *SF* = 0.96. Less reliable processes may require more down time, hence lower *SF* values. The "stream factor" represents the fraction of time that the process unit is on line and operating at design capacity. When estimating the size of equipment, you must be careful to use the design flowrate for a typical stream day and *not* a calendar day. The following example illustrates the use of the stream factor.

Example 3.5

   a. Determine the yearly cost of toluene for the process given in Chapter 1.
   b. What is the yearly consumption of toluene?
   c. What is the yearly revenue from the sale of benzene?

Assume a stream factor of 0.95 and note that the flowrates given on the PFD are in kg per stream hour.

     From Table 1.5: Flowrate of toluene = 10,000 kg/h (Stream 1)
     From Table 1.5: Flowrate of benzene = 8,210 kg/h (Stream 15)
     From Table 3.5: Cost of toluene = $0.24/kg
     From Table 3.5: Cost of benzene = $0.27/kg

   a. Yearly Cost of Toluene = (24)(365)(10,000)(0.24)(0.95) = $19,973,000/yr
   b. Yearly Consumption of Toluene = (24)(365)(10,000)(0.95) / 1000 = 83,200 tonnes/yr
   c. Yearly Revenue from Benzene Sales = (24)(365)(8210)(0.27)(0.95) = $18,447,000/yr

Comparing the results from parts a and c, we can see that with the current prices for these two chemicals it is not economical to produce benzene from toluene. Historically, the price differential between benzene and toluene has been greater than the $0.03/kg shown in Table 3.5, and this is the reason why this process has been used and is currently being used to produce benzene. Clearly, if this low price differential were to exist for a long period of time, this process might have to be shut down.

3.6   ESTIMATING UTILITY COSTS FROM THE PFD

Most often, utilities do not directly contact process streams. Instead, they exchange heat energy (fuel gas, steam, cooling water, and boiler feed water) in equipment such as heat exchangers and process heaters, or they supply work (electric power or steam) to pumps, compressors, and other rotating equipment. In most cases, the flowrate can be found either by inspection or by doing a simple heat balance around the equipment. Table 3.6 provides efficiencies that can be used for a variety of thermal process units.

Steam can be used to drive a piece of rotating equipment such as a compressor. In this case, both the theoretical steam requirement and efficiency are required. Table 3.7 provides the theoretical steam requirements as a function of the steam inlet pressure and the exhaust pressure along with the mechanical efficiency of various drives.

To illustrate the techniques used to estimate the utility flowrates and utility costs for various types of equipment, the following example is provided.

Example 3.6

Estimate the quantities and yearly costs of the appropriate utilities for the following pieces of equipment on the toluene hydrodealkylation PFD (Figure 1.5). You may assume a stream factor of 0.95 and that all the numbers on the PFD are on a stream time basis. The duty on all of the units can be found in Table 1.7.

  a. E-101, Feed Preheater
  b. E-102, Reactor Effluent Cooler
  c. H-101, Heater
  d. C-101, Recycle Gas Compressor, assuming electric drive

Table 3.6   Typical Thermal Efficiencies of Fired Heaters and Furnaces (from Ulrich, G. D., *A Guide to Chemical Engineering Process Design and Economics,* Wiley , New York, 1984, Copyright © 1984 John Wiley & Sons, Inc., reprinted by permission of John Wiley & Sons Inc.)

Type of Heater/Furnace	Typical Range of Thermal Efficiencies (based on % of Lower Heating Value of fuel transmitted to process stream or utility)
Industrial Boiler (Water-Tube)	85–90 %
Thermal Fluid Heater	
(i) Hot Water	85–90 %
(ii) Diphenyls (Dowtherm™)	80–85 %
(iii) Molten Salt	80–85 %
(iv) Mineral or Silicon Oil	80–85 %
Reactive Process Heaters	80–85 %
Non-Reactive Process Heaters	90–92 %

Table 3.7   Theoretical Steam Requirements (kg steam/kWh) (from Perry, R. H., and D. W. Green, *Perry's Chemical Engineers Handbook,* **6th ed., McGraw-Hill, New York, 1984. Reprinted by permission McGraw-Hill Companies.) and Efficiencies for Pump and Compressor Drives (from Ulrich, G. D.,** *A Guide to Chemical Engineering Process Design and Economics,* **Wiley , New York, 1984, Copyright © 1984 John Wiley & Sons, Inc., reprinted by permission of John Wiley & Sons Inc.)**

Theoretical Steam Requirements	Inlet Condition of Steam (barg) (Superheat in °C)							
Exhaust Pressure	10.0 sat'd	13.8 sat'd	17.2 50	27.6 170	41.4 145	41.4 185	58.6 165	58.6 205
2" Hg abs	4.77	4.54	4.11	3.34	3.22	3.07	2.98	2.85
4" Hg abs	5.33	5.04	4.54	3.62	3.47	3.30	3.20	3.05
0 barg	8.79	7.94	6.88	5.08	4.72	4.45	4.22	4.00
0.69 barg	10.87	9.57	8.11	5.77	5.28	4.97	4.67	4.40
2.07 barg	15.24	12.72	10.40	6.91	6.18	5.78	5.35	5.02
3.45 barg	20.86	16.32	12.79	7.97	6.97	6.49	5.93	5.54
4.14 barg	24.45	18.32	14.11	8.50	7.34	6.83	6.20	5.78
4.82 barg	28.80	20.68	15.47	9.05	7.71	7.16	6.45	6.01

Drive Efficiencies (%)	Shaft Power (kW)								
Type of Drive	1	5	10	50	100	500	1000	5000	10^4
Electric Drive	72	82	85	91	92	95	96	96	97
Steam Turbine or Gas Expanders	—	—	—	—	—	57	63	74	77
Internal Combustion Engines* or Gas Turbines	—	—	—	—	28	32	34	38	38

*Based on lower heating value of fuel, all others based on theoretical performance of ideal machine.

    e. C-101, Recycle Gas Compressor, assuming steam drive using 10 barg steam discharging to atmospheric pressure.

    f. P-101, Toluene Feed Pump

    a. E-101: Duty is 15.1 GJ/h. From Table 3.4: Cost of High Pressure Steam = $5.09/GJ

        Energy Balance: $Q = 15.1 \text{ GJ/h} = (\dot{m}_{steam})(\Delta H_{vap}) = (\dot{m}_{steam})(1699.3) \text{ kJ/kg}$

$$\dot{m}_{steam} = 8886 \text{ kg/h} = 2.47 \text{ kg/s}$$

Yearly Cost $= (Q)(C_{steam})(t) = (15.1 \text{ GJ/h})(\$5.09/\text{GJ})(24)(365)(0.95) = \$639,600/\text{yr}$

Alternatively: Yearly Cost = (Yearly flowrate)(Cost per unit mass)

Yearly Cost $= (2.47)(3600)(24)(365)(0.95)(8.65/1000) = \$640,100/\text{yr}$

b. E-102: Duty is 46.66 GJ/h

From Table 3.4 Cost of Cooling Water = \$0.16/GJ

$Q = 46.66 \text{ GJ/h} = (\dot{m}_{cw})(C_{pcw})(\Delta T_{cw} = (\dot{m}_{cw})(4.18)(10) = 41.8 \dot{m}_{cw}$

$\dot{m}_{cw} = (46.66)(10^9/41.8)(10^3) = 1116270 \text{ kg/h} = 310 \text{ kg/s}$

Yearly Cost $= (46.66 \text{ GJ/h})(24)(365)(0.95)(\$0.16/\text{GJ}) = \$62,100/\text{yr}$

c. H-101: Duty is 27 GJ/h (7510 kW). From Table 3.6, an indirect, non-reactive process heater has a thermal efficiency (ξ_{th}) of 90% From Table 3.4, natural gas cost \$2.50/GJ and the heating value is 0.0377 GJ/m³.

$Q = 27 \text{ GJ/h} = (\dot{v}_{gas})(\Delta H_{natural\ gas})(\text{efficiency}) = (\dot{v}_{gas})(0.0377)(0.9)$

$\dot{v}_{gas} = 796 \text{ std m}^3/\text{h} (0.22 \text{ std m}^3/\text{sec})$

Yearly Cost $= (27)(2.5)(24)(365)(0.95) = \$566,000/\text{yr}$

d. C-101: Shaft power is 49.1 kW and from Table 3.7 the efficiency of an electric drive (ξ_{dr}) is 91 %.

Electric Power $= P_{dr} = \text{Output power}/\xi_{dr} = (49.1)/0.91 = 54.0 \text{ kW}$

Yearly Cost $= (54.0)(0.06)(24)(365)(0.95) = \$27,000/\text{yr}$

Table E3.7   Summary of Utility Requirements for the Equipment in the Toluene Hydrodealkylation Process (data from Figure 1.5, Table 1.7, and Example 3.6)

Equipment	Electric Power (kW)	Steam High Pressure (kg/s)	Steam Med Pressure (kg/s)	Steam Low Pressure (kg/s)	Cooling Water (m³/s)	Fuel Gas (std m³/s)
E-101	—	2.47	—	—	—	—
E-102	—	—	—	—	0.31	—
E-103	—	—	—	0.14	—	—
E-104	—	—	—	—	0.055	—
E-105	—	—	—	—	0.007	—
E-106	—	—	1.26	—	—	—
H-101	—	—	—	—	—	0.22
C-101	54.0	—	—	—	—	—
P-101	16.7	—	—	—	—	—
P-102	4.9	—	—	—	—	—
Totals	75.6	2.47	1.26	0.14	0.372	0.22
Total Yearly Cost \$/yr	37,700	640,100	275,900	27,800	74,700	566,000

e. Same as part d with steam driven compressor. For 10 barg steam with exhaust at
   0 barg Table E3.7 provides a steam requirement of 8.79 kg-steam/kWh of power.
   The shaft efficiency is about 50% (extrapolating from Table E3.7)

   Steam required by drive = (49.1)(8.79/0.5) = 863 kg/h (0.24 kg/s)

   Cost of Steam = (863)(24)(365)(0.95)(7.3 × 10⁻³) = $52,400/yr

f. P-101: Shaft Power is 14.2 kW. From Table E3.7 the efficiency of an electric drive is
   about 85%.

   Electric Power = 14.2/0.85 = 16.7 kW

   Yearly Cost = (16.7)(0.06)(24)(365)(0.95) = $8,300/yr

Note : The cost of using steam to power the compressor is greater than the cost of electricity even though the cost per unit energy is much lower for the steam. The reasons for this
are: (1) The thermodynamic efficiency is low, and (2) the efficiency of the drive is low for a
small compressor. Usually steam drives are only used for compressor duties greater than
100 kW.

3.7   COST OF TREATING LIQUID AND SOLID WASTE STREAMS

As environmental regulations continue to tighten, the problems and costs associated with the treatment of waste chemical streams will increase. In recent years
there has been a trend to try to reduce or eliminate the volume of these streams
through waste minimization strategies. Such strategies involve utilizing alternative process technology or using additional recovery steps in order to reduce or
eliminate waste streams. Although waste minimization will become increasingly
important in the future, the need to treat waste streams will continue. Some typical costs associated with this treatment are given in Table 3.4, and flowrates can
be obtained from the PFD. It is worth noting that the costs associated with the
disposal of solids waste streams, especially hazardous wastes, have grown im-
mensely in the past few years and the values given in Table 3.4 are only approximate average numbers. Escalation of these costs should be done with extreme
caution.

3.8   EVALUATION OF COST OF MANUFACTURE FOR THE PRODUCTION
OF BENZENE VIA THE HYDRODEALKYLATION OF TOLUENE

Example 3.7

Calculate the Cost of Manufacture (*COM*) for the toluene hydrodealkylation process using
the PFD in Figure 1.5 and the flow table given in Table 1.5.

A Utility Summary for all the equipment is given in Table E3.7, from which we find
the total yearly utility costs for this process are:

Steam = $ 943,800/yr
Cooling water = $ 74,700/yr

Fuel gas = $566,000/yr
Electricity = $37,700/yr

Total Utilities = $1,622,200 = $1.622 × 10^6/yr

Raw Material Costs from the PFD and Table 3.5 are

Toluene = $19.973 × 10^6/yr

Hydrogen = $1.605 × 10^6/yr (based on a value of $0.0286/std m^3)

Total Raw Materials = $21.578 × 10^6/yr

There are no waste streams shown on the PFD so

Waste Treatment = $0.0/yr

From Example 3.1 the Cost of Operating Labor (C_{OL}) is

C_{OL} = **(10)(46,800) = $468,000/yr**

From Problem 2.11 (using CAPCOST©), we find that the fixed capital investment (C_{GR}) for the process is $5.6 × 10^6.

FCI = **$5.6 × 10^6/yr**

Finally, using Eq. (3.1) we estimate the total manufacturing cost as:

COM = **0.304FCI + 2.73C_{OL} + 1.23(Utilities + Raw Materials + Waste Treatment)**

COM = **$31.52 × 10^6/yr**

3.9   SUMMARY

In this chapter, we have seen that the cost of manufacturing a chemical product depends on the fixed capital investment, the cost of operating labor, the cost of utilities, the cost of waste treatment, and the cost of raw materials. In most cases, the cost of raw materials is, by far, the biggest cost.

The amount of the raw materials and utilities can be obtained directly from the PFD. The cost of operating labor can be estimated from the number of pieces of equipment given on the PFD. Finally, the fixed capital investment may again be estimated from the PFD using the techniques given in Chapter 2.

REFERENCES

1. Ulrich, G. D., *A Guide to Chemical Engineering Process Design and Economics,* Wiley, New York, 1984.

2. Peters, M. S., and K. D. Timmerhaus, *Plant Design and Economics for Chemical Engineers,* 4th ed., McGraw-Hill, New York, 1990.

3. Valle-Riestra J. F., *Project Evaluation in the Chemical Process Industries,* McGraw-Hill, New York, 1983.

4. *Chemical Marketing Reporter,* Published weekly by Schnell Publishing Co., New York.

5. *Annual Energy Review,* Energy Information Administration, 1991.

6. Perry, R. H., and D. W. Green, *Perry's Chemical Engineers Handbook,* 6th ed., McGraw-Hill, New York, 1984.

PROBLEMS

1. Estimate the *COM* (cost of manufacture) for the light gas plant given in Problem 2.4. The duties of the four heat exchangers are given below:

   E-201 $Q = -3.80 \times 10^6$ kJ/h
   E-202 $Q = 5.55 \times 10^6$ kJ/h
   E-203 $Q = -4.20 \times 10^6$ kJ/h
   E-204 $Q = 3.09 \times 10^6$ kJ/h

   Assume a stream factor of 0.913 (8000 h/yr) for this process.

2. Estimate the cost of operating labor, C_{OL}, for the dimethyl ether process shown in Figure B.1.

3. Estimate the cost of operating labor, C_{OL}, for the acrylic acid process shown in Figure B.2.

4. Estimate the cost of operating labor, C_{OL}, for the acetone process shown in Figure B.3.

5. Estimate the cost of operating labor, C_{OL}, for the heptenes process shown in Figure B.4.

6. Estimate the cost of utilities for the dimethyl ether process shown in Figure B.1.

7. Estimate the cost of utilities for the acrylic acid process shown in Figure B.2.

8. Estimate the cost of utilities for the acetone process shown in Figure B.3.

9. Estimate the cost of utilities for the heptenes process shown in Figure B.4.

10. Estimate the cost of manufacture, *COM*, for the dimethyl ether process shown in Figure B.1.

11. Estimate the cost of manufacture, *COM*, for the acrylic acid process shown in Figure B.2.

12. Estimate the cost of manufacture, *COM*, for the acetone process shown in Figure B.3.

13. Estimate the cost of manufacture, *COM*, for the heptenes process shown in Figure B.4.

4

Engineering
Economic Analysis

The goal of any manufacturing industry is to make money. This is realized by producing products with a high market value from raw materials with a low market value. The chemical process industry produces high-value chemicals from low-value raw materials.

In the previous chapters, a process flow diagram (Chapter 1), an estimate of the capital cost (Chapter 2), and an estimate of operating costs (Chapter 3) were provided for the production of benzene. From this material an economic evaluation can be carried out to determine:

1. If the process generates money
2. If the process is attractive compared to other processes (such as those for the production of dimethyl ether, acrylic acid, heptenes, and acetone given in Appendix B)

In the next two chapters, the necessary background to perform this economic analysis is provided.

The principles of economic analysis are covered in this chapter. The material presented covers all of the major topics required for completion of the Fundamentals of Engineering (FE) examination. This is the first requirement for becoming a Registered Professional Engineer in the United States.

It is important for you, the graduating student, to understand the principles presented in this chapter at the beginning of your professional career in order to manage your money skillfully. As a result, we have elected to integrate discussions and examples of personal money management throughout the chapter.

The evaluation of profitability and comparison of alternatives for proposed projects are covered in Chapter 5.

4.1   INVESTMENTS AND THE TIME VALUE OF MONEY

The ability to profit from investing money is the key to our economic system. In this text, we introduce investment in terms of personal financing and then apply the concepts to chemical process economics.

There are various ways to distribute one's personal income. The first priority is to maintain a basic (no-frills) standard of living. This includes necessary food, clothing, housing, transportation, and expenses such as taxes imposed by the government. The remaining money, termed discretionary money, can then be distributed. It is wise to distribute this money in a manner that will realize both your short-term and long-term goals.

Generally, there are two classifications for spending discretionary money:

1. Consume money as received.   This provides immediate personal gratification and/or satisfaction. We experience this use for money early in life.
2. Retain money for future consumption.   This is money put aside to meet future needs. These may result from hard-to-predict causes such as sickness and job layoffs or from a more predictable need for long-term retirement income. It is unlikely that you have considered these types of financial needs and have little experience in investing to secure a comfortable life style after you stop working.

There are two approaches to setting money aside for use at a later date:

- Simple Savings: Put money in a safety deposit box, sugar bowl or other such container.
- Investments: Put money into an investment.

These two approaches are considered in the following example.

Example 4.1

Upon graduation, you start your first job at $50,000/yr. You decide to set aside 10% or $5,000/yr for retirement in 40 years time, and you assume that you will live 20 years after retiring. You have been offered an investment that will pay you $67,468/year during your retirement years for the money you invest.

   a. How much money would you have per year in retirement if you had saved the money, but not invested it, until retirement?
   b. How does this compare to the investment plan offered?

c. How much money was produced from the investment?

a. Money Saved: ($5,000)(40) = $200,000.
   Income during retirement: $200,000/20 = $10,000/yr.

b. Comparison: (Income from savings)/(Income from investments) = $10,000/$67,468 = 0.15

c. Money Produced = Money received − Money Invested = ($67,468)(20) − $200,000 = $1,149,360

The value of the investment is clear. The income in retirement from savings amounts to only 15% of the investment income. The amount of money provided during retirement, by setting $200,000 aside, was over a million dollars. We will show later that this high return on investment resulted from two factors: the long time period for the investment and the interest rate earned on the savings.

Money, when invested, makes money.

We will now define what is meant by an investment.

An *Investment* is an agreement between two parties, whereby, one party, the *Investor*, provides money, P, to a second party, the *Producer*, with the expectation that the *Producer* will return money, F, to the *Investor* at some future specified date, where $F > P$. The terms used in describing the investment are:

P – Principal or Present Value
F – Future Value
n – Years between F and P

The amount of money earned from the investment

$$E = F - P \qquad (4.1)$$

The yearly earnings rate is

$$i_s = \frac{E}{Pn} = \frac{(F - P)}{Pn} \qquad (4.2)$$

where i_s is termed the simple interest rate.

From Eq. (4.2), we have:

$$\frac{F}{P} = (1 + ni_s) \text{ or, in general, } \frac{F}{P} = f(n,i) \qquad (4.3)$$

Example 4.2

You decide to put $1000 into a bank that offers a special rate if left in for two years. After two years you will be able to withdraw $1,150.

  a. Who is the producer?
  b. Who is the investor?
  c. What are the values of P, F, i_s, and n?

  a. Producer: The bank has to produce $150.00 after two years.
  b. Investor: You invest $1,000 in an account at the beginning of the two-year period.
  c. $P = \$1,000$ (given)

   $F = \$1,150$ (given)

   $n = 2$ years (given)

   From Eq. (4.2),

   $i_s = (\$1,150 - \$1,000)/(\$1,000)/(2) = 0.075$ or 7.5% per year

---

In Example 4.2, you were the investor and invested in the bank. The bank was the "money producer" and had to return to you more dollars ($1,150) than you invested (1,000). This bank transaction is an investment commonly termed as "savings." In the reverse situation, termed "loan," the bank becomes the investor. You must produce money during the time of the investment.

Eqs. (4.1) through (4.3) apply to a single transaction between the investor and the producer that covers n years and uses simple interest. There are other investment schedules and interest formulations in practice, and these will be covered later in this chapter.

Figure 4.1 illustrates a possible arrangement to provide the funds necessary to build a new chemical plant such as the one introduced in the narrative in Chapter 1.

In this arrangement, a bank invests in a company, which in turn invests in a project to produce a chemical. There are two agreements in this project (see Figure 4.1).

  1. The bank is an investor and the company is the producer.
  2. The company is an investor and the project is the producer.

In this illustration, all of the money produced in the project is sent to the company. The company pays its investor, the bank, and draws off the rest as profit. The bank also makes a profit from its investment loan to the company. The project is the source of money to provide profits to both the company and the bank. The project converts a low-value, raw chemical into chemicals of higher value. Without the investor, the plant would not be built, and without the plant, there would be no profits for either the company or the bank.

Money is a measure of the value of products and services. The value of a chemical material is the price it can be exchanged for in dollars. Investments can

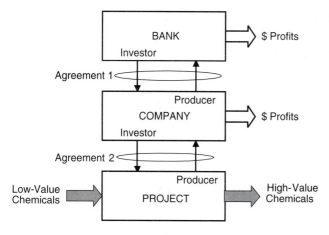

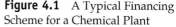

Figure 4.1   A Typical Financing Scheme for a Chemical Plant

be made in units other than dollars such as stocks, bonds, grain, oil, or gold. We will often refer to value, or value added, in describing investments. The term value is a general one and, in our case, can be assigned a dollar figure for economic calculations.

Figure 4.1 shows that all profits were produced from an operating plant. The role of engineers in our economy should be clear. This is to assure efficient production of high-value products, including current as well as new and improved products.

The economic analysis of processes, in almost all cases, will be made from the point of view of the company as the investor in a project. The project may be the construction of a new plant or a modification to an existing plant.

Consider the decisions involved in the investment in a new plant (the project) from the point of view of the company. The company must invest the money to build the plant before any income resulting from production can begin. Once the plant has been built and is operational, it is expected to operate for many years. During this time, the plant produces a profit and the company receives income from its investment. It is necessary to be able to determine if this future income is sufficiently attractive to make the investment worthwhile.

The *time value of money* refers to a concept that is fundamental to evaluating an investment. We illustrate this in the next example:

Example 4.3

You estimate that in two years time you will need $1,150 in order to replace the linoleum in your kitchen. Consider two choices.

1. Wait two years to take action, or
2. Invest $1,000 now (Assume that interest is offered by the bank at the same rate as given in Example 4.1.)

What would you do (Explain your answer)?

Solution: Consider investing the $1,000 today as it will provide the $1,150 in two years. The key is that the dollar I have today is worth 15% more than a dollar I will have in two years' time.

From the above example, we conclude that today's dollar is worth more than tomorrow's dollar because it can be invested to earn more dollars. This must not be confused with inflation that erodes purchasing power and is discussed in Section 4.6.

Money today is worth more than money in the future.

You will find in the upcoming sections, when we compare capital investments made at different times, that we *must* account for the timing of each investment. In other words, we must take into account the time value of money.

4.2   DIFFERENT TYPES OF INTEREST

Two types of interest are used when calculating the future value of an investment. They are referred to as *simple* and *compound interest*. Simple interest calculations are rarely used today. Unless specifically noted, all interest calculations will be carried out using compound interest methods.

4.2.1   Simple Interest

In simple interest calculations, the amount of interest paid is based solely on the initial investment.

Interest paid in any year $= Pi_s$

For an investment period of n years, the total interest paid $= Pi_s n$

Total value of investment in n years $= F_n = P + Pi_s n$

$$F_n = P(1 + i_s n) \tag{4.4}$$

If, instead of setting the earned interest aside, it was reinvested, the total amount of interest earned would be greater. When earned interest is reinvested, the interest is referred to as compound interest.

4.2.2   Compound Interest

Let us determine the future value of an investment, F_n, after n years at an interest rate of i per year for an initial investment of P when the interest earned is reinvested each year.

a. At the start we have our initial investment $= P$
b. In year 1, we earn Pi in interest.
    For year 2, we invest $P + Pi$ or $P(1 + i)$.
c. In year 2, we earn $P(1 + i)i$ in interest.
    For year 3, we invest $P(1 + i) + P(1 + i)i$ or $P(1 + i)^2$.
d. In year 3, we earn $P(1 + i)^2 i$ in interest.
    For year 4, we invest $P(1 + i)^2 + P(1 + i)^2 i$ or $P(1 + i)^3$.
e. By induction we find that after n years the value of our investment is $P(1 + i)^n$

Thus, for compound interest we may write

$$F_n = P(1 + i)^n \qquad\qquad (4.5)$$

We can reverse the process and ask how much would I have to invest now, P, in order to receive a certain sum F_n, in n years time? The solution to this problem is found by rearranging Eq. (4.5):

$$P = \frac{F_n}{(1 + i)^n} \qquad\qquad (4.6)$$

We illustrate the use of these equations in the following examples. The letters, p.a., following the interest refers to per year, ({p}er {a}nnum).

Example 4.4

For an investment of $500 at an interest rate of 8% p.a. for 4 years, what would be the future value of this investment, assuming compound interest?

From Eq. (4.5) for $P = 500$, $i = 0.08$ and $n = 4$ we obtain

$F_4 = P(1 + i)^n = 500(1 + 0.08)^4 = \$680.24$

Note: Simple interest would have yielded $F_4 = 500(1 + (4)(0.08)) = \$660$ ($20.24 less)

Example 4.5

How much would I need to invest in a savings account, yielding 6% interest p.a. to have $5,000 in five years time?

From Eq. (4.6) using $F_5 = \$ 5,000$, $i = 0.06$ and $n = 5$ we get

$$P = F_n/(1 + i)^n = 5000/(1.06)^5 = \$3736.29$$

If we invest $ 3736.29 into the savings account today, we will have $5,000 in 5 years time.

Example 4.6

I need to borrow a sum of money (P) and have two loan alternatives:

a. I borrow from my local bank who will lend me money at an interest rate of 7% p.a. and pay compound interest.

b. I borrow from "Honest Sam" who offers to loan me money at 7.3% p.a. using simple interest.

In both cases, I need the money for 3 years. How much money would I need in 3 years to pay off this loan? Consider each option separately.

Bank: From Eq. (4.5) for $n = 3$ and $i = 0.07$ we get

$$F_3 = (P)(1 + 0.07)^3 = 1.225\ P$$

Sam: From Eq. (4.4) for $n = 3$ and $i = 7.3$ we get

$$F_3 = (P)(1 + (3)(0.073)) = 1.219\ P$$

Even though Sam stated a higher interest rate to be paid, I would borrow the money from Sam since $1.219P < 1.225P$. This was because Sam used simple interest and the bank used compound interest.

4.2.3 Interest Rates Changing with Time

If we have an investment over a period of years, and the interest rate changes each year, then the appropriate calculation for compound interest is given by

$$F_n = P \prod_{j=1}^{n} (1 + i_j) = P(1 + i_1)(1 + i_2)\cdots(1 + i_n) \tag{4.7}$$

4.3 TIME BASIS FOR COMPOUND INTEREST CALCULATIONS

In industrial practice, the length of time assumed when expressing interest rates is one year. However, we are sometimes confronted with terms such as 6% p.a. compounded monthly. In this case, the 6% is referred to as a "nominal annual interest rate," i_{nom}, and the number of compounding periods per year is m (12 in this case). The nominal rate is not used directly in any calculations. The actual rate is the interest rate per compounding period, r. The relationship needed to evaluate r is

$$r = \frac{i_{nom}}{m} \tag{4.8}$$

Example 4.7

For the case of 12% p.a. compounded monthly, what are m, r, and i_{nom}?

Given: $m = 12$ (months in a year), $i_{nom} = 12\% = 0.12$

From Eq. (4.8)

$r = 0.12/12 = 0.01$ (or 1% per month)

4.3.1   Effective Annual Interest Rate

We can use an effective annual interest rate, i_{eff}, that will allow us to make interest calculations on an annual basis and obtain the same result as using the actual compounding periods. If we look at the value of an investment after one year, we may write

$$F_1 = P(1 + i_{eff}) = P\left(1 + \frac{i_{nom}}{m}\right)^m$$

which, upon rearrangement, gives:

$$i_{eff} = \left(1 + \frac{i_{nom}}{m}\right)^m - 1 \tag{4.9}$$

Example 4.8

What is the effective annual interest rate for a nominal rate of 8% p.a. when compounded monthly?

From Eq. (4.9) for $i_{nom} = 0.08$, and $m = 12$ we obtain

$i_{eff} = (1 + 0.08/12)^{12} - 1 = 0.083$ (or 8.3 % p.a.)

The effective annual interest rate is greater than the nominal annual rate. This indicates that the effective interest rate will continue to increase as the number of compounding periods per year increases. For the limiting case, the interest is compounded continuously.

4.3.2   Continuously Compounded Interest

For the case of continuously compounded interest, we must look at what happens to Eq. (4.9) as $m \to \infty$:

$$i_{eff} = \lim_{m \to \infty}\left[\left(1 + \frac{i_{nom}}{m}\right)^m - 1\right]$$

rewriting the left-hand side as $\lim\limits_{m \to \infty} \left[\left\{ \left(1 + \dfrac{i_{nom}}{m} \right)^{\frac{m}{i_{nom}}} \right\}^{i_{nom}} - 1 \right]$

and noting that $\lim\limits_{n \to \infty} \left[1 + \dfrac{x}{n} \right]^{\frac{n}{x}} = e$

we find that for continuous compounding:

$$i_{eff} = e^{i_{nom}} - 1 \tag{4.10}$$

Eq. (4.10) represents the maximum effective annual interest rate for a given nominal rate.

Example 4.9

What is the effective annual interest rate for an investment made at a nominal rate of 8% p.a. compounded continuously?

From Eq. (4.10) for $i_{nom} = 0.08$ we obtain

$i_{eff} = e^{0.08} - 1 = 0.0833$ or 8.33% p.a.

Note: We can see, by comparison with Example 4.8, that by compounding continuously little was gained over monthly compounding.

In comparing alternatives, the effective annual rate and not the nominal annual rate of interest must be used.

4.4 CASH FLOW DIAGRAMS

Up to this point, we have considered only an investment made at a single point in time at a known interest rate, and we learned to evaluate the future value of this investment. More complicated transactions involve several investments and/or payments of differing amounts made at different times. For more complicated investment schemes, we must keep careful track of the amount and time of each transaction. An effective way to track these transactions is to utilize a *Cash Flow Diagram* or *CFD*. Such a diagram offers a visual representation of each investment. Figure 4.2 is the cash flow diagram for Example 4.10. It is used to introduce the basic elements of the "discrete" CFD.

Figure 4.2 shows that cash transactions were made periodically. The values given represent payments made at the end of the year. Figure 4.2 shows that $1,000, $1,200, and $1,500 were received at the end of the first, second, and third year, respectively. In the fifth and seventh year, $2,000 and $ X were paid out. There were no transactions in the fourth and sixth years.

Each cash flow is represented by a vertical line with length proportional to the cash value of the transaction. The sign convention uses a downward-pointing

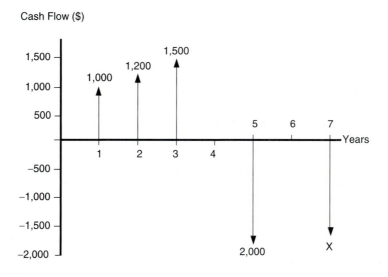

Figure 4.2   An Example of a Representative Discrete Cash Flow Diagram (CFD)

arrow when cash flows outward and an upward-pointing arrow representing inward cash flows. When a company invests money in a project, the company CFD shows a negative cash flow (outward flow) while the project CFD shows a positive cash flow (inward flow). Lines are placed periodically in the horizontal direction to represent the time axis. Most frequently, we perform our analysis from the point of view of the investor.

The cash flow diagram shown in Figure 4.2 can be presented in a simplified format, using the following simplifications:

1. The y-axis is not shown.
2. Units of monetary transactions are not given for every event.

In addition to the discrete CFD described above, we can show the same information in a "cumulative" CFD. This type of CFD presents the accumulated cash flow at the end of each period.

4.4.1   Discrete Cash Flow Diagram

The discrete CFD provides a clear, unambiguous pictorial record of the value, type, and timings of each transaction occurring during the life of a project. In order to avoid making mistakes and to save time, it is recommended that prior to doing any calculations, you sketch a cash flow diagram.

Example 4.10

I borrow $1,000, $1,200, and $1,500 from a bank (at 8% p.a. effective interest rate) at the end of years 1, 2, and 3, respectively. At the end of year 5, I make a payment of $2,000, and at the end of year 7, I pay off the loan in full. The CFD for this exchange from my point of view (producer) is given to the right.

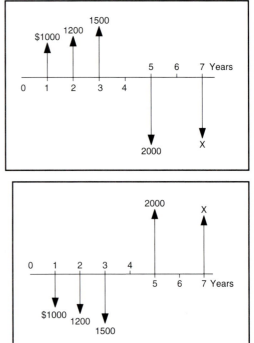

Note: This figure is the short-hand version of the one presented in Figure 4.2 used to introduce the CFD.

Draw a discrete cash flow diagram for the investor.

The bank represents the investor. From the investor's point of view, the initial three transactions are negative and the last two are positive.

The figure to the right represents the CFD for the bank. It is the mirror image of the one given above in the problem statement.

The value of X in Example 4.10 depends upon the interest rate. Its value is a direct result of the time value of money. The effect of interest rate and the calculation of the value of X (in Example 4.13) will be determined in the next section.

Example 4.11

You borrow $10,000 from a bank to buy a new car and agree to make 36 equal monthly payments of $320 each to repay the loan. Draw the discrete CFD for the investor in this agreement.

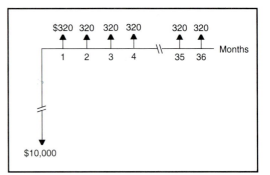

The bank is the investor. The discrete CFD for this investment is shown on the right.

Notes:

1. There is a break in both the time scale and in the investment at time = 0 (the initial investment).

2. From your point of view, the cash flow diagram would be the mirror image of the one shown.

The cash flow diagram constructed in Example 4.11 is typical of those you will encounter throughout this text. The investment (negative cash flow) is made early in the project during design and construction before there is an opportunity for the plant to produce product and generate money to repay the investor. In Example 4.11, the payback was made in a series of equal payments over three years to repay the initial investment by the bank. In Section 4.5, you will learn how to calculate the interest rate charged by the bank in this example.

4.4.2   Cumulative Cash Flow Diagram

As the name suggests, the cumulative CFD keeps a running total of the cash flows occurring in a project. To illustrate how to construct a cumulative CFD, consider Example 4.12. This example illustrates the cash flows associated with the construction and operation of a new chemical plant.

Example 4.12

The yearly cash flows estimated for a project involving the construction and operation of a chemical plant producing a new product are provided in the discrete CFD given on the right. Using this information, construct a cumulative CFD.

The numbers shown in the worksheet below were obtained from this diagram.

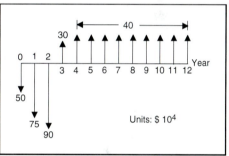

Year	Cash Flow ($) (from discrete CFD)	Cumulative Cash Flow (calculated)
0	−500,000	−500,000
1	−750,000	−1,250,000
2	−900,000	−-2,150,000
3	300,000	−1,850,000
4	400,000	−1,450,000
5	400,000	−1,050,000
6	400,000	−650,000
7	400,000	−250,000
8	400,000	150,000
9	400,000	550,000
10	400,000	950,000
11	400,000	1,350,000
12	400,000	1,750,000

The cumulative cash flow diagram is plotted below.

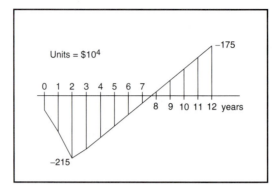

4.5   CALCULATIONS FROM CASH FLOW DIAGRAMS

To compare investments that take place at different times, it is necessary to account for the time value of money.

> **When cash flows occur at different times, each cash flow must be brought forwards (or backwards) to the same point in time and then compared.**

The point in time that we choose is arbitrary. This is illustrated in the next example.

Example 4.13

The CFD obtained from Example 4.10 (for the borrower) is copied below. The annual interest rate paid on the loan is 8% p.a..

In year 7, the remaining money owed on the loan is paid off.

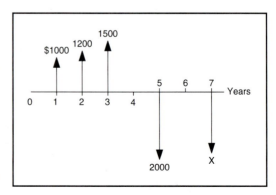

- **a.** Determine the amount, X, of the final payment.
- **b.** Compare the value of X to the value that would be owed if there were no interest paid on the loan.

With the final payment at the end of year 7, no money is owed on the loan. If we sum all the positive and negative cash flows adjusted for the time of the transactions, this adjusted sum must equal zero.

We select as the base time the date of the final payment.

a. From Eq. (4.5) for $i = 0.08$ we obtain:

For withdrawals:

$1,000 end of year 1: $F_6 = (\$1000)(1 + 0.08)^6 = \$1586.87$
$1,200 end of year 2: $F_5 = (\$1200)(1 + 0.08)^5 = \$1763.19$
$1,500 end of year 3: $F_4 = (\$1500)(1 + 0.08)^4 = \$2040.73$

$$\text{Total withdrawals} = \$5390.79$$

For repayments:

$2,000 end of year 5: $F_2 = -(\$2000)(1 + 0.08)^2 = -\$2332.80$
$X   end to year 7: $F_0 = -(\$X)(1 + 0.08)^0 = -\$X$
Total repayments $= -\$(2332.80 + X)$

Summing the cash flows and solving for X yields

$0 = \$5390.79 - \$(2332.80 + X)$
$X = \$3057.99 \approx \$3058$

b. For $i = 0.00$

Withdrawals $= \$1,000 + \$1,200 + \$1,500 = \$3,700$
Repayments $= -\$(2,000 + X)$

$0 = \$3,700 - \$(2,000 + X)$
$X = \$1,700$

Note: Because of the interest paid to the bank, the borrower repaid a total of $1,358 ($3,058 − $1,700) more than was borrowed from the bank seven years earlier.

To demonstrate that any point in time could be used as a basis, let us compare the amount repaid based on the end of year 1. We use Eq. (4.6) and move backwards in time (exponents become negative). This gives

$$0 = 1000 + \frac{1200}{1.08} + \frac{1500}{1.08^2} - \frac{2000}{1.08^4} - \frac{X}{1.08^6}$$

and solving for X yields

$$X = (1.08)^6 \left[1000 + \frac{1200}{1.08} + \frac{1500}{1.08^2} - \frac{2000}{1.08^4} \right] = \$3058 \text{ (the same answer as before!)}$$

Usually, we are interested in comparing investments at the start or at the end of a project, but the conclusions that we draw will be independent of where we make that comparison.

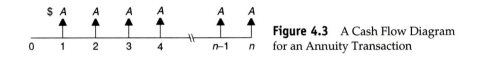

Figure 4.3   A Cash Flow Diagram for an Annuity Transaction

4.5.1   Annuities—A Uniform Series of Cash Transactions

We often encounter problems that involve a series of uniform cash transactions, each of value A, taking place at the end of each year for n consecutive years. This pattern is called an *annuity* and the discrete CFD for an *annuity* is shown in Figure 4.3.

To avoid the need to do a year-by-year analysis, like the one done in Example 4.13, we will develop an equation to provide the future value of an annuity.

The future value of an annuity at the end of time period n is found by bringing each of the investments forward to time n as we did in Example 4.13.

$$F_n = A(1 + i)^{n-1} + A(1 + i)^{n-2} + A(1 + i)^{n-3} + \cdots + A$$

This equation is a geometric series of the form $a, ar, ar^2, ..., ar^{n-1}$ with sum $S_n = F_n$.

$$S_n = a\left[\frac{r^n - 1}{r - 1}\right]$$

For our case, we have: $a = A$; $r = 1 + i$; $n = n$. Therefore,

$$F_n = A\left[\frac{(1 + i)^n - 1}{i}\right] \tag{4.11}$$

It is important to notice that Eq. (4.11) is correct when the annuity starts at the end of the *first* time period not at time zero. In the next section, we provide a shorthand notation that we will find useful in CFD calculations.

4.5.2   Discount Factors

We introduce the shorthand notation for the future value of an annuity starting with Eq. (4.11). We shorten the term F_n by simply calling it F, and then by dividing through by A, we get

$$F/A = [(1 + i)^n - 1]/i = f(i,n)$$

This ratio of F/A is a function of i and n, that is, $f(i,n)$. It can be evaluated when both the interest rate, i, and the time duration, n, are known. The value of $f(i,n)$ is referred to as a discount factor. If either A or F is known, the remaining unknown can be evaluated.

In general terms, a discount factor is designated as

Discount Factor for $X/Y = (X/Y, i, n) = f(i, n)$

Discount factors represent simple ratios and can be multiplied or divided by each other to give additional discount factors. For example, assume that we need to know the present worth, P, of an annuity, A, that is, the discount factor for P/A, but do not have the needed equation. The only available formula containing the annuity term, A, is the one for F/A derived above. We can eliminate the future value, F, and introduce the present value, P, by multiplying by the ratio of P/F, from Eq. (4.6), as shown below.

$$\text{Discount factor for } P/A = (P/A, i, n)$$

$$= (F/A, i, n)(P/F, i, n)$$

Substituting for F/A and P/F gives:

$$P/A = \frac{(1 + i)^n - 1}{i} \frac{1}{(1 + i)^n}$$

$$= \frac{(1 + i)^n - 1}{1} \frac{1}{i(1 + i)^n} = (P/A, i, n)$$

Table 4.1 lists the most frequently used discount factors in this text with their common names and corresponding formulae.

The key to performing any economic analysis is the ability to evaluate and compare equivalent investments. In order to understand that the equations presented in Table 4.1 provide a comparison of alternatives, we suggest that you re-

Table 4.1 Commonly Used Factors for Cash Flow Diagram Calculations

Conversion	Symbol	Common Name	Eq. No.	Formula
P to F	$(F/P,i,n)$	Single Payment Compound Amount Factor	(4.5)	$(1 + i)^n$
F to P	$(P/F,i,n)$	Single Payment Present Worth Factor	(4.6)	$\dfrac{1}{(1 + i)^n}$
A to F	$(F/A,i,n)$	Uniform Series Compound Amount Factor, Future Worth of Annuity	(4.11)	$\dfrac{(1 + i)^n - 1}{i}$
F to A	$(A/F,i,n)$	Sinking Fund Factor	(4.12)	$\dfrac{i}{(1 + i)^n - 1}$
P to A	$(A/P,i,n)$	Capital Recovery Factor	(4.13)	$\dfrac{i(1 + i)^n}{(1 + i)^n - 1}$
A to P	$(P/A,i,n)$	Uniform Series Present Worth Factor, Present Worth of Annuity	(4.14)	$\dfrac{(1 + i)^n - 1}{i(1 + i)^n}$

place the equal sign with the words "is equivalent to." As an example, consider the equation given for the value of an annuity, A, needed to provide a specific future worth, F. From Table 4.1, Eq. (4.11) can be expressed as:

$$F \text{ (Future value) is equivalent to } \{f(i,n) \, A(\text{Annuity value})\}$$

where

$$f(i,n) = (F/A, i, n)$$

Example 4.14

You have just won \$2,000,000 in the Texas Lottery as one of seven winners splitting up a jackpot of \$14,000,000. It has been announced that each winner will receive \$100,000/year for the next 20 years. What is the equivalent present value of your winnings if you have a secure investment opportunity providing 7.5% p.a.?

From Table 4.1, Eq. (4.14), for $n = 20$ and $i = 0.075$

$$P = (\$100,000)[(1 + 0.075)^{20} - 1]/[(0.075)(1 + 0.075)^{20}]$$

$$P = \$1,019,000$$

A present value of \$1,019,000 is equivalent to a 20-year annuity of \$100,000/yr when the effective interest rate is 7.5%.

Some examples to illustrate how to use these discount factors and how to approach problems involving discrete CFDs are given next.

Example 4.15

Consider Example 4.11, involving a car loan. The discrete CFD from the bank's point of view was shown previously.

What interest rate is the bank charging for this loan?

You have agreed to make 36 monthly payments of \$320. The time selected for evaluation is the time at which the final payment is made. At this time, the loan will be fully paid off. This means that the future value of the \$10,000 borrowed is equivalent to a \$320 annuity over 36 payments.

$$(\$10,000)(F/P, i, n) = (\$320)(F/A, i, n)$$

Substituting the equations for the discount factors given in Table 4.1, with $n = 36$ months, we get:

$$0 = -(10,000)(1 + i)^{36} + (320)[(1 + i)^{36} - 1]/i$$

This equation cannot be solved explicitly for i. We solve this equation by plotting the value of the right-hand

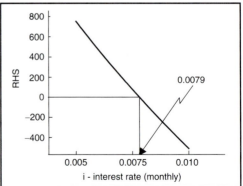

side of the equation shown above for various interest rates. This equation could also be solved using a numerical technique. From the graph, the interest rate that gives a value of zero represents the answer. From the graph on the previous page the rate of interest is $i = 0.0079$.

The nominal annual interest rate is $(12)(0.00786) = 0.095$ (9.5%).

Example 4.16

I invest money in a savings account that pays a nominal interest rate of 6% p.a. compounded monthly. I open the account with a deposit of $1000 and then deposit $50 at the end of each month for a period of 2 years followed by a monthly deposit of $100 for the following 3 years. What will the value of my savings account be at the end of the 5-year period?

First, draw a discrete CFD (shown to the right).

Although this CFD looks rather complicated, we can break it down into 3 easy sub-problems:

1. The initial investment
2. The 24 monthly investments of $50
3. The 36 monthly investments of $100

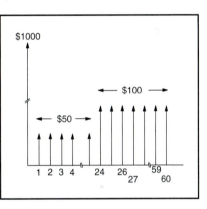

Each of these investments is brought forward to the end of month 60.

$F = (\$1000)(F/P, 0.005, 60) + (\$50)(F/A, 0.005, 24)(F/P, 0.005, 36) + (\$100)(F/A, 0.005, 36)$

Note: the effective monthly interest rate is $0.06/12 = 0.005$

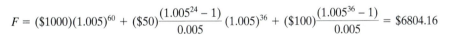

$$F = (\$1000)(1.005)^{60} + (\$50)\frac{(1.005^{24} - 1)}{0.005}(1.005)^{36} + (\$100)\frac{(1.005^{36} - 1)}{0.005} = \$6804.16$$

There are many ways to solve most complex problems. No one method is more or less correct than another. For example, we could regard the discrete CFD to be made up of a single investment of $1000 at the start, a $50 monthly annuity for the next 60 months and another $50 annuity for the last 36 months. Evaluating the future worth of the investment gives:

$$F = (\$1000)(F/P, 0.005, 60) + (\$50)(F/A, 0.005, 60) + (\$50)(F/A, 0.005, 36)$$

$$F = (\$1000)(1.005)^{60} + (\$50)\frac{(1.005^{60} - 1)}{0.005} + (\$50)\frac{(1.005^{36} - 1)}{0.005} = \$6804.16.$$

This is the same result as before.

Example 4.17

In Example 4.1, we introduced an investment plan for retirement. It involved investing $5,000/year for 40 years leading to retirement. The plan then provided $67,468/year for 20 years of retirement income.

   a. What yearly interest rate was used in this evaluation?

   b. How much money was invested in the retirement plan before withdrawals began?

   a. The evaluation is performed in two steps:

Step 1: Find the value of the $5,000 annuity investment at the end of the 40 years.

Step 2: Evaluate the interest rate of an annuity that will pay out this amount in 20 years at $67,468/year.

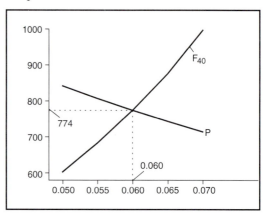

Step 1: From Eq. (4.11), Table 4.1, for A = $5,000 and n = 40,

$$F_{40} = (A)(F/A, n, i) = (\$5,000) [(1+i)^{40}-1]/i$$

Step 2: From Eq. (4.14), Table 4.1, for A = $67,468 and n = 20,

$$P = (A)(P/A, n, i) = (\$67,468)[(1+i)^{20} - 1]/[(i)(1+i)^{20}]$$

Set $F_{40} = P$ and solve for i. From the graph above, we get $i = 0.060$

   b. With $i = 0.060$, we have from the graph $F_{40} = \$774,000$

Note: The interest rate of 6.0% p.a. represents a relatively low interest rate, involving small risk.

4.6   INFLATION

As a result of inflation, a dollar set aside (not invested) will purchase fewer goods and services in the future than the same dollar would today. In Chapters 2 and 3, we have seen that we can track the inflation of equipment, labor, and fuel costs by the use of cost indexes. We sometimes wish to express these trends in terms of rates of inflation (f). This can be done using the cost indexes as follows:

$$CEPCI(j + n) = (1 + f)^n\, CEPCI(j) \tag{4.15}$$

where n = time span in years
       f = average inflation rate over the time span
       j = arbitrary year

We illustrate the use of Eq. (4.15) to estimate the inflation rate in the following example.

Example 4.18

What was the average rate of inflation for the costs associated with building a chemical plant over the following periods?

   a. 1980 through 1986
   b. 1986 through 1992

From Table 2.4, the values of the Chemical Engineering Plant Cost Index (CEPCI) are

   CEPCI (1980) = 261
   CEPCI (1986) = 318
   CEPCI (1992) = 358

   Eq. (4.15) yields

   a. $318 = (261)(1 + f)^6$,
       $f = 0.033$ (3.3% p.a.)
   b. $358 = 318 (1 + f)^6$
       $f = 0.020$ (2.0% p.a.)

To understand inflation, we must be able to distinguish between cash and the purchasing power (for the purchase of goods and services) of cash. Inflation decreases this purchasing power with time. All of the previous discussions on A, P, and F are given in terms of cash and not in terms of the relative purchasing power of this cash. We now introduce the term F', which represents the purchasing power of future cash. This purchasing power can be estimated using Eq. (4.16).

$$F' = \frac{F}{(1 + f)^n} \tag{4.16}$$

Substituting the equation for F in terms of P, from Eq. (4.5), gives

$$F' = P\frac{(1 + i)^n}{(1 + f)^n} = P\left[\frac{1 + i}{1 + f}\right]^n \tag{4.17}$$

We now write Eq. (4.17) in terms of an effective interest rate, i', that includes the effect of inflation.

$$F' = P(1 + i')^n \tag{4.18}$$

By comparing Eq. (4.18) with Eq. (4.17) we see that i' is given by

$$i' = \frac{1+i}{1+f} - 1 = \frac{i-f}{1+f} \qquad (4.19)$$

For small values of $f < 0.05$, we can approximate Eq. (4.19) by

$$i' \approx (i - f) \qquad (4.20)$$

Example 4.19

In this example, we consider the effect of inflation on the purchasing power of the money set aside for retirement in Example 4.17. Previously, we calculated the amount of cash available at the time of retirement in 40 years to be $774,000. This provided an income of $67,468 for twenty years.

a. Assuming an annual inflation rate of 2%, what is the purchasing power of the cash available at retirement?

b. What is the purchasing power of the retirement income in the first and 20th year of retirement?

c. How does Part a compare with the total annuity payments of $5,000/yr for 40 yrs.?

a. Using Eq. (4.16) for $f = 0.02$, $n = 40$, and $F = \$774,000$

$$F' = \$774,000/(1 + 0.02)^{40} = \$351,000$$

b. At the end of the 41st year (1st year of retirement)
Purchasing Power $= \$67,468/(1 + 0.02)^{41} = \$29,956/\text{yr}$
At the end of the 60th year (20th year of retirement)
Purchasing Power $= \$67,468/(1 + 0.02)^{60} = \$20,563/\text{yr}$

c. Amount invested $= (\$5,000/\text{yr})(40 \text{ yr}) = \$200,000$, compared to $351,000

Example 4.19 reveals the consequences of inflation. It showed that the actual income received in retirement of $67,468/yr had a purchasing power equivalent to between $29,956 and $20,563 at the time that the initial investment was made. This does not come close to the $50,000/yr base salary at that time. To increase this value, you would have to increase one or more of the following

a. the amount invested

b. the interest rate for the investment

c. the time over which the investment was made

The effects of inflation should not be overlooked in any decisions involving investments. Because inflation is influenced by politics, future world events, and so on, it is hard to predict. In this book, we will not consider inflation directly, and we will consider the cash flows to be in uninflated dollars.

4.7   DEPRECIATION OF CAPITAL INVESTMENT

When a company builds and operates a chemical process plant, the physical plant (equipment and buildings) associated with the process has a finite life. The value or worth of this physical plant decreases with time. Some of the equipment wears out and has to be replaced during the life of the plant. Even if the equipment is seldom used and is well maintained, it becomes obsolete and of little value. When the plant is closed, the plant equipment can be salvaged and sold for only a fraction of the original cost.

The cash flows associated with the purchase and installation of equipment are expenses that occur before the plant is operational. This results in a negative cash flow on a discrete CFD. When the plant is closed, equipment is salvaged, and this results in a positive cash flow at that time. The difference between these costs represents capital depreciation.

For tax purposes, the government does not allow companies to charge the full costs of the plant as a one-time expense when the plant is built. Instead, it allows only a fraction of the capital depreciation to be charged as an operating expense each year until the total capital depreciation has been charged.

The amount and rate at which equipment may be depreciated is set by the federal government (Internal Revenue Service of the U.S. Treasury Department). The regulations that cover the capital depreciation change quite regularly. Rather than outline the current method of depreciation suggested by the IRS, we choose to present several of the techniques that have been used in the past to depreciate capital investment. We leave the calculation of the current depreciation technique to Problems 4.18 and 4.19 at the back of the chapter.

4.7.1   Fixed Capital, Working Capital, and Land

When we talk about the depreciation of capital investment, we must be careful to distinguish between what can and cannot be depreciated. In general, the total capital investment in a chemical process is made up of the two components:

$$\text{Total Capital Investment} \ = \ \text{Fixed Capital} \ + \ \text{Working Capital} \qquad (4.21)$$

Fixed capital is all the costs associated with building the plant and was covered in Chapter 3 (either total module cost or grass roots cost). The only part of the fixed capital investment that cannot be depreciated is the *land*, which usually represents only a small fraction of the total.

Working capital is the amount of capital required to start up the plant and finance the first few months of operation before revenues from the process start. Typically, this money is used to cover salaries, raw material inventories, and any contingencies. The working capital will be recovered at the end of the project and represents a float of money to get the project started. This concept is similar to that of paying the first and last month's rent on an apartment. The last month's rent is fully recoverable at the end of the lease but must be paid at the beginning.

Since the working capital is fully recoverable, it cannot be depreciated. Typical values for the working capital are between 15% and 20% of the fixed capital investment.

4.7.2   Different Types of Depreciation

First of all, we will introduce and define terms that we will use to evaluate depreciation:

> **Fixed Capital Investment, FCI_L:** This represents the fixed capital investment to build the plant less the cost of land and represents the depreciable capital investment.
>
> **Salvage Value, S:** This represents the fixed capital investment of the plant, less the value of the land, evaluated at the end of the plant life. Usually, the equipment salvage (scrap) value represents a small fraction of the initial fixed capital investment. Often the salvage value of the equipment is assumed to be zero.
>
> **Life of the equipment, n:** This is specified by the United States Internal Revenue Service (IRS). It does not reflect the actual working life of the equipment but rather the time allowed by the IRS for equipment depreciation. Chemical process equipment currently has a depreciation class life of 9.5 years (see Problem 18).
>
> **Total Capital for Depreciation:** The total amount of depreciation allowed is the difference between the fixed capital investment and the salvage value.

$$D = FCI_L - S$$

> **Yearly Depreciation:** The amount of depreciation varies from year to year. The amount allowed in the kth year is denoted d_k.
>
> **Book Value:** The amount of the depreciable capital that has not yet been depreciated.

$$BV_k = FCI_L - \sum_1^k d_j$$

We provide a discussion of three representative depreciation methods that have been widely used to determine the depreciation allowed each year. Currently, only the straight line and double declining balance methods are approved by the IRS. The sum of the years digits method has been used previously and is included here for completeness.

> **Straight Line Depreciation Method, SL:** An equal amount of depreciation is charged each year over the depreciation period allowed. This is shown as

$$d_k^{SL} = \frac{[FCI_L - S]}{n} \qquad (4.22)$$

Sum of the Years Digits Depreciation Method, *SOYD*: The formula for calculating the depreciation allowance is given below:

$$d_k^{soyd} = \frac{[n + 1 - k][FCI_L - S]}{\frac{n}{2}[n + 1]} \tag{4.23}$$

The method gets its name from the denominator of Eq. (4.23), which is equal to the sum of the number of years over which the depreciation is allowed

$$1 + 2 + 3... + n = (n)(n + 1)/2$$

for example, if $n = 7$, then the denominator equals 28.

Double Declining Balance Depreciation Method, *DDB*: The formula for calculating the depreciation allowance is given below:

$$d_k^{DDB} = \frac{2}{n}\left[FCI_L - \sum_{j=0}^{j=k-1} d_j\right] \tag{4.24}$$

In the declining balance method, the amount of depreciation each year is a constant fraction of the book value, BV_{k-1}. The word double in *DDB* refers to the factor 2 in Eq. (4.24). Values other than 2 are sometimes used; for example, for the 150% declining balance method, 1.5 is substituted for the 2 in Eq. (4.24).

In this method, the salvage value does not enter into the calculations. We cannot, however, depreciate more than the value of *D*. To avoid this problem, we reduce the final year's depreciation to obtain this limiting value.

The following example illustrates the use of each of the above formulas to calculate the yearly depreciation allowances.

Example 4.20

The fixed capital investment (excluding the cost of land) of a new project is estimated to be $150.0 million, and the salvage value of the plant is $10.0 million. Assuming a 7-year equipment life, estimate the yearly depreciation allowances using:

  a. The straight line method
  b. The sum of the years digits method
  c. The double declining balance method.

   We have $FCI_L = \$150 \times 10^6$, $S = \$10.0 \times 10^6$, and $n = 7$ years

   Sample calculations for year 2 give the following:

   For straight line depreciation, using Eq. (4.22),

   $d_2 = (\$150 \times 10^6 - \$10 \times 10^6)/7 = \$20 \times 10^6$

Table E4.20   Calculations and Results for Example 4.20: The Depreciation of Capital Investment for a New Chemical Plant (all values in 10^7).

Year (k)	d_k^{SL}	d_k^{SOYD}	d_k^{DDB}	Book Value $FCI_L - Sd_k^{DDB}$
0				$(15 - 0) = 15$
1	$\dfrac{(15 - 1)}{7} = 2$	$\dfrac{(7 + 1 - 1)(15 - 1)}{28^a} = 3.5$	$\dfrac{(2)(15)}{7} = 4.29$	$(15 - 4.29) = 10.71$
2	$\dfrac{(15 - 1)}{7} = 2$	$\dfrac{(7 + 1 - 2)(15 - 1)}{28^a} = 3.0$	$\dfrac{(2)(10.71)}{7} = 3.06$	$(10.71 - 3.06) = 7.65$
3	$\dfrac{(15 - 1)}{7} = 2$	$\dfrac{(7 + 1 - 3)(15 - 1)}{28^a} = 2.5$	$\dfrac{(2)(7.65)}{7} = 2.19$	$(7.65 - 2.19) = 5.46$
4	$\dfrac{(15 - 1)}{7} = 2$	$\dfrac{(7 + 1 - 4)(15 - 1)}{28^a} = 2.0$	$\dfrac{(2)(5.46)}{7} = 1.56$	$(5.46 - 1.56) = 3.90$
5	$\dfrac{(15 - 1)}{7} = 2$	$\dfrac{(7 + 1 - 5)(15 - 1)}{28^a} = 1.5$	$\dfrac{(2)(3.90)}{7} = 1.11$	$(3.90 - 1.11) = 2.79$
6	$\dfrac{(15 - 1)}{7} = 2$	$\dfrac{(7 + 1 - 6)(15 - 1)}{28^a} = 1.0$	$\dfrac{(2)(2.79)}{7} = 0.80$	$(2.79 - 0.80) = 1.99$
7	$\dfrac{(15 - 1)}{7} = 2$	$\dfrac{(7 + 1 - 7)(15 - 1)}{28^a} = 0.5$	$1.99 - 1.0 = 0.99^b$	$(1.99 - 0.99) = 1.00$
Tot.	14.0	14.0	14.0	$1.0 = $ Salvage Value^b

^{a}Sum of Digits: $[n + 1]n/2 = [7 + 1]\, 7/2 = 28$

^{b}The depreciation allowance in the final year of the double declining balance method is adjusted to give a final book value equal to the salvage value.

For SOYD depreciation, using Eq. (4.23),

$d_2 = (7 + 1 - 2)\ (\$150 \times 10^6 - \$10 \times 10^6)/28 = \$30 \times 10^6$

For double declining balance depreciation, using Eq. (4.24),

$d_2 = (2/7)\ (\$150 \times 10^6 - \$42.86 \times 10^6) = \$30.6 \times 10^6$

A summary of all the calculations is given in Table E4.20 and presented graphically in Figure E4.20.

From Figure E4.20 we see that:

1. The depreciation values obtained from the sum of the years digits and the double declining balance methods are similar.
2. The double declining balance method has the largest depreciation in the early years.

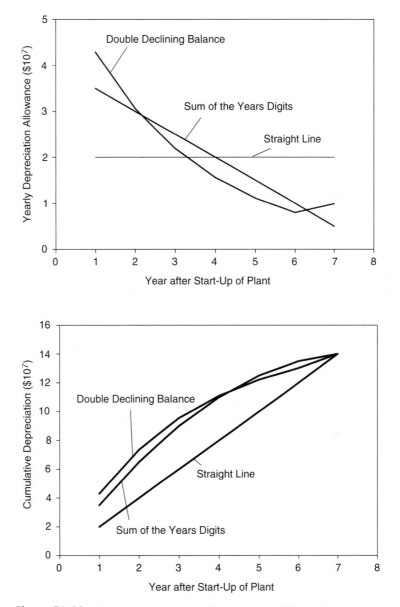

Figure E4.20   Yearly Depreciation Allowances and Cumulative Depreciation Amounts for Example 4.20.

3. The straight line method represents the slowest depreciation in the early years.

We characterize the SOYD and the DDB methods as examples of accelerated depreciation schemes (relative to the straight line). It will be shown in Section 4.8 that accelerated depreciation has significant economic advantages over the straight line method. However, we must remember that:

We can only depreciate capital investment in accordance with current tax regulations.

4.8 TAXATION, CASH FLOW, AND PROFIT

Taxation has a direct impact on the profits realized from building and operating a plant. Tax regulations are complex and companies have tax accountants and attorneys to assure compliance and to maximize the benefit from these laws. When we are considering individual projects or comparing similar projects, we must account for the effect of taxes. Taxation rates for companies and the laws governing taxation change frequently, and for the purposes of this book, we will assume a fixed taxation rate of 30%, unless specified differently.

$$\text{Taxation Rate } (t) = 0.30 \ (30\%)$$

Table 4.2 provides the definition of important terms and equations used to evaluate the cash flow and the profits produced from a project.

We use the equations from Table 4.2 in the next example.

Example 4.21

For the project given in Example 4.20, the manufacturing costs, excluding depreciation, are $30 million per year, and the revenues from sales are $75 million per year. Given the depreciation values calculated in Example 4.20, calculate the following for a 10-year period after start-up of the plant.

a. The after tax profit (net profit).
b. The after tax cash flow, assuming a taxation rate of 30%.

    From Eqs. (4.27) and (4.28) (all numbers are in 10^6),
    After tax profit $= (75 - 30 - d_k)(1 - 0.3) = 31.5 - (0.7)(d_k)$
    After tax cash flow $= (75 - 30 - d_k)(1 - 0.3) + d_k = 31.5 + (0.3)(d_k)$
    We provide a sample calculation for year 1 ($k = 1$):
    From Example 4.20, we know $d_1{}^{SL} = 20$, $d_1{}^{SOYD} = 35$ and $d_1{}^{DDB} = 42.9$

Table 4.2 Evaluation of Cash Flows[1] and Profits[1] in Terms of Revenue (*R*), Cost of Manufacturing (*COM*), Depreciation (*d*), and Tax Rate (*t*)

	Description	Formula	Equation
Expenses	= Manufacturing Costs + Depreciation	$= COM_d + d$	(4.25)
Income Tax	= (Revenue − Expenses)(Tax Rate)	$= (R - COM_d - d)(t)$	(4.26)
After Tax (Net) Profit	= Revenue − Expenses − Income Tax	$= (R - COM_d - d)(1 - t)$	(4.27)
After Tax Cash Flow	= Net Profit + Depreciation	$= (R - COM_d - d)(1 - t) + d$	(4.28)
Variables:			
t	Tax Rate		Constant
COM_d	Cost of Manufacture Excluding Depreciation		(3.2)
d	Depreciation: Depends upon method used		(4.22) (4.23) (4.24)
R	Revenue from Sales		

[1]To obtain before tax values set the tax rate (*t*) to zero

	SL	SOYD	DDB
Profit after Tax	17.5	7.0	1.47
Cash Flow after Tax	37.5	42.0	44.37

The calculations for years 1 through 10 are plotted in Figure E4.21. From this plot, we can see that the cash flow at the start of the project is greatest for the DDB method and lowest for the SL method.

The sum of the profits and cash flows over the 10-year period are $217 and $357 million, respectively. These totals are the same for each of the depreciation schedules used. The difference between the cash flows and the profits is seen to be the depreciation ($357 − 217 = $140 million).

Example 4.21 demonstrated how different depreciation schedules affect the after tax cash flow. The accelerated schedules for depreciation provided the greatest cash flows in the early years. Since money earned in early years has a greater value than money earned in later years, the accelerated schedule of depreciation is the most desirable alternative.

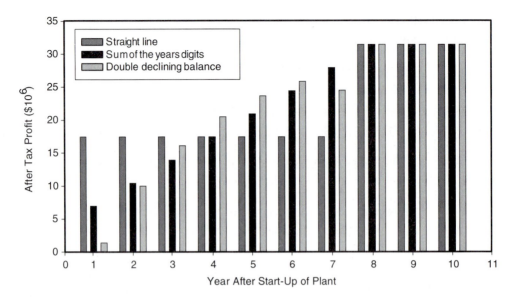

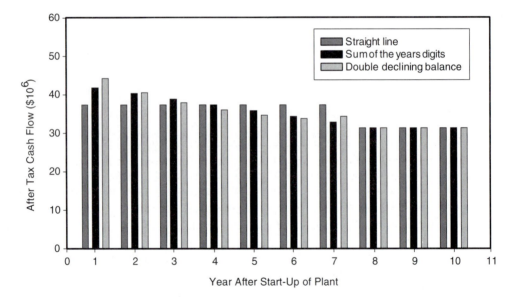

Figure E4.21 Comparison of the After-Tax Profit and Cash Flow Using Different Depreciation Schedules from Example 4.21.

4.9 SUMMARY

In this chapter, we have covered the basics of economic analysis required to evaluate project profitability. The material presented in this chapter is founded on the principle that

Money + Time = More Money

To benefit from this principle, it is necessary to have resources to make an investment and the time to allow the investment to grow. When this principle is applied to chemical processes, the revenue or additional money is generated when low-valued materials and services are converted into high-valued materials and services.

A central concept identified as the "time value of money" grows out of this principle. This principle is applied to a wide range of applications from personal financial management to the analysis of new chemical plants.

The use of cash flow diagrams to visualize the timing of cash flows and to manage cash flows during a project was illustrated. A shorthand notation for the many discount factors involved in economic calculations (that account for the time value of money) were introduced to simplify cash flow calculations. Other items necessary for a comprehensive economic evaluation of a chemical plant were covered and included depreciation, taxation, and the evaluation of profit and cash flow.

Applications involving these principles and concepts directly relating to chemical plants will be pursued in the next chapter.

PROBLEMS

1. I need to borrow $100 in a hurry! I have two alternatives: either borrow from my old friend "Blank" Frank or from my local bank. Frank failed first grade math and has never recovered. He insists on the simplest method of repayment and says that for every month I keep the money I owe him an extra $1.10 (some friend!). The bank, on the other hand, will charge me a nominal interest rate of 12% p.a. compounded monthly.
   a. If I intend to repay the loan in a lump sum at the end of one year, from whom should I borrow?
   b. Calculate, to the closest month, how long it will take until I owe the bank and Frank the same amount (assuming that I borrow the same amount from each at the same time).

2. Which investment scheme is the most profitable, assuming that the initial investment (principal) is the same for each case given below?
   a. 11.0% p.a. (nominal rate) compounded continuously.

  b. 11.5% p.a. (nominal rate) compounded monthly.
  c. 12.0% p.a. (nominal rate) compounded yearly.

3. At an investment seminar that I recently attended I learned about something called the "Rule of 72." According to the person in charge of the seminar "a good estimate for finding how long it takes for an investment to double is given by the following equation":

$$\text{number of years to double investment} = \frac{72}{\text{effective annual interest rate (in \%)}}$$

  Using what you know about the time value of money calculate the error in using the above equation to estimate how long it takes to double an investment made at the following effective annual interest rates: 6%, 9%, and 12%.

  Express your answers as % error to two significant figures.

  Comment on the Rule of 72 and its accuracy for typical financial calculations today.

4. I invested $1,000 six years ago and I want to know how much this investment is worth now. During the past six years, the nominal interest rate has fluctuated as shown below:

  through year 1 nominal interest rate = 6 % p.a.
  through year 2 nominal interest rate = 7 % p.a.
  through year 3 nominal interest rate = 6 % p.a.
  through year 4 nominal interest rate = 5 % p.a.
  through year 5 nominal interest rate = 6 % p.a.
  through year 6 nominal interest rate = 5 % p.a.

  If my investment is compounded monthly, how much is it worth today? (Assume that we are now at the end of the sixth year of this investment.)

5. I was going to invest $20,000 with a bank that offered a nominal interest rate of 10% p.a., compounded continuously. A friend of mine suggested that I go with another bank that offers the same nominal rate but compounds every two weeks. She says that the second bank will be more convenient for me and that the difference between the two compounding schemes will not amount to more than $5 per year. Is my friend giving me accurate advice?

6. You wish to invest an amount now that will provide you with enough capital for the first two years of your daughter's education in 8 and 9 years, respectively. These costs you estimate to be $7,000 for the first year and $8,500 for the second year. You can invest your money in tax-free bonds that will yield an effective annual interest rate of 13.5%. How much should you invest now so that you can meet the two payments of $7,000 and $8,500 in 8 and 9 years,

respectively. (Hint: Treat the two withdrawals separately and then add the principals together.)

7. You have the choice of receiving $1,100 in 3 years' time or $1,000 now. Over what range of effective annual interest rates would you accept the offer of $1,100 in 3 years' time?

8. A small manufacturer of pharmaceutical products currently produces its top-selling drug (SPRAIN) in a batch process. In order to meet the projected increase in sales over the next five years, the company has been considering an investment to upgrade its facility to a continuous process. The company estimates that this upgrade will require an investment of $5 million dollars, and that the bank that they are dealing with will lend them the money at an effective annual interest rate of 10%. For the following four repayment schemes, calculate the loan repayment schedule for the five years that the loan will be made.

   a. Repay the loan as a lump sum at the end of five years.
   b. Repay the loan in five equal payments at the end of each year.
   c. Repay the loan on a sliding scale making five end-of-year payments such that the second payment is 50% greater than the first payment, the third payment is 50% greater than the second payment, and so on.
   d. Repay the loan with 60 equal monthly payments. (Remember that the 10% p.a. is an effective annual interest rate.)

9. An investment of $500,000 by a small chemical company in a process improvement project is expected to yield the following annual cash flows over the six-year project life.

Year	Annual Cash Flow
1	$ 25,000
2	$ 50,000
3	$ 150,000
4	$ 250,000
5	$ 100,000
6	$ 75,000

The company can alternatively get a 6% p.a. effective interest rate in an investment account at a bank. Do you recommend that the company invest the money in the process improvement project or go with the investment account? (Hint: Evaluate both alternatives using the 6% rate and choose the most favorable one.)

10. The cash flows (all figures in $) to and from my savings account are described below in the discrete CFD.

If the savings account yields an effective annual interest rate of 5.5 %, calculate the following:
a. The future value of all the cash flows at the end of 10 years
b. The future value of all the cash flows at the end of 15 years, assuming that I make no more investments or withdrawals after the end of year 10

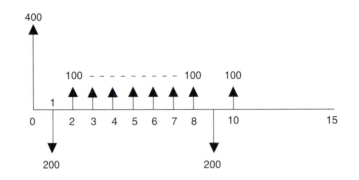

11. I plan to borrow $150,000 from a bank to buy a new house. The interest rate is a nominal 8.75 % p.a., compounded monthly. If I plan to repay the loan in 15 years, calculate:
a. My monthly payment (assume that I have 180 equal monthly payments)
b. The total amount of interest that I will pay on the loan

12. For the following cash flow diagram (all figures in $), find the net value of the sum of all cash flows brought forward or backward to time = 0 (Assume $i = 6\%$ p.a.)

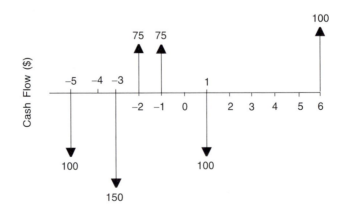

13. I intend to make an initial investment of $1000 into a savings account, and then I will add $100 per month (starting at the end of the first month) for the first year. After that, I will invest $200 per month for the next four years. Assume that I get a nominal interest rate of 6% p.a., and the interest is compounded monthly.
    a. Draw a discrete CFD indicating the cash flows over the 5-year period.
    b. How much money will I have at the end of five years?
    c. What is the total amount of interest that I will have earned over the 5-year investment period?
    d. At the end of five years, I stop investing money but leave the accumulated investment in the account. How much will I have in the account at the end of ten years (five years after I stop investing)?

14. Repeat Problem 11 for the case when the mortgage is spread out over 30 years instead of 15 years. Comment on the effect that this longer payment period has on the monthly installment and the total interest paid on the loan.

15. A new plant is projected to cost $100 million to build, and the money will be paid in two equal lump sums with the first payment at the start of the project and the second at the end of construction, which is scheduled to occur at the end of year 2. The plant, when operating (after the second year), is expected to generate after-tax annual cash flows of $25 million, and the projected plant life is 15 years after start up. Assuming the effective annual interest rate or compounding rate is 7%, answer the following questions :
    a. Draw a discrete and cumulative CFD for this project for the 17 year duration of the investment. (Assume that all the yearly revenue occurs at the end of the year in which it was generated.)
    b. What is the future value of the project at the end of year 17?
    c. What is the equivalent value of this investment in today's dollars (i.e., bring all cash flows back to zero)?
    d. What would the effective annual interest rate have to be such that the value of the investment in today's dollars was zero? (This interest rate is known as the discounted cash flow rate of return, DCFROR.)

16. A family wishes to set up a fund to pay for the education of its child (who is now 5 years old). This account will be arranged so that at the end of the child's 4-year college education, all the money in the account will have been spent. The investment scheme will consist of 17 equal payments. The last 16 will be made at the end of each year plus the first payment, which is due now. The money in the account will accumulate interest at a nominal rate of 8% p.a.

    Four withdrawals of $5,000 each will be made from the fund at the end of years 13, 14, 15, and 16 and these will be used to defray the costs of the child's education. For this fund calculate the following:

a. Draw a discrete CFD to represent the fund (Remember that although there are 17 payments, the fund is only to last for 16 years since the first payment will be made at time = 0)

b. What should the annual premium be so that at the end of year 16 there is no money remaining in the fund?

17. My friend recently obtained a bank loan for $75,000 to buy a house. Her monthly payment for the 20-year loan was $689. What is the nominal interest rate that she is being charged if the bank uses a monthly compounding scheme?

18. The current federal tax law (1987–1996) is based on a Modified Accelerated Cost Recovery System (MACRS) using a half-year convention. All equipment is assigned a class life, which is the period over which the depreciable portion of the investment may be discounted. Most equipment in a chemical plant has a class life of 9.5 years with no salvage value. This means that the capital investment may be depreciated using a straight line method over 9.5 years. Alternatively, we may use a MACRS method over a shorter period of time, which is 5 years for this class life.

The MACRS method uses a double declining balance method and switches to a straight line method when the straight line method yields a greater depreciation allowance for that year (The straight line method is applied to the remaining depreciable capital over the remaining time allowed for depreciation.) The half-year convention assumes that the equipment is bought midway through the first year for which depreciation is allowed. In the first year, the depreciation is only half that for a full year. Likewise in the sixth (and last) year the depreciation is again for half a year. Confirm that the depreciation schedule for equipment with a 9.5-year class life and 5-year recovery period, using the MACRS method, yields the following yearly percentage depreciation allowances.

Year	Depreciation Allowance (% of Capital Investment)
1	20.00
2	32.00
3	19.20
4	11.52
5	11.52
6	5.76

19. Show that the MACRS depreciation allowances for recovery periods of 3, 7, and 10 years are as follows:

	Recovery Period (years)		
	3	**7**	**10**
Year	**Depreciation allowance (% of Capital Investment)**		
1	33.3	14.3	10.0
2	44.5	24.5	18.0
3	14.8	17.5	14.4
4	7.4	12.5	11.5
5		8.9	9.2
6		8.8	7.4
7		8.8	6.6
8		4.5	6.6
9			6.5
10			6.5
11			3.3

Hint: Follow the same logic described in Problem 4.18 to determine the depreciation allowances.

CHAPTER

5

Profitability Analysis

In this chapter, we will see how to apply the techniques of economic analysis developed in Chapter 4. These techniques will be used to assess the profitability of projects involving both capital expenditures and yearly operating costs. We will look at a variety of projects ranging from large multimillion dollar ventures to much smaller process improvement projects. Several criteria for profitability will be discussed and applied to the evaluation of process and equipment alternatives. We start with the profitability criteria for new large projects.

5.1 A TYPICAL CASH FLOW DIAGRAM FOR A NEW PROJECT

A typical cumulative, after-tax cash flow diagram for a new project is illustrated in Figure 5.1. We find it convenient to relate profitability criteria to the cumulative CFD rather than the discrete CFD. The timings of the different cash flows are explained below.

In the economic analysis of the project, it is assumed that any new land purchases required are done at the start of the project, that is, at time zero. After the decision has been made to build a new chemical plant or expand an existing facility, the construction phase of the project starts. Depending on the size and scope of the project, this construction may take anywhere from 6 months to 3 years to complete. In the example shown in Figure 5.1, a typical value of 2 years for the time from project initiation to the start-up of the plant has been assumed. Over the 2-year construction phase, there is a major capital outlay. This represents the fixed capital expenditures for purchasing and installing the equipment and auxiliary facilities required to run the plant (see Chapter 2). The distribution of this

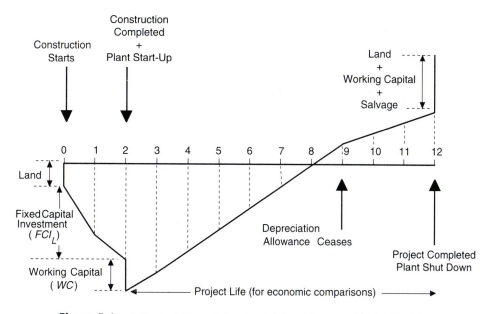

Figure 5.1   A Typical Cumulative Cash Flow Diagram for the Evaluation of a New Project

fixed capital investment is usually slightly larger towards the beginning of construction, and this is reflected in Figure 5.1. At the end of the second year, construction has finished and the plant is started up. At this point, the additional expenditure for working capital required to float the first few months of operations is shown. This is a one-time expense at the start-up of the plant and will be recovered at the end of the project.

After start-up, the process begins to generate finished products for sale, and the yearly cash flows become positive. This is reflected in the positive slope of the cumulative CFD in Figure 5.1. Usually the revenue for the first year after start-up is less than subsequent years due to "teething" problems in the plant; this is also reflected in Figure 5.1. The cash flows for the early years of operation are larger than those for later years due to the effect of the depreciation allowance discussed in Chapter 4. The time used for depreciation in Figure 5.1 is 7 years. The time over which the depreciation is allowed depends on the IRS regulations and the method of depreciation used.

In order to evaluate the profitability of a project, we must assume a life for the process. This is not usually the working life of the equipment nor is it the time over which depreciation is allowed. It is a specific length of time over which the profitability of different projects are to be compared. Lives of 10, 12, and 15 years are commonly used for this purpose. It is necessary to standardize the project life when comparing different projects. This is because profitability is directly related to project life, and comparing projects using different lives biases the results.

Usually, chemical processes have anticipated operating lives much greater than 10 years. If much of the equipment in a specific process is not expected to last for a 10-year period, then the operating costs for that project should be adjusted. These operating costs should reflect a much higher maintenance cost to include the periodic replacement of equipment necessary for the process to operate the full 10 years. We will use a project life of 10 years for the examples in the next section.

From Figure 5.1 we can see a steadily rising cumulative cash flow over the 10 operating years of the process, that is, years 2 through 12. At the end of the 10 years of operation, that is, at the end of year 12, we assume that the plant is closed down and that all the equipment is sold for its salvage or scrap value, that the land is also sold, and that we recover our working capital. This additional cash flow, received on closing down the plant, is shown by the upward pointing vertical line in year 12. Remember that in reality, the plant will most likely *not* be closed down—we only assume that it will be in order to perform our economic analysis.

The question that we must now address is how to evaluate the profitability of a new project. Looking at Figure 5.1, we can see that at the end of the project the cumulative CFD is positive. Does this mean that the project will be profitable? The answer to this question depends on whether the value of the income earned during the time the plant operated was smaller or greater than the investment made at the beginning of the project. Therefore, we must consider the time value of money when evaluating profitability. In the following sections, we will look at different ways to evaluate project profitability.

5.2   PROFITABILITY CRITERIA FOR PROJECT EVALUATION

There are essentially three bases used for the evaluation of profitability. They are:

1. Time
2. Cash
3. Interest rate

For each of these bases, we can consider discounted or nondiscounted techniques. The nondiscounted techniques do not take into account the time value of money and are *not* recommended for evaluating new, large projects. Traditionally, however, such methods have been and are still used to evaluate smaller projects, such as process improvement schemes. We will present examples of both types of methods for all the three bases.

5.2.1 Nondiscounted Profitability Criteria

Four nondiscounted profitability criteria are illustrated in Figure 5.2. The graphical interpretation of these profitability criteria are shown in the figure. We explain each of the four criteria below.

Time Criterion.   The term used for this criterion is the *Payback Period (PBP)* (also known by a variety of other names, such as payout period, payoff period, and cash recovery period). We define the payback period as follows:

PBP = Time required, after start-up, to recover the fixed capital investment, FCI_L, for the project.

The payback period is shown as a length of time on Figure 5.2. Clearly, the shorter the payback period, the more profitable the project.

Cash Criterion.   The criterion used here is the *Cumulative Cash Position (CCP)*, which is simply the worth of the project at the end of its life. For criteria using cash or monetary value, it is difficult to compare projects with dissimilar fixed capital investments and sometimes it is more useful to use the *Cumulative Cash Ratio (CCR)* defined as:

$$CCR = \frac{\text{Sum of All Positive Cash Flows}}{\text{Sum of All Negative Cash Flows}}$$

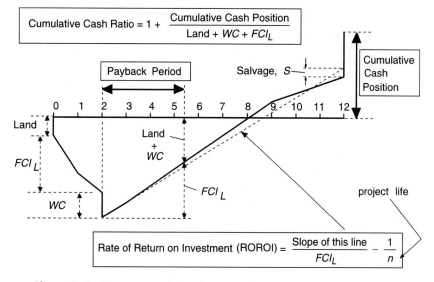

Figure 5.2 Illustration of Nondiscounted Profitability Criteria

The above definition effectively gives the cumulative cash position normalized by the initial investment. Projects with cumulative cash ratios greater than unity are potentially profitable while those with ratios less than unity cannot be profitable.

Interest Rate Criterion.   The criterion used here is called the *Rate of Return on Investment (ROROI)* and represents the nondiscounted rate at which we make money from our fixed capital investment. The definition is given as:

$$ROROI = \frac{\text{Average Annual Net Profit}}{\text{Fixed Capital Investment } (FCI_L)}$$

The annual net profit in the above definition is an average over the life of the plant after start-up.

The use of fixed capital investment, FCI_L, in the calculations for payback period and rate of return on investment given above seems reasonable, since this is the capital that must be recovered by project revenue. Many alternative definitions for these terms can be found, and the reader will find that total capital investment (FCI_L + WC + Land) is often used instead of fixed capital investment. When the plant has a salvage value (S), the fixed capital investment minus the salvage value ($FCI_L - S$) could be used instead of FCI_L. However, since the salvage value is usually very small, we prefer to use FCI_L alone.

Example 5.1

A new chemical plant is going to be built and will require the following capital investments (all figures are in $ million):

Cost of land, L = $10.0
Total fixed capital investment, FCI_L = $ 150.0
Fixed capital investment during year 1 = $90.0
Fixed capital investment during year 2 = $60.0
Plant start-up at end of year 2
Working capital = 20% of FCI_L = (0.20)($150) = $30.0 at end of year 2

The sales revenues and costs of manufacturing are given below:
Yearly sales revenue (after start-up), R = $75.0 per year
Cost of manufacturing excluding depreciation allowance (after start-up),
    COM_d = $30.0 per year

Taxation rate, t = 30%
Salvage value of plant, S = $10.0
Depreciation use double declining balance over 7 years, d_k
Assume a project life of 10 years.

Calculate each nondiscounted profitability criteria given in this section for this plant.

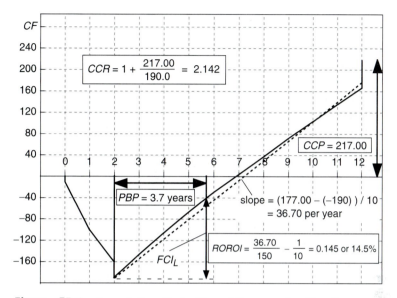

Figure E5.1 Cumulative Cash Flow Diagram for Nondiscounted After-Tax Cash flows for Example 5.1

The discrete and cumulative nondiscounted cash flows for each year are given in Table E5.1. The fixed capital investment, revenue, and cost of manufacture figures are the same as used in Examples 4.20 and 4.21 and may be reviewed to verify the numbers in Table E5.1. Using this data, the cumulative cash flow diagram is drawn, Figure E5.1.

The method of evaluation for each of the criteria is given on Figure E5.1 and in Table E5.1.

Payback Period (PBP) = 3.7 years
Cumulative Cash Position (CCP) = $ 217 \times 10^6$
Cumulative Cash Ratio (CCR) = 2.142
Rate of Return on Investment ($ROROI$) = 14.5 %

All of these criteria indicate that the project cannot be eliminated as unprofitable. They all fail to take into account the "time value of money" that is necessary for a thorough measure of profitability. We consider the effects of the time value of money on profitability in the next section.

5.2.2   Discounted Profitability Criteria

The main difference between the nondiscounted and discounted criteria is that for the latter we discount each of the yearly cash flows back to time zero. We then

Table E5.1 Nondiscounted After-Tax Cash Flows for Example 5.1 (all numbers in 10^6)

End of Year (k)	Investment	d_k	$FCI_L - \Sigma d_k$	R	COM_d	$(R - COM - d_k)*(1 - t) + d_k$	Cash Flow	Cumulative Cash Flow
0	(10)†	—	150.00	—	—	—	(10.00)	(10.00)
1	(90)	—	150.00	—	—	—	(90.00)	(100.00)
2	(60 + 30) = (90)	—	150.00	—	—	—	(90.00)	(190.00)
3	—	42.86	107.14	75	30	44.36	44.36	(145.64)
4	—	30.61	76.53	75	30	40.68	40.68	(104.96)
5	—	21.87	54.66	75	30	38.06	38.06	(66.90)
6	—	15.62	39.04	75	30	36.19	36.19	(30.71)
7	—	11.16	27.89	75	30	34.85	34.85	4.13
8	—	7.97	19.92	75	30	33.89	33.89	38.02
9	—	9.92	10.00	75	30	34.48	34.48	72.50
10	—	—	10.00	75	30	31.50	31.50	104.00
11	—	—	10.00	75	30	31.50	31.50	135.50
12	10 + 30 + 10 = 50	—	10.00	75	30	31.50	81.50	217.00

†numbers in ( ) are negative cash flows

Nondiscounted Profitability Criteria

Payback Period (PBP)

Land + WC = 10 + 30 = $40 × 10^6$ Find time after start-up for which cumulative cash flow = −$40 × 10^6$

$PBP = 3 + (-66.90 + 40)/(-66.90 + 30.71) = $ **3.7 years**

Cumulative Cash Position (CCP) and Cumulative Cash Ratio (CCR)

$CCP = $217.00 × 10^6$ and $CCR = \Sigma$ positive cash flows/Σ negative cash flows = $(44.36 + 40.68 + 38.06 + \ldots + 31.50 + 81.50)/(10 + 90 + 90) = $ **2.142**

Rate of Return on Investment (ROROI)

$ROROI = (44.36 + 40.68 + 38.06 + \ldots + 31.50 + 10)/10/150 - 1/10 = $ **0.145 or 14.5% p.a.**

use the resulting discounted cumulative cash flow diagram to evaluate profitability. The three different types of criteria are given below.

Time Criterion.   The *Discounted Payback Period (DPBP)* is defined in a manner similar to the nondiscounted version given above.

> $DPBP$ = Time required, after start-up, to recover the fixed
> capital investment, FCI_L, required for the project, with all cash
> flows discounted back to time zero.

The project with the shortest discounted payback period is the most desirable.

Cash Criterion.   The *Discounted Cumulative Cash Position,* more commonly known as the *Net Present Value (NPV)* or *Net Present Worth (NPW)* of the project, is defined as:

> NPV = Cumulative discounted cash position at the end of the project

Again, the NPV of a project is greatly influenced by the level of fixed capital investment and a better criterion for comparison of projects with different investment levels may be the *Present Value Ratio (PVR):*

$$PVR = \frac{\text{Present Value of All Positive Cash Flows}}{\text{Present Value of All Negative Cash Flows}}$$

A present value ratio of unity for a project represents a break-even situation. Values greater than unity indicate profitable processes while those less than unity represent unprofitable projects.

Example 5.2

For the plant described in Example 5.1 determine the following discounted profitability criteria:

    a. Discounted Payback Period (*DPBP*)
    b. Net Present Value (*NPV*)
    c. Present Value Ratio (*PVR*).

Assume a discount rate of $0.1 - (10\%$ p.a.)

The procedure used is similar to the one used for the nondiscounted evaluation shown in Example 5.1. The discounted cash flows replace the actual cash flows. For the discounted case, we must first discount all the cash flows in Table E5.1 back to the beginning of the project (time = 0). We do this simply by multiplying each cash flow by the discount factor $(P/F,i,n)$, where n is the number of years after the start of the project. These discounted cash flows are shown along with the cumulative discounted cash flows in Table E5.2.

Table E5.2   Discounted Cash Flows for Example 5.2 (all numbers are in $ millions)

End of Year	Non-Discounted Cash Flow	Discounted Cash Flow	Cumulative Discounted Cash Flow
0	(10)	(10)	(10.00)
1	(90)	$(90)/1.1 = (81.82)$	(91.82)
2	(90)	$(90)/1.1^2 = (74.38)$	(166.20)
3	44.36	$44.37/1.1^3 = 33.33$	(132.87)
4	40.68	$40.70/1.1^4 = 27.79$	(105.08)
5	38.06	$38.07/1.1^5 = 23.63$	(81.45)
6	36.19	$36.18/1.1^6 = 20.43$	(61.03)
7	34.85	$34.86/1.1^7 = 17.88$	(43.14)
8	33.89	$33.90/1.1^8 = 15.81$	(27.33)
9	34.48	$34.56/1.1^9 = 14.62$	(12.71)
10	31.50	$31.50/1.1^{10} = 12.14$	(0.57)
11	31.50	$31.50/1.1^{11} = 11.04$	10.47
12	81.50	$81.50/1.1^{12} = 25.97$	36.44

Discounted Profitability Criteria
Discounted Payback Period (*DPBP*)
Discounted value of $L + WC = 10 + 30/1.1^2 = \$34.8 \times 10^6$. Find time after start-up when cumulative cash flow = $-\$34.8 \times 10^6$
DPBP = 5 + (−43.14 + 34.8)/(−43.14 + 27.33) = **5.5 yr**
Net Present Value (*NPV*) and Present Value Ratio (*PVR*)
NPV = \$36.44 × 10⁶
$PVR = \Sigma$ positive discounted cash flows/Σ negative discounted cash flows = (33.33 + 27.79 + 20.43 + . . . +25.97)/(10 + 81.82 + 74.38)
PVR = (202.64)/(166.2) = 1 + 36.44/166.2 = **1.22**

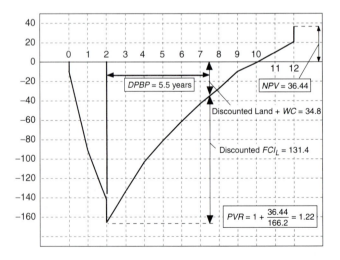

Figure E5.2   Cumulative Cash Flow Diagram for Discounted After Tax Cash Flows for Example 5.2

The cumulative discounted cash flows are shown on Figure E5.2 and the calculations are given in Table E5.2 from these sources the profitability criteria are given as:

   a. Discounted Payback Period ($DPBP$) = 5.5 years
   b. Net Present Value (NPV) = $\$36.4 \times 10^6$
   c. Present Value Ratio (PVR) = 1.22

We can see from these examples that there are significant effects of discounting the cash flows to account for the time value of money. From these results, we can make the following observations:

   1. In terms of the time basis, the payback period increases as the discount rate increases. In the above examples, it increased from 3.7 to 5.5 years.
   2. In terms of the cash basis, replacing the cash flow with the discounted cash flow decreases the value at the end of the project. In the above examples, it dropped from \$217 to \$36.4 million dollars.
   3. In terms of the cash ratios, discounting the cash flows gives a lower ratio. In the above examples, the ratio dropped from 2.14 to 1.22.

As the discount rate increases, all of the discounted profitability criteria will show a reduction in profitability.

Interest Rate Criterion. The *Discounted Cash Flow Rate of Return (DCFROR)* is defined to be the interest rate at which all the cash flows must be discounted in order for the net present value of the project to be equal to zero. Thus, we may write:

$$DCFROR = \text{Interest or Discount Rate for which the Net}$$
$$\text{Present Value of the project is equal to zero}$$

Therefore, the *DCFROR* represents the highest, after-tax interest or discount rate at which the project can just break even.

For the discounted payback period and the net present value calculations, the question arises as to what interest rate we should use to discount the cash flows. This "internal" interest rate is usually determined by corporate management and represents the minimum acceptable rate of return that the company will accept for any new investment. Many factors influence the determination of this discount interest rate, and for our purposes, we will assume that it is always given.

It is worth noting that for evaluation of the discounted cash flow rate of return, no interest rate is required since this is what we calculate. Clearly, if the *DCFROR* is greater than the internal discount rate, then the project is considered to be profitable.

Example 5.3

For the problem presented in Examples 5.1 and 5.2, determine the discounted cash flow rate of return (*DCFROR*).

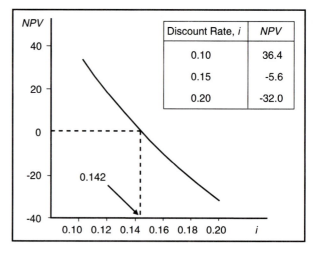

Discount Rate, *i*	*NPV*
0.10	36.4
0.15	-5.6
0.20	-32.0

The *NPV*s for several different discount rates were calculated and the results are shown in the figure to the right. The value of the *DCFROR* is found at *NPV* equals 0. From the graph we obtain

$$DCFROR = 0.142 \ (14.2\%)$$

An alternate method for obtaining the *DCFROR* is to solve for the value of *i* in an implicit, nonlinear algebraic expression. This is illustrated in Example 5.4.

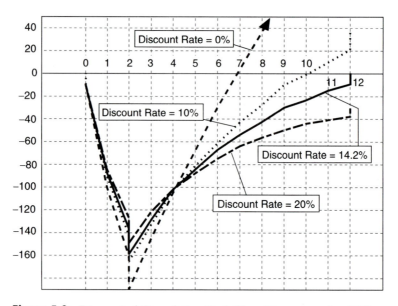

Figure 5.3   Discounted Cumulative Cash Flow Diagrams using Different Discount Rates for Example 5.3

Figure 5.3 provides the cumulative discounted cash flow diagram for Example 5.3 for several discount factors. It shows the effect of changing discount factors on the profitability and shape of the curves. It includes a curve for the *DCFROR* found in Example 5.3. For this case, it can be seen that the *NPV* for the project is zero. In Example 5.3, if the acceptable rate of return for our company were set at 20%, then the project would not be considered an acceptable investment. This is indicated by a negative *NPV* for $i = 20\%$.

Each method described above used to gauge the profitability of a project has advantages and disadvantages. For projects having a short life and small discount factors, the effect of discounting is small and nondiscounted criteria may be used to give an accurate measure of profitability. However, it is fair to say that for large projects, involving many millions of dollars of capital investment, discounting techniques should always be used.

Since all of the above techniques are commonly used in practice, you must be familiar with and be able to use each technique.

5.3   COMPARING SEVERAL LARGE PROJECTS—
INCREMENTAL ECONOMIC ANALYSIS

In this section, we will compare and select among investment alternatives. When comparing project investments, the *DCFROR* tells us how efficiently we are using our money. The higher the *DCFROR*, the more attractive the individual investment. However, when comparing investment alternatives, it may be better to choose a project that does not have the highest *DCFROR*. The rationale for comparing projects and choosing the most attractive alternative is discussed in this section.

In order to make a valid decision regarding alternative investments (projects), it is necessary to know a base line rate of return that must be attained in order for an investment to be attractive. A company that is considering whether to invest in a new project always has the option to reject all alternatives offered and invest the cash (or resources) elsewhere. The baseline or benchmark investment rate is related to these alternative investment opportunities, such as investing in the stock market.

Example 5.4

Our company is seeking to invest approximately $\$120 \times 10^6$ in new projects. After extensive research and preliminary design work, three projects have emerged as candidates for construction. The minimum acceptable internal discount (interest) rate, after tax, has been set at 10%. The after-tax cash flow information for the three projects using a 10-year operating life is as follows (values in $ million):

	Initial Investment	After-Tax Cash Flow in Year i	
		$i = 1$	$i = 2\text{-}10$
Project A	$60	$10	$12
Project B	$120	$22	$22
Project C	$100	$12	$20

For this example it is assumed that the cost of land, working capital, and salvage are zero. Furthermore, it is assumed that the initial investment occurs at time = 0, and the yearly annual cash flows occur at the end of each of the 10 years of plant operation.
Determine:

a. The *NPV* for each project.
b. The *DCFROR* for each project.

For Project A we get:

$$NPV = -\$60 + (\$10)(P/F,0.10,1) + (\$12)(P/A,0.10,9)(P/F,0.10,1)$$

$$= -\$60 + \frac{(\$10)}{1.1} + (\$12)\frac{1.1^9 - 1}{(0.1)(1.1^9)}\frac{1}{1.1} = \$11.9$$

The *DCFROR* is the value of i that results in an $NPV = 0$.

$$NPV = 0 = -\$60 + (\$10)(P/F,i,1) + (\$12)(P/A,i,9)(P/F,i,1)$$

Solving for i yields $i = DCFROR = 14.3\,\%$.
Values obtained for *NPV* and *DCFROR* are:

	NPV (i=10%)	DCFROR
Project A	11.9	14.3%
Project B	15.2	12.9%
Project C	15.6	13.3%

Note: Projects A, B, and C are mutually exclusive since we cannot invest in more than one of them, due to our cap of 120×10^6. The analysis that follows is limited to projects of this type. For the case when projects are not mutually exclusive, the analysis becomes somewhat more involved and is not covered here.

While all of the projects in Example 5.4 showed a positive *NPV* and a *DCFROR* of over 10%, at this point it is not clear how to select the most attractive option with this information. We will see later that the choice of the project with the highest *NPV* will be the most attractive. However, let us consider the following alternative analysis. If Project B is selected, a total of 120×10^6 is invested and yields 12.9%, whereas the selection of Project A yields 14.3% on the

60×10^6 invested. To compare these two options, we would have to consider a situation in which the same amount is invested in both cases. In Project A, this would mean that 60×10^6 in invested in the project and the remaining 60×10^6 is invested elsewhere, whereas in Project B, a total of 120×10^6 is invested in the project.

It is necessary in our analysis that we are sure that the last dollar invested earns at least 10%. To do this we must perform an incremental analysis on the cash flows and establish that at least 10% is made on each additional increment of money invested in the project.

Example 5.5

This is a continuation of Example 5.4.

    a. Determine the *NPV* and the *DCFROR* for each increment of investment.
    b. Recommend the best option.

    a. Project A to Project C:

Incremental investment is $40 \times 10^6 = (\$100 - \$60) \times 10^6$
Incremental cash flow for $i = 1$ is $2 \times 10^6/\text{yr} = (\$12/\text{yr} - \$10/\text{yr}) \times 10^6$
Incremental cash flow for $i = 2$ to 10 is $8 \times 10^6/\text{yr} = (\$20/\text{yr} - \$12/\text{yr}) \times 10^6$

$$NPV = -\$40 \times 10^6 + (\$2 \times 10^6)(P/F,0.10,1) + (\$8 \times 10^6)(P/A,0.10,9)(P/F,0.10,1)$$
$$NPV = \$3.7 \times 10^6$$

Setting $NPV = 0$ yields $DCFROR = 0.119$ (11.9%)

Project C to Project B:

Incremental investment is $20 \times 10^6 = (\$120 - \$100) \times 10^6$
Incremental cash flow for $i = 1$ is $10 \times 10^6/\text{yr} = (\$22/\text{yr} - \$12/\text{yr}) \times 10^6$
Incremental cash flow for $i = 2$ to 10 is $2 \times 10^6/\text{yr} = (\$22/\text{yr} - \$20/\text{yr}) \times 10^6$

$$NPV = -\$0.4 \times 10^6 \text{ and } DCFROR = 0.094 \text{ (9.4\%)}$$

    b. It is recommended that we move ahead on Project C.

From Example 5.5, it is clear that the rate of return on the 20×10^6 incremental investment required to go from Project C to Project B did not return the 10% required, and gave a negative *NPV*.

The information from Example 5.4 shows that an overall return on investment of over 10% is obtained for each of the three projects. However, the correct choice, Project C, also has the highest *NPV* using a discount rate of 10%, and it is this criterion that should be used to compare alternatives.

> **When comparing mutually exclusive investment alternatives, choose the alternative with the greatest positive net present value.**

When carrying out an incremental investment analysis on projects that are mutually exclusive, the following four-step algorithm is recommended:

Step 1: Establish the minimum acceptable rate of return on investment for such projects.

Step 2: Calculate the *NPV* for each project using the interest rate from Step 1, above.

Step 3: Eliminate all projects with negative *NPV* values.

Step 4: Of the remaining projects, select the project with the highest *NPV*.

5.4   ESTABLISHING ACCEPTABLE RETURNS FROM INVESTMENTS: THE CONCEPT OF RISK

Most comparisons of profitability will involve the rate of return of an investment. Company management usually provides several "benchmarks" or "hurdle rates" for acceptable rates of return that must be used in comparing alternatives.

A company Vice President (V.P.) has been asked to recommend which of the following two alternatives to pursue.

Option 1: A new product is to be produced that has never been made before on a large scale. Pilot plant runs have been made and the products sent to potential customers. Many of these customers have expressed an interest in the product but need more material to evaluate it fully. The calculated return on the investment for this new plant is 33%.

Option 2: A second plant is to be built in another region of the country to meet increasing demand in the region. The company has a dominant market position for this product. The new facility would be similar to other plants. It would involve more computer control and attention will be paid to meeting pending changes in environmental regulations. The rate of return is calculated to be 12%.

The recommendation of the V.P. and the justifications are given below:

Items that favor Option 2

- The market position for Option 2 is well established. The market for the new product has not been fully established.
- The manufacturing costs are well known for Option 2 but are uncertain for the new process since only estimates are available.

- Transportation costs will be less than current values due to the proximity of plant.
- The technology used in Option 2 is mature and well known to us. For the new process we have no guarantee that it will work.

Items that favor Option 1 if pursued

- High return on the investment.
- Opens new product possibilities.

The closing statement from the V.P. included the following summary:

"We have little choice but to expand our established product line. If we fail to build these new production facilities, our competitors are likely to build a new plant in the region to meet the increasing demand. They could undercut our regional prices, and this would put at risk our market share and dominant market position in the region."

Clearly, the high return on investment for Option 1 was associated with a high risk. This is usually the case. There are often additional business reasons that must be considered prior to making the final decision. The concern for lost market position is a serious one and weighs heavily in any decision. The relatively low return on investment of 12% given in this example would probably not be very attractive had it not been for this concern. It is the job of company management to weigh all of these factors, along with the rate of return, in order to make the final decision.

In this chapter, we often refer to "internal interest/discount rates" or "internal rates of return." This deals with benchmark interest rates that are to be used to make profitability evaluations. There are likely to be different values that reflect dissimilar conditions of risk, that is, the value for mature technology would differ from that for unproven technology. For example, the internal rate of return for mature technology might be set at 12%, while that for very new technology might be set at 40%. Using these values the decision by the V.P. given above seems more reasonable. The analysis of risk is not covered in this text. Instead, we choose to concentrate on the evaluation of the different criteria of profitability and assume that an acceptable internal interest rate is known.

5.5 EVALUATION OF EQUIPMENT ALTERNATIVES

Often during the design phases of a project, it will be necessary to evaluate different equipment options. Each alternative piece of equipment performs the same process function. However, the capital cost, operating cost, and equipment life may be different for each, and we must determine which is the "best choice" using some economic criterion.

Clearly, if we have two pieces of equipment, each with the same expected operating life that can perform the desired function with the same operating cost, then *common sense* tells us that we should choose the *least expensive alternative!* When the expected life and operating expenses vary, the selection becomes more difficult. Techniques available to make the selection are discussed in this section.

5.5.1   Equipment with the Same Expected Operating Lives

When the operating costs and initial investments are different but the equipment lives are the same, then we can make our choice based on *NPV*. The choice with the least negative *NPV* will be the best choice.

Example 5.6

In the final design stage of a project the question has arisen as to whether to use a water-cooled exchanger or an air-cooled exchanger in the overhead condenser loop of a distillation tower. The information available on the two pieces of equipment are provided below:

	Initial Investment	Yearly Operating Cost
Air-Cooled	$23,000	$1,200
Water-Cooled	$12,000	$3,300

Both pieces of equipment have service lives of 12 years. For an internal rate of return of 8% p.a., which piece of equipment represents the better choice?

The *NPV* for each exchanger is evaluated below.

$$NPV = -[\text{Initial Investment} + (\text{Operating Cost})(P/A, 0.08, 12)]$$

	NPV
Air-Cooled	-$32,040
Water-Cooled	-$36,870

The air-cooled exchanger represents the better choice.

Despite the higher capital investment for the air-cooled exchanger in Example 5.6, it was the recommended alternative. The lower operating cost more than compensated for the higher initial investment.

5.5.2   Equipment with Different Expected Operating Lives

When process equipment have different expected operating lives, we must be careful how we determine the best choice. When we talk about expected equipment life, we assume that this is less than the expected working life of the plant.

Therefore, during the normal operating life of the plant, we can expect to replace the equipment at least once. This requires that we apply different profitability criteria. We present three commonly used methods to evaluate this situation. The effect of inflation is not considered in these methods. All of the methods consider both the capital and operating cost in minimizing expenses, thereby maximizing our profits.

Capitalized Cost Method.   In this method we establish a fund for each piece of equipment that we wish to compare. This fund provides the amount of cash that we would need

a. to purchase the equipment initially
b. to replace it at the end of its life
c. to continue replacing it forever.

The size of the initial fund and the logic behind the capitalized cost method are illustrated in Figure 5.4.

From Figure 5.4, we can see that if the equipment replacement cost is P, then the total fund set aside (called the capitalized cost) is $P + R$, where R is termed the residual. The purpose of this residual is to earn sufficient interest during the life of the equipment to pay for its replacement. At the end of the equipment life, n_{eq}, the amount of interest earned is P, the equipment replacement cost. As Figure 5.4 shows, we may continue to replace the equipment every time it wears out. Refer-

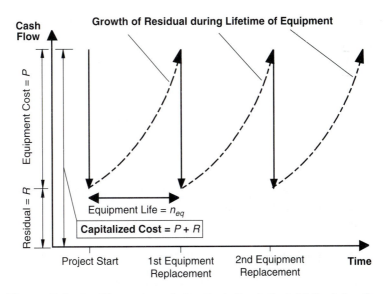

Figure 5.4   An Illustration of the Capitalized Cost Method for the Analysis of Equipment Alternatives.

ring to Figure 5.4, we develop the equation for the capitalized cost defined as $(P + R)$:

$$R(1 + i)^{n_{eq}} - R = P$$

and

$$\text{Capitalized Cost} = P + R = P + \frac{P}{(1 + i)^{n_{eq}} - 1} = P\left[\frac{(1 + i)^{n_{eq}}}{(1 + i)^{n_{eq}} - 1}\right] \quad (5.1)$$

The term in square brackets in Eq. (5.1) is commonly referred to as the capitalized cost factor.

The capitalized cost obtained from Eq. (5.1) does not include the operating cost and is useful in comparing alternatives only when the operating costs of the alternatives are the same. When the operating costs vary, it is necessary to capitalize the operating cost. An equivalent capitalized operating cost, that converts the operating cost into an equivalent capital cost, is added to the capitalized cost calculated from Eq. (5.1) to provide the Equivalent Capitalized Cost (ECC),

$$\text{Equivalent Capitalized Cost} = \text{Capitalized Cost} + \text{Capitalized Operating Cost}$$

$$\text{Equivalent Capitalized Cost} = \left[\frac{P(1 + i)^{n_{eq}} + YOC(F/A,i,n_{eq})}{(1 + i)^{n_{eq}} - 1}\right] \quad (5.2)$$

This cost considers both the capital cost of equipment and the yearly operating cost (YOC) needed to compare alternatives. The extra terms in Eq. (5.2) represent the effect of taking the yearly cash flows for operating costs from the residual, R.

By using Eqs. (5.1) or (5.2), we correctly account for the different operating lives of the equipment by calculating an effective capitalized cost for the equipment and operating cost. The following example illustrates the use of these equations.

Example 5.7

During the design of a new project, we are faced with a decision regarding which type of pump we should use for a corrosive service. Our options are as follows:

	Capital Cost	Operating Cost (per year)	Equipment Life (years)
Carbon Steel Pump	$8,000	$1,800	4
Stainless Steel Pump	$16,000	$1,600	7

Assume a discount rate of 8%, p.a.
Using Eq. (5.2) for the carbon steel pump:

$$\text{Capitalized Cost} = \frac{(8000)(1.08)^4 + (1800)\dfrac{[1.08^4 - 1]}{0.08}}{1.08^4 - 1} = \$52,700$$

For the stainless steel pump:

$$\text{Capitalized Cost} = \frac{(16000)(1.08)^7 + (1600)\dfrac{[1.08^7 - 1]}{0.08}}{1.08^7 - 1} = \$58,400$$

Since the carbon steel pump has the lower capitalized cost, it is recommended.

In Example 5.7, the stainless steel pump costs twice that of the carbon steel pump and, because of its superior resistance to corrosion, will last nearly twice as long. In addition, the operating cost for the stainless steel pump is lower due to lower maintenance costs. In spite of these advantages, the carbon steel pump was still judged to offer a cost advantage.

Equivalent Annual Operating Cost (*EAOC*) Method. In the previous method, we lumped both capital cost and yearly operating costs into a single cash fund or equivalent cash amount. An alternative method is to amortize (spread out) the capital cost of the equipment over the operating life to establish a yearly cost. This is added to the operating cost to yield the Equivalent Annual Operating Cost (*EAOC*).

Figure 5.5 illustrates the principles behind this method. From the figure, we can see that we spread the cost of the initial purchase out over the operating life of the equipment. The *EAOC* is expressed by Eq. (5.3).

$$EAOC = (\text{Capital Investment})(A/P, i, n_{eq}) + YOC \tag{5.3}$$

Example 5.8

Compare the stainless steel and carbon steel pumps in Example 5.7 using the *EAOC* method.

For the carbon steel pump:

$$EAOC = \frac{(8000)(0.08)(1.08)^4}{1.08^4 - 1} + 1800 = \$4,220 \text{ per year}$$

For the stainless steel pump:

$$EAOC = \frac{(16000)(0.08)(1.08)^7}{1.08^7 - 1} + 1600 = \$4,670 \text{ per year}$$

The carbon steel pump is shown to be the preferred equipment using the *EAOC* method as it was in Example 5.7 using the *ECC* method.

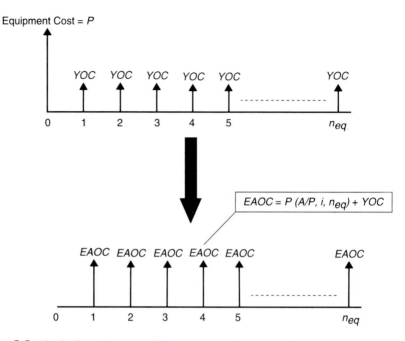

Figure 5.5   Cash Flow Diagrams Illustrating the Concept of Equivalent Annual Operating Cost

Common Denominator Method.   Another method for comparing equipment with unequal operating lives is the common denominator method. This method is illustrated in Figure 5.6 in which two pieces of equipment with operating lives of n and m years are to be compared. This comparison is done over a period of nm years during which the first piece of equipment will need m replacements and the second will require n replacements. Each piece of equipment has an integer number of replacements, and the time over which the comparison is made is the same for both pieces of equipment. For these reasons the comparison can be made using the net present value of each alternative. In general, an integer number of replacements can be made for both pieces of equipment in a time N, where N is the smallest number into which m and n are both exactly divisible, that is, N is the common denominator.

Example 5.9

Compare the two pumps given in Example 5.7 using the common denominator method. The discrete cash flow diagrams for the two pumps are shown in Figure E5.9. The minimum time over which the comparison can be made is 4(7) = 28 years.

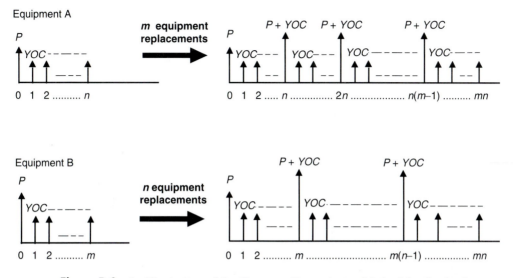

Figure 5.6   An Illustration of the Common Denominator Method for the Analysis of Equipment Alternatives

NPV for carbon steel pump:

$$NPV = -(\$8{,}000)(1 + 1.08^{-4} + 1.08^{-8} + 1.08^{-12} + 1.08^{-16} + 1.08^{-20} + 1.08^{-24})$$
$$-(\$1{,}800)(P/A,0.08,28)$$
$$= -\$46{,}580$$

NPV for stainless steel pump:

$$NPV = -(\$16000)(1 + 1.08^{-7} + 1.08^{-14} + 1.08^{-21}) - (\$1600)(P/A,0.08, 28) = -\$51{,}643$$

The carbon steel pump has a less negative *NPV* and is recommended.

As found for the previous two methods, the common denominator method favors the carbon steel pump.

Choice of Methods.   Since all three methods of comparison correctly take into account the time value of money, the results of all the methods are equivalent. In most problems, the common denominator method becomes unwieldy. We favor the use of the *EAOC* or the capitalized cost methods for our calculations.

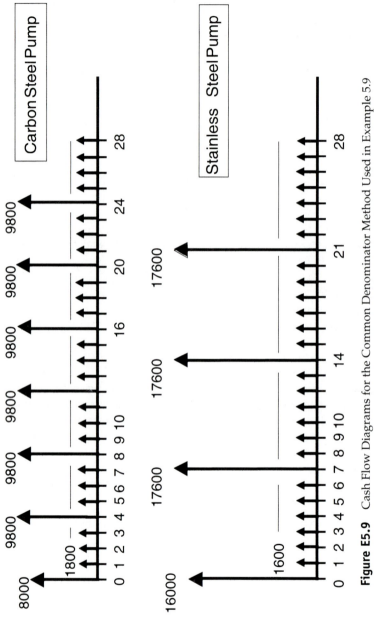

Figure E5.9 Cash Flow Diagrams for the Common Denominator Method Used in Example 5.9

5.6   INCREMENTAL ANALYSIS FOR RETROFITTING FACILITIES

This topic involves profitability criteria used for analyzing situations where a piece of equipment is added to an existing facility. The purpose of adding the equipment is to improve the profitability of the process. Such improvements are often referred to as "retrofitting." Such retrofits may be extensive, requiring millions of dollars of investment, or small, requiring an investment of only a few thousand dollars.

The decisions involved in retrofitting projects may be of the discrete type, the continuous type, or a combination of both. An example of a discrete decision is whether to add an on-line monitoring and control system to a wastewater stream. The decision is a simple yes or no. An example of a continuous decision is what size of heat recovery system should we add to an existing process heater to improve fuel efficiency. This type of decision would involve sizing the optimum heat exchanger where the variable of interest (heat exchanger area) is continuous.

Since retrofit projects are carried out on existing operating plants, it becomes necessary to identify all of the costs and savings associated with the retrofit. When comparing alternative schemes, we focus our attention on the profitability of the incremental investment required. We will consider simple discrete choices in this section. The problem of optimizing a continuous variable is covered in Chapter 19.

The initial step in an incremental analysis of competing alternatives is to identify the potential alternatives to be considered and to specify the increments over which the analysis is to be performed. Our first step is to rank the available alternatives by the magnitude of the capital cost. We will identify the alternatives as $A_1, A_2, \ldots, A_n$. There are n possible alternatives. The first alternative, A_1, which is always available, is the "do nothing" option. It requires no capital cost (and achieves no savings). For each of the available alternatives, the project cost (capital cost), PC, and the yearly savings generated (yearly cash flow), YS, must be known.

For larger retrofit projects, discounted profitability criteria should be used. The algorithm to compare alternatives using discounted cash flows is essentially the same as the four-step method outlined in Section 5.3. For small retrofit projects, nondiscounted criteria may often be sufficiently accurate for comparing alternatives. Both types of criteria are discussed in the next sections.

5.6.1   Nondiscounted Methods for Incremental Analysis

For nondiscounted analyses, two methods are provided below:

1. Rate of Return on Incremental Investment (*ROROII*)

$$ROROII = \frac{\text{Incremental Yearly Savings}}{\text{Incremental Investment}}$$

2. Incremental Payback Period (*IPBP*)

$$IPBP = \frac{\text{Incremental Investment}}{\text{Incremental Yearly Savings}}$$

The following examples illustrate the method of comparison of projects using these two criteria.

Example 5.10

A circulating heating loop for an endothermic reactor has been in operation for several years. Due to an oversight in the design phase, a certain portion of the heating loop piping was left uninsulated. The consequence is a significant energy loss. We have two types of insulation that can be used to reduce the heat loss. They are both available in two thicknesses. The estimated cost of the insulation and the estimated yearly savings in energy costs are given below. (The ranking has been added to the alternatives and is based on increasing project cost).

Ranking (option #)	Alternative Insulation	Project Cost (*PC*)	Savings Generated by Project (*YS*)
1	No Insulation	0	0
2	B-1″ Thick	$3,000	$1,400
3	B-2″ Thick	$5,000	$1,900
4	A-1″ Thick	$6,000	$2,000
5	A-2″ Thick	$9,700	$2,400

Assume an acceptable internal rate of return for a nondiscounted profitability analysis to be 15% (0.15).

a. For the four types of insulation determine the rate of return on incremental investment (*ROROII*) and the incremental payback period (*IPBP*).

b. Determine the value of the incremental payback period equivalent to the 15% internal rate of return.

a. Evaluation of *ROROII* and *IPBP*

Option #-Option 1	*ROROII*	*IPBP* (years)
2-1	$1400/$3000 = 0.47 (47%)	$3000/$1400 = 2.1
3-1	$1900/$5000 = 0.38 (38%)	$5000/$1900 = 2.6
4-1	$2000/$6000 = 0.33 (33%)	$6000/$2000 = 3.0
5-1	$2400/$9700 = 0.25 (25%)	$9700/$2400 = 4.0

b. $IPBP = 1/(ROROII) = 1/0.15 = 6.67$ yrs

Note that in Part a of Example 5.10, the incremental investment and savings are given by the difference between installing the insulation and doing nothing. All of the investments considered in Example 5.10 satisfied the internal benchmark for investment of 15%, which means that the "do nothing" option (Option 1) can be discarded. However, which of the remaining options is the best can only be determined using pairwise comparisons.

Example 5.11

Which of the options in Example 5.10 is the best based on the nondiscounted *ROROII* of 15%?

Step 1: Choose Option 2 as the base case, since it has the lowest capital investment.

Step 2: Evaluate incremental investment and incremental savings in going from the base case to the case with the next higher capital investment, Option 3.

Incremental Investment = ($5000 − $3000) = $2000
Incremental Savings = ($1900/yr − $1400/yr) = $ 500/yr
ROROII = 500/2000 = 0.4 or 40% per year

Step 3: Since the result of Step 2 gives an *ROROII* > 15%, we use Option 3 as the base case and compare it with the option with the next higher capital investment, Option 4.

Incremental Investment = ($6000 − $5000) = $1000
Incremental Savings = ($2000/yr − $1900/yr) = $100/yr
ROROII = 100/1000 = 0.1 or 10% per year

Step 4: Since the result of Step 3 gives an *ROROII* < 0.15 we reject Option 4 and compare Option 3 with the option with the next higher capital investment, Option 5.

Incremental Investment = ($9700 − $5000) = $4700
Incremental Savings = ($2400/yr − $1900/yr) = $500/yr
ROROII = 500/4700 = 0.106 or 10.6%.

Step 5: Again the *ROROII* from Step 4 is less than 15%, hence we reject Option 5. Since Option 3 (Insulation B-2″ thick) is the current base case and no more comparisons remain, we accept Option 3 as the "Best Option."

It is important to note that in Example 5.11 we reject Options 4 and 5 even though they give *ROROII* greater that 15% when compared to the "do nothing" option (see Example 5.10). The key here is that in going from Option 3 to either Option 4 or 5 the incremental investment loses money, that is, *ROROII* < 15%.

Example 5.12

Repeat the comparison of options in Example 5.10 using a nondiscounted incremental payback period of 6.67 years.

The steps are the similar to those used in Example 5.11 and are given below without further explanation.

Step 1: (Option 3 – Option 2) $IPBP = 2000/500 = 4$ years < 6.67
Reject Option 2 – Option 3 becomes the base case

Step 2: (Option 4 – Option 3) $IPBP = 1000/10 = 10$ years > 6.67
Reject Option 4

Step 3: (Option 5 – Option 3) $IPBP = 4700/500 = 9.4$ years > 6.67
Reject Option 5

Option 3 is the best option.

5.6.2   Discounted Methods for Incremental Analysis

Incremental analyses taking into account the time value of money should always be used when large capital investments are being considered. We may make comparisons either by discounting the operating costs to yield an equivalent capital investment or by amortizing the initial investment to give an equivalent annual operating cost. Both these techniques are considered in the following sections.

Capital Cost Methods.   The Incremental Net Present Value ($INPV$) for a project is given by

$$INPV = -PC + YS(P/A,i,n) \tag{5.4}$$

When comparing investment options, we will always compare a given case with a "do nothing" option. Thus, we may consider these comparisons as incremental investments. In order to use $INPV$, it is necessary to know the internal discount rate and the time over which the comparison is to be made.

Example 5.13

Based on the information provided in Example 5.10 for an acceptable internal interest rate of 15% and time $n = 5$ yrs, determine the most attractive alternative, using the $INPV$ criterion to compare options.

For $i = 0.15$ and $n = 5$ the value for $(P/A, i, n) = 3.352$. (See Eq. (4.14).)

Eq. (5.4) becomes

$$NPV = -PC + 3.352\, YS$$

Option	INPV
2-1	1693
3-1	1369
4-1	704
5-1	−1655

From the results above, it is clear that Options 2, 3, and 4 are all potentially profitable since $INPV > 0$. However, the best option is Option 2 because it has the highest $INPV$ when compared with the "do nothing" case, Option 1. Note that other pairwise comparisons are unnecessary. We simply choose the option that yields the highest $INPV$. The reason that the $INPV$ gives the best option directly is because by knowing i and n, each dollar of incremental investment is correctly accounted for in the calculation of $INPV$. Thus, if the incremental investment in going from Option A to Option B is profitable, then the $INPV$ will be greater for Option B and vice versa. It should also be pointed out that by using discounting techniques, the best option has changed from Option 3 (in Example 5.11) to Option 2.

Operating Cost Methods.   In the previous section, the yearly savings were converted to an equivalent present value using the present value of an annuity, and this was measured against the capital cost. An alternative method is to convert all the investments to annual costs using the capital recovery factor and measure them against the yearly savings.

We develop the needed relationship from Eq. (5.4), giving

$$INPV/(P/A,i,n) = -PC/(P/A,i,n) + YS$$

It can be seen that the capital recovery factor $(A/P,i,n)$ is the reciprocal of the present worth factor $(P/A,i,n)$. Substituting this relationship and multiplying by -1 gives

$$-(INPV)(A/P,i,n) = (PC)(A/P,i,n) - YS$$

The term on the left is identified as the Equivalent Annual Operating Cost $(EAOC)$. Thus, we may write:

$$EAOC = (PC)(A/P,i,n) - YS \tag{5.5}$$

When we substitute an acceptable rate for i and n, a negative $EAOC$ indicates the investment is acceptable (since a negative cost is the same as a positive savings).

Example 5.14

Repeat Example 5.13 using *EAOC* in place of *NPV*.

For $i = 0.15$ and $n = 5$ the value for $(A/P, i, n) = 1/3.352$.

Eq. (5.5) becomes

$EAOC = PC/3.352 - YS$

Option	EAOC
2–1	−505
3–1	−408
4–1	−210
5–1	494

The best alternative is Option 2 since it has the most negative *EAOC*.

5.7   PROFIT MARGIN ANALYSIS

All the techniques that have been discussed in this chapter use the fixed capital cost and the operating costs in order to evaluate the profitability of a process. Clearly, the accuracy of such predictions will depend on the accuracy of the estimates for the different costs. When screening alternative processes, it is sometimes useful to evaluate the difference between the revenue from the sale of products and the cost of raw materials. This difference is called the profit margin or sometimes just the margin.

$$\text{Profit Margin} = \Sigma(\text{Revenue Products}) - \Sigma(\text{Cost of Raw Materials}) \quad (5.6)$$

If the profit margin is negative, the process will not be profitable. This is because no capital cost, utility costs, and other ancillary operating costs have been taken into account. A positive profit margin does not guarantee that the process will be profitable but does suggest that further investigation may be warranted. Therefore, the profit margin is a useful, but limited, tool for the initial screening of process alternatives.

Example 5.15

Consider the DME process shown in Appendix B.1. Estimate the profit margin for this process using the costs of raw materials and products from Table 3.5.

From Tables 3.5 and B.2 the following flowrates and costs are found:

Cost of methanol = $ 0.17/kg
Cost of dimethyl ether (DME) = $ 1.23/kg
Feed rate of methanol to process (Stream 1, Figure B.1) = 8370 kg/h
Product rate of DME (Stream , Figure B.1) = 5970 kg/h
Profit Margin = (5970)(1.23)– (8370)(0.17) = $5920/h

Clearly, from an analysis of the profit margin, further investigation of the DME process is warranted.

5.8   SUMMARY

In this chapter, we have covered the basics of profitability analysis for projects involving large capital expenditures. The concepts of nondiscounted and discounted profitability criteria were introduced as were the three bases for these criteria namely, time, money, and interest rate.

We also considered how to choose the economically optimum piece of equipment among a group of alternatives using the capitalized cost, equivalent annual operating cost, and the common denominator methods.

The concept of incremental economic analysis was introduced and applied to an example involving large capital budgets and also to a retrofit project. It was shown that both the net present value (NPV) and equivalent annual operating cost ($EAOC$) methods were particularly useful when comparing alternatives using discounted cash flows.

PROBLEMS

For the following problems, unless stated otherwise, you may assume that the cost of land, L, and the salvage value, S, of the plant are both zero.

1. The projected costs for a new plant are given below (all numbers are in 10^6):

    Land Cost = $5
    Fixed Capital Investment = $100 ($60 at end of year 1 and $40 at end of
        year 2)
    Working Capital = $30 (at start-up)
    Start up at end of year 2
    Revenue from sales = $35
    Cost of manufacturing (without depreciation) = $13
    Tax rate = 35%
    Depreciation method = Current MACRS (see Problem 4.18)

Length of time over which profitability is to be assessed = 10 years after start-up

Internal rate of return = 8 % p.a.

For this project, do the following:
a. Draw a cumulative (nondiscounted) after-tax cash flow diagram.
b. From Part a, calculate the following nondiscounted profitability criteria for the project:
   (i) cumulative cash position and cumulative cash ratio
   (ii) payback period
   (iii) rate of return on investment
c. Draw a cumulative (discounted) after-tax cash flow diagram.
d. From Part c, calculate the following discounted profitability criteria for the project:
   (i) net present value and net present value ratio
   (ii) discounted payback period
   (iii) discounted cash flow rate of return (*DCFROR*)

2. Repeat Problem 5.1 using a straight line depreciation method over 7 years. Compare the results with those obtained in Problem 5.1. If you were allowed to choose, which depreciation method would you use?

3. The following expenses and revenues have been estimated for a new project:

Revenues from sales = $2.9 × 10^6$/yr
Cost of manufacturing (excluding depreciation) = $1.3 × 10^6$/yr
Taxation rate = 35%
Fixed Capital Investment = $5 × 10^6$
(two payments of $3 × 10^6$ and $2 × 10^6$ at the end of years 1 and 2, respectively)
Start-up at the end of year 2
Working Capital = $1 × 10^6$ at the end of year 2
Land Cost = $0.5 × 10^6$ at the beginning of the project (time = 0)
Project life (for economic evaluation) = 10 yr after start-up

For this project, estimate the *NPV* of the project assuming an after-tax internal hurdle rate of 10% p.a., using the following depreciation schedules.
a. MACRS method (see Problem 4.18)
b. Straight line depreciation with an equipment life (for depreciation) of 9.5 years.

Comment on the effect of accelerated depreciation schedules on the overall profitability of large capital projects.

4. In reviewing currently operating processes, the company accountant has provided you with the following information about a small chemical process that was built ten years ago.

Year	Process 1	Process 2
1	1	4
2	3	4
3	5	4
4	6	4
5	6	4

a. Calculate the *NPV* of both plants for interest rates of 5% and 20%. Which plant do you recommend? Explain any unusual results that you find.

b. Calculate the *DCFROR* for each plant. Based on the efficiency of resource usage, which plant do you recommend?

c. Calculate the payback period (*PBP*) for each plant. Which plant do you recommend?

d. Explain any differences between your answers to Parts a, b, and c.

8. In a design, you have the choice of purchasing either of the following pieces of equipment:

Equipment	A	B	C
Material of Construction	CS	SS	Ni Alloy
Installed Cost	$ 6,000	$12,000	$15,000
Equipment Life	4 yrs	7 yrs	8 yrs
Yearly Maintenance Cost	1,700	$1,300	$1,000

If the internal rate of return for such comparisons is set at 10% p.a., which of the alternatives is the least costly?

9. Two pieces of equipment are being considered for an identical service. The installed costs and yearly operating costs associated with each equipment are given below:

Costs	Equipment A	Equipment B
Installed Cost	$ 5,000	$10,000
Operating Cost	$750/yr	$450/yr
Equipment Life	4 years	7 years

If the internal hurdle rate for comparison of alternatives is set at 12% p.a., which piece of equipment do you recommend we purchase?

10. We are considering implementing a bypass arrangement around an existing exchanger that is required for heat integration purposes. The new equipment that needs to be installed consists of an exchanger, a pump, a control valve, and some piping. All the equipment is made of carbon steel and the installed costs, operating costs and equipment lives are tabulated below:

Equipment	Installed Cost ($)	Operating Cost ($/yr)	Equipment life (yr)
Exchanger	20,000	7,500	15
Piping	8,500	—	20
Pump	17,500	9,300	8
Control valve instrumentation	14,500	—	8
Total	60,500	16,800	

It has been suggested that a variable speed pump could be used instead of the conventional pump and control valve. This will save operating expenses but the variable speed pump is more expensive and has a slightly shorter life. The cost information for this option is given below:

Equipment	Installed Cost ($)	Operating Cost ($/yr)	Equipment life (yr)
Exchanger	20,000	7,500	15
Piping	8,500	—	20
Variable speed motor and pump	42,000	6,000	6
Total	70,500	13,500	

The current interest rate for comparisons of investments has been set at 10% p.a. Which option (variable speed pump or conventional pump and control valve) do you recommend?

11. Because of rapid corrosion, a pump, in a large production plant must be replaced every four years. What is the capitalized cost for the pump given the following information?

Purchased cost of pump = $ 15,000
Cost of installing new pump = 50% of purchased cost
Internal hurdle rate for study = 8% p.a.
Life of the pump = 4 yr

12. Two alternative pieces of equipment are being considered for the separation of solids from a liquid slurry. Details are given below:

Equipment Type	Service Life	Operating Cost	Capital Investment
Rotary Vacuum Filter	5yr	$1,900/yr	$13,000
Hydroclone + Centrifuge	8yr	$1,300/yr	$26,000

If the discount rate for this project is 10% p.a., which alternative would you recommend?

13. An improvement project for a wastewater treatment process is being considered. The project will involve installing an advanced "smart" control system for the metering of neutralizing chemicals. With this system in place, the existing neutralizer can be scrapped, which will save a significant amount of money due to decreased maintenance costs and wasted chemicals. The costs and savings are listed below:

Cost of installing new "smart" control system = $375,000
Projected savings = $85,000 per year
Time over which cost comparison should be made = 5 years
If the internal rate of return for such projects is set at 8%, should the new control system be implemented?

14. In considering investments in large capital projects, a company is faced with the decision of which of the following projects to invest in.

	Project A	Project B	Project C
Capital Required (in yr 0)	90×10^6	120×10^6	180×10^6
After Tax Yearly Cash Flow (yr 1–10)	15×10^6	20×10^6	26×10^6

The company can always invest its money in long-term bonds, which currently yield 8% p.a. (after tax). In which, if any, of the projects should the company invest assuming the capital ceiling for investment is 220×10^6, and a project life of 10 years is assumed?

15. The house that I live in is very hot in the summer and very cold in the winter. As a consequence I pay a lot of money for utilities. A friend of mine suggested that I should insulate my roof and walls and install a heat pump that can heat the house in the winter and cool it in the summer. I do not have to do either of these suggested improvements, but I would like to make a rational choice as to which of the following options I should implement. Should I:
a. do nothing?
b. buy and install the insulation?
c. buy and install the heat pump?
d. buy and install both the insulation and heat pump?

Using the following information please advise me as to what option makes the most sense from an economic standpoint. Explain the rationale for your recommendation.

Data

Purchase and installation cost of insulation for roof and walls =	$ 3,000
Purchase and installation cost of heat pump =	$ 6,000
Current heating and cooling bill =	$ 2,000/yr
Heating and cooling bill with insulation added =	$ 1,500/yr
Heating and cooling bill with heat pump added =	$ 1,200/yr
Heating and cooling bill with heat pump and insulation added =	$ 700/yr
Cost of house (1982) =	$62,000
Water and sewage bill =	$ 1,200/yr
Maintenance costs for house =	$ 1,700/yr
Interest rate at which I can invest my money (before tax) =	10% p.a.
Estimated life of heat pump and insulation =	15 years

How would your decision change if the life of the heat pump and insulation were 10 years instead of 15 years?

16. You have been asked to evaluate several investment opportunities for the chemical company for which you work. These potential investments concern a new chemical plant to manufacture a certain chemical. The financial information for these investments is presented below:

Case	Capital Investment (10^6)	Yearly Cash Flow (after tax) (10^6/yr)
Base	25	7

Alternative	Incremental Capital Investment (10^6)	Incremental Yearly Cash Flow (after tax) (10^6/yr)
Option 1	5	2
Option 2	10	4
Option 3	5	2.5

The alternative investments (Options 1, 2, and 3) are changes to the base case. The investments shown are incremental to those for the base case. For example, for Option 1, the total capital investment is ($25 + $5) × 10^6, and the total yearly cash flow (after tax) is ($7 + $2) × 10^6/yr. Answer the following:

a. If an acceptable nondiscounted rate of return on investment (*ROROI*) is 30% p.a., which is the best option?

b. If an acceptable after-tax discount rate is 15% p.a., which option is best? Assume a plant life of 10 years with all the capital investment occurring at time = 0.

17. The following information has been calculated for two competing, mutually exclusive projects, each with the same project life (all figures in $ millions per year).

	Project A	Project B
NPV ($i = 10\%$)	105	75
NPV ($i = 15\%$)	55	45
NPV ($i = 20\%$)	−62	1

a. For each project, estimate the *DCFROR*.
b. If the internal hurdle rate for your company is set at 15%, which project would you choose? Would your answer change if the economic climate were changing rapidly, and the internal hurdle rate for alternative investments was expected to increase?

18. A company is considering an investment in a process improvement scheme that would require an initial capital investment of $250,000. The projected increase in revenues from this project over the next five years are given below:

Year	Incremental Revenue
1	$75,000/yr
2	$65,000/yr
3	$65,000/yr
4	$55,000/yr
5	$55,000/yr

a. If the company could invest its money in a mutual fund yielding 10% p.a. interest, which investment should it choose, the project or the mutual fund?
b. What is the minimum effective annual percentage rate that the mutual fund could yield so that it would be more profitable to invest in the mutual fund? What is this rate called?

19. During the design of a new chemical plant (manufacture of acetone), several alternative heat recovery schemes are being considered. There are essentially three options and the details of each option are given below:

Case	Capital Investment (10^4)	Annual Cash Flow After Tax (10^4/yr)	DCFROR
Base	300	50	14.5%
Base + Option 1	305	51	14.5%
Base + Option 1 + Option 2	307	51.2	14.5%

The base case represents the main heat recovery scheme, while Options 1 and 2 are minor modifications to each scheme.

An acceptable (nondiscounted) rate of return for investment is set at 12% p.a. (after tax). What heat recovery scheme (if any) should we use?

20. Consider a chemical process containing the following equipment:

Equipment	Purchased Cost (10^3)
Pump P-901A/B	5
Compressor C-901	750
Heat Exchanger E-901	75
Heat Exchanger E-902	45
Reactor R-901	250
Flash V-901 (contains vessel + heat exchanger)	125
Tower T-901	450
Condenser E-903	70
Reboiler E-904	105
Total Purchased Cost of Equipment	**1,875**

The annual cash flows start at the end of year two and are as follows:

Source of Cash Flow	Amount of Cash Flow (10^6/yr)
Product revenue	36.00
Raw material costs	21.43
Utilities + waste treatment costs	2.00
Operating labor	0.41

The criterion for profitability in your company has been set at 15% rate of return (after tax) and a 10-year plant life should be assumed for economic calculations. Assume that all the capital investment occurs at time 0; that the tax rate is 35%; that you will use MACRS depreciation (see Problem 4.18), and a working capital equal to 0.1 (COM_d) at the end of year 1. Answer the following:

a. Do you recommend construction of the above plant? Justify your answer quantitatively.
b. A plant built to the above specifications is now in operation. The following two modifications have been suggested. What is your recommendation regarding these two modifications? The same profitability measures as used above apply for this comparison.

Modification	Installed Capital Cost ($10³)	Yearly Savings ($10³/yr)
New reboiler	56.5	11.2
Second flash unit to recover product	72.5	251.3

21. The installation of a new heat exchanger is proposed for an existing formalin plant. It costs $150,000 (installed) and saves $30,000/yr in operating costs. The tax rate is 35%, and you may use a straight line depreciation method over 5 years. The service life of the heat exchanger is estimated to be 10 years. The internal hurdle rate for capital in our company for these types of projects is set at 10% p.a., before tax. Calculate the before-tax effect of this plant modification on the $EAOC$. Is this a good investment?

22. A process to produce formalin from methanol has been designed. Pertinent cost data for this base case are in the table below (all figures are in 10^6 or $10^6/yr$). For all parts of this problem, assume that the required before-tax return on investment is 15% p.a., and that the equipment life is considered to be 12 years.
a. Do you recommend construction of this plant, that is, is the base case design a profitable venture?
b. There are two alternative processes that require more expensive equipment, as indicated in the table. What do you recommend?
c. Suppose that there were another alternative requiring an additional 2×10^6 capital investment. How much savings in operating costs would be required to make this an attractive investment?

	Base Case Design	Alternative 1	Alternative 2
Capital investment	11.9	—	—
Additional investment	—	1.02	0.5
Annual product revenue	11.1	11.1	11.1
Annual raw material costs	3.16	3.16	3.16
All other annual operating costs	0.299	0.206	0.258

2

Technical Analysis of a Chemical Process

The purpose of this section is to provide a technical basis and rationale for examining and understanding a process flow diagram. In this section we show that all processes have a similar input/output structure where raw materials enter a process and are reacted to form products and by-products. These products are separated from unreacted feed, which is usually recycled. The streams are then purified to yield products acceptable to the marketplace. All equipment in a process can be categorized into one of the six elements of the Generic Block Flow Process Diagram.

In addition, methods for tracing chemical species through a process flow diagram are given. By following the paths of feed chemicals and reactants, we obtain a much clearer picture of what is happening in the process and of the interaction between the equipment. Furthermore, the conditions at which different equipment operates are discussed and explained. The concept of conditions of special concern are explained and examples of such conditions are identified and explained in the context of the Toluene hydrodealkylation process. Finally, the role that experienced-based guidelines play in the checking of design calculations is discussed. A comprehensive set of tables covering process engineering guidelines is presented.

This material is treated in the following chapters:

Chapter 6: Structure of Chemical Process Flow Diagrams
The basic input/output structure of a chemical process is presented, and it is shown that all chemical processes can be described in terms of a Generic Block Flow Process Diagram, which has six basic elements. These six elements are Reactor Feed Preparation, Reactor, Separator Feed Preparation, Separator, Recycle, and Environmental Control.

Chapter 7: Tracing Chemicals through the Process Flow Diagram

In order to gain a better understanding of a PFD, it is often necessary to follow the flow of key chemical components through the diagram. This chapter presents two different methods to accomplish this. The tracing of chemicals through the process reinforces our understanding of the role that each piece of equipment plays. In most cases, the major chemical species can be followed throughout the flow diagram using simple logic without referring to the flow summary table.

Chapter 8: Understanding Process Conditions

Once the connectivity or topology of the PFD has been understood, it is necessary to understand why a piece of equipment is operated at a given pressure and temperature. The idea of conditions of special concern are introduced. These conditions are either expensive to implement (due to special materials of construction and/or the use of thick-walled vessels) or use expensive utilities. The reasons for using these conditions are introduced and explained.

Chapter 9: Utilizing Experience-Based Principles to Confirm the Suitability of a Process Design

When the design of a process is reviewed, it is prudent to compare the design with well established guidelines. Although the design will most often differ from the results of these shortcut methods, it does allow large differences and potential errors to be identified quickly. This, in turn, allows engineers to check their work as the design progresses.

Structure of Chemical Process Flow Diagrams

A chemical process represents a logical arrangement of unit operations that, working together, transforms raw chemical material(s) into useful chemical product(s). To understand a chemical process it is important to be able to recognize:

1. Why each unit operation was chosen
2. The logic behind the placement of each unit in the process

The Process Flow Diagram (PFD) provides the foundation upon which to build a solid understanding of a chemical process. In the next three chapters, we will illustrate the use of the PFD to understand a chemical process. We use the process flow diagram introduced in Chapter 1 for the production of benzene from toluene (presented in Figures 1.3 and 1.5). In the next three chapters, the function of each piece of process equipment shown in the PFD will be analyzed. The logic behind the placement of each operation in the process and a basis for the choice of operating conditions (T,P) will also be discussed. We will show that all flow diagrams have certain common features that provide a useful pattern and give further understanding of chemical processes, regardless of the chemicals involved.

6.1 THE INPUT-OUTPUT STRUCTURE OF A FLOW DIAGRAM

Process flow diagrams, by convention, show the process feed stream(s) entering from the left and the process product stream(s) leaving to the right. This illustrates the input-output structure of the overall process.

There are other auxiliary streams shown on the PFD, such as utility streams, which are necessary for operations but are not part of the basic input-output structure. Ambiguities between process streams and utility streams may be eliminated by starting our analysis of the process with an over all input-output Block Flow Diagram (BFD). The first step required to construct such a diagram is to identify the chemical reaction or reactions taking place within the process. In most cases, this information is known. If it is not, certain assumptions must be made. We will assume in the following discussion that the chemical reactions taking place are known. The balanced chemical reaction(s) forms the basis for the overall process input-output BFD. Figure 6.1 presents the overall input-output BFD for the PFD shown in Figure 1.3 (We refer to the skeleton PFD for most of the discussion in this section). It should be noted that only process streams are identified on this diagram, utility streams will be identified later on in this chapter. The steps used to create this diagram are:

1. A single block is drawn. Within this block the stoichiometry for all reactions that take place in the process is written. The normal convention of the reactants on the left and products on the right is used.
2. The reactant streams are drawn entering from the left. The number of streams corresponds to the number of reactants (2 in this case). Each stream is labeled with the name of the reactant (toluene and hydrogen in this case).
3. Product streams are drawn leaving to the right. The number of streams corresponds to the number of products (2 in this case). Each stream is labeled with the name of the product (benzene and methane in this case).

Figure 6.1 unmasks the basic input-output structure from the detailed process diagram presented in the PFD (Figure 1.3). This also can be observed from Figure 6.2 where the input and output streams for the Toluene hydrodealkylation PFD are shown in bold. Both figures have the same overall input-output structure. The input streams labeled toluene and hydrogen shown on the left in Figure 6.1 appear on the left on the PFD in Figure 1.3. They are easy to identify since they have been labeled as toluene and hydrogen. These streams are identified as Streams 1 and 3, respectively. Likewise, the output streams, benzene and methane, must appear on the right on the PFD. The benzene leaving the process, Stream 15, is clearly labeled, but there is no clear identification for the methane. Here, it becomes necessary to

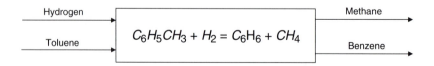

Figure 6.1   The Input-Output Block Flow Diagram for Toluene Hydrodealkylation Process

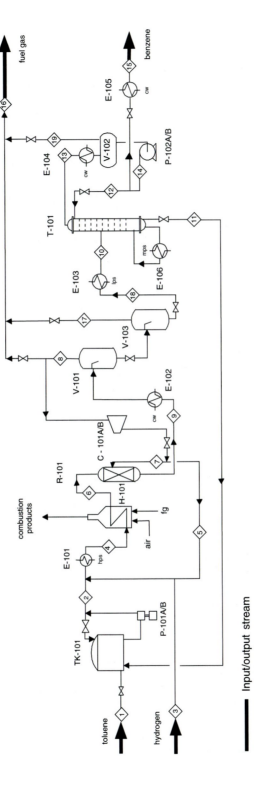

Figure 6.2 Input and Output Streams on Toluene Hydrodealkylation PFD

use a little deductive reasoning to determine where the methane leaves the process in the PFD. There are only two choices, the stream labeled benzene and the stream labeled "fuel gas." It is not possible for the methane, which is a gas, to leave with the liquid benzene from a distillation operation. Methane, therefore, cannot be part of the stream labeled benzene. In addition, if the stream labeled benzene contained the methane (that has equal number of moles leaving the process), it would not be labeled as benzene on the PFD. Since methane is the major constituent of natural gas, it is reasonable to assume that it would be found in the other exit stream labeled fuel gas. All of these reasons suggest that the methane produced in the reaction must leave in the stream labeled fuel gas, Stream 16.

Deductions, such as the ones made above for the location of the methane, must often be made in analyzing flow diagrams. Fortunately, these deductions are mostly straightforward. At times, it may become necessary to use the process conditions given on the PFD to determine where a chemical is to be found. As a last resort, the flow table that is part of the PFD may be used to locate a specific chemical. We recommend that the flow table should be used as a last resort or to verify a thoughtful assessment of the location of any chemical. A greater understanding of the process will be obtained when you utilize your knowledge and experience to deduce the location or content of a process stream. The flow table then gives an independent check to verify your reasoning.

There are several important factors to consider in analyzing the overall input-output structure of a process. We list some of these factors below.

1. Seldom does a single reaction occur, and unwanted side reactions must be considered. Be certain to include all reactions and the reaction stoichiometry taking place. The unwanted products are treated as by-products and must leave along with the product streams shown on the right of the diagram.

2. Chemicals entering the PFD from the left that are not consumed in the chemical reactor are either required to operate a piece of equipment or are inert materials that simply pass through the process. Examples of chemicals required but not consumed include catalyst make-up, solvent make-up, and inhibitors. Inert chemicals may enter with feed materials that are not pure. Alternatively, they may be added in order to control reaction rates, to keep the reactor feed outside of the explosive limits, or to act as a heat sink or heat source to control temperatures.

3. Any chemical leaving a process must have either entered in one of the feed streams or have been produced by a chemical reaction within the process.

4. Utility streams are treated differently from process streams. Utility streams, such as cooling water, steam, fuel, and electricity, rarely contact directly the process streams. They usually provide or remove thermal energy or work.

Figure 6.3 identifies, with bold lines, the utility streams in the benzene process. You will see that two streams, fg (fuel gas) and air, enter the fired heater.

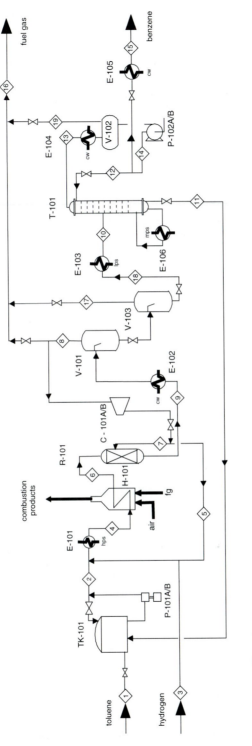

Figure 6.3 Identification of Utility Streams on Toluene Hydrodealkyhlation PFD

utility streams

These are burned to provide heat to the process, but never come in direct contact (i.e., mix) with the process streams. Other streams, such as cooling water and steam, are also highlighted in Figure 6.3. All of these streams are considered to be utility streams and are not extended to the left or right boundaries of the diagram, as were the process streams.

6.2   THE GENERIC BLOCK FLOW PROCESS DIAGRAM

There are features, other than the input-output structure, that are common to all chemical processes. Figure 6.4 provides a generic Block Flow Process Diagram that shows a chemical process broken down into six basic areas or blocks. Each block provides insight into why the operations are included in the process. These common elements or blocks are:

   1. Reactor Feed Preparation
   2. Reactor
   3. Separator Feed Preparation
   4. Separator
   5. Recycle
   6. Environmental Control

   Each of these blocks may contain several unit operations. For example, a separation section might contain four distillation columns, two flash units, and a

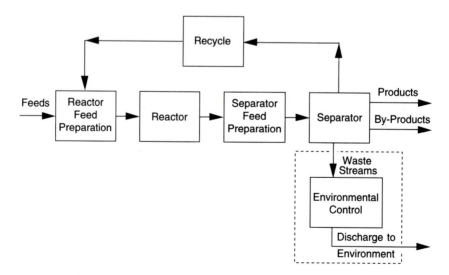

Figure 6.4   The Six Elements of the Generic Block Flow Process Diagram

liquid-liquid decanter, while a reaction section might contain three staged reaction vessels. Each unit operation found on a PFD can be placed into one of these blocks, depending upon the reason for including the unit. It is sometimes difficult to resolve where a given piece of equipment or unit operation should be included. For example, should a separator that removes a trace chemical be listed in the separator block or in the environmental control block? Without further information, it is not possible to answer this question, and you will have to make an assumption to locate the unit. All units can be placed into one of the six blocks listed above. While all processes do not include operations in each block, all processes involve some of these blocks.

You can see that we have drawn a dotted line around the block containing the Environmental Control Operations. This identifies the unique role of environmental control operations in a chemical plant complex. A single environmental control unit may treat the waste from several processes. For example, the waste water treatment facility for an oil refinery might treat the waste water from as many as twenty separation processes. In addition, the refinery may contain a single stack and incinerator to deal with gaseous wastes from these processes. Often, we find that this common environmental control equipment is not shown in the PFD for an individual process. Just because the environmental units do not appear on the PFD does not indicate that they do not exist nor that they are unimportant.

An explanation of the function of each block in Figure 6.4 is given below.

1. **Reactor Feed Preparation Block:** In most cases, the feed chemicals entering a process come from storage. These chemicals are most often not at a suitable concentration, temperature, and pressure for the best performance in the reactor. The purpose of the reactor feed preparation section is to change the conditions of these process feed streams as required in the reactor.

2. **Reactor Block:** All chemical reactions take place in this block. The streams leaving this block contain the desired product(s), any unused reactants, and a variety of undesired by-products produced by competing reactions.

3. **Separator Feed Preparation Block:** The output stream from the reactor, in general, is not at a condition suitable for the effective separation of products, by-products, waste streams, and unused feed materials. The units contained in the separator feed preparation block alter the temperature and pressure of the reactor output stream to provide the conditions required for the effective separation of these chemicals.

4. **Separator Block:** The separation of products, by-products, waste streams, and unused feed materials is accomplished via a wide variety of physical processes. The most common of these techniques are typically taught in unit operations and/or separations classes, for example, distillation, absorption, and extraction.

5. **Recycle Block:** The recycle block represents the return of unreacted feed chemicals, separated from the reactor effluent, to the reactor for further reaction. Since the feed chemicals are not free, it most often makes economic sense to separate the unreacted reactants and recycle them back to the Reactor Feed Preparation Block. Normally, the only equipment in this block is a pump or compressor and perhaps a heat exchanger.

6. **Environmental Control Block:** Virtually all chemical processes produce waste streams. These include gases, liquids, and solids that must be treated prior to being discharged into the environment. These waste streams may contain unreacted materials, chemicals produced by side reactions, fugitive emissions, and impurities coming in with the feed chemicals and the reaction products of these chemicals. Not all of the unwanted emissions come directly from the process streams. An example of an indirect source of pollution results when the energy needs of the plant are met by burning high sulfur oil. The products of this combustion include the pollutant sulfur dioxide that must be removed before the gaseous combustion products can be vented to the atmosphere. The purpose of the Environmental Control Block is to reduce significantly the waste emissions from a process and to render all nonproduct streams harmless to the environment.

6.3   CATEGORIZING EQUIPMENT INTO GENERIC FUNCTION BLOCKS

With deductive reasoning and a well-presented PFD, we can classify all of the units on a PFD into the generic classification blocks identified above. Since the PFD is prepared and presented to provide a clear, unambiguous description of the process, it should not be difficult to identify the block into which each operation falls.

Example 6.1

Starting with Figure 1.3, the PFD for the production of benzene from toluene, identify and classify each piece of equipment into one of the six generic blocks described above. Provide the reasons for the selection of the category chosen.

Since the mass flows on a PFD, in general, move from left to right, the units to the left of the reactor, R-101, are likely to be associated with the Reactor Feed Preparation Block. This observation suggests that the pump P-101, feed heater E-101 and heater H-101 are used to increase the pressure and temperature of the feed to the reactor, and are so classified in Table E6.1.

Applying similar logic, the units to the right of the reactor include the Separator Feed Preparation Block, the Separator Block and the Recycle Blocks. From the PFD we identify these units as compressor C-101, two phase separators V-101 and V-103, the reactor effluent cooler E-102, the tower feed heater E-103, distillation tower T-101, condenser

Table E6.1   Grouping of Unit Operations into Generic Function Blocks

Equipment Identification	Generic Function Block
TK-101	Recycle
P-101 A/B	Reactor Feed Preparation
P-102 A/B	Separator
C-101 A/B	Recycle
E-101	Reactor Feed Preparation
E-102	Separator Feed Preparation
E-103	Separator Feed Preparation
E-104	Separator
E-105	Separator
E-106	Separator
H-101	Reactor Feed Preparation
R-101	Reactor
V-101	Separator
V-102	Separator
V-103	Separator
T-101	Separator

E-104, reflux drum V-102, reboiler E-106, and reflux pump P-102. The distillation unit consists of units T-101, E-104, E-105, E-106, V-102, and P-102. All of these are part of the distillation unit that separates the product benzene and, therefore, are listed as belonging to the Separator Block. The function of heat exchangers E-102 and E-103 is to change the temperature of a stream prior to a physical separation and they are therefore placed in the Separator Feed Preparation Block. Units V-101 and V-103 are identified on the flow sheet as phase separators, and they are included as part of the Separator Block.

The compressor C-101 operates on a portion of Stream 8 that has come indirectly from the exit of the reactor and is returned to the Reactor Feed Preparation section. Therefore, C-101 is part of the Recycle Section.

Table E6.1 lists all sixteen pieces of equipment shown on the PFD for the benzene process presented in Figure 1.3. We have also identified the generic block into which each unit belongs.

This example uncovered no unit operations identified as part of the Environmental Control Block. Upon close inspection of the PFD, you find a stream leaving the process labeled "Combustion Products," that we have shown as a utility stream in Figure 6.3. Depending upon the fuel chosen for the heater, this stream may have to be treated to remove pollutants, for example, sulfur oxides (SO_x) and nitrogen oxides (NO_x).

6.4 SUMMARY

In this chapter you were shown that all process flow diagrams possess an input-output structure with the feeds entering from the left and products leaving to the right. The process stoichiometry is the basis of a block flow process diagram for the overall process. This block flow process diagram distinguishes the utility streams on a PFD from the process streams.

The generic chemical block flow process diagram divides any chemical process into six functional elements that identify the purpose of each unit operation in the process. The reasoning used to break down a chemical process into these elements was demonstrated for the production of benzene. By breaking down a chemical process into the generic elements, it is easier to understand and follow the overall flow of process and utility streams throughout the process. The only information required is the PFD and the chemical reactions involved.

PROBLEMS

1. For the Acrylic Acid PFD given in Appendix B, Figure B.2, the following reactions take place:

$$C_3H_6 + 1.5O_2 = C_2H_3COOH + H_2O$$

$$C_3H_6 + 2.5O_2 = CH_3COOH + CO_2 + H_2O$$

$$C_3H_6 + 4.5O_2 = 3CO_2 + 3H_2O$$

   With this information do the following:
   a. Draw the overall input-output block diagram for the process.
   b. From Figure B.2, identify all the process streams that enter or leave the process that contain chemicals that do not take part in the reactions given above.
   c. Identify all utility streams on the PFD.

2. For the Acrylic Acid PFD flow diagram given in Appendix B, Figure B.2, and the stoichiometry provided in Problem 6.1, do the following:
   a. Prepare a table that places all of the unit operations shown into the appropriate blocks in the generic block flow process diagram.
   b. State your logic for placing each item in the table in Part a.

3. For the dimethyl ether (DME) production process, Figure B.1, Appendix B, do the following:

   a. Identify all the chemical reactions taking place (read introduction for DME process given in Appendix B).

   b. Draw the overall input-output block diagram for the process.

   c. Identify all the process streams that enter or leave the process that contain chemicals that do not take part in the reactions given in Part a.

   d. Identify all utility streams on the PFD.

   e. Place all the equipment in the PFD into one (or more) of the six generic process function blocks of the generic block flow process diagram.

4. Repeat Problem 3 for the acetone production process, Appendix B, Figure B.3.

5. Repeat Problem 3 for the heptenes production process, Appendix B, Figure B.4.

6. Repeat Problem 3 for the allyl chloride production process, Appendix C, Figure C.1 and C.3.

7. Repeat Problem 3 for the phthalic anhydride production process, Appendix C, Figure C.5.

8. Repeat Problem 3 for the cumene production process, Appendix C, Figure C.8.

CHAPTER 7

Tracing Chemicals through the Process Flow Diagram

In the previous chapter, we classified the unit operations from a PFD into one of the six blocks of a generic block flow process diagram. In this chapter, you will gain a deeper understanding of a chemical process by learning how to trace the paths taken by chemical species through a chemical process.

7.1 GUIDELINES AND TACTICS FOR TRACING CHEMICALS

We provide guidelines and outline some useful tactics to help you to trace chemical components through a process.

We introduce two important operations for tracing chemical pathways in PFDs. These are the adiabatic mixer and adiabatic splitter described below.

> **Mixer**: Two or more input streams are combined to form a single stream. This single output stream has a well-defined composition, phase(s), pressure, and temperature.
>
> **Splitter**: A single input stream is split into two or more output streams with the same temperature, pressure, and composition as the input stream. All streams involved differ only in flowrate.

These operations are found where streams meet or a stream divides on a PFD. They are little more than tees in pipelines in the plant. These operations involve little design and involve minimal cost. Hence, they are not important in estimating the capital cost of a plant and would not appear on a list of major equipment. However, you will find in Chapter 18 that these units are included in the design of flowsheets for implementing and using chemical process simulators.

We have highlighted the mixers and splitters as "shadow boxes" on the flow diagrams presented in this chapter. They carry an "m" and "s" designation, respectively.

7.2   TRACING PRIMARY PATHS TAKEN BY CHEMICALS IN A CHEMICAL PROCESS

Chemical species identified in the overall block flow process diagram (those associated with chemical reaction(s)) are termed primary chemicals. The paths followed by primary chemicals between the reactor and the boundaries of the process are termed primary flow paths. Two general guidelines should be followed when tracing these primary chemicals.

1. **Reactants:** Start with the feed (left-hand side of the PFD) and trace chemicals forward toward the reactor.
2. **Products:** Start with the product (right-hand side of the PFD) and trace chemicals backward toward the reactor.

The following tactics for tracing chemicals apply to all unit operations *except for* chemical reactors.

Tactic 1: Any unit operation, or group of operations, that has a single, or multiple, input stream(s) and a single output stream is traced in a forward direction . If chemical A is present in any input stream, it must appear in the single output stream (see Figure 7.1a).

Tactic 2: Any unit operation, or group of operations, that has a single input stream and single, or multiple, output stream(s) is traced in a backward direction. If chemical A is present in any output stream, it must appear in the single input stream (see Figure 7.1b).

Tactic 3: Systems such as distillation columns are composed of multiple unit operations with a single input or output stream. It is sometimes necessary to consider such equipment combinations as blocks before implementing Tactics 1 and 2.

When tracing chemicals through a PFD, it is important to remember the following:

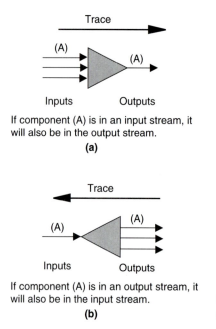

If component (A) is in an input stream, it
will also be in the output stream.

(a)

If component (A) is in an output stream, it
will also be in the input stream.

(b)

Figure 7.1   Tactics for Tracing
Chemical Species

You may occasionally encounter situations where both reactions and physical separations take place in a single piece of equipment. In most cases, this is undesirable but unavoidable. In such situations, it will be necessary to divide the
unit into two imaginary, or phantom, units. The chemical reaction(s) takes place
in one phantom unit and the separation in the second phantom unit. These phantom units are never shown on the PFD, but we will see that such units are useful
when building a flowsheet for a chemical process simulator (see Chapter 18).

We demonstrate these guidelines by determining the paths of the primary
chemicals in the benzene hydrodealkylation process. The only information used
is that provided in the skeleton process flow diagram given in Figure 1.3.

Example 7.1

For the toluene hydrodealkylation process, establish the primary flow pathway for:

- **a.** toluene between the feed (Stream 1) and the reactor
- **b.** benzene between the reactor and the product (Stream 15)

    Hint: Consider only one unit of the system at a time . Refer to Figure E7.1.
    Toluene Feed: Tactic 1 is applied to each unit operation in succession.

- **a.** Toluene feed Stream 1 mixes with Stream 11 in TK-101. A single unidentified stream
    leaves Tank, TK-101, and goes to pump P-101. All of the toluene feed is in this
    stream.

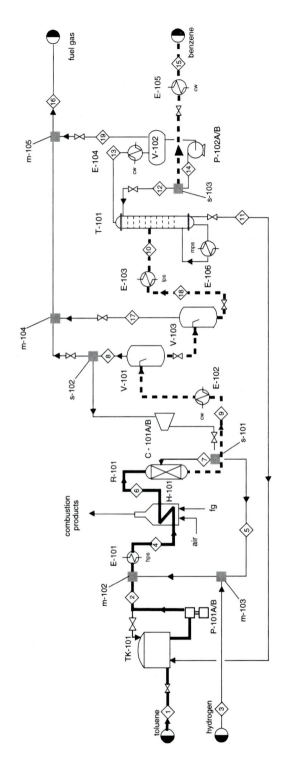

Figure E7.1 Primary Chemical Pathways for Benzene and Toluene in the Toluene Hydrodealkylation Process (Figure 1.3)

■ m-mixer / s-splitter

– – – – Benzene

———— Toluene

   b. Stream 2 leaves pump P-101 and goes to mixer m-102. All of the feed toluene is in this stream.

   c. A single unidentified stream leaves mixer m-102, and goes to exchanger E-101. All of the feed toluene is in this stream.

   d. Stream 4 leaves exchanger E-101, and goes to heater H-101. All of the feed toluene is in this stream.

   e. Stream 6 leaves heater H-101, and goes to reactor R-101. All of the feed toluene is in this stream.

Benzene Product: Tactic 2. Applied to each unit operation in succession.

   a. Product Stream 15 leaves exchanger E-105.

   b. Entering exchanger E-105, is an undesignated stream from s-103 of the distillation system. It contains all of the benzene product.

   c. Apply Tactic 3 and treat the tower T-101, pump P-102, exchangers E-104 and E-106, vessel V-102, and splitter s-103 as a system.

   d. Entering this distillation unit system is Stream 10 from exchanger E-103. It contains all of the benzene product.

   e. Entering exchanger E-103, is Stream 18 from vessel V-103. It contains all of the benzene product.

   f. Entering vessel V-103, is an undesignated stream from vessel V-101. It contains all of the benzene product.

   g. Entering vessel V-101, is an undesignated stream from exchanger E-102. It contains all of the benzene product.

   h. Entering exchanger E-102, is Stream 9 from reactor R-101. It contains all of the benzene product.

The path for toluene was identified in Example 7.1 and is shown in Figure E7.1 as a solid enhanced line. For this case, it was not necessary to apply any additional information about the unit operations to establish this path. The two streams that joined the toluene path did not change the fact that all of the feed toluene remained as part of the stream. All of the toluene fed to the process in Stream 1 entered the reactor, and this path represents the primary path for toluene.

The path for benzene identified in Example 7.1 is shown in Figure E7.1 as an enhanced dotted line. The equipment that makes up the distillation system was considered as an operating system and treated as a single unit operation. The fact that, within this group of process units, some streams were split with some of the flow returning upstream did not change the fact that the product benzene always remained in the part of the stream that continued to flow toward the product discharge. All of the benzene product followed this path, and it represents the primary path for the benzene. The flow path taken for the benzene through the distillation column section is shown in more detail in Figure 7.2. The concept of

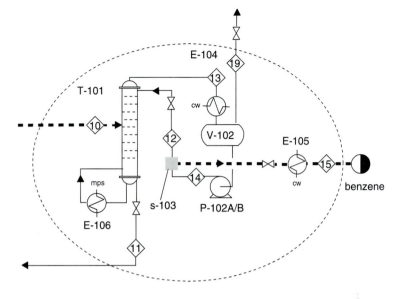

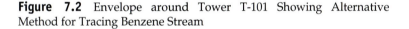

—————— benzene path through tower T-101 and related equipment

Figure 7.2 Envelope around Tower T-101 Showing Alternative Method for Tracing Benzene Stream

drawing envelopes around groups of equipment in order to carry out material and energy balances is introduced early into the chemical engineering curriculum. This concept is essentially the same as the one used here to trace the path of benzene through the distillation column. The only information needed about unit operations used in this analysis was the identification of the multiple units that made up the distillation system. This procedure can be used to trace chemicals throughout the PFD and forms an alternative tracing method that is illustrated by the following example.

Example 7.2

Establish the primary flow pathway for:

   a. hydrogen between its introduction as a feed and the reactor
   b. methane between its generation in the reactor and the discharge from the process as a product.

   In order to determine the primary flow paths, we develop systems (by drawing envelopes around equipment) that progressively include additional unit operations. Refer-

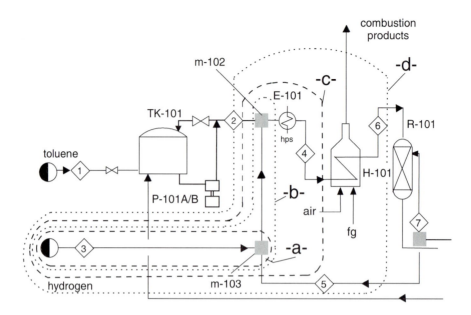

Figure E7.2A   Tracing Primary Chemical Pathways Using the Envelope Method

ence should be made to Figure E7.2A for viewing and identifying systems for tracing hydrogen.

Hydrogen Feed: Tactic 1 is applied to each system in a forward progression. Each system includes the hydrogen feed Stream 3, and the next piece of equipment to the right.

System -a-: This system illustrates the first step in our analysis. The system includes the first unit into which the hydrogen feed stream flows. The unidentified stream leaving mixer m-103 contains the feed hydrogen.

System -b-: Includes mixers m-103 and m-102. The exit stream for this system includes the feed hydrogen.

System -c-: Includes mixers m-103, m-102, and exchanger E-101. The exit stream for this system, Stream 4, includes the feed hydrogen.

System -d-: Includes mixers m-103, m-102, exchanger E-101, and heater 101. The exit stream for this system, Stream 6, includes the feed hydrogen. Stream 6 goes to the reactor.

The four steps described above are illustrated in Figure E7.2A. A similar analysis is possible for tracing methane, and the steps necessary to do this are illustrated in Figure E7.2B. These steps are discussed briefly below.

Methane Product: The methane produced in the process leaves in the fuel gas, Stream 16. Tactic 2 is applied to each system containing the fuel gas product, in backward progression.

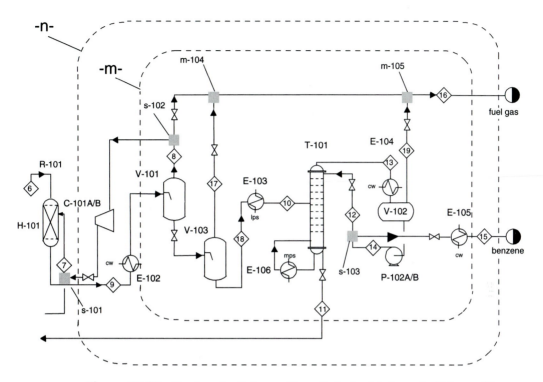

Figure E7.2B Tracing the Primary Flow Path for Methane in Toluene Hydrodealkylation PFD

System -m-: Consists of: m-105, m-104, E-105, T-101, V-102, P-102, s-103, E-104, E-106, E-103, V-101, V-103, and s-102. This is the smallest system that can be found that contains the fuel gas product stream and has a single input.

System -n-: Includes the system identified above plus exchanger E-102 and compressor C-101. The inlet to E-102 contains all the methane in the fuel stream. This is Stream 9, which leaves the reactor.

In the first step of tracing methane, including only m-105 was attempted. This unit had two input streams, and it was not possible to determine which of these streams (or both streams) carried the methane that made up the product stream. Thus, Tactic 2 could not be used. In order to move ahead, additional units were added to m-105 to create a system that had a single input stream. The resulting system, System -m-, has a single input, the unidentified stream coming from exchanger E-102. The identical problem would arise if the procedure used in Example 7.1 were implemented. Figure 7.3 shows the primary paths for the hydrogen and methane.

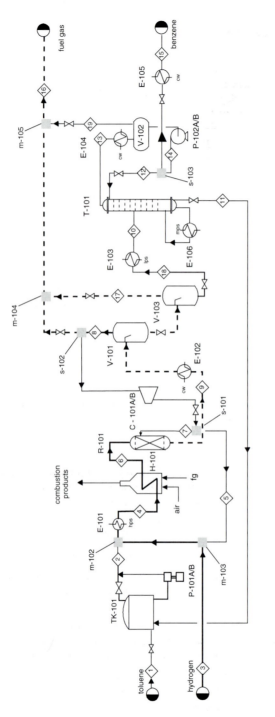

Figure 7.3 Primary Chemical Pathways for Methane and Hydrogen in the Toluene Hydrodealkylation PFD

7.3   RECYCLE AND BYPASS STREAMS

It is important to be able to recognize recycle and bypass streams in a chemical processes. When identifying recycle and bypass streams, we look for flow loops in the PFD. Any time we can identify a flow loop, we have either a recycle or a bypass stream. The direction of the streams, as indicated by the direction of the arrow heads, determines whether the loop contains a recycle or a bypass. The following tactics are applied to flow loops:

Tactic 4:   If the streams in a loop flow so that the flow path forms a complete circuit back to the point of origin, then this is a recycle loop.

Tactic 5:   If the streams in a loop flow so that the flow path does not form a complete circuit back to the place of origin, then there is bypass.

It is worth noting that certain pieces of equipment normally contain recycle streams. In particular, distillation columns nearly always have top and bottom product reflux streams, which are essentially recycle loops. When identifying recycle loops, we can easily determine which loops contain reflux streams and which do not. The following example illustrates the procedure for identifying recycle and bypass streams in the toluene hydrodealkylation PFD.

Example 7.3

For the toluene hydrodealkylation PFD given in Figure E7.1, identify all recycle and bypass streams.

The recycle loops are identified in Figures E7.3A and E7.3B. The main toluene recycle loop is highlighted in Figure E7.3A, while the hydrogen recycle loops are shown in Figure E7.3B(a) and E7.3B(b). There are two reflux loops associated with T-101, and these are shown in Figures E7.3B(c) and E7.3B(d). Finally, there is a second toluene recycle loop identified in Figure E7.3B(e). This recycle loop is used for control purposes (see Chapter 15) and will not be discussed further here. The logic used to deduce what chemical is being recycled in each loop is discussed in the next example.

The bypass streams are identified in Figure E7.3C. These bypass streams contain mostly hydrogen and methane and are combined to form the fuel gas stream, Stream 16.

It is important to remember that flow diagrams represent the most meaningful and useful documents to describe and understand a process. Although PFDs contain a lot of process information, it is sometimes necessary to apply additional knowledge about a unit operation to determine which chemicals are contained in a recycle stream. We demonstrate this idea in the next example.

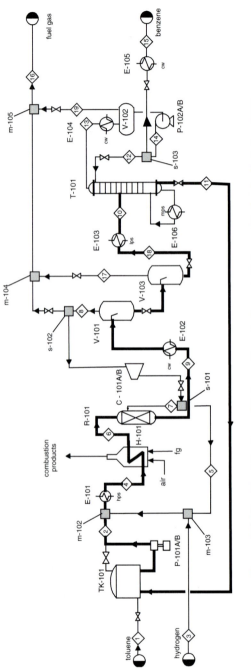

Figure E7.3A   Identification of Toluene Recycle Loop in Toluene Hydrodealkylation PFD

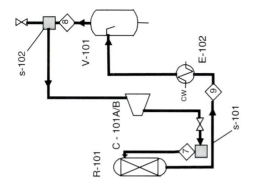

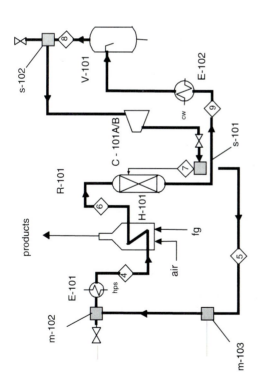

(a)

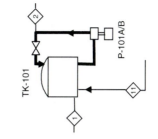

(b)

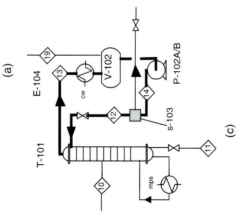

(c)

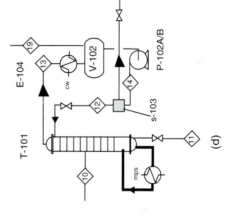

(d)

(e)

Figure E7.3B Identification of other Recycle Loops in Toluene Hydrodealkylation PFD

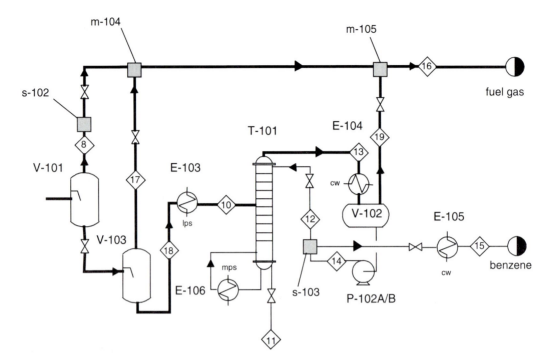

Figure E7.3C   Identification of Bypass Streams in Toluene Hydrodealkylation PFD

Example 7.4

Provide preliminary identification of the important chemical species in each of the three recycle streams identified in Example 7.3 (see Figures E7.3A, E7.3B(a), and E7.3B(b)).

Figure E7.3A
Stream 11: This is the bottom product stream out of the distillation tower that provides the product benzene as distillate. The bottom product stream must have a lower volatility than benzene. The only possible candidate is toluene. Stream 11 is essentially all toluene.

Figure E7.3B(a) and E7.3B(b)
Two undesignated streams leave splitter s-102: One stream leaves as part of a product stream and joins with other streams to form Stream 16. The other stream passes through C-101 to splitter s-101. The input and the two streams leaving s-101 have the same composition. If we know any of the stream compositions, we know them all. In addition, methane is a reaction product and must leave the process. There are only two streams that leave the process, namely Streams 15 and 16. Since the methane is unlikely to be part of the benzene stream, it must, therefore, be in the stream identi-

fied as fuel gas, Stream 16. We make the following assumption regarding the composition of the fuel gas stream.

ASSUMPTION: The product stream leaving is gaseous and not pure methane. If it were pure, it would be labeled methane.

The only other gas that could be present is hydrogen. Therefore, the fuel gas stream is a mixture of methane and hydrogen and all three streams associated with s-101 have the same composition of methane and hydrogen.

The stream that leaves splitter s-102 and goes through compressor C-101 to splitter s-101 is split further into Streams 5 and 7. All streams have the same composition.

Stream 5 then mixes with additional hydrogen from Stream 3 in mixer m-103. The stream leaving m-103 contains both hydrogen and methane, but with a composition of hydrogen greater than that in the other gas streams discussed.

Finally, Stream 7, which also leaves splitter s-101, flows back to the reactor and forms the third recycle stream.

Before the analysis in Example 7.4 can be accepted, it is necessary to check out the assumption used to develop the analysis. Up to this point, we have used the skeleton flow diagram that did not provide the important temperatures, pressures, and flowrates that are seen in the completed PFD (Figure 1.5). Figure 1.5 gives the following information for the flowrates of reactants:

hydrogen (Stream 3): 572 kg/h (286.0 kmol/h)
toluene (Stream 1):   10,000 kg/h (108.7 kmol/h)

Based on the information given in Table 1.5, only 108 kmol/h of hydrogen reacts to form benzene, and 178 kmol/h is excess reactant that leaves in the fuel gas. The fuel gas content is about 40 mole % methane and 60 mole % hydrogen. This confirms the assumption made in Example 7.4.

7.4   TRACING NON-REACTING CHEMICALS

Chemical processes may contain chemicals other than the primary chemicals analyzed in the previous sections. These chemicals must appear in both the input and output streams and are neither created nor destroyed in the process. Unlike the primary chemicals, it makes no difference in what direction we choose to trace these chemicals. You can trace in the forward direction, the backward direction, or start in the middle and trace in both directions. Other than this additional flexibility, the tactics provided above can be applied to all non-reacting chemicals.

7.5   LIMITATIONS

When the tracing procedure resorts to combining several unit operations into a single system that provides a single stream, the path is incomplete. This can be seen in the paths of both product streams, methane and benzene, in Figure E7.1.

> Benzene: The benzene flows into and out of the distillation system as the figure shows. There is no indication how it moves through the internal units consisting of V-102, s-103, E-104, E-106, and T-101.
>
> Methane: The methane flows into and out of a system composed of V-101, V-103, s-102, and m-104. Again, there is no indication of the methane path.

In order to determine the performance and the flows through these compound systems, you need more information than provided in the skeleton PFD, and you must know the function of each of the units.

The development given in the previous sections used only the information provided on the skeleton PFD (without the description of the unit operation) and did not include the important flows, temperatures, and pressures that were given in the full PFD, Figure 1.5, and the flow table, Table 1.5. With this additional information and knowledge of the unit operations, you will be able to fill in some of the paths that are yet unknown.

Each step in tracing the flow paths increases our understanding of the process for the production of benzene represented in the PFD. As a last resort, reference should be made to the flow table to determine the composition of the streams, but this fails to develop analytical skills that are essential to understand the process.

7.6   WRITTEN PROCESS DESCRIPTION

A process description, like a flow table, is often included with a PFD. When a description is not included, it is necessary to provide a description based upon the PFD. Based on the techniques developed in this and the previous chapter, you should be able to write a detailed description of the toluene hydrodealkylation process. Table 7.1 provides such a description. You should read this description carefully and make sure you understand it fully. You should find it useful, if not essential, to refer to the PFD in Figure 1.5 during your review. It is a good idea to have the PFD in front of you while you follow the process description.

The process description should capture all of the knowledge that you have developed in the last two chapters and represents a culmination of our understanding of the process up to this point.

Table 7.1   Process Description of the Toluene Hydrodealkylation Process (refer to Figures 7.2 and 1.5)

Fresh toluene, Stream 1, is combined with recycled toluene, Stream 11, in the storage tank, TK-101. Toluene from the storage tank is pumped, via P-101, up to a pressure of 25.8 bar and combined with the recycled and fresh hydrogen streams, Streams 3 and 5. This two-phase mixture is then fed through the feed preheater exchanger, E-101, where its temperature is raised to 225°C, and the toluene is completely vaporized. Further heating is accomplished in heater, H-101, where the temperature of the stream is raised to 600°C. The stream leaving the heater, Stream 6, enters the reactor, R-101, at 600°C and 25.0 bar. The reactor consists of a vertical packed bed of catalyst, down through which the hot gas stream flows. The hydrogen and toluene react catalytically to produce benzene and methane according to the following exothermic reaction:

$$C_7H_8 + H_2 = C_6H_6 + CH_4$$
$$\textit{toluene} \qquad \textit{benzene}$$

The reactor effluent, Stream 9, consisting of benzene and methane produced from the reaction, along with the unreacted toluene and hydrogen, is quenched in exchanger E-102 where the temperature is reduced to 38°C, using cooling water. Most of the benzene and toluene condenses in E-102, and the two-phase mixture leaving this exchanger is then fed to the high pressure phase separator, V-101, where the liquid and vapor streams are allowed to disengage.

The liquid stream leaving V-101 is flashed to a pressure of 2.8 bar and is then fed to the low-pressure phase separator, V-103. The liquid leaving V-103, Stream 18, contains toluene and benzene with only trace amounts of dissolved methane and hydrogen. This stream is heated in exchanger E-103 to a temperature of 90°C prior to being fed to the benzene purification column, T-101. The benzene column, T-101, contains 42 sieve trays and operates at approximately 2.5 bar. The overhead vapor, Stream 13, from the column is condensed using cooling water in E-104 and the condensate is collected in the reflux drum, V-102. Any methane and hydrogen in the column feed accumulates in V-102 and these non-condensables, Stream 19, are sent to fuel gas. The condensed overhead vapor stream is fed from V-102 to the reflux pump P-102. The liquid stream leaving P-102, Stream 14, is split into two, one portion of which, Stream 12, is returned to the column to provide reflux. The other portion of the condensed liquid is cooled to 38°C in E-105, prior to being sent to storage as benzene product, Stream 15. The bottoms product from T-101, Stream 11, contains virtually all of the toluene fed to the column and is recycled back to TK-101 for further processing.

The vapor stream leaving V-101 contains most of the methane and hydrogen in the reactor effluent stream plus small quantities of benzene and toluene. This stream is split into two, with one portion being fed to the recycle gas compressor, C-101. The stream leaving C-101 is again split into two. The major portion is contained in Stream 5, which is recycled back to the front end of the process where it is combined with fresh hydrogen feed, Stream 3, prior to being mixed with the toluene feed upstream of E-101. The remaining gas leaving C-101, Stream 7, is used for temperature control in the reactor, R-101. The second portion of the vapor leaving V-101 constitutes the major portion of the fuel gas stream. This stream is first reduced in pressure and then combined with the flashed vapor from V-103, Stream 17, and with the non-condensables from the overhead reflux drum, Stream 19. The combination of these three streams is the total fuel gas product from the process, Stream 16.

7.7 SUMMARY

This chapter showed how to trace many of the chemical species through a PFD, based solely upon the information shown on the skeleton PFD. It introduced operations involving splitting and mixing, not explicitly shown on the PFD, that were helpful in tracing these streams.

For situations where there was no single input or output stream, systems containing multiple unit operations were created. The tracing techniques for these compound systems did not provide the information needed to determine the internal flows for these systems. In order to determine reflux ratios for columns, for example, the process flow table must be consulted.

With the information provided, an authoritative description of the process can be prepared.

PROBLEMS

1. Identify the main reactant and product chemical pathways for the dimethyl ether (DME) process shown in Figure B.1, Appendix B.

2. Identify the main reactant and product chemical pathways for the acrylic acid production facility shown in Figure B.2, Appendix B.

3. Identify the main reactant and product chemical pathways for the acetone production facility shown in Figure B.3, Appendix B.

4. Identify the main reactant and product chemical pathways for the heptenes production process shown in Figure B.4, Appendix B.

5. Identify the main recycle and bypass streams for the dimethyl ether (DME) process shown in Figure B.1, Appendix B.

6. Identify the main recycle and bypass streams for the acrylic acid production facility shown in Figure B.2 Appendix B.

7. Identify the main recycle and bypass streams for the acetone production facility shown in Figure B.3, Appendix B.

8. Identify the main recycle and bypass streams for the heptenes production process shown in Figure B.4, Appendix B.

9. Write a process description for the dimethyl ether (DME) process shown in Figure B.1, Appendix B.

10. Write a process description for the acrylic acid production facility shown in Figure B.2, Appendix B.

11. Write a process description for the acetone production facility shown in Figure B.3, Appendix B.

12. Write a process description for the heptenes production process shown in Figure B.4, Appendix B.

CHAPTER 8

Understanding Process Conditions

In previous chapters, process flow diagrams (PFDs) were accepted without evaluating the technical features of the process. The process topology and process operating conditions were provided but were not examined. Economic evaluations were carried out, but without confirming that the process would operate as indicated by the flow diagram.

It is not uncommon to investigate process economics based upon assumed process performance. For example, in order to justify spending the capital to develop a new catalyst, the economics of a process using a hypothetical catalyst with assumed characteristics, such as no unwanted side reactions, might be calculated.

> **The ability to make an economic analysis of a chemical process based on a PFD is not proof that the process will actually work.**

In this chapter, you will learn how to analyze the reasons why the specific temperatures, pressures, and compositions selected for important streams and unit operations have been chosen. Stream specifications and process conditions are influenced by physical processes as well as economic considerations and are *not chosen arbitrarily*. The conditions used in a process most often represent an economic compromise between process performance and the capital and operating costs of the process equipment. Final selection of operating conditions should not be made prior to the analysis of the process economics. In this chapter, we concentrate on analyzing process conditions that require special consideration. As an example, we do not address why a reactor is run at 600°C instead of 580°C, but rather concentrate on the reasons why the reactor is not run at a much lower temperature, 200°C, for example. This type of analysis leads us to question how process conditions are chosen and makes us consider the consequences of changing these conditions.

8.1   CONDITIONS OF SPECIAL CONCERN FOR THE OPERATION OF SEPARATION AND REACTOR SYSTEMS

Rarely are process streams available at conditions most suitable for reactor and separation units. Temperatures, pressures, and stream compositions must be adjusted to provide conditions that allow effective process performance. This was discussed in Chapter 6, where the Generic PFD was introduced (see Figure 6.4). This figure showed two feed preparation blocks, one associated with the reactor and the second with the separation section.

We provide two generalizations to assist you in analyzing and understanding the selection of process conditions:

- It is usually easier to adjust the temperature and/or pressure of a stream than it is to change its composition. In fact, often the concentration of a compound in a stream (for a gas) is a dependent variable and is controlled by the temperature and pressure of the stream.
- In general, pressures between 1 and 10 bar and temperatures between 40°C and 250°C do not cause severe processing difficulties.

The rationale for the conditions given in the second generalization are explained below.

8.1.1   Pressure

There are economic advantages associated with operating equipment at above ambient pressure when gases are present. These result from the increase in gas density (decrease in gas volume) with increasing pressure. All other things being equal, in order to maintain the same fluid (gas) residence time in a piece of equipment, the size of the equipment through which the gas stream flows need not be as large when the pressure is increased.

Most chemical processing equipment can withstand pressures up to 10 bar without additional capital investment (see the cost curves in Appendix A). At pressures above 10 bar, thicker walled, more expensive equipment is necessary. Likewise, operating below ambient pressure (vacuum conditions) tends to make equipment large and may require special construction techniques, thus increasing the cost of equipment.

> **A decision to operate outside the pressure range of 1 to 10 bar must be justified.**

8.1.2   Temperature

There are several critical temperature limits that apply to chemical processes. At elevated temperatures, common construction materials (primarily carbon steel), suffer a significant drop in physical strength and must be replaced by more costly materials.

Example 8.1

The maximum allowable tensile strength for a typical carbon steel and stainless steel, at ambient temperature, 400°C, and 550°C is provided below (from Walas [1])

Temperature	Tensile Strength of Material at Temperature Indicated (bar)		
	Ambient	400°C	550°C
Carbon Steel (grade 70)	1190	970	170
Stainless Steel (Type 302)	1290	1290	430

Determine the fractional decrease in the maximum allowable tensile strength (relative to the strength at ambient conditions) for the temperature intervals: (a) ambient to 400°C and (b) 400°C to 550°C.

  a. Interval ambient to 400°C:
     Carbon Steel: (1190-970)/1190 =   0.18
     Stainless Steel: (1290-1290)/1290 = 0.0

  b. Interval 400°C to 550°C:
     Carbon Steel: (970-170)/1190 =   0.67
     Stainless Steel: (1290-430)/1290 = 0.67

Example 8.1 has shown that carbon steel suffers a loss of 18% and the stainless steel suffers no loss in tensile strength when heated to 400°C. With an additional temperature increase of 150°C to 550°C, the stainless steel suffers a 67% loss while the carbon steel suffers an additional 67% loss in strength. At operating temperatures of 550°C, the carbon steel has a maximum allowable tensile strength of about 15% of its value at ambient conditions. For the stainless steel, the maximum allowable strength at 550°C is about 33% of its ambient value. For this example, it is clear that carbon steel is unacceptable for service temperatures above about 400°C, and that the use of stainless steel is severely limited. For higher service temperatures, more exotic (and expensive) alloys are required and/or equipment may have to be refractory lined.

A decision to operate above 400°C must be justified.

Thus, if we specify higher temperatures, we must be able to justify the economic penalty associated with more complicated processing equipment, such as refractory-lined vessels or exotic materials of construction. In addition to the critical temperature of 400°C, there are temperature limits associated with the availability of common utilities for heating and cooling a process stream.

> Steam: High pressure steam between 40 and 50 bar is commonly available and provides heat at 250 to 265°C. Above this temperature additional costs are involved.
>
> Water: Water from a cooling tower is commonly available at about 30°C (and is returned to the cooling tower at around 40°C). For utilities below this temperature, costs increase due to refrigeration. As the temperature decreases, the costs increase dramatically (see Table 3.4).

If cryogenic conditions are necessary, there may be an additional need for expensive materials of construction.

A decision to operate outside the range of 40°C to 260°C, thus requiring special heating/cooling media, must be justified.

8.2   REASONS FOR OPERATING AT CONDITIONS OF SPECIAL CONCERN

When you review the PFD for different processes, you are likely to encounter conditions in reactors and separators that lie outside the temperature and pressure ranges presented in Section 8.1. This does not mean to say that these are "bad" processes, but rather that these conditions had to be used, despite the additional costs involved, to have the process operate effectively. These conditions, outside the favored temperature and pressure ranges, are identified as "conditions of special concern."

When you encounter these conditions, you should seek a rational explanation for their selection. If no explanation can be identified, the condition used may be unnecessary. In this situation, the condition may be changed to a less severe one that provides an economic advantage.

A list of possible justifications for using temperature and pressure conditions outside the ranges given above are identified in Tables 8.1 through 8.3. The material provided in Tables 8.1 to 8.3 is based upon elementary concepts presented in undergraduate texts covering thermodynamics and reactor design.

We describe below the rationale used to justify operating at temperatures that are of special concern presented in Table 8.1. The justification for entries in Table 8.2 and 8.3 have similar rationale.

For chemical reactors, we considered the following items in our justification:

1. **Favorable equilibrium conversion:** If the reaction is endothermic and approaches equilibrium, it benefits from operating at high temperatures. We recall Le Chatelier's principle, which states "for a reacting system at equilibrium, the extent of the reaction will change so as to oppose any changes in temperature or pressure." For an endothermic reaction, an increase in temperature tends to push the reaction equilibrium to the right (toward products). Conversely, low temperatures decrease the equilibrium conversion.

2. **Increase reaction rates:** All chemical reaction rates are strongly dependent upon temperature through an Arrhenius type equation:

$$k_{reaction} = k_0 e^{-\frac{E_{act}}{RT}} \qquad (8.1)$$

As temperature increases, so does the reaction rate constant, $k_{reaction}$, for both catalytic and non-catalytic reactions. Therefore, temperatures above 250°C may be required to obtain a high enough reaction rate in order to keep the size of the reaction vessel reasonable.

3. **Maintain a gas phase:** Many catalytic chemical reactions used in processes today require both reactants and products to be in the gas phase. For high boiling point materials or operations where high pressure is used, a temperature in excess of 400°C may be required in the reactor in order to maintain all the species in the vapor phase.

4. **Improve selectivity:** If competing reactions (series, parallel or a combination of both) occur and the different reactions have different activation energies then the production of the desired product may be favored by using a high temperature. Schemes for competing reactions are covered in greater detail in many of the well-known texts on chemical reaction engineering and as well as in Chapter 14.

Table 8.1   Possible Reasons for Operating Reactors and Separators Outside the Temperature Ranges of Special Concern

Stream Condition	Process Justification for Operating at This Condition	Penalty for Operating at This Condition
High Temperature [T > 250°C]	**REACTORS** (i) Favorable equilibrium conversion for endothermic reactions (ii) Increase reaction rates (iii) Maintain a gas phase (iv) Improve selectivity (v) (vi) **SEPARATORS** (i) Obtain a gas phase required for vapor-liquid equilibrium (ii) (iii)	(i) Use of special process heaters (ii) T > 400°C requires special materials of construction (iii) (iv) (v)
Low Temperature [T < 40°C]	**REACTORS** (i) Favorable equilibrium conversion for exothermic reactions (ii) Temperature sensitive materials (iii) Improve selectivity (iv) Maintain a liquid phase (v) (vi) **SEPARATORS** (i) Obtain a liquid phase required for vapor-liquid or liquid-liquid equilibrium (ii) Obtain a solid phase for crystal-lization (iii) Temperature sensitive materials (iv) (v)	(i) Uses expensive refrigerant (ii) May require special materials of construction for very low temperatures (iii) (iv) (v)

Table 8.2   Possible Reasons for Operating Reactors and Separators Outside the Pressure Range of Special Concern

Stream Condition	Process Justification for Operating at This Condition	Penalty for Operating at This Condition
High Pressure (P > 10 bar)	**REACTORS** (i) Favorable equilibrium conversion (ii) Increase reaction rates for gas phase reactions (due to higher concentration) (iii) Maintain a liquid phase (iv) (v) **SEPARATORS** (i) Obtain a liquid phase for vapor-liquid or liquid-liquid equilibrium (ii) (iii)	(i) Requires thicker-walled equipment (ii) Requires expensive compressors if gas streams must be compressed (iii) (iv) (v)
Low Pressure (P < 1 bar)	**REACTORS** (i) Favorable equilibrium conversion (ii) Maintain a gas phase (iii) (iv) **SEPARATORS** (i) Obtain a gas phase for vapor-liquid equilibrium (ii) Temperature-sensitive materials (iii) (iv)	(i) Requires large equipment (ii) Special design for vacuum operation (iii) Air leaks into equipment that may be dangerous and expensive to prevent (iv) (v) (vi)

Table 8.3   Possible Reasons for Non-Stoichiometric Reactor Feed Compositions of Special Concern

Stream Condition	Process Justification for Operating at This Condition	Penalty for Operating at This Condition
Inert Material in Feed to Reactor	(i) Acts as a diluent to control the rate of reaction and/or to ensure that the reaction mixture is outside the explosive limits (exothermic reactions) (ii) Inhibits unwanted side reactions (iii) (iv)	(i) Causes reactor and downstream equipment to be larger since inert takes up space (ii) Requires separation equipment to remove inert material (iii) May cause side reactions (material is no longer inert) (iv) Decreases equilibrium conversion (v) (vi)
Excess Reactant	(i) Increases the equilibrium conversion of the limiting reactant. (ii) Inhibits unwanted side reactions. (iii) (iv)	(i) Requires separation equipment to remove excess reactant. (ii) Requires recycle. (iii) Added feed material costs (due to losses in separation and/or no recycle) (iv) (v)
Product present in feed to reactor	(i) Product cannot easily be separated from recycled feed material. (ii) Recycled product retards the formation of unwanted by-products formed from side reactions. (iii) Product acts as a diluent to control the rate of reaction and/or to ensure that the reaction mixture is outside the explosive limits, for exothermic reactions. (iv) (v)	(i) Causes reactor and downstream equipment to be larger (ii) Requires larger recycle loop (iii) Decreases equilibrium conversion. (iv) (v)

For separators, we considered the following item of justification:

> Obtain a vapor phase for vapor-liquid equilibrium: This situation arises quite frequently when high boiling point materials need to be distilled. An example is the distillation of crude oil in which the bottom of the atmospheric column is typically operated in the region of 310°C to 340°C (590°F to 645°F).

You would benefit by spending time to acquaint yourself with the information presented in Tables 8.1 to 8.3 and to convince yourself that you understand the justifications given in these tables. These tables should not be considered an exhaustive list of possible reasons for operating in the ranges of special concern. Instead, they represent a starting point in analyzing process conditions. You should add your own explanations as you discover them.

8.3 CONDITIONS OF SPECIAL CONCERN FOR THE OPERATION OF OTHER EQUIPMENT

Additional equipment (such as pumps, compressors, heaters, exchangers, and valves) produce the temperature and pressure required by the feed streams entering the reactor and separation sections. When initially choosing the stream conditions for the reactor and separator sections, it is worthwhile using certain guidelines or heuristics. These technical heuristics are useful guidelines for doing design. Comprehensive lists of heuristics are described and applied in the next chapter. In this chapter, we present some of the more general guidelines that apply to streams passing through process equipment. These are presented in Table 8.4. Some of these guidelines are explored in Example 8.2 below.

Example 8.2

It is necessary to provide a nitrogen stream at 80°C and a pressure of 6 bar. The source of the nitrogen is at 200°C and 1.2 bar. Determine the work and cooling duty required for three alternatives.

    a. Compress in a single compression stage and cool the compressed gas.

    b. Cool the feed gas to 80°C and then repeat Part a, above.

    c. Repeat Part b, above, except use two stages of compression with an intercooler.

    d. Identify any conditions of special concern that occur.

Nitrogen can be treated as an ideal diatomic gas for this comparison. Use as a basis 1 kmol of nitrogen and assume that the efficiency, ε, of each stage of compression is 70%.

    For ideal diatomic gas: $C_p = 3.5R$, $C_v = 2.5R$, $\gamma = C_p/C_v = 1.4$, $R = 8.314$ kJ/kmol K, $\varepsilon = 0.70$

    Equations used: $q = C_p\Delta T$, $w = RT_{in}\gamma/(\gamma - 1)[(P_{out}/P_{in})^{(\gamma-1)/\gamma} - 1]/\varepsilon$,

$$T_{out} = T_{in}(1 + 1/\varepsilon)[(P_{out}/P_{in})^{(\gamma-1)/\gamma} - 1]$$

    Figure E8.2 gives the process flow diagrams for the three alternatives and identifies stream numbers and utilities.

    The results of the calculations for Parts a, b, and c are provided in Table E8.2, which shows stream conditions and utility requirements. To keep the calculations simple, the pressure drops across and between equipment have been ignored.

Table 8.4   Changes in Process Conditions that are of Special Concern for a Stream Passing through a Single Piece of Equipment

Type of Equipment	Change in Stream Condition Causing Concern	Justification or Remedy	Penalty for Operating Equipment in this Manner
1. Compressors	$P_{out}/P_{in} > 3$	**Remedy:** Use multiple stages and inter-coolers	High theoretical work requirement due to large temperature rise of gas stream
	High temperature inlet gas	**Remedy:** Cool the gas before compression.	High theoretical work requirement and special construction materials required
2. Heat Exchangers	$\Delta T_{ln} > 100°C$	**Remedy:** Integrate heat better within process (see Chapter 19)	Large temperature driving force means we are wasting valuable high temperature energy
		Justification: Heat integration not possible or not profitable	
3. Process Heaters	$T_{out} < T_{steam\ available}$	**Remedy:** Use high-pressure steam to heat process stream	Process heaters are expensive and unnecessary if heating may be accomplished by using an available utility
		Justification: Heater may be needed during start-up	
4. Valves	Large ΔP across valve	**Remedy:** For gas streams install a turbine to recover lost work	Wasteful expenditure of energy due to throttling
		Justification: (a) Valve used for control purposes (b) Installation of turbine not profitable (c) Liquid is being throttled	
5. Mixers (streams mixing)	Streams of greatly differing temperatures mix	**Remedy:** Bring temperatures of streams closer together using heat integration	Wasteful expenditure of high temperature energy
	Streams of greatly differing composition mix	**Justification:** (a) Quenching of reaction products (b) Provides driving force for mass transfer	Causes extra separation equipment and cost

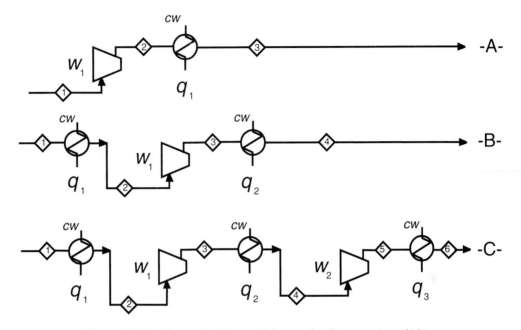

Figure E8.2   Alternative Process Schemes for Compression of Nitrogen.

Part d: Alternative -A- requires a compressor exit temperature of 595°C that is a condition of special concern. Note also that although the intermediate temperature of the gas (stream) in Alternative -B- was 374°C, since this stream is to be cooled there are no concerns about utility requirements.

Example 8.2 showed three alternatives of differing complexity for achieving the same final conditions. The amount of work (w) and cooling utilities (q) required for each alternative were calculated. Based solely upon process complexity, Alternative -A- is the most desirable. However, this alternative has several disincentives that should be considered before final selection. Namely, Alternative -A- has:

1. The highest electric utility demand and cost (assuming that the compressor is electrically driven).
2. The highest cooling utility demand and cost.
3. A "condition of special concern," i.e., T > 400°C (see Table 8.1).
   Note: Compressors are high-speed rotating devices where the loss of material strength and thermal expansion is critical. It would be expected that the purchased cost of the compressor would undergo a quantum jump for high-temperature operations.
4. Exceeds the 3/1 pressure ratio provided as a guideline (see Table 8.4).

Table E8.2   Flow Summary Table for Example 8.2 and Figure E8.2

Stream No. in Figure E8.2	System -A-		System -B-		System -C-	
	$T(°C)$	$P(bar)$	$T(°C)$	$P(bar)$	$T(°C)$	$P(bar)$
1	200	1.2	200	1.2	200	1.2
2	595	6.0	80	1.2	80	1.2
3	80	6.0	374	6.0	210	2.68
4	—	—	80	6.0	80	2.68
5	—	—	—	—	210	6.0
6	—	—	—	—	80	6.0
Work: kJ/kmol						
w_1	11,470		8,560		3,780	
w_2	—		—		3,780	
w_{total}	11,470		8,560		7,560	
Heat: kJ/kmol						
q_1	14,970		3,490		3,490	
q_2	—		8,550		3,780	
q_3	—		—		3,780	
q_{total}	14,970		12,040		11,050	

Alternative -B- is more complex than Alternative -A- because it requires an additional heat exchanger, but it avoids the condition of special concern in Item 3, above. The result of using this extra exchanger is a significant decrease in utilities over Alternative -A-. As a result, it is likely that Alternative -B- would be preferred to Alternative -A-.

Alternative -C- requires an extra stage of compression and an additional cooler before the second compressor that is not required by Alternative -B-. However, Alternative -C- results in an additional savings in utilities over Alternative -B-.

The qualitative analysis given above suggests that both Alternatives -B- and -C- are superior to Alternative -A-. This conclusion is consistent with the two heuristics for compressors in Table 8.4: Namely, it is better to cool a hot gas prior to compressing it and it is usually desirable to keep the compression ratio below 3:1. Before a final selection is made, an economic analysis, which must include both the capital investment and the operating costs, should be carried out on each of the competing schemes. The Equivalent Annual Operating Cost (EAOC), described in Chapter 5, would be a suitable criterion to make such a comparison.

You should review the information given in Table 8.4 and convince yourself that you understand the rules, along with the penalties, remedies, and justifications for operating equipment under these conditions. You may also be able to provide additional reasons why operating the equipment in this way would be justified. You should add these reasons to the list provided along with additional heuristics that are uncovered as you work problems and gain experience.

8.4 ANALYSIS OF IMPORTANT PROCESS CONDITIONS

In this section, we begin to analyze and to justify the conditions of special concern found in a process flow diagram. To help with this analysis, it is beneficial to prepare a Process Conditions Matrix (PCM). In the PCM, all the equipment is listed vertically and the conditions of special concern are listed horizontally. Each unit is reviewed for conditions of special concern, and a check mark is used to identify which pieces of equipment have been identified. The PCM for the toluene hydrodealkylation process is shown in Table 8.5. The information for this PCM was obtained from Chapter 1 and you should verify that none of the areas of special concern have been missed.

We will now consider and justify all of the special conditions identified in Table 8.5.

8.4.1 Evaluation of Reactor R-101

Three conditions of concern have been identified for the reactor. They are high temperature, high pressure, and non-stoichiometric feed conditions.

In order to understand why these conditions are needed, additional information about the toluene dealkylation reaction is required. Table 8.6 provides additional but limited information. This information is divided into two groups.

Thermodynamic Information: This is information found in most chemical engineering thermodynamic textbooks. It includes:
a. information required to perform energy balances: This includes heats of reaction and phase change, heat capacities, and so on.
b. information required to determine equilibrium conversion: This includes heats of formation, free energy of formation, and so on.

Figure 8.1 is a plot of the heat of reaction and the equilibrium constant as a function of temperature, evaluated from the information provided in Table 8.6. From these plots it is evident that the chemical reaction is slightly exothermic, causing the equilibrium constant to decrease with temperature.

Table 8.5   Process Conditions Matrix for the PFD of the Toluene Hydrodealklyation Process Shown in Figure 1.5

| Equipment | Reactors and Separators Tables 8.1–8.3 | | | | | Other Equipment Table 8.4 | | | | |
	High Temp	Low Temp	High Pres.	Low Pres.	Non-Stoich. Feed	Comp	Exch.	Htr.	Valve	Mix
R-101	X		X		X					
V-101			X							
V-102										
V-103										
T-101										
H-101										
E-101							X			
E-102							X			
E-103										
E-104										
E-105										
E-106										
C-101										
P-101										
P-102										
TK-101										
PCV on Stream 8									X	
PCV on Stream from V-101 to V-103									X	

Reaction Kinetics Information: This information is reaction specific and must be obtained experimentally. The overall kinetics may involve homogeneous and heterogeneous reactions both catalytic and non-catalytic. The expressions are often complex.

Before a process is commercialized, reaction kinetics information, such as space velocity and residence times, must be obtained for different temperatures and pressures from pilot plant studies. Such data are necessary to deign the reactor. At this point, we are not interested in the reactor design, and hence specific

Table 8.6   Equilibrium and Reaction Kinetics Data for the Toluene Hydrodealkylation Process

Reaction Stoichiometry

$$C_6H_5CH_3 + H_2 = C_6H_6 + CH_4$$

toluene   benzene

Equilibrium Constant (T is in units of K)

$$\ln(K_p) = 57.18 + \frac{21{,}080}{T} - 8.677\ln(T) + 7.319 \times 10^{-4}T + 5.389 \times 10^{-8}T^2 + \frac{12{,}620}{T^2}$$

Heat of Reaction

$$\Delta H_{reaction} = -37{,}190 - 17.24T + 29.09 \times 10^{-4}T^2 + 0.6939 \times 10^{-6}T^3 + \frac{50{,}160}{T} \quad \frac{kJ}{kmol}$$

At the Reaction Conditions of 600°C (873 K)

Equilibrium Constant, $K_p = 262$

Heat of Reaction, $\Delta H_{reaction} = -49{,}500 \quad \dfrac{kJ}{kmol}$

Information on Reaction Kinetics

No side reactions
Reaction is kinetically controlled

kinetics expressions have not been included in Table 8.6 and are not necessary for the following analysis.

The analysis of the reactor takes place in two parts.

a. Evaluation of the special conditions from the thermodynamic point of view. This assumes that chemical equilibrium is reached and provides a limiting case.

b. Evaluation of the special conditions from the kinetics point of view. This accounts for the limitations imposed by reaction kinetics, mass transfer, and heat transfer.

If a process is unattractive under equilibrium (thermodynamic) conditions, analysis of the kinetics is not necessary. For processes in which the equilibrium conditions give favorable results, further study is necessary. The reason for this is that conditions that favor high equilibrium conversion may be unfavorable from the standpoint of reaction kinetics.

Thermodynamic Considerations. We consider the use of high-temperature, high-pressure and non-stoichiometric feed conditions separately.

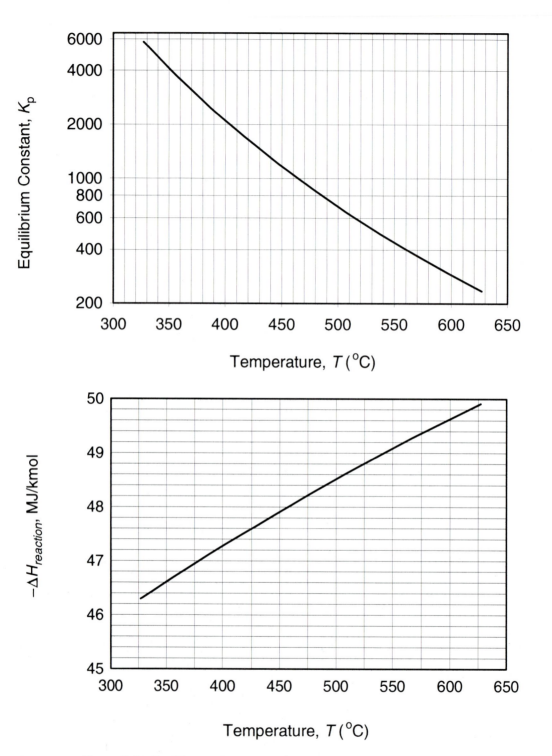

Figure 8.1 Equilibrium Constant and Heat of Reaction as a Function of Temperature for the Toluene Hydrodealkylation Reaction

High-Temperature Concern (see Table 8.1).   Figure 8.1 provided the important information that the reaction is exothermic. Table 8.1 notes that for an exothermic reaction the result of increasing temperature is a reduction in equilibrium conversion. This is confirmed by the plot of the equilibrium constant vs. temperature given in Figure 8.1. The decrease in the equilibrium conversion is undesirable.

Example 8.3

For the PFD presented in Figure 1.5:

    a. Calculate the actual conversion
    b. Evaluate the equilibrium conversion at 600°C.

    Assuming ideal gas behavior: $K_p = (N_{benzene} N_{methane}) / (N_{toluene} N_{hydrogen})$
    where N represents the moles of each species at equilibrium

    Information on the feed stream to the reactor from Table 1.5 (Stream 6 on Figure 1.5).

Hydrogen	735.4 kmol/h
Methane	317.3
Benzene	7.6
Toluene	144.0
Total	1204.3

    a. Actual Conversion: Toluene in exit stream (Stream 9) = 36 kmol/h

    Conversion = $(144 – 36)/144 = 0.75$ (75%)

    b. Equilibrium Conversion at 600°C. From Table 8.6 @600°C K_p = 262

    Let N = kmol/h of benzene formed
    $262 = [(N + 7.6)(N + 317.3)]/[(735.4 – N)(144 – N)]$
    $N = 143.5$
    Equilibrium Conversion = $143.5/144 = 0.997$ (99.7%)

The equilibrium conversion for the hydrodealkylation reaction remained high in spite of the high temperature. Although, there is no real problem with using the elevated temperature in the reactor, it cannot be justified from a thermodynamic point of view.

High Pressure Concern (see Table 8.2).   From the reaction stoichiometry, we see that there are equal numbers of reactant and product moles in the hydrodealkylation reaction. For this case, there is no effect of pressure on equilibrium conversion. From a thermodynamic point of view there is no reason for the high pressure in the reactor.

Non-Stoichiometric Feed (see Table 8.3):   The component feed rates to the reactor (see Example 8.3) show that:

1. toluene is the limiting reactant.
2. hydrogen is an excess reactant (over 400% excess).
3. methane, a reaction product, is present in significant amounts.

Reaction Products (Methane) in Feed. The presence of reaction product in the feed results in a reduction in the equilibrium conversion (see Table 8.3). However, Example 8.3 shows that at the conditions selected for the reactor, the equilibrium conversion remained high despite the presence of the methane in the feed.

Excess Reactant (Hydrogen) in Feed. The presence of excess reactants in the feed results in an increase in equilibrium conversion (see Table 8.3). Example 8.4 explores the effect of this excess hydrogen on conversion.

Example 8.4

(Reference Example 8.3) Reduce the amount of hydrogen in the feed to the reactor to the stoichiometric amount, i.e., 144 kmol/h, and determine the effect on the equilibrium conversion at 600°C.

The calculations are not shown. They are similar to those in Example 8.3(b). The total moles of hydrogen in the feed were changed from 790.6 kmol/h to the stoichiometric value of 144 kmol/h.

The results obtained were: N = 128.8 kmol/h, Equilibrium Conversion = 0.894 (89.4%)

Example 8.4 reveals that the presence of the large excess of hydrogen had a noticeable effect on the equilibrium conversion.

We conclude that thermodynamic considerations do not explain the selection of the high temperature, high pressure, and the presence of reaction products in the feed. The presence of a large excess of hydrogen is the only positive effect predicted by thermodynamics.

Consideration of Reaction Kinetics. The information on reaction kinetics is limited in this chapter. We will present a more detailed description of the kinetics rate expression in a case study in Chapter 14, and investigate the predictions made in this chapter with this limited information. However, you will find that a great deal of understanding can be extracted from the limited information presented here.

From the information provided in Table 8.6 and Chapter 1 we know that:

1. The reaction takes place in the gas phase.
2. The reaction is kinetically controlled.
3. There are no side reactions.

High Temperature Concern (see Table 8.1).   In a region where the reaction kinetics control, the reaction rate increases rapidly with temperature.

Example 8.5

The activation energy for the rate of reaction for the hydrodealkylation of toluene is equal to 148.1 kJ/mol (Tarhan [2]). What is the reaction rate at 400°C relative to that at 600°C?

$$\text{Ratio of Reaction Rates} = \exp[-E/R\{1/T_2 - 1/T_1\}]$$
$$= \exp[148100/8.314\{1/673 - 1/873\}] = 430$$

The size of a reactor would increase by nearly three orders of magnitude if the reaction were carried out at 400°C (the critical temperature for materials selection Table 8.1) rather than 600°C. Clearly the effect of temperature is significant.

Most reactions are not kinetically controlled as is the case here. In most cases the rate is controlled by heat or mass transfer considerations. These are not as sensitive to temperature changes as chemical reaction rates. For more detail, see Chapter 14.

High Pressure Concern (see Table 8.2).   For gas phase reactions the concentration of reactants is proportional to the pressure. For a situation where the reaction rate is directly proportional to the concentration, operation at 25 bar rather than at 1 bar would increase the reaction rate by a factor of 25 (assuming ideal gas behavior). While we do not know that the rate is directly proportional to the concentration, we can predict that the effect of pressure is likely to be substantial, and the reactor size will be substantially reduced.

Non-Stoichiometric Feed (see Table 8.3).   The reactor feed contains both excess hydrogen and the reaction product methane.

Methane in the Feed.   The effect of methane is to reduce the reactant concentrations. This decreases the reaction rate and represents a negative impact. The methane could possibly reduce the formation of side products, but you have no information to suggest that this is the case.

Excess Hydrogen in the Feed.   The large amount of excess hydrogen in the feed assures that the concentration of hydrogen will remain large throughout the reactor. This increases the reaction rate. While there is no information provided regarding the decision to maintain the high hydrogen levels, it may be linked to reducing the formation of side products.

With the exception of the presence of methane product in the feed, the high-temperature operation, the excess hydrogen, and the elevated pressure all support an increase in reaction rate and a reduction in reactor volume. This suggests that the catalyst is not "hot," that is, the catalyst is still operating in the reaction controlled regime and mass transfer effects have not started to intrude. For these conditions, the manipulation of temperatures and pressures are essential to limit the reactor size.

There is a significant economic penalty for using over 400% excess hydrogen in the reactor feed. The raw material cost of hydrogen would be reduced significantly if excess hydrogen were not used. The fact that this large excess is used in spite of the economic penalty involved suggests that the hydrogen plays an important role in the prevention of side products. The concept of selectivity is discussed further in Chapter 14.

The presence of methane in the feed has not yet been resolved. At best it behaves as an inert and occupies volume that must be handled downstream of the reactor, thus making all the equipment larger and more expensive. This question is considered in more detail in the next example.

Example 8.6

It has been proposed that we handle the hydrogen/methane stream in the same manner that we handled the toluene/benzene stream. We recall that the unreacted toluene was separated from the benzene product and then recycled. It is proposed that the methane be separated from the hydrogen. The methane would then become a process by-product and the hydrogen would be recycled. Discuss this proposal using the arguments provided in Tables 8.1 and 8.2.

To use distillation for the separation of methane from hydrogen, as was used with the toluene/benzene, requires a liquid phase. For methane/hydrogen systems, this requires extremely high pressures together with cryogenic temperatures.

If the hydrogen could be separated from the methane and recycled, the reactor feed would not contain significant quantities of methane, and the large excess of hydrogen could be maintained without the steep cost of excess hydrogen feed. Note that the overall conversion of hydrogen in the process is only 37%, while for toluene it is 99%.

Alternative separation schemes that do not require a liquid phase (e.g., a membrane separator) should be considered. The use of alternative separation technologies is addressed further in Chapter 19.

8.4.2   Evaluation of High-Pressure Phase Separator V-101

This vessel separates toluene and benzene as a liquid from the non-condensable gases hydrogen and methane. The reactor product is cooled and forms a vapor and a liquid stream that are in equilibrium. The vapor-liquid equilibrium is that at the temperature and pressure of the stream entering V-101. From Tables 8.1 and 8.2, we conclude that the lower temperature (38°C) was provided to obtain a liquid phase for the vapor-liquid equilibrium. The pressure was maintained to support the formation of the liquid phase. Since the separation can be affected relatively easily at high pressure, it is worthwhile maintaining V-101 at this high pressure.

8.4.3   Evaluation of Large Temperature Driving Force in Exchanger E-101

There is a large temperature driving force in this exchanger, since the heating medium is at a temperature of approximately 250°C, and the inlet to the exchanger is only 30°C. This is greater than the 100°C suggested in Table 8.4. This is an example of poor heat integration, and we will take a closer look at improving this in Chapter 19 (also see the case study presented in Chapter 23).

8.4.4   Evaluation of Exchanger E-102

Stream 9 is cooled from 654°C to 40°C using cooling water at approximately 35°C. Again this is greater than the 100°C suggested in Table 8.4, and the process stream has a lot of valuable energy that is being wasted. Again, we can save a lot of money by using heat integration.

8.4.5   Pressure Control Valve on Stream 8

The purpose of this control valve is to reduce the pressure of the stream entering the fuel gas line from 23.9 bar to 2.5 bar. This reduction in pressure represents a potential loss of useful work due to the throttling action of the valve. Referring to Table 8.4, we can see that when we throttle a gas, we can recover work by using a turbine, although this may not be economically attractive. The operation of this valve is justified because of its control function.

8.4.6   Pressure Control Valve on Stream from V-101 to V-103

The purpose of this valve is to reduce the pressure of the liquid leaving V-101. This reduction in pressure causes some additional flashing and recovery of dissolved methane and hydrogen from the toluene/benzene mixture. The flashed gas is separated in V-103 and sent to the fuel gas line. The purpose of this valve is to control the pressure of the material fed to the distillation column T-101. Since the stream passing through the valve is essentially all liquid, little useful work could be recovered from this stream.

This completes our review of the conditions of special concern for the toluene hydrodealkylation process.

8.5   SUMMARY

In this chapter, you learned to identify process conditions that are of special interest or concern in the analysis of the PFD. A series of tables were presented in which justifications for using process conditions of special concern were given. We introduced the Process Conditions Matrix (PCM) for the toluene hydrodealkylation process and identified all the equipment in which process conditions of special concern existed. Finally, by comparing the process conditions from the PFD to those given in the tables, we learned to analyze why these conditions were selected for the process and where improvements may be made.

REFERENCES

1. Walas, S. M., *Chemical Process Equipment: Selection and Design*, Butterworth, Stoneham, MA, 1988.

2. Tarhan, M. O., *Catalytic Reactor Design*, McGraw-Hill, New York, 1983.

PROBLEMS

For Problems 8.1–8.3, refer to Figure B.2 in Appendix B.

1. Acrylic Acid Process—Reaction Section

    Acrylic acid is produced by the catalytic oxidation of propylene with air

    $$C_3H_6 + 1.5\ O_2 \rightarrow C_3H_4O_2 + H_2O$$
    $$\text{acrylic acid}$$

    Along with the main reaction shown above, several additional oxidation reactions occur to yield a variety of by-products, including acetic acid:

    $$C_3H_6 + 2.5\ O_2 \rightarrow C_2H_4O_2 + CO_2 + H_2O$$
    $$\text{acetic acid}$$

    $$C_3H_6 + 4.5\ O_2 \rightarrow 3CO_2 + 3H_2O$$

    The reactor section of a process to produce acrylic acid is shown in Appendix B, Figure B.2. Using the information given on this PFD, answer the following questions:
    a. Is the reaction endothermic or exothermic?
    b. Why is steam fed into the reactor?
    c. Why is the reactor operated at the elevated temperature of 310°C?
    d. What would be the effect of using a higher pressure and temperature in the reactor? Hint: Kinetics information for this reaction sequence is provided in Appendix B.

2. Acrylic Acid Process—Separation Products
    The stream leaving the acrylic acid reactor contains a variety of products, by-products and unused reactants. In order to separate the non-condensable fraction of this stream, the reactor effluent is sent to two absorbers. In the first absorber, T-301, the reactor effluent is quenched by a cool recirculating stream of product. In the second absorber, T-302, the non-condensed portion of the reactor effluent is scrubbed with water into which the remaining condensables are absorbed. The combined stream leaving the bottom of both ab-

sorbers, Stream 9, is then fed to a liquid-liquid extractor tower, T-303, where the non-aqueous portion of the stream is transferred to an organic solvent and then sent for further processing.

    a. In the process illustrated in Figure B.2, the organic components (acrylic and acetic acid) are first absorbed into water and then absorbed into an organic solvent. Explain why this separation sequence is used.

    b. Can you suggest alternatives to this separation sequence? What are the consequences of these alternative separation schemes, that is, what process conditions might have to be changed and what savings or additional expenses might be obtained?

    c. It has been proposed that a portion of the off-gas, Stream 11, be recycled back to the reactor while the amount of steam feed is reduced. Can you foresee any problems or areas of special concern if this were implemented?

3. Acrylic Acid Process— Product Purification

In the purification of the products, several distillation columns are used. First the organic solvent containing the acids, Stream 13, is separated in T-304, and the solvent is recycled back to the liquid-liquid extraction column, T-303. The bottoms product from T-304 contains the organic acids, Stream 14, and is sent to T-305 for final purification. Finally, the aqueous stream from the extractor is sent to T-306, where a small amount of solvent is recovered and recycled to T-303. The conditions used for the purification process are given in Figure B.2 and Table B.4. For the purification section, answer the following questions:

    a. What conditions of special concern are evident from this section of the PFD?

    b. What reasons can you give for justifying the conditions identified in part a, above?

    c. Why is the water leaving T-306 sent to a waste water treatment plant?

4. Separation of Propane from Higher Hydrocarbons—Depropanizer

A depropanizer is a column that separates nearly pure propane from a mixture of propane, butanes, and other straight chain hydrocarbons. This column is typically operated at a pressure of approximately 220 psig (16.2 bar). What justification can you give for using this high pressure condition of special concern? Try to quantify your reasoning.

5. Vacuum Distillation of Crude Oil Bottoms

The front end of a typical crude oil refinery is illustrated in the process block flow diagram shown in Figure P8.5. Crude oil, after desalting, is sent to the main fractionating column which operates at a pressure of between 1.5 and 2.0 bar. Many side streams are taken from the column, and each side stream or "cut" has a characteristic boiling point range. As we move toward the bottom of the column the cuts become heavier and the corresponding boiling points of the streams become higher (these are shown on the diagram). The material that leaves the bottom of the main fractionator is called atmospheric residuum. This stream is further treated in a second column that operates at a

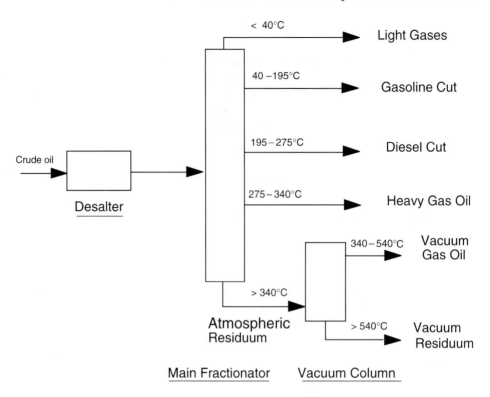

Note : Temperatures indicate the approximate range of boiling points for the stream

Figure P8.5 Process Block Flow Diagram for the Front End of a Typical Oil Refinery

relatively high vacuum (top pressure may be as low as 20mm of Hg or 0.027 bar). Why does this second column operate at such a low pressure?

For Problems 8.6–8.8, refer to Figure B.1 in Appendix B.

6. Areas of Special Concern in DME Process

The production of dimethyl ether (DME) is achieved by the catalytic dehydration of methanol, via the following reaction:

$$2 \, CH_3OH \rightarrow (CH_3)_2O + H_2O$$

methanolDME

A PFD for this process is given in Figure B.1. Using the information shown on the PFD and given in Table B.2, identify all the areas of special concern. Try to justify these conditions.

7. Pressure Profile in the Purification Sequence
   The pressures at which the two distillation columns (T-201 and T-201) operate are approximately 10.4 and 7.4 bar, respectively.
   a. What changes would be required if it were decided to operate both columns at 1.1 bar?
   b. What changes would be required if it were decided to operate both columns at 10.4 bar?

8. Process Alternatives for DME Process
   a. The methanol recycle stream could be recycled as a vapor if we operated E-207 as a partial condenser. What are the advantages and disadvantages of making this change?
   b. During the initial start-up of the process, a flaw was found in the design shown in Figure B.1. Can you identify this flaw and suggest how it may be overcome? Hint: Look carefully at E-202 during the start-up sequence!

9. For the production of acetone from isopropyl alcohol shown in Appendix B, Figure B.3, do the following.
   a. Construct a process conditions matrix (PCM) for the process in order to identify any conditions of special concern.
   b. For each condition of special concern identified in part a, suggest at least one reason why such a condition was used.
   c. For each condition of special concern identified in part a, suggest at least one process alternative to eliminate the condition.

10. It was suggested in Section 8.1 that operation above 250°C requires the use of an expensive fired heater. In addition, there are concerns about the materials of construction that can be used at temperatures greater than 400°C. It has been suggested that a simple remedy to this problem is to produce higher temperature steam, for example, at 300°C or 350°C.
   What do you think of this suggestion? Try to quantify your reasoning.
   Hint: Look at the steam tables!!

9

Utilizing Experience-Based Principles to Confirm the Suitability of a Process Design

Experienced chemical engineers possess the skills necessary to perform detailed and accurate calculations for the design, analysis, and operation of equipment and chemical processes. In addition, these engineers will have formulated a number of experienced-based short-cut calculation methods and guidelines that are useful for:

1. Checking new process designs
2. Providing equipment size and performance estimates
3. Helping to troubleshoot problems with operating systems
4. Verifying the reasonableness of results of computer calculations and simulations
5. Providing reasonable initial values for the input to a process simulator required to achieve program convergence
6. Obtaining approximate costs for process units
7. Developing preliminary process layouts

These short-cut methods are forms of heuristics that are helpful to the practicing engineer. All heuristics are, in the final analysis, fallible and sometimes difficult to justify. They are merely plausible aids or directions toward the solution of a problem [1]. Especially for the heuristics described in this chapter, we need to keep in mind the four characteristics of any heuristic:

1. A heuristic does not guarantee a solution.
2. It may contradict other heuristics.

3. It can reduce the time to solve a problem.

4. Its acceptance depends on the immediate context instead of on an absolute standard.

The fact that one cannot precisely follow all heuristics all the time is to be expected, as it is with any set of technical heuristics. However, despite the limitations of heuristics, they are nevertheless valuable guides for the process engineer.

In Chapter 8, you analyzed process units and stream conditions that were identified as areas of special concern. These areas were highlighted in a series of informational tables. In this chapter, you will complete the analysis of chemical processes by checking the equipment parameters and stream conditions in the PFD for agreement with observations and experiences in similar applications.

The required information to start an analysis is provided in a series of "informational tables" containing short-cut calculation techniques. In this chapter, we demonstrate the use of these resources by checking the conditions given in the basic toluene hydrodealkylation PFD.

9.1   THE ROLE OF EXPERIENCE IN THE DESIGN PROCESS

The short narrative given below illustrates a situation that could be encountered early in your career as an engineer.

> *You are given an assignment that involves writing a report that is to be completed and presented in two weeks. You work diligently and feel confident you have come up with a respectable solution. You present the written report personally to your director (boss), who asks you to summarize only your final conclusions. Immediately after providing this information, your boss declares that "your results must be wrong" and returns your report unopened and unread.*
>
> *You return to your desk angry. Your comprehensive and well-written report was not even opened and read. Your boss did not tell you what was wrong, and you did not receive any "partial credit" for all your work. After a while, you cool off and review you report. You find that you had made a "simple" error, causing your answer to be off by an order of magnitude. You correct the error and turn in a revised report.*
>
> *What remains is the nagging question, "How could your boss know you made an error without having reviewed your report or asking any questions?"*

The answer to this nagging question is probably a direct result of your director's experience with a similar problem or knowledge of some guideline that contradicted your answer. The ability of your boss to transfer personal experience to new situations is one reason why he or she was promoted to that position.

> **It is important to be able to apply knowledge gained through experience to future problems.**

9.1.1   Introduction to Technical Heuristics and Short-Cut Methods

A heuristic is a statement concerning equipment sizes, operating conditions, and equipment performance that reduces the need for calculations. A *short-cut method* replaces the need for extensive calculations in order to evaluate equipment sizes, operating conditions, and equipment performance. These are aptly referred to as "back-of-the-envelope calculations." In this text we refer to both these experienced-based tools as guidelines or heuristics.

The guidelines provided in this chapter are limited to materials specifically covered in this text (including problems at the end of the chapters). All such material is likely to be familiar to final year B.S. chemical engineering students and new graduates as a result of their education. Upon entering the work force, engineers will develop guidelines that apply specifically to their area of responsibility.

Guidelines and heuristics must be applied with an understanding of their limitations. In most cases, a novice chemical engineer should have sufficient background to apply the rules provided in this text.

The narrative started earlier is now revisited. The assignment remains the same; however, the approach to solving the problem changes.

> *Before submitting your report, you applied a heuristic that highlighted an inconsistency in your initial results. You then reviewed your calculations, found the error, and made corrections before submitting your report. Consider two possible responses to this report.*
>
> 1. *Your boss accepts the report and notes that the report appears to be excellent and looks forward to reading it.*
> 2. *Your boss expresses concern and the returns report as before. In this case, you have available a reasoned response. You point out and show that your solution is consistent with the heuristic you used to check your work. With this supporting evidence your boss would have to rethink his or her response and provide you with an explanation regarding his or her concern.*
>
> *In either case, your work will have made a good impression.*

Guidelines and heuristics are frequently used to make quick estimates during meetings and conferences and are valuable in refreshing one's memory with important information.

9.1.2   Maximizing the Benefits Obtained from Experience

No printed article, lecture, or text is a substitute for the perceptions that result from experience. An engineer must be capable of transferring knowledge gained from one or more experiences to resolve future problems successfully.

To benefit fully from experience, it is important to make a conscious effort to use each new experience to build a foundation upon which to increase your ability to handle and to solve new problems.

> **An experienced engineer retains a body of information, made up largely of heuristics and short-cut calculation methods, that is available to help solve new problems.**

The process by which an engineer uses information and creates new heuristics consists of three steps. These three steps are **Predict**, **Authenticate**, and **Re-evaluate** and they form the basis of the **PAR** Process. The elements of this process are presented in Table 9.1. The following example illustrates the steps used in the PAR process.

Example 9.1

Evaluate the heat transfer coefficient for water at 93°C (200°F) flowing at 3.05 m/s (10 ft/s) inside a 38mm (1.5″) diameter tube. From previous experience, you know that the heat transfer coefficient for water, at 21°C (70°F) and 1.83 m/s (6 ft/s), in these tubes is

---

Table 9.1:   PAR Process to Maximize Benefits of Experience—(P)redict, (A)uthenticate, (R)e-evaluate

1. **Predict:** This is a precondition of the PAR process. It represents your "best prediction" of the solution. It often involves making assumptions and applying heuristics based on experience. Calculations should be limited to "back-of-the-envelope" or short-cut techniques.

2. **Authenticate/Analyze:** In this step you seek out equations and relationships, do research relative to the problem, and perform the calculations that lead toward a solution. The ability to carry out this activity provides a necessary but not sufficient condition to be an engineer. When possible, information from actual operations are included in order to achieve "the best possible solution."

3. **Re-evaluate/Rethink:** The "best possible solution" from Step 2 is compared to the predicted solution in Step 1. When the prediction is not acceptable it is necessary to correct the reasoning that lead to the poor prediction. It becomes necessary to remove, revise, and replace assumptions made in Step 1. This is the critical step in learning from experience.

5250 W/m²°C. Follow the PAR process to establish the heat transfer coefficient at the new conditions.

Step 1—Predict:
Assume that the velocity and temperature have no effect
Predicted Heat Transfer Coefficient = 5250 W/m²°C.

Step 2—Authenticate/Analyze:

Using the properties given below we find that the Reynolds Number for the water in the tubes is

$$Re = u\rho D_{pipe}/\mu = (1.83)(997.4)(1.5)(0.0254)/(9.8 \times 10^{-4}) = 71 \times 10^3 \rightarrow \text{Turbulent Flow}$$

Use the Sieder-Tate Equation [2] to check the prediction

$$hD/k = (0.023)(Du\rho/\mu)^{0.8}(C_p\mu/k)^{1/3} \tag{9.1}$$

Property	21°C (70°F)	93°C (200°F)	Ratio of (new/old)
ρ (kg/m³)	997.4	963.2	0.966
k (W/m°C)	0.604	0.678	1.12
C_p (kJ/kg°C)	4.19	4.20	1.00
μ (kg/m.s)	9.8×10^{-4}	3.06×10^{-4}	0.312

Take the ratio of Eq. (9.1) for the two conditions given above, rearrange and substitute numerical values. Using ' to identify the new condition at 93°C, we get:

$$h'/h = (D/D)^{0.2}(u'/u)^{0.8}(\rho'/\rho)^{0.8}(\mu/\mu')^{0.47}(C_p'/C_p)^{0.33}(k'/k)^{0.67} \tag{9.2}$$

$$= (1)(1.50)(0.973)(1.73)(1.00)(1.08) = 2.725 \tag{9.3}$$

$$h' = (2.725)(5250) \text{ W/m}^2°C = 14300 \text{ W/m}^2°C$$

The initial assumption that the velocity and temperature do not have a significant effect is incorrect. Eq. (9.3) reveals a velocity effect of a factor of 1.5 and a viscosity effect of a factor of 1.73. All other factors are close to 1.0.

Step 3—(R)e-evaluate/Rethink: The original assumptions that velocity and temperature had no effect on the heat transfer coefficient have been rejected. Improved assumptions for future predictions are:

1. The temperature effect on viscosity must be evaluated.
2. The effects of temperature on C_p, ρ, and k are negligible.
3. Pipe diameter has a small effect on h (all other things being equal).
4. Results are limited to the range where the Sieder-Tate equation is valid.

With these assumptions, the values for water at 21°C are substituted into Eq. (9.2). This creates a useful heuristic for evaluating the heat transfer coefficients for water.

$$h'(\text{W/m}^2°C) = 125u'^{0.8}/\mu'^{0.47} \text{ for } u'(\text{m/s}), \mu' \text{ (kg/m s)}$$

While it takes longer to obtain a solution when you start to apply the PAR process, the development of the heuristic and the addition of a more in-depth understanding of the factors that are important, offers substantial long term advantages.

There are hundreds of heuristics covering areas in chemical engineering, some general and some specific to a given application, process or material. In the next section, we have gathered a number of these rules that you can use to make predictions to start the PAR analysis.

9.2   PRESENTATION OF TABLES OF TECHNICAL HEURISTICS AND GUIDELINES

We provide a number of these guidelines for you in this section. The information given is limited to operations most frequently encountered in this text. Most of the information was extracted from a collection that is presented in Walas [3]. In addition, this excellent reference also includes additional guidelines for the following equipment:

1. Conveyors for Particulate Solids
2. Cooling Towers
3. Crystallization from Solution
4. Disintegration
5. Drying of Solids
6. Evaporators
7. Size Separation of Particles

The heuristics or rules are contained in a number of tables and apply to operating conditions that are most often encountered. The information provided is used in the following example and should be used to work problems at the end of the chapter and to check information on any PFD.

Example 9.2

Refer to the information given in Chapter 1 for the toluene hydrodealkylation process, namely Figure 1.7 and Tables 1.5 and 1.7. Using the information provided in the tables in this chapter, estimate the size of the equipment and other operating parameters for the following units:

a. V-101
b. E-105
c. P-101
d. C-101

  e. T-101

  f. H-101

Compare your findings with the information given in Chapter 1.

a. V-101—High-Pressure Phase Separator

From Table 9.6, we use the following heuristics

Rule 3 → Vertical Vessel
Rule 4 → L/D between 2.5 and 5 with optimum at 3.0
Rule 5 → liquid hold-up time is 5 min based on 1/2 volume of vessel
Rule 9 → Gas velocity u is given by

$$u = k\sqrt{\frac{\rho_l}{\rho_v} - 1}\ m/s$$

where $k = 0.0305$ for vessels without mesh entrainers

Rule 12 → Good performance obtained at 30–100% of u from Rule 9; typical value is
            75%

From Table 1.5, we have
Vapor flow = Stream 8 = 9200 kg/h, P = 23.9 bar, T = 38°C
Liquid flow = Stream 17 + 18 = 11570 kg/h, P = 2.8 bar, T = 38°C
ρ_v = 8 kg/m^3 and ρ_l = 850 kg/m^3 (estimated from Table 1.7)

From Rule 9, we get $u = 0.0305[850/8 - 1]^{0.5} = 0.313$ m/s
Use $u_{act} = (0.75)(0.313) = 0.23\ m/s$
Now mass flowrate of vapor $= u\rho_v\pi D^2/4 = 9200/3600 = 2.56$ kg/s
Solving for D, we get $D = 1.33$ m

From Rule 5, we have volume of liquid $= 0.5\ L\pi D^2/4 = 0.726L$ m^3
5 minutes of liquid flow $= (5)(60)(11570)/850/3600 = 1.13$ m^3
Equating the two results above, we get $L = 1.56$ m

From Rule 4, we have L/D should be in range 2.5 to 5. For our case $L/D = 1.56/1.33 =$
            1.17
Since this is out of range we should change to $L = 2.5D = 3.3$ m
Heuristics from Table 9.6 suggest that V-101 should be a vertical vessel with $D =$
1.33 m, $L = 3.3$ m

From Table 1.7, we see that the actual V-101 is vertical vessel with D = 1.1 m, $L =$
3.5 m

We should conclude that the design of V-101 given in Chapter 1 is consistent with the
heuristics given in Table 9.6. The small differences in L and D are to be expected in a comparison such as this one.

b. E-105 Product Cooler

From Table 9.11 we use the following heuristics

Rule 1: Set $F = 0.9$
Rule 6: min. $\Delta T = 10$°C
Rule 7: Water enters at 30°C and leaves at 40°C
Rule 8: $U = 850$ W/m^2°C

We note immediately from Table 1.5 and Figure 1.5 that Rule 6 has been violated since $\Delta T_{min} = 8°C$

For the moment ignore this and return to the heuristic analysis:

$\Delta T_{lm} = [(105 - 40) - (38 - 30)]/\ln[(105 - 40)/(38 - 30)] = 27.2°C$
$Q = 1085 \text{ MJ}/h = 301 \text{ kW (from Table 1.7)}$
$A = Q/U\Delta T_{lm}F = (301000)/(850)/(27.2)/(0.90) = 14.46 \text{ m}^2$
From Rule 9, Table 9.11, this heat exchanger should be a double-pipe or multiple pipe design.
Comparing our analysis with the information in Table 1.7 we get:

Heuristic: Double-pipe design, Area $= 14.5 \text{ m}^2$
Table 1.7: Multiple-pipe design, Area $= 12 \text{ m}^2$

Again the heuristic analysis is close to the actual design. The fact that the minimum approach temperature of 10°C has been violated should not cause us too much concern, since the actual minimum approach is only 8°C and the heat exchanger is quite small suggesting that a little extra area (due to a smaller overall temperature driving force) is not very costly.

c. P-101

From Table 9.9, use the following heuristics:
Rule 1: Power(kW) $= (1.67)[\text{Flow}(\text{m}^3/\text{min})]\Delta P(\text{bar})/\varepsilon$
Rules 4–7: Type of pump based on head

From Figure 1.5 and Tables 1.5 and 1.7, we have:
Flowrate (Stream 2) $= 13300 \text{ kg}/h$
Density of fluid $= 870 \text{ kg}/\text{m}^3$
$\Delta P = 25.8 - 1.2 = 24.6 \text{ bar} = 288 \text{ m of liquid (head} = \Delta P/\rho g)$
Volumetric flowrate $= (13300)/(60)/(870) = 0.255 \text{ m}^3/\text{min}$
Fluid pumping power $= (1.67)(0.255)(24.6) = 10.5 \text{ kW}$

From Rules 4–7, pump choices are multistage centrifugal, rotary, and reciprocating. Choose reciprocating to be consistent with Table 1.7. Typical $\varepsilon = 0.75$

Power (shaft power) $= 10.5/0.75 = 14.0 \text{ kW} \rightarrow$ compares with 14.2 kW from Table 1.7

d. C-101

From Table 9.10, use the following heuristics:

Rule 2: $W_{rev\ adiab} = mz_1RT_1[(P_2/P_1)^a - 1]/a$

From Table 1.7, we have, flow $= 6770 \text{ kg}/h$, $T_1 = 38°C = 311 \text{ K}$, $mw = 8.45$, $P_1 = 23.9$ bar, $P_2 = 25.5$
$k = 1.41$ (assume) and $a = 0.2908$
$m = (6770)/(3600)/(8.45) = 0.223 \text{ kmol/s}$
$W_{rev\ adiab} = (223)(1.0)(8.314)(311)\{ (25.5/23.9)^{0.2908} - 1)/0.2908 = 37.7 \text{ kW}$
using a compressor efficiency of 75%
$W_{actual} = (37.7)/(0.75) = 50.3 \text{ kW} \rightarrow$ This checks with the shaft power requirement given in Table 1.7.

e. T-101

From Table 9.13, we use the following heuristics

Rule 5: Optimum reflux in the range of $1.2 - 1.5\ R_{min}$
Rule 6: Optimum number of stages approximately $2N_{min}$
Rule 7: $N_{min} = \ln\{\ [x/(1-x)]_{ovhd}/[x/(1-x)]_{bot}\}/\ln\ \alpha$
Rule 8: $R_{min} = \{F/D\}/(\alpha - 1)$
Rule 9: Use a safety factor of 10% on number of trays
Rule 14: $L_{max} = 53$ m and $L/D < 30$

From Table 9.14, we use the following heuristics
Rule 2: $F_s = u\rho_v^{0.5} = 1.2 \rightarrow 1.5$ m/s(kg/m^3)$^{0.5}$
Rule 3: $\Delta P_{tray} = 0.007$ bar
Rule 4: $\varepsilon_{tray} = 60 - 90$ %

$x_{ovhd} = 0.9962,\ x_{ovhd} = 0.0308,\ \alpha_{ovhd} = 2.44,\ \alpha_{bot} = 2.13,\ \alpha_{geom\ ave} = (\alpha_{ovhd}\alpha_{bot})^{0.5} = 2.28$
$N_{min} = \ln\{\ [0.9962/(1 - 0.9962)]/[0.0308/(1 - 0.0308)]\}\ /\ln\ (2.28) = 10.9$
$R_{min} = \{142.2/105.6\}/(2.28 - 1) = 1.05$
Range of $R = (1.2 \rightarrow 1.5)R_{min} = 1.26 \rightarrow 1.58$

$N_{theoretical} \approx (2)(10.9) = 21.8$
$\varepsilon_{tray} = 0.6$
$N_{actual} \approx (21.6/0.6)(1.1) = 40$ trays

$\rho_v = 6.1$ kg/m^3
$u = (1.2 \rightarrow 1.5)/6.1^{0.5} = 0.49 \rightarrow 0.60$ m/s
Vapor flow rate (stream 13) = 22,700 kg/h
Vol. flow rate, $v = 1.03$ m^3/s
$D_{tower} = [4v/\pi u]^{0.5} = [(4)(1.03)/(3.142)/(0.49 \rightarrow 0.60)]^{0.5} = 1.64 - 1.48$ m

$\Delta P_{tower} = (N_{actual})(\Delta P_{tray}) = (40)(0.007) = 0.28$ bar

A comparison of the actual equipment design and the predictions of the heuristic methods are given below:

	From Tables 1.5, 1.7 and Figure 1.5	**From Heuristics**
Tower Diameter	1.5 m	1.48 → 1.64 m
Reflux Ratio, R	1.75	1.26 →1.58
Number of Trays	42	40
Pressure drop, ΔP_{tower}	0.30 bar	0.28 bar

f. H-101

From Table 9.11, we use the following heuristics
Rule 13: Equal heat transfer in radiant and convective sections
        radiant rate = 37.6 kW/m^2, convective rate = 12.5 kW/m^2

$$\text{Duty} = 27,040\ \text{MJ/h} = 7511\ \text{kW}$$

Area radiant section = $(0.5)(7511)/(37.6) = 99.9$ m^2 (106.8 m^2 in Table 1.7)
Area convective section = $(0.5)(7511)/(12.5) = 300.4$ m^2 (320.2 m^2 in Table 1.7)

From the worked examples above, it is clear that the sizing of the equipment in Table 1.7 agrees well with the predictions of the heuristics presented in this chapter. Exact agreement is not to be expected. Instead, the heuristics should be used to check calculations performed using more rigorous methods and to flag any inconsistencies.

LIST OF INFORMATIONAL TABLES

Table	Description
9.2a	Physical Property Heuristics
9.2b	Typical Physical Property Variations with Temperature and Pressure
9.3	Capacities for Processes Units in Common Usage
9.4	Effect of Typical Materials of Construction on Product Color, Corrosion, Abrasion, and Catalytic Effects
9.5	Heuristics for Drivers and Power Recovery Equipment
9.6	Heuristics for Process Vessels (Drums)
9.7	Heuristics for Vessels (Pressure and Storage)
9.8	Heuristics for Piping
9.9	Heuristics for Pumps
9.10	Heuristics for Compressors, Fans, Blowers, and Vacuum Pumps
9.11	Heuristics for Heat Exchangers
9.12	Heuristics for Thermal Insulation
9.13	Heuristics for Towers (Distillation and Gas Absorption)
9.14	Heuristics for Tray Towers (Distillation and Gas Absorption)
9.15	Heuristics for Packed Towers (Distillation and Gas Absorption)
9.16	Heuristics for Liquid-Liquid Extraction
9.17	Heuristics for Reactors
9.18	Heuristics for Refrigeration and Utility

Table 9.2a   Physical Property Heuristics

	Units	Liquids	Liquids	Gases	Gases	Gases
		Water	Organic Material	Steam	Air	Organic Material
Heat Capacity	kJ/kg°C	4.2	1.0–2.5	2.0	1.0	2.0–4.0
Density	kg/m³	1000	700–1500		1.29@STP	
Latent Heat	kJ/kg	1200–2100	200–1000			
Thermal Conductivity	W/m°C	0.55–0.70	0.10–0.20	0.025–0.07	0.025–0.05	0.02–0.06
Viscosity	kg/m s	0°C 1.8×10^{-3} 50°C 5.7×10^{-4} 100°C 2.8×10^{-4} 200°C 1.4×10^{-4}	Wide Range	$10\text{–}30 \times 10^{-6}$	$20\text{–}50 \times 10^{-6}$	$10\text{–}30 \times 10^{-6}$
Prandtl No.		1–15	10–1000	1.0	0.7	0.7–0.8

Table 9.2b   Typical Physical Property Variations with Temperature and Pressure

	Liquids	Liquids	Gases	Gases
Property	Temperature	Pressure	Temperature	Pressure
Density	$\rho_l \propto (T_c - T)^{0.3}$	Negligible	$\rho_g = MW.P/ZRT$	$\rho_g = MW.P/ZRT$
Viscosity	$\mu_l = Ae^{B/T}$	Negligible	$\mu_g \ \alpha \ \dfrac{T^{1.5}}{(T + 1.47T_b)}$	Significant only for $P > 10$ bar
Vapor Pressure	$P^* = ae^{b/(T+c)}$	—	—	—

T is temperature (K), T_c is the critical Temperature (K), T_b is the normal boiling point (K), MW is molecular weight, P is pressure, Z is compressibility, R is the gas constant, and P^* is the vapor pressure.

Table 9.3 Capacities of Process Units in Common Usage[a]

Process Unit	Capacity Unit	Max. Value	Min. Value	Comment
Horizontal Vessel	Pressure (bar)	400	Vacuum	L/D typically 2–5, see Table 9.6
	Temper. (°C)	400[b]	−200	
	Height (m)	10	2	
	Diameter (m)	2	0.3	
	L/D	5	2	
Vertical Vessel	Pressure (bar)	400	400	L/D typically 2–5, see Table 9.6.
	Temper. (°C)	400[b]	−200	
	Height (m)	10	2	
	Diameter (m)	2	0.3	
	L/D	5	2	
Towers	Pressure (bar)	400	Vacuum	Normal Limits
	Temper. (°C)	400[b]	−200	Diameter L/D
	Height (m)	50	2	0.5 3.0–40[c]
	Diameter (m)	4	0.3	1.0 2.5–30[c]
	L/D	30	2	2.0 1.6–23[c]
				4.0 1.8–13[c]
Pumps				
Reciprocating	Power[e](kW)	250	< 0.1	
	Pressure (bar)	1,000		
Rotary and Positive	Power[e](kW)	150	< 0.1	
Displacement	Pressure (bar)	300		
Centrifugal	Power[e](kW)	250	< 0.1	
	Pressure (bar)	300		
Compressors				
Axial, Centrifugal + Recipr.	Power[e](kW)	8,000	50	
Rotary	Power[e](kW)	1,000	50	
Drives for Compressors				
Electric	Power[f](kW)	15,000	< 1	
Steam Turbine	Power[f](kW)	15,000	100	
Gas Turbine	Power[f](kW)	15,000	10	
Internal Combustion Eng.	Power[f](kW)	15,000	10	
Process Heaters	Duty (MJ/h)	500,000	10,000	Duties different for reactive heaters/furnaces.
Heat Exchangers	Area (m²)	1,000	10	For Area < 10 m² use double pipe exchanger.
	Tube Dia. (m)	0.0254	0.019	
	Length (m)	6.5	2.5	
	Pressure (bar)	150	Vacuum	For 150 < P < 400 bar need special design.
	Temp. (°C)	400[b]	−200	

[a]Most of the limits for equipment sizes shown here correspond to the limits used in the costing program (CAPCOST.BAS) introduced in Chapter 2.
[b]Maximum temperature and pressure are related to the materials of construction and may differ from values shown here.
[c]For 20 < L/D < 30 special design may be required. ~Diameters up to 7m possible but over 4m must be fabricated on site.
[d]Power values refer to fluid/pumping power.
[e]Power values refer to shaft power

Table 9.4   Effect of Typical Materials of Construction on Product Color, Corrosion[a], Abrasion and Catalytic Effects

Metals		
Material	*Advantages*	*Disadvantage*
Carbon Steel	Low cost, readily available, resists abrasion, standard fabrication, resists alkali	Poor resistance to acids and strong alkali, often causes discoloration and contamination
Stainless Steel	Resists most acids, reduces discoloration, available with a variety of alloys, abrasion less than mild steel	Not resistant to chlorides, more expensive, fabrication more difficult, alloy materials may have catalytic effects
Monel-Nickel	Little discoloration, contamination, resistant to chlorides	Not resistant to oxidizing environments, expensive
Hasteloy	Improved over Monel-Nickel	More expensive than Monel-Nickel
Other Exotic Metals	Improves specific properties	Can be very high cost
Non-metals		
Material	*Advantages*	*Disadvantage*
Glass	Useful in laboratory and batch system, low diffusion at walls	Fragile, not resistant to high alkali, poor heat transfer, poor abrasion resistance
Plastics	Good at low temperature, large variety to select from with various characteristics, easy to fabricate, seldom discolors, minor catalytic effects possible	Poor at high temperature, low strength, not resistant to high alkali conditions, low heat transfer, low cost
Ceramics	Withstands high temperatures, variety of formulations available, modest cost	Poor abrasion properties, high diffusion at walls (in particular hydrogen), low heat transfer, may encourage catalytic reactions

[a]In addition, see Table 2.10 for preliminary selection of materials of construction.

Table 9.5   Heuristics for Drivers and Power Recovery Equipment (Adapted from Walas S. M., *Chemical Process Equipment: Selection and Design,* Butterworths, Stoneham, MA, 1988, copyright © 1988 by Butterworth Publishers, adapted by permission of Butterworth Publishers, Stoneham, MA, all rights reserved)

1. Efficiency is greater for larger machines. Electric Motors are 85–95%; steam turbines are 42–78%; gas engines and turbines are 28–38% efficient (see Table 3.7)

2. For under 74.6kW (100hp), electric motors are used almost exclusively. They are made for services up to 14,900kW (20,000hp).

3. Steam turbines are competitive above 76.6kW (100hp). They are speed controllable. They are frequently used as spares in case of power failure.

4. Combustion engines and turbines are restricted to mobile and remote locations.

5. Gas expanders for power recovery may be justified at capacities of several hundred hp; otherwise any pressure reduction in process is done with throttling valves.

6. The following useful definitions are given:

$$\text{shaft power} = \frac{\text{theoretical power to pump fluid (liquid or gas)}}{\text{efficiency of pump or compressor, } \varepsilon_{sh}}$$

$$\text{drive power} = \frac{\text{shaft power}}{\text{efficiency of drive, } \varepsilon_{dr}}$$

Overall Efficiency, $\varepsilon_{ov} = \varepsilon_{sh}\, \varepsilon_{dr}$

ε_{dr} values are given in this Table and Table 3.7

ε_{sh} values are given in Tables 9.9 and 9.10. Usually ε_{sh} are given on PFD.

Table 9.6   Heuristics for Process Vessels (Drums) (Adapted from Walas S. M., *Chemical Process Equipment: Selection and Design,* Butterworths, Stoneham, MA, 1988, copyright ©️ 1988 by Butterworth Publishers, adapted by permission of Butterworth Publishers, Stoneham, MA, all rights reserved)

1. Drums are relatively small vessels that provide surge capacity or separation of entrained phases.

2. Liquid drums are usually horizontal.

3. Gas/liquid phase separators are usually vertical.

4. Optimum length/diameter = 3, but the range 2.5 to 5 is common.

5. Holdup time is 5 min for half-full reflux drums and gas/liquid separators, 5-10 min for a product feeding another tower.

6. In drums feeding a furnace, 30 min for half-full drum is allowed.

7. Knockout drums placed ahead of compressors should hold no less than 10 times the liquid volume passing per minute.

8. Liquid/liquid separations are designed for settling velocity of 0.085–0.127 cm/s (2–3 in/min)

9. Gas velocity in gas/liquid separators, $u = k \sqrt{\rho_l/\rho_v - 1}$ m/s (ft/sec) $k = 0.11$ (0.35) for systems with mesh deentrainer and $k = 0.0305$ (0.1) without mesh deentrainer.

10. Entrainment removal of 99% is attained with 10.2–30.5 cm (4–12 in) mesh pad thickness; 15.25 cm (6 in) thickness is popular.

11. For vertical pads, the value of the coefficient in Step 9 is reduced by a factor of 2/3.

12. Good performance can be expected at velocities of 30–100% of those calculated with the given k; 75% is popular.

13. Disengaging spaces of 15.2–45.7 cm (6–18 in) ahead of the pad and 30.5 cm (12 in) above the pad are suitable.

14. Cyclone separators can be designed for 95% collection at 5 μm particles, but usually only droplets greater than 50 μm need be removed.

Table 9.7 Heuristics for Vessels (Pressure and Storage) (Adapted from Walas, S. M., *Chemical Process Equipment: Selection and Design*, **Butterworths, Stoneham, MA, 1988, copyright © 1988 by Butterworth Publishers, adapted by permission of Butterworth Publishers, Stoneham, MA, all rights reserved)**

Pressure Vessels

1. Design temperature between −30 and 345°C is 25°C above maximum operating temperature; higher safety margins are used outside the given temperature range.

2. The design pressure is 10% or 0.69-1.7 bar (10-25 psi) over the max. operating pressure, whichever is greater. The max. operating pressure, in turn, is taken as 1.7 bar (25 psi) above the normal operation.

3. Design pressures of vessels operating at 0–0.69 bar (0-10 psig) and 95–540°C (200–1000°F) are 2.76 barg (40 psig).

4. For vacuum operation, design pressures are 1 barg (15 psig) and full vacuum.

5. Minimum wall thickness for rigidity; 6.4 mm (0.25 in) for 1.07 m (42 in) dia. and under, 8.1 mm (0.32 in) for 1.07–1.52 m (42-60 in) dia., and 9.7 mm (0.38 in) for over 1.52 m (60 in) dia.

6. Corrosion allowance 8.9 mm (0.35 in) for known corrosive conditions, 3.8 mm (0.15 in) for noncorrosive streams, and 1.5 mm (0.06 in) for steam drums and air receivers.

7. Allowable working stresses are one-fourth of the ultimate strength of the material.

8. Maximum allowable stress depends sharply on temperature.

Temperature (°F)		−20-650	750	850	1,000
(°C)		−30-345	400	455	540
Low alloy steel SA 203	(psi)	18,759	15,650	9,950	2,500
	(bar)	1,290	1,070	686	273
Type 302 stainless steel	(psi)	18,750	18,750	15,950	6,250
	(bar)	1,290	1,290	1,100	431

Storage Vessels

1. For less than 3.8 m³ (1000 gal), use vertical tanks on legs.

2. Between 3.8-38 m³ (1000–10,000 gal), use horizontal tanks on concrete supports.

3. Beyond 38 m³ (10,000 gal) use vertical tanks on concrete pads.

4. Liquids subject to breathing losses may be stored in tanks with floating or expansion roofs for conservation.

5. Freeboard is 15% below 1.9 m³ (500 gal) and 10% above 1.9 m³ (500 gal) capacity.

6. Thirty days capacity often is specified for raw materials and products, but depends on connecting transportation equipment schedules.

7. Capacities of storage tanks are at least 1.5 times the size of connecting transportation equipment; for instance, 28.4 m³ (7500 gal) tanker trucks, 130 m³ (34,500 gal) rail cars, and virtually unlimited barge and tanker capacities.

Table 9.8   Heuristics for Piping (Adapted from Walas, S. M., *Chemical Process Equipment: Selection and Design*, Butterworths, Stoneham, MA, 1988, copyright © 1988 by Butterworth Publishers, adapted by permission of Butterworth Publishers, Stoneham, MA, all rights reserved)

1. Line velocities(u) and pressure drop (ΔP): (a) For liquid pump discharge; $u = (5 + D/3)$ ft/sec and $\Delta P = 2.0$ psi/100ft (b) For liquid pump suction; $u = (1.3 + D/6)$ ft/sec and $\Delta P = 0.4$ psi/100 ft (c) For steam or gas flow: $u = 20D$ ft/sec and $\Delta P = 0.5$ psi/100ft, D = diameter of pipe in inches.

2. Gas/steam line velocities = 61 m/s (200ft/sec) and pressure drop = 0.1 bar/100m (0.5 psi/100ft).

3. In preliminary estimates set line pressure drops for an equivalent length of 30 m (100 ft) of pipe between each piece of equipment.

4. Control valves require at least 0.69 bar (10 psi) drop for good control.

5. Globe valves are used for gases, control, and wherever tight shut-off is required. Gate valves for most other services.

6. Screwed fittings are used only on sizes 3.8 cm (1.5 in) or less, flanges or welding used otherwise.

7. Flanges and fittings are rated for 10, 20, 40, 103, 175 bar (150, 300, 600, 1500 or 2500 psig).

8. Approximate schedule number required = $1000 \, P/S$, where P is the internal pressure in psig and S is the allowable working stress {about 690 bar (10,000 psi)} for A120 carbon steel at 260° (500°F). Schedule 40 is most common.

Table 9.9   Heuristics for Pumps (Adapted from Walas, S. M., *Chemical Process Equipment: Selection and Design*, Butterworths, Stoneham, MA, 1988, copyright © 1988 by Butterworth Publishers, adapted by permission of Butterworth Publishers, Stoneham, MA, all rights reserved)

1. Power for pumping liquids: kW = (1.67)[Flow(m^3/min)][ΔP(bar)]/ε [hp = Flow(gpm) ΔP(psi)/1714/ε] ε = Fractional Efficiency = ε_{sh} (see Table 9.5)

2. Net positive suction head (NPSH) of a pump must be in excess of a certain number, depending upon the kind of pumps and the conditions, if damage is to be avoided. *NPSH* = (pressure at the eye of the impeller −vapor pressure)/(ρg). Common range is 1.2–6.1 m of liquid (4–20 ft).

3. Specific speed $N_s = $ (rpm)(gpm)$^{0.5}$/(head in feet)$^{0.75}$. Pump may be damaged if certain limits on N_s are exceeded, and the efficiency is best in some ranges.

4. Centrifugal pumps: Single stage for 0.057–18.9 m^3/min (15–5000 gpm), 152 m (500 ft) maximum head; multistage for 0.076–41.6 m^3/min (20–11,000 gpm), 1675 m (5500 ft) maximum head. Efficiency 45% at 0.378 m^3/min (100 gpm), 70% at 1.89 m^3/min (500 gpm), 80% at 37.8 m^3/min (10,000 gpm).

5. Axial pumps for 0.076–378m^3/min (20–100,000 gpm), 12 m (40 ft) head, 65–85% efficiency.

6. Rotary pumps for 0.00378–18.9 m^3/min (1–5000 gpm), 15,200 m (50,000 ft head), 50–80% efficiency.

7. Reciprocating pumps for 0.0378–37.8 m^3 (10–10,000 gpm), 300 km (1,000,000 ft) head max. Efficiency 70% at 7.46 kW (10 hp), 85% at 37.3 kW (50 hp) and 90% at 373 kW (500 hp).

Table 9.10   Heuristics for Compressors, Fans, Blowers and Vacuum Pumps (Adapted from Walas, S. M., *Chemical Process Equipment: Selection and Design,* **Butterworths, Stoneham, MA, 1988, copyright © 1988 by Butterworth Publishers, adapted by permission of Butterworth Publishers, Stoneham, MA, all rights reserved)**

1. Fans are used to raise the pressure about 3% {12 in (30 cm) water}, blowers raise to less than 2.75 barg (40 psig) and compressors to higher pressures, although the blower range is commonly included in the compressor range.

2. Theoretical reversible adiabatic power $= mz_1RT_1[(\{P_2/P_1\}^a-1)]/a$
   where T_1 is inlet temperature, R = Gas Constant, z_1 = compressibility, m = molar flow rate, $a = (k-1)/k$ and $k = C_p/C_v$
   Values of R: = 8.314 J/mol K = 1.987 Btu/lbmol R = 0.7302 atm ft^3/lbmol R

3. Outlet temperature for reversible adiabatic process $T_2 = T_1 (P_2/P_1)^a$

4. Exit temperatures should not exceed 167–204°C (350–400°F); for diatomic gases $(C_p/C_v = 1.4)$ this corresponds to a compression ratio of about 4.

5. Compression ratio should be about the same in each stage of a multistage unit, ratio $= (P_n/P_1)^{1/n}$, with n stages.

6. Efficiencies of reciprocating compressors: 65% at compression ratios of 1.5, 75% at 2.0 and 80–85% at 3–6.

7. Efficiencies of large centrifugal compressors, 2.83–47.2 m^3/s (6,000–100,000 acfm) at suction, are 76–78%.

8. For vacuum pumps use the following:

Reciprocating piston type	- down to 1 Torr
Rotary piston type	- down to 0.001 Torr
Two lobe rotary type	- down to 0.0001 Torr
Steam jet ejectors	- 1 stage down to 100 Torr
	- 3 stage down to 1 Torr
	- 5 stage down to 0.05 Torr

9. A 3 stage ejector needs 100 kg steam/kg air to maintain a pressure of 1 Torr.

10. In-leakage of air to evacuated equipment depends on the absolute pressure, Torr, and the volume of the equipment, V in m^3 (ft^3) according to $W = kV^{2/3}$ kg/h (lb/hr) with $k = 0.98$ (0.2) when $P > 90$ Torr, $k = 0.39$ (0.08) between 3 and 20 Torr, and $k = 0.12$ (0.025) at less than 1 Torr.

Table 9.11   Heuristics for Heat Exchangers (Adapted from Walas, S. M., *Chemical Process Equipment: Selection and Design*, Butterworths, Stoneham, MA, 1988, copyright © 1988 by Butterworth Publishers, adapted by permission of Butterworth Publishers, Stoneham, MA, all rights reserved)

1. For conservative estimate set $F = 0.9$ for shell and tube exchangers with no phase changes, $q = UAF\Delta T_{lm}$. When ΔT at exchanger ends differ greatly then check F, reconfigure if F is less than 0.85.

2. Standard tubes are 1.9 cm (3/4 in) OD, on a 2.54 cm (1 in) triangle spacing, 4.9 m (16 ft) long.

   A shell 30 cm (1 ft) dia., accommodates 9.3 m^2 (100ft^2);

         60 cm (2 ft) dia., accommodates 37.2 m^2 (400ft^2),

         90 cm (3 ft) dia., accommodates 102 m^2 (1,100ft^2).

3. Tube side is for corrosive, fouling, scaling, and high pressure fluids.

4. Shell side is for viscous and condensing fluids.

5. Pressure drops are 0.1 bar (1.5 psi) for boiling and 0.2–0.62 bar (3–9 psi) for other services.

6. Minimum temperature approach are 10°C (20°F) for fluids and 5°C (10°F) for refrigerants.

7. Cooling water inlet is 30°C (90°F), maximum outlet 45°C (115°F).

8. Heat transfer coefficients for estimating purposes, W/m^2°C (Btu/hr ft^2°F): water to liquid, 850 (150); condensers, 850 (150); liquid to liquid, 280 (50); liquid to gas, 60 (10); gas to gas 30 (5); reboiler 1140 (200). Maximum flux in reboiler 31.5 kW/m^2 (10,000 Btu/hr ft^2).

   When phase changes occur, use a zoned analysis with appropriate coefficient for each zone.

9. Double-pipe exchanger is competitive at duties requiring 9.3–18.6 m^2 (100–200ft^2).

10. Compact (plate and fin) exchangers have 1150 m^2/m^3 (350 ft^2/ft^3), and about 4 times the heat transfer per cut of shell-and-tube units.

11. Plate and frame exchangers are suited to high sanitation services, and are 25–50% cheaper in stainless steel construction than shell-and-tube units.

12. Air coolers: Tubes are 0.75–1.0 in. OD., total finned surface 15–20 m^2/m^2 (ft^2/ft^2 bare surface), U = 450–570 W/m^2°C (80–100 Btu/hr ft^2 (bare surface) °F). Minimum approach temperature = 22°C (40°F). Fan input power = 1.4–3.6 kW/(MJ/h) [2–5 hp / (1000 Btu/hr)]

13. Fired heaters: radiant rate, 37.6 kW/m^2 (12,000 Btu/hr ft^2); convection rate, 12.5 kW/m^2 (4,000 Btu/hr ft^2); cold oil tube velocity = 1.8 m/s (6 ft/sec); approximately equal transfer in the two sections; thermal efficiency see Table 3.6; flue gas temperature 140–195°C (250–350°F) above feed inlet; stack gas temperature 345–510°C (650–950°F).

Table 9.12   Heuristics for Thermal Insulation (Adapted from Walas, S. M., *Chemical Process Equipment: Selection and Design*, Butterworths, Stoneham, MA, 1988, copyright © 1988 by Butterworth Publishers, adapted by permission of Butterworth Publishers, Stoneham, MA, all rights reserved)

1. Up to 345°C (650°F) 85% magnesia is used.
2. Up to 870–1040°C (1600–1900°F) a mixture of asbestos and diatomaceous earth is used.
3. Ceramic refractories at higher temperature.
4. Cryogenic equipment −130°C (−200°F) employs insulations with fine pores of trapped air e.g. Perlite™.
5. Optimal thickness varies with temperature: 1.27 cm (0.5 in) at 95°C (200°F), 2.54 cm (1.0 in) at 200°C (400°F), 3.2 cm (1.25 in) at 315°C (600°F).
6. Under windy conditions 12.1 km/h (7.5 miles/hr), 10–20% greater thickness of insulation is justified.

Table 9.13 Heuristics for Towers (Distillation and Gas Absorption) (Adapted from Walas, S. M., *Chemical Process Equipment: Selection and Design*, Butterworths, Stoneham, MA, 1988, copyright © 1988 by Butterworth Publishers, adapted by permission of Butterworth Publishers, Stoneham, MA, all rights reserved)

1. Distillation is usually the most economical method for separating liquids, superior to extraction, absorption crystallization, or others.

2. For ideal mixtures, relative volatility is the ratio of vapor pressures $\alpha_{12} = P_1^* / P_2^*$.

3. Tower operating pressure is most often determined by the temperature of the condensing media, 38–50°C (100–120°F) if cooling water is used; or by the maximum allowable reboiler temperature to avoid chemical decomposition/degradation.

4. Sequencing of columns for separating multi-component mixtures:[a]

   a. Perform the easiest separation first, that is, the one least demanding of trays and reflux, and leave the most difficult to the last

   b. When neither relative volatility nor feed composition vary widely, remove components one by one as overhead products

   c. When the adjacent ordered components in the feed vary widely in relative volatility, sequence the splits in order of decreasing volatility

   d. When the concentrations in the feed vary widely but the relative volatilities do not, remove the components in order of decreasing concentration.

5. Economical optimum reflux ratio is in the range of 1.2–1.5 times the minimum reflux ratio, R_{min}.

6. The economically optimum number of theoretical trays is near twice the minimum value N_{min}.

7. The minimum number of trays is found with the Fenske-Underwood equation
   $N_{min} = \ln\{[x/(1-x)]_{ovhd}/[x/(1-x)]_{btms}\}/\ln \alpha$.

8. Minimum reflux for binary or pseudobinary mixtures is given by the following when separation is essentially complete ($x_D \approx 1$) and D/F is the ratio of overhead product to feed rate:
   $R_{min}D/F = 1/(\alpha-1)$, when feed is at the bubble point
   $(R_{min}+1) D/F = \alpha/(\alpha-1)$, when feed is at the dew point.

9. A safety factor of 10% of the number of trays calculated by the best means is advisable.

10. Reflux pumps are made at least 10% oversize.

11. The optimum value of the Kremser absorption factor $A = (L/mV)$ is in the range of 1.25 to 2.0.

12. Reflux drums usually are horizontal, with a liquid holdup of 5 min half full. A take-off pot for a second liquid phase, such as water in hydrocarbon systems, is sized for a linear velocity of that phase of 1.3 m/s (0.5 ft/sec), minimum diameter is 0.4 m (16 in).

13. For towers about 0.9 m (3 ft dia), add 1.2 m (4 ft) at the top for vapor disengagement and 1.8 m (6 ft) at bottom for liquid level and reboiler return.

14. Limit the tower height to about 53 m (175 ft) max. because of wind load and foundation considerations. An additional criterion is that L/D be less than 30 (20 < L/D < 30 often will require special design)

[a]Additional information on sequencing is given in Table 17.2

Table 9.14   Heuristics for Tray Towers (Distillation and Gas Absorption) (Adapted from Walas, S. M., *Chemical Process Equipment: Selection and Design*, Butterworths, Stoneham, MA, 1988, copyright © 1988 by Butterworth Publishers, adapted by permission of Butterworth Publishers, Stoneham, MA, all rights reserved)

1. For reasons of accessibility, tray spacings are made 0.5–0.6 m (20–24 in).

2. Peak efficiency of trays is at values of the vapor factor $F_s = u\rho^{0.5}$ in the range of 1.2–1.5 m/s $\{kg/m^3\}^{0.5}$ [1–1.2 ft/s $\{lb/ft^3\}^{0.5}$]. This range of F_s establishes the diameter of the tower. Roughly, linear velocities are 0.6 m/s (2 ft/sec) at moderate pressures and 1.8 m/s (6 ft/sec) in vacuum.

3. Pressure drop per tray is on the order of 7.6 cm (3 in) of water or 0.007 bar (0.1 psi).

4. Tray efficiencies for distillation of light hydrocarbons and aqueous solutions are 60–90%; for gas absorption and stripping, 10–20%.

5. Sieve trays have holes 0.6–0.7 cm (0.25–0.5 in) dia., area being 10% of the active cross section.

6. Valve trays have holes 3.8 cm (1.5 in) dia. each provided with a liftable cap, 130–150 caps/m² (12–14 caps/ft²) of active cross section. Valve trays are usually cheaper than sieve trays.

7. Bubblecap trays are used only when a liquid level must be maintained at low turndown ratio; they can be designed for lower pressure drop than either sieve or valve trays.

8. Weir heights are 5 cm (2 in), weir lengths are about 75% of tray diameter, liquid rate—a maximum of 1.2 m³/min m of weir (8 gpm/in of weir); multi-pass arrangements are used at higher liquid rates.

Table 9.15   Heuristics for Packed Towers (Distillation and Gas Absorption), adapted from Walas S. M., *Chemical Process Equipment: Selection and Design,* Butterworths, Stoneham, MA, 1988, copyright © 1988 by Butterworth Publishers, adapted by permission of Butterworth Publishers, Stoneham, MA, all rights reserved.

1. Structured and random packings are suitable for packed towers less than 0.9 m (3 ft) when low pressure drop is required.

2. Replacing trays with packing allows greater throughput and separation in existing tower shells.

3. For gas rates of 14.2 m³/min (500 ft³/min), use 2.5 cm (1 in) packing; for 56.6 m³/min (2,000 ft³/min) or more use 5 cm (2 in) packing.

4. Ratio of tower diameter/packing diameter should be >15/1.

5. Because of deformability, plastic packing is limited to 3–4 m (10–15 ft) and metal to 6.0–7.6 m (20–25 ft) unsupported depth.

6. Liquid distributors are required every 5–10 tower diameters with pall rings and at least every 6.5 m (20 ft) for other types of dumped packing.

7. Number of liquid distributors should be >32–55/m² (3–5/ft²) in towers greater that 0.9 m (3 ft) diameter and more numerous in smaller columns.

8. Packed tower should operate near 70% of flooding (evaluated from Sherwood and Lobo correlation)

9. Height Equivalent to Theoretical Stage (HETS) for vapor-liquid contacting is 0.4–0.56 m (1.3–1.8 ft) for 2.5 cm (1 in) pall rings and 0.76–0.9 m. (2.5–3.0 ft) for 5 cm (2 in) pall rings.

10. Generalized pressure drops

	Design Pressure Drops (cm of H_2O/m of packing)	Design Pressure Drops (inches of H_2O/ft of packing)
Absorbers and Regenerators (non-foaming systems)	2.1–3.3	0.25–0.40
Absorbers and Regenerators	0.8–2.1	0.10–0.25
Atmospheric/Pressure Stills and Fractionators	3.3–6.7	0.40–0.80
Vacuum Stills and Fractionators	0.8–3.3	0.10–0.40
Maximum value	8.33	1.0

Table 9.16   Heuristics for Liquid-Liquid Extraction (Adapted from Walas, S. M., *Chemical Process Equipment: Selection and Design*, Butterworths, Stoneham, MA, 1988, copyright © 1988 by Butterworth Publishers, adapted by permission of Butterworth Publishers, Stoneham, MA, all rights reserved)

1. The dispersed phase should be the one that has the higher volumetric flow rate except in equipment subject to back-mixing where it should be the one with the smaller volumetric rate. It should be the phase that wets material of construction less well. Since the holdup of continuous phase is greater, that phase should be made up of the less expensive or less hazardous material.

2. There are no known commercial applications of reflux to extraction processes, although the theory is favorable.

3. Mixer-settler arrangements are limited to at most five stages. Mixing is accomplished with rotating impellers or circulation pumps. Settlers are designed on the assumption that droplet sizes are about 150 μm dia. In open vessels, residence times of 30–60 min or superficial velocities of 0.15–0.46 m/min (0.5–1.5 ft/min) are provided in settlers. Extraction stage efficiencies commonly are taken as 80%.

4. Spray towers as tall as 6–12 m (20–40 ft) cannot be depended on to function as more than a single stage.

5. Packed towers are employed when 5–10 stages suffice. Pall rings 2.5–3.8 cm (1–1.5 in) size are best. Dispersed phase loadings should not exceed 10.2 m^3/min m^2 (25 gal/min ft^2). HETS of 1.5–3.0 m (5–10 ft) may be realized. The dispersed phase must be redistributed every 1.5–2.1 m (5–7 ft). Packed towers are not satisfactory when the surface tension is more than 10 dyne/cm.

6. Sieve tray towers have holes of only 3–8 mm dia. Velocities through the holes are kept below 0.24 m/s (0.8 ft/sec) to avoid formation of small drops. Redispersion of either phase at each tray can be designed for. Tray spacings are 15.2–60 cm (6 to 24 in). Tray efficiencies are in the range of 20–30%.

7. Pulsed packed and sieve tray towers may operate at frequencies of 90 cycles/min. and amplitudes of 6–25 mm. In large diameter towers, HETS of about 1 m have been observed. Surface tensions as high as 30–40 dyne/cm have no adverse effect.

8. Reciprocating tray towers can have holes 1.5 cm (9/16 in) dia., 50–60% open area, stroke length 1.9 cm (0.75 in), 100–150 strokes/min, plate spacing normally 5 cm (2 in) but in the range of 2.5–15 cm (1–6 in). In a 76 cm (30 in) diameter tower, HETS is 50–65 cm (20–25 in) and throughput is 13.7 m^3/min m^2 (2,000 gal/hr ft^2). Power requirements are much less than that of pulsed towers.

9. Rotating disk contactors or other rotary agitated towers realize HETS in the range of 0.1–0.5 m (0.33–1.64 ft). The especially efficient Kuhni with perforated disks of 40% free cross section has HETS of 0.2 m (.66 ft) and a capacity of 50 m^3/m^2 h (164 ft^3/ft^2 hr).

Table 9.17   Heuristics for Reactors (Adapted from Walas, S. M., *Chemical Process Equipment: Selection and Design*, Butterworths, Stoneham, MA, 1988, copyright © 1988 by Butterworth Publishers, adapted by permission of Butterworth Publishers, Stoneham, MA, all rights reserved)

1. The rate of reaction in every instance must be established in the laboratory, and the residence time or space velocity and product distribution eventually must be found from a pilot plant.

2. Dimensions of catalyst particles are 0.1 mm (0.004 in) in fluidized beds, 1 mm in slurry beds, and 2–5mm (0.078–0.197 in) in fixed beds.

3. The optimum proportions of stirred tank reactors are with liquid level equal to the tank diameter, but at high pressures slimmer proportions are economical.

4. Power input to a homogeneous reaction stirred tank is 0.1–0.3 kW/m^3 (0.5–1.5 hp/1,000 gal), but three times this amount when heat is to be transferred.

5. Ideal CSTR (continuous stirred tank reactor) behavior is approached when the mean residence time is 5–10 times the length needed to achieve homogeneity, which is accomplished with 500–2,000 revolutions of a properly designed stirrer.

6. Batch reactions are conducted in stirred tanks for small daily production rates or when the reaction times are long or when some condition such as feed rate or temperature must be programmed in some way.

7. Relatively slow reactions of liquids and slurries are conducted in continuous stirred tanks. A battery of four or five in series is most economical.

8. Tubular flow reactors are suited to high production rates at short residence times (sec. or min.) and when substantial heat transfer is needed. Embedded tubes or shell-and-tube construction then are used.

9. In granular catalyst packed reactors, the residence time distribution is often no better than that of a five-stage CSTR battery.

10. For conversion under about 95% of equilibrium, the performance of a five-stage CSTR battery approaches plug flow.

11. The effect of temperature on chemical reaction rate is to double the rate every 10°C.

12. The rate of reaction in a heterogeneous system is more often controlled by the rate of heat or mass transfer than by the chemical reaction kinetics.

13. The value of a catalyst may be to improve selectivity more than to improve the overall reaction rate.

Table 9.18   Heuristics for Refrigeration and Utility Specifications (Adapted from Walas, S. M., *Chemical Process Equipment: Selection and Design*, Butterworths, Stoneham, MA, 1988, copyright © 1988 by Butterworth Publishers, adapted by permission of Butterworth Publishers, Stoneham, MA, all rights reserved)

1.	A ton of refrigeration is the removal of 12,700 kJ/h (12,000 Btu/hr) of heat.
2.	At various temperature levels: -18 to $-10°C$ (0 to 50°F), chilled brine and glycol solutions; -45 to $-10°C$ (-50 to $-40°F$), ammonia, freon, butane; -100 to $-45°C$ (-150 to $-50°F$) ethane or propane.
3.	Compression refrigeration with 38°C (100°F) condenser requires kW/tonne (hp/ton) at various temperature levels; 0.93 (1.24) at $-7°C$ (20°F); 1.31 (1.75) at $-18°C$ (0°F); 2.3 (3.1) at $-40°C$ ($-40°F$); 3.9 (5.2) at $-62°C$ ($-80°F$).
4.	Below $-62°C$ ($-80°F$), cascades of two or three refrigerants are used.
5.	In single stage compression, the compression ratio is limited to 4.
6.	In multistage compression, economy is improved with interstage flashing and recycling, so-called economizer operation.
7.	Absorption refrigeration: ammonia to $-34°C$ ($-30°F$), lithium bromide to 7°C (+45°F) is economical when waste steam is available at 0.9 barg (12 psig).
8.	Steam: 1–2 barg (15–30 psig), 121–135°C (250–275°F); 10 barg (150 psig), 186°C (366°F); 27.6 barg (400 psig), 231°C (448°F); 41.3 barg (600 psig), 252°C (488°F) or with 55–85°C (100–150°F) superheat.
9.	Cooling water: For design of cooling tower use—supply at 27–32°C (80–90°F) from cooling tower, return at 45–52°C (115–125°F); return seawater at 43°C (110°F); return tempered water or steam condensate above 52°C (125°F).
10.	Cooling air supply at 29–35°C (85–95°F); temperature approach to process, 22°C (40°F).
11.	Compressed air 3.1 (45), 10.3 (150), 20.6 (300), or 30.9 barg (450 psi) levels.
12.	Instrument air at 3.1 barg (45 psig), $-18°C$ (0°F) dewpoint.
13.	Fuels: gas of 37,200 kJ/m^3 (1000 Btu/SCF) at 0.35–0.69 barg (5–10 psig), or up to 1.73 barg (25 psig) for some types of burners; liquid at 39.8 GJ/m^3 (6 million Btu/bbl).
14.	Heat transfer fluids: petroleum oils below 315°C (600°F), Dowtherms below 400°C (750°F), fused salts below 600°C (1100°F), direct fire or electricity above 450°F.
15.	Electricity: 0.75–74.7 kW. (1–100 hp), 220–550 V; 149–1864 kW (200–2500 hp), 2300–4000 V.

9.3 SUMMARY

In this chapter, we have introduced a number of heuristics that allow us to check the reasonableness of the results of engineering calculations. These heuristics or guidelines cannot be used to determine whether a particular answer is correct or incorrect. However, they are still useful guides that allow the engineer to flag possible errors and help focus attention on areas of the process that may require special attention. Several heuristics, provided in the tables at the end of this chapter, were used to check the designs provided in Table 1.5 for the toluene hydrodealkylation process.

REFERENCES

1. Koen, B. V., *Definition of the Engineering Method*, American Society for Engineering Education, Washington, DC, 1985.

2. Sieder, E. N., and G. E. Tate, "Heat Transfer and Pressure Drop of Liquids in Tubes," *Ind. Eng. Chem.*, **28,** 1429 (1936).

3. Walas S. M., *Chemical Process Equipment: Selection and Design*, Butterworths, Stoneham, MA, 1988.

PROBLEMS

1. For the DME process shown in Appendix B.1, check the design specifications for the following three pieces of equipment against the appropriate heuristics, P-201, T-201, and E-202. Comment on any significant differences that you find.

2. For the acrylic acid process shown in Appendix B.2, check the design specifications for the following three pieces of equipment against the appropriate heuristics, C-301, P-305, and V-302. Comment on any significant differences that you find.

3. For the acetone process shown in Appendix B.3, check the design specifications for the following three pieces of equipment against the appropriate heuristics, H-401, E-401, and V-402. Comment on any significant differences that you find.

3

Analysis of System Performance

In the previous two sections, we focused on problems associated with the design and economics of a new chemical process, where there was freedom to select equipment. In this section, we explore problems associated with an existing chemical process. Two important factors that must be understood in dealing with existing equipment are:

1. Changes are limited by the performance of the existing equipment.
2. Any changes in operation of the process cannot be considered in isolation. The impact on the total process must always be considered.

Over the 10 to 30 years or more a plant is expected to operate, process operations vary. A plant seldom operates at the original process conditions provided on the design PFD. This is due to:

- Design/Construction: Installed equipment is often oversized. This reduces risks resulting from inaccuracies in design correlations, uncertainties in material properties, and so on.
- External Effects: Feed materials, product specifications and flowrates, environmental regulations, and costs of raw materials and utilities all are likely to change during the life of the process.
- Replacement of Equipment: New and improved equipment (or catalysts) may replace existing units in the plant.
- Changes in Equipment Performance: In general, equipment effectiveness degrades with age. For example, heat transfer surfaces foul, packed towers

develop channels, catalysts lose activity, and bearings on pumps and compressors become worn. Plants are shut down periodically for maintenance to restore equipment performance.

To remain competitive, it is necessary to be able to alter process operations in response to changing conditions. Therefore, it is necessary to understand how equipment performs over its complete operating range to quantify the effects of changing process conditions on process performance.

The material provided in this section involves several categories of performance problems:

1. Predictive Problems: An examination of the changes that take place for a change in process or equipment input and/or a change in equipment effectiveness.

2. Diagnostic/Troubleshooting Problems: If a change in process output (process disturbance or upset) is observed, the cause (change in process input, change in equipment performance) must be identified.

3. Control Systems Problems: If a change in process output is undesirable or a change in process input or equipment performance is anticipated, compensating action that can be taken to maintain process output must be identified.

4. Debottlenecking Problems: Often, a process change is necessary or desired, such as scale-up (increasing production capacity) or allowing for a change in product or raw material specifications. Identification of the equipment that limits the ability to make the desired change or constrains the change is necessary.

This section introduces the basic principles by which existing equipment and processes can be evaluated, operated, controlled, and subjected to changes in operating conditions. This material is treated in the following chapters:

Chapter 10: Process Input/Output Models
The basic structure of performance problems is considered in the context of an input, an output, and a system.

Chapter 11: Tools for Evaluating System Performance
Tools needed to analyze performance problems, such as ratios, limiting resistances, and base cases, are presented.

Chapter 12: Performance Curves for Individual Unit Operations
The performance of single pieces of equipment is analyzed for changes in process conditions, flowrates, utility flowrates, and degradation of equipment. It is assumed that the equipment has been designed and built and that the physical parameters of the equipment cannot change.

Chapter 13: Multiple Unit Performance

The performance of multiple pieces of equipment is analyzed. It is shown how a change in one unit affects the performance of another unit.

Chapter 14: Reactor Performance

Evaluation of the performance of different types of reactors is illustrated. The choice of process conditions to change selectivity is addressed.

Chapter 15: Regulating Process Conditions

Using examples from earlier chapters in this section, it is shown how a deviation in output from a piece of equipment can be controlled by altering an input. This is different from, and complementary to, what is treated in a typical process control class.

Chapter 16: Process Troubleshooting and Debottlenecking

Case studies are presented to introduce the philosophy and methodology for process troubleshooting and debottlenecking.

10

Process Input/Output Models

Imagine that you are in charge of operations for a portion of a chemical plant when you are informed that the pressure of a distillation column has begun to rise slowly. You know that if the pressure continues to rise the structural integrity of the column may be in jeopardy or that a relief valve will open and release valuable product to the stack. Both of these scenarios have serious negative consequences. In order to solve the problem without shutting down the plant, which might be very costly, it is necessary to understand how the distillation column performs in order to diagnose the problem and determine a remedy.

Dealing with the day-to-day performance of a chemical process differs from design of a new process. When designing a new process, there is freedom to choose equipment specifications as long as it produces the desired performance. However, once a piece of equipment has been designed and constructed, it performs in a unique manner. The specific performance of a piece of equipment must be considered when operating such equipment.

> **When designing equipment, alternative specifications can yield the same operating results. When dealing with existing equipment, day-to-day operation is constrained by the fixed equipment characteristics.**

While it may be tempting to proceed immediately to the analysis of performance of individual pieces of equipment, it is important to develop an intuitive understanding of how equipment performs. This chapter introduces a framework by

263

which individual equipment and complete process performance may be understood. The following chapters develop and use tools for analysis of system performance. Having an intuitive understanding of process and equipment performance is a necessary complement to the ability to do numerous, repetitive, high-speed calculations.

To determine the outputs of a unit operation or chemical process we must know the process inputs and understand the performance for each unit of equipment involved in the process. The relationship between input and output can be described as

$$Output = f(Input, \ Unit \ Performance) \qquad (10.1)$$

Input changes are the driving force for change.

Unit performance defines the characteristics of fixed equipment by which inputs are changed to outputs.

To change process output, process input and/or equipment performance must be altered. Conversely, the cause of a process output disturbance is a change in process input, equipment performance, or both.

Changes in process output result from changes in process inputs and/or equipment performance.

10.1   REPRESENTATION OF PROCESS INPUTS AND OUTPUTS

Figure 10.1 shows us how outputs are connected to inputs through the process system. The process system may consist of a single piece of equipment or several unit operations with the output from one unit becoming the input to another unit. As a result, a change of input to one unit operation affects all of the downstream units. For systems containing recycle stream(s) the output from a unit operation returns to affect its input.

Figure 10.1 provides the input/output representation of a chemical process. It shows,

    a. Process System (shown in the center box): This may consist of a single process unit, or a collection of process units (such as a distillation system) that, taken together, perform a specific function, or a complete chemical process.

    b. Process Flow Streams: These are divided into

        i. Input streams (shown on the left): They consist of process flow streams entering a unit operation or process.

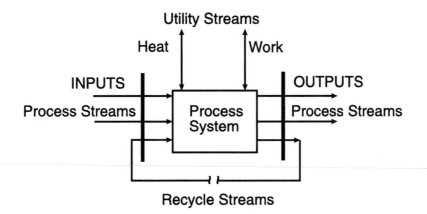

Figure 10.1   Input/Output Diagram for Chemical Process

    ii. Output streams (shown on the right): They consist of process flow streams exiting a unit operation or process.

  c. Utility Streams: Utility streams provide for the transfer of energy to or from process streams or process units. They are used to regulate temperatures and pressures required in the process. Utility stream inputs and outputs are shown at the top of Figure 10.1.

    i. Heat: Heat is most often provided by the flow of a heating or cooling medium, but can be provided by other sources such as electrical energy.

    ii. Work: This represents shaft work for a pump or compressor doing work on the fluid.

  d. Recycle Streams: These streams are internal to the process. They are critical to the operation of the process system. Recycle streams must be identified and their impact on the process understood. A recycle stream is illustrated by "cutting or tearing" the recycle stream (see Chapter 18). The "cut ends" create a set of pseudo-input and pseudo-output streams. They are shown along the bottom of the process system block, connected by a broken line. The pseudo-output stream is fed back to the process as a pseudo-input stream.

  Recycle streams fall into two categories:

    i. Process recycle streams recover raw materials and/or energy to reduce the cost of raw materials, waste disposal, and energy.

    ii. Equipment recycle streams affect equipment performance. As an example, increasing reflux to a distillation tower improves separations.

The following example illustrates how to represent process inputs and outputs for individual pieces of equipment.

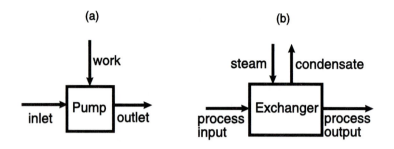

Figure E10.1a and b   Solutions to Example 10.1

Example 10.1

Draw an input/output model for

  a. a pump
  b. a heat exchanger in which a process stream is heated using condensing steam
  c. a distillation column

  a. The result is shown in Figure E10.1a. The mass balance on a pump has one input and one output, as shown. The energy balance on a pump (or one's intuitive understanding of how a pump works) indicates that there is energy input to the process stream, as shown.
  b. The result is shown in Figure E10.1b. A heat exchanger using a utility has one process stream input and one process stream output. The utility also flows in (steam) and out (condensate) of the heat exchanger.
  c. The result is shown in Figure E10.1c. The process streams are the feed, bottoms, and distillate. The utility streams are steam for the reboiler and cooling water for the condenser. Finally, the reflux from the condenser and the boil-up from the reboiler are represented as recycle streams, which both leave and enter across system boundaries.

Using the input/output diagram in Figure 10.1, two classifications are identified for process analysis problems encountered in this text.

  1. Process Design Analysis: Input and output streams are fixed. A process system is designed to transform the input into the output.
  2. Process Performance Analysis: Input and equipment are fixed. The outputs are determined by the process system.

(c)

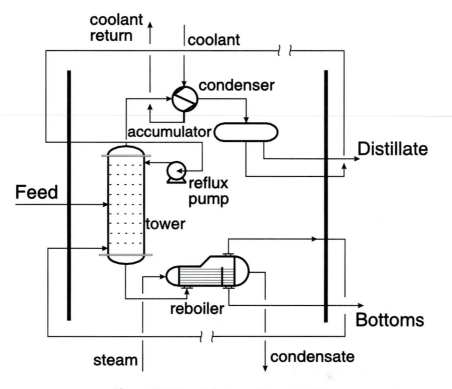

Figure E10.1c   Solution to Example 10.1c

10.2   ANALYSIS OF THE EFFECT OF PROCESS INPUTS ON PROCESS OUTPUTS

An intuitive understanding of the effect of process inputs on process outputs can be obtained by categorizing the relationships used to analyze equipment. We break down these relationships into two groups.

1. Equipment Independent Relationships: These relationships are independent of equipment specifications. Material balance, energy balance, kinetics, and equilibrium relationships are examples of equipment independent relationships.

2. Equipment Dependent Relationships: These are often called design equations, and involve equipment specifications. The heat transfer equation (contains area, A) and the frictional pressure relationship (contains pipe

size, D_p, and the equivalent length, L_{eq}) are examples of equipment dependent relationships.

> **Unless equipment dependent relationships are known and used, the impact of a change of process input on the output cannot be determined correctly.**

The following examples illustrate equipment independent and equipment dependent relationships.

Example 10.2

What relationships are used to analyze the following pieces of equipment? Classify these as equipment independent and equipment dependent.

   a. heat exchanger

   b. adiabatic reactor

   c. multistage extraction

   a. The energy balance, which for a temperature change without phase change is $Q = \dot{m}C_p\Delta T$, and for a phase change is $Q = \dot{m}\lambda$, are equipment independent relationships. The design (or performance) equation for a heat exchanger is $Q = UA\Delta T_{lm}F$. It is observed that only the equipment dependent relationship contains equipment specifications, in this case, the area for heat transfer, A.

   b. In a reactor, equipment independent relationships may include kinetics and/or the definition of equilibrium. If there are no equilibrium limitations, the kinetics may take the form $r = kC_AC_B$, where A and B are the reactants. All kinetic expressions are equipment independent and are only functions of temperature, pressure, and concentration. The energy balance, which is necessary to determine the outlet temperature of the adiabatic reactor, is another relevant equipment independent relationship. One equipment dependent relationship is the design equation for the reactor, and it depends upon the type of reactor. For example, for a CSTR, this relationship is $V/F_{Ao} = X/(-r_A)$. It is observed that this relationship contains an equipment specification, in this case, the reactor volume, V. Another equipment dependent relationship is for the pressure drop through the reactor, which is important in analyzing changes in flowrates.

   c. In any multistage separation involving a mass separating agent, the mass balance is the key equipment independent relationship. For example, it is understood that increasing the solvent rate results in a better separation. Another equipment independent relationship is the statement of phase equilibrium. An equipment-dependent relationship does not exist as a simple, closed-form equation, except in certain limiting cases. However, an intuitive understanding of the equipment dependent relationship is possible. For example, it is understood that increasing the number of stages will (usually) improve the separation. For the limiting case of dilute solutions, all of the above relationships can be expressed as a single, closed-form equation, the Kremser equation. This will be discussed in Chapter 12. Another equipment dependent relationship is the efficiency of the equipment, which is dependent on the liquid and vapor flowrates.

10.3   A PROCESS EXAMPLE

We assume that both the equipment independent and dependent relationships are understood, and we focus our attention on integrating these relationships to analyze complete processes. In a chemical process, a change in input to one piece of equipment or a change in performance of that piece of equipment affects more than just the output from that piece of equipment. Since the output from the piece of equipment in question usually becomes the input to another piece of equipment, the disturbance caused by the change in the original piece of equipment can propagate through the process.

This is important in two types of problems that will be discussed in later chapters. In one type of problem, the effect of a planned disturbance, such as a 10% scale-up in production, can only be quantified by considering the effect of a changed output from one unit on the subsequent unit. Similarly, in a second type of problem, an observed disturbance in the output of a particular unit may trace back to a disturbance in a unit far removed from the observed disturbance.

In order to analyze these problems, the input/output model for an entire process must be understood, as must the equipment independent and equipment dependent relationships for each piece of equipment in the process. This is illustrated in the following example.

Example 10.3

Consider the toluene hydrodealkylation process illustrated in Figure 1.5. Draw and label an input/output diagram, similar to Figure 10.1, for this process. Label all process inputs and outputs, all utility streams, and all recycles. Then, pick any two adjacent units and draw their input/output diagrams together, showing how the output from one unit affects the adjacent unit.

Figure E10.3a is the input/output diagram for the entire process. It is observed that there are two recycles, one for hydrogen and one for toluene, both unreacted reactants. This is a particularly useful representation because all of the inputs and outputs also represent cash flows. By showing all of the utilities, it is less likely that one could be omitted from a cash flow analysis.

Figure E10.3b is an input/output diagram for the toluene feed pump (P-101) and the heat exchanger (E-101). This diagram makes it evident that any disturbance in the feed to

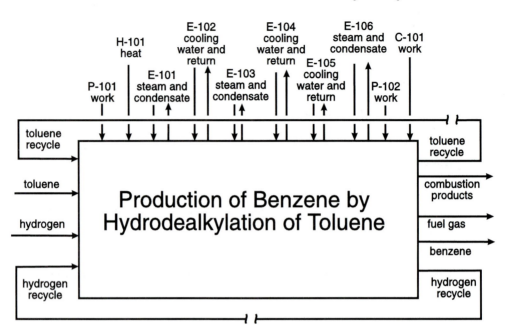

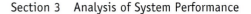

Figure E10.3a   Solution to Example 10.3a

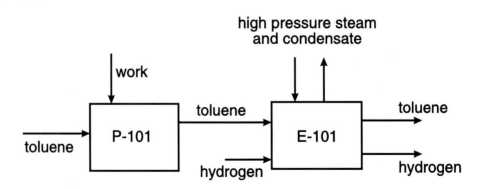

Figure E10.3b   One Possible Solution to Example 10.3b

or the operation of the pump not only affects the pump output, but also affects the heat exchanger output. Clearly, this disturbance would also propagate to other downstream equipment.

10.4 SUMMARY

This chapter introduced and demonstrated the importance of the input/output characteristics of chemical processes. In the input/output structure, process inputs are the driving force for change and the unit operations are the mechanism for change.

Recycle streams are of special importance, and create pseudo-input/output streams that are critical to analysis of process changes. Utility streams act as "servants" of the process units and streams and provide the desired temperature and pressure for the process units. They exchange energy with the process stream and process units.

As a consequence of the linkage between process units, a change in a single input or unit affects the total system. The impact on the process cannot be evaluated solely from analysis of isolated units.

PROBLEMS

1. Draw input/output diagrams for the following pieces of equipment:
    a. a pipe with frictional losses
    b. a turbine
    c. a compressor
    d. an absorber
    e. a reactor with heat removal or addition
    f. a membrane separation unit
    g. a heat exchanger between two process streams (no utilities)
2. Write down the equipment independent and equipment dependent relationships for each piece of equipment listed in Problem 1.
3. Write down the equipment independent and equipment dependent relationships for a distillation column.
4. Draw input/output diagrams for the following processes. Pick any two adjacent pieces of equipment and draw input/output diagrams for these, clearly showing how the output from one affects the input to the other.
    a. dimethyl ether, Figure B.1 (Note: These are PFDs located in the appendix.)
    b. acrylic acid, Figure B.2
    c. acetone, Figure B.3
    d. heptenes, Figure B.4

11

Tools for Evaluating System Performance

Several important computational and graphical tools used to analyze plant operations and changes in process design are developed and demonstrated in this chapter. Many of these tools are keyed to actual system performance. Tools presented are used extensively in future chapters.

Intuitive understanding is emphasized over computational complexity. All problems can be solved using a calculator without special functions. Graphical representations are also emphasized. They illustrate the characteristic behavior of a unit or process over a wide range of operations. Graphical presentations reveal critical operating regions that demand careful assessment.

11.1 KEY RELATIONSHIPS

In analyzing equipment performance, there are certain key relationships that are used over and over again. These are shown in Table 11.1. While these should be familiar to the reader, they are important enough to be reviewed here.

Table 11.1   Typical Key Performance Relationships

Situation	Equation	Trends	Comments
frictional loss for fluid flow	$$\Delta P = \frac{2\rho f L_{eq} u^2}{D}$$	$\lvert \Delta P \rvert \propto u^2$ $\lvert \Delta P \rvert \propto D^{-5}$ $\lvert \Delta P \rvert \propto L$	assumes fully developed turbulent flow, i.e., constant friction factor; for laminar flow $\Delta P \propto D^{-4}$
heat transfer	of form $$\left(\frac{hD}{k}\right) = c\left(\frac{Du\rho}{\mu}\right)^a\left(\frac{\mu C_p}{k}\right)^b$$	$h_i \propto v^{0.8}$ inside closed channels $h_o \propto v^{0.6}$ cross flow outside pipes	equations given are for no phase change; if phase change, weak flow dependence, but some ΔT dependence
kinetics	$$r = k\Pi c_i^{a_i}$$ $$k = k_o e^{-\left(\frac{E}{RT}\right)}$$	$\ln k$ vs $1/T$ is linear	as $T\uparrow$, $k\uparrow$ for ideal gases $P\uparrow$, $c_i\uparrow$, so $r\uparrow$
reactor	mixed flow: $$\frac{V}{F_{A0}} = \frac{\tau}{C_{A0}} = \frac{X}{-r_A}$$ plug flow: $$\frac{V}{F_{A0}} = \frac{\tau}{C_{A0}} = \int_0^X \frac{dX}{-r_A}$$	$\tau \propto V$ as $\tau\uparrow$ or $V\uparrow$, $X\uparrow$	τ is space time V is reactor volume X is conversion of limiting reactant, A assumes one reaction and constant volumetric flowrate
separator using mass separating agent	not necessarily described by a single equation	as flow of mass separating agent $\uparrow$, or as number stages or height of packed tower $\uparrow$, degree of separation $\uparrow$	for certain cases, there are limitations to the effect of increasing number of stages or packed tower height
distillation	not necessarily described by a single equation	as reflux ratio $\uparrow$, degree of separation $\uparrow$	complicated analysis, see Chapter 12

The first entry deals with the pressure drop due to frictional losses in a pipe. The key relationship is the proportionality between pressure drop and the square of the velocity, and the proportionality between pressure drop and length (or equivalent length) of pipe. While it is intuitive that the pressure drop should increase with length of pipe and with increasing velocity, the quantitative relationship reveals the square dependence on velocity. Similarly, while it is intuitive that a larger pipe diameter should reduce frictional losses, the negative fifth power dependence is only revealed by the equation describing the actual physical situation. Similarly, the general trends for heat transfer coefficients and for rate constants may be intuitive, but the equation describing the actual physical situation shows that quantitative dependence. For reactors and separators, only the intuitive trends can be shown because the exact, quantitative dependence is either situation specific or not easily quantifiable using a closed-form equation.

The relationships in Table 11.1 will be used extensively in Chapters 12, 13, and 16.

11.2   THINKING WITH EQUATIONS

It is possible to quantify equipment performance without resorting to extensive, detailed calculations. This involves using equations to understand trends. The first step is to identify the equations necessary to quantify a given situation. Wales and Stager [1] have termed this *thinking with equations,* and have used the acronym GENI to describe the associated problem-solving strategy. The second step involves predicting trends from equations. These methods are described and then illustrated in an example.

11.2.1   GENI

GENI is a method for solving quantitative problems. The name GENI is an acronym for the four steps in the method. The steps are:

1. Goal: Identify the goal. This is usually the unknown that needs to be calculated.
2. Equation: Identify the equation that relates the unknown to known values or properties.
3. Need: Identify additional relationships that are needed to solve the equation in #2, above.
4. Information: List additional information that is available to determine if what is needed in #3 is known. If the correct information is not known, the need becomes the new goal and the process is repeated.

11.2.2   Predicting Trends

The following method can be used to predict trends from equations known to apply to a given physical situation. In this method, there are four possible modifiers to a term in an equation.

1. ¢ value remains constant
2. ↑: value increases
3. ↓: value decreases
4. ?: value change not known

Each variable in an expression has one of these identifiers appended. If x is a constant then $x \rightarrow x(\cent)$. The symbol in parenthesis identifies the effect the variable has on the term, that is, reducing a value in the denominator of a term increases the term.

The application of the techniques described in Sections 11.2.1 and 11.2.2 is illustrated in the following example.

Example 11.1

For a bimolecular, elementary, gas-phase reaction, the rate expression is known to be $-r_A = kC_AC_B$.

a. What is the effect on the reaction rate of increasing the reaction pressure by 10% while maintaining constant temperature?

b. What is the effect on the reaction rate of increasing the reaction temperature by 10% while maintaining constant pressure?

The **goal** is to determine the effect of increasing pressure on the reaction rate. The **equation** containing the unknown value, which is the reaction rate, r_A, is given. We **need** a relationship between pressure and at least one variable in the equation. We have **information,** from the ideal gas law, that tells us that $C_i = P_i/RT$, where i is A or B.

Now that we have the necessary relationships, an intuitive understanding can be obtained by predicting trends.

a. First, let us examine the ideal gas relationship. If the pressure increases with the temperature remaining constant, then the ideal gas relationship can be written as

$$C_i(\uparrow) = y_i(\cancel{\cent})P(\uparrow)/R(\cancel{\cent})T(\cancel{\cent}) = P_i(\uparrow)/R(\cent)T(\cent)$$

Therefore, the concentration increases because the pressure increases while the temperature and the gas constant remain constant. Intuitively, if the pressure increases by 10%, the concentration must increase by 10%.

Moving to the rate expression, since the rate constant is not affected by pressure, the resulting trend is predicted by

$$r_A(\uparrow) = k(\cent)C_A(\uparrow)C_B(\uparrow)$$

which means that the reaction rate increases. Intuitively, since each concentration increases by 10%, then the reaction rate changes by a factor of $(1.1)^2 = 1.21$, an increase of 21%.

b. In this case, from the ideal gas relationship

$$C_i(\downarrow) = y_i(\cent)P(\cent)/R(\cent)T(\uparrow) = P_i(\cent)/R(\cent)T(\uparrow)$$

From the rate expression, since the rate constant is a function of temperature

$$r_A(?) = k(\uparrow)C_A(\downarrow)C_B(\downarrow)$$

Since the concentrations decrease and the rate constant increases, it cannot be determined, *a priori*, whether the rate increases or decreases. If the activation energy in the rate constant were known, a quantitative evaluation could be made.

11.3   BASE CASE RATIOS

The calculation tool provided in this section combines use of fundamental relationships with plant operating data to form a basis for predicting changes in system behavior.

Your ability to predict changes in a process design or in plant operations is improved by anchoring an analysis to a base case. For design changes, we would *18* like to identify a design proven in practice as the base case. For operating plants, actual data are available and are chosen as the base case. It is important to put this base case into perspective. Assuming that there are no instrument malfunctions and the operating data are correct, then these data represent a real operating *84* point at the time the data were taken. As the plant ages, the effectiveness of process units change and operations are altered to account for these changes. As a consequence, recent data on plant operations should be used in setting up the base case.

> **Establish predictions of process changes on known operating data, not on design data.**

The base case ratio integrates the "best available" information from the operating plant with design relationships to predict process changes. It is an important and powerful technique with wide application. The base case ratio, X, is defined as the ratio of a new case system characteristic, x_2, to the base case system characteristic, x_1.

$$X \equiv x_2/x_1 \qquad (11.1)$$

Using a base case ratio often reduces the need for knowing actual values of physical and transport properties (physical properties refer to thermodynamic and transport properties of fluids), equipment, and equipment characteristics. The values identified in the ratios fall into three major groups. They are defined below and applied in Example 11.2.

1. Ratios related to equipment sizes (L_{eq}, equivalent length; diameter, D; surface area, A): Assuming that the equipment is not modified, these values are constant, the ratios are unity, and these terms cancel out.

2. Ratios related to physical properties (such as density, ρ; viscosity, μ): These values can be functions of material composition, temperature, and pressure. Absolute values are not needed, only the functional relationships. Quite often, for small changes in composition, temperature, or pressure, the properties are unchanged, and the ratio is unity and cancels out. An exception to this is gas-phase density.

3. Ratios related to stream properties: These usually involve velocity, flowrate, concentration, temperature, and pressure.

Using the base case ratio eliminates the need to know equipment characteristics and reduces the amount of physical property data needed to predict changes in operating systems.

The base case ratio is a powerful and straightforward tool to analyze and predict process changes. This is illustrated in the following example.

Example 11.2

It is necessary to scale up production in an existing chemical plant by 25%. It is your job to determine whether a particular pump has sufficient capacity to handle the scale-up. The pump's function is to provide enough pressure to overcome frictional losses between the pump and a reactor.

The relationship for frictional pressure drop is given in Table 11.1. This relationship is now written as the ratio of two base cases as follows:

$$\frac{\Delta P_2}{\Delta P_1} = \frac{2\rho_2 f_2 L_{eq2} u_2^2 D_1}{2\rho_1 f_1 L_{eq1} u_1^2 D_2}$$

Since the pipe has not been changed, the ratios of diameters (D_2/D_1) and lengths (L_{eq2}/L_{eq1}) are unity. Since a pump is only used for liquids, and liquids are (practically) incompressible, the ratio of densities is unity. If the flow is assumed to be fully turbulent, which is usually true for process applications, the friction factor is not a function of Reynolds number. (This fact should be checked for a particular application.) Therefore, the friction factor is constant, and the ratio of friction factors is unity. The above ratio reduces to

$$\frac{\Delta P_2}{\Delta P_1} = \frac{u_2^2}{u_1^2} = \frac{\dot{m}_2^2/A_2^2\rho_2^2}{\dot{m}_1^2/A_1^2\rho_1^2} = \frac{\dot{m}_2^2}{\dot{m}_1^2}$$

where the second equality is obtained by substituting for u_i in numerator and denominator using the mass balance $\dot{m}_i = \rho_i A_i u_i$, canceling the ratio of densities for the same reason as above, and canceling the ratio of cross-sectional areas since the pipe has remained unchanged. Therefore, by assigning the base case mass flow to have a value of 1, for a 25% scale-up, the new case has a mass flow of 1.25, and the ratio of pressure drops becomes

$$\frac{\Delta P_2}{\Delta P_1} = \left(\frac{\dot{m}_2}{\dot{m}_1}\right)^2 = \left(\frac{1.25}{1}\right)^2 = 1.56$$

and the pump must be able to deliver enough head to overcome 56% additional frictional pressure drop while pumping 25% more material.

It is important to observe that Example 11.2 was solved without knowing any details of the system. The pipe diameter, length, and number of valves and fittings were not known. The liquid being pumped, its temperature, and its density were not known. Yet, the use of base case ratios along with simple assumptions permitted a solution to be obtained. This illustrates the power and simplicity of base case ratios.

11.4   ANALYSIS OF SYSTEMS USING CONTROLLING RESISTANCES

Design relationships for many operations such as fluid flow, heat transfer, mass transfer, and chemical reactors all involve rate equations of the general form,

$$\text{Rate} = \text{Driving Force/Resistance} \tag{11.2}$$

For resistances in series,

$$R_T \,(\text{Total Resistance}) = R_1 + R_2 + R_3 + \cdots + R_N \tag{11.3}$$

For certain situations, one resistance dominates all other resistances. For example, if resistance R_1 dominates, then

$$R_1 \gg R_2 + R_3 + \cdots + R_N \tag{11.4}$$

and

$$R_T \approx R_1 \tag{11.5}$$

where R_1 represents the "controlling resistance." Other resistances have little impact on the rate. Only those factors that impact R_1 have a significant impact on the rate.

As an example, an overall heat transfer coefficient for a clean (non-fouling) service, U_o, can be expressed as

$$\frac{1}{U_o} = \frac{1}{h_o} + \frac{D_o \ln\left(\dfrac{D_o}{D_i}\right)}{2k} + \frac{D_o}{D_i h_i} \tag{11.6}$$

where D_o and D_i are the outer and inner tube radii, respectively, k is the thermal conductivity of the tube material, and h_o and h_i are the outer and inner heat transfer coefficients. We will assume that Eq. (11.6) can be simplified by assuming $D_o \approx D_i$, and that the conduction resistance is negligible, so that

$$\frac{1}{U_o} = \frac{1}{h_o} + \frac{1}{h_i} \qquad (11.7)$$

Suppose that we wanted to use a base case ratio of a new overall heat transfer coefficient, U_{o2}, to an original overall heat transfer coefficient, U_{o1}. For a situation when no phase change is occurring on either side of a shell and tube heat exchanger, from Eq. (11.7), this ratio would be

$$\frac{\left(\dfrac{1}{U_{o1}}\right)}{\left(\dfrac{1}{U_{o2}}\right)} = \frac{U_{o2}}{U_{o1}} = \frac{\dfrac{1}{\alpha u_{i1}^{0.8}} + \dfrac{1}{\beta u_{o1}^{0.6}}}{\dfrac{1}{\alpha u_{i2}^{0.8}} + \dfrac{1}{\beta u_{o2}^{0.6}}} \qquad (11.8)$$

where the individual heat transfer coefficients have been expressed as indicated in Table 11.1, and the proportionality factors α and β contain the lead constant and all of the properties contained in the dimensionless groups other than the heat transfer coefficient and the velocity. From Eq. (11.8), it would not be possible, as it was in Example 11.2, to determine quantitatively how the overall heat transfer coefficient changes with changing velocity (i.e., mass flowrate) without knowing values for all of the physical properties. However, if it could be assumed that one resistance were dominant, the problem would be greatly simplified. For example, let us assume that a heat transfer fluid in the shell is heating a gas in the tubes. It is likely that the resistance in the tubes would dominate, given the low film heat transfer coefficient for gases. Therefore, the base case ratio would reduce to

$$\frac{U_{o2}}{U_{o1}} = \frac{h_{i2}}{h_{i1}} = \left(\frac{u_{i2}}{u_{i1}}\right)^{0.8} \qquad (11.9)$$

and it would be possible to predict a new heat transfer coefficient without having values for the physical properties.

The reduction of the heat transfer coefficient ratio using a limiting resistance described above is very powerful but must be used with care. It is not valid for all situations. It is most likely valid when one resistance is for a boiling liquid or condensing steam and the other resistance is for a liquid or gas without phase change. In other situations, the base case ratio can only be determined from Eq. (11.8) if the relative magnitude for each resistance is known for the base case, as is illustrated in Example 11.4. It is also important to understand that Eqs. (11.8) and (11.9) do not represent all possible base case ratios arising in heat transfer. Each situation must be analyzed individually to ensure that a correct solution is obtained. Problems of this type are given at the end of the chapter.

Situations in which the simplifications leading to Eq. (11.9) are valid, along with situations that require a form for the base case ratio similar to Eq. (11.8), will be applied to heat exchanger performance in Chapter 12. Examples of how Eqs. (11.8) and (11.9) can be applied are presented here.

Example 11.3

It is desired to scale down process capacity by 25%. In a particular heat exchanger, process gas in the tubes is heated by condensing steam in the shell. By how much will the overall heat transfer coefficient for the heat exchanger change after scale-down?

Using the subscript 2 for the new, scaled-down operation and the subscript 1 for the original conditions, a scale-down of the mass flow rate of the process stream by 25% means that $u_{i2} = 0.75u_{i1}$. For a process gas heated by condensing steam, it is certain that the heat transfer coefficient for condensing steam will be at least 100 times larger than that for the process gas. Although Eq. (11.9) is applicable, it is not derived from Eq. (11.8) because of the phase change. The correct derivation is assigned in Problem 11.5 at the end of the chapter. The result is $U_{o2}/U_{o1} = (u_{i2}/u_{i1})^{0.8} = (0.75)^{0.8} = 0.79$, and the overall heat transfer coefficient is reduced by 21%.

Example 11.4

In a similar process as in Example 11.3, there is a heat exchanger between two process streams, neither of which involves a phase change. It is known that the resistances on the shell and tube sides are approximately equal before scale-down. By how much will the overall heat transfer coefficient for the heat exchanger change after scale-down by 25%?

Here, Eq. (11.7) is required. Since both original resistances are equal, both original heat transfer coefficients will be denoted h_1. In the tubes, $h_{i2}/h_1 = (u_{i2}/u_{i1})^{0.8}$, and in the shell $h_{o2}/h_1 = (u_{o2}/u_{o1})^{0.6}$. Both velocity ratios are 0.75. Therefore, Eq. (11.7) reduces to

$$\frac{U_2}{U_1} = \frac{\dfrac{1}{h_1} + \dfrac{1}{h_1}}{\dfrac{1}{h_1\left(\dfrac{u_{i2}}{u_{i1}}\right)^{0.8}} + \dfrac{1}{h_1\left(\dfrac{u_{o2}}{u_{o1}}\right)^{0.6}}} = \frac{2}{\dfrac{1}{(0.75)^{0.8}} + \dfrac{1}{(0.75)^{0.6}}} = 0.82$$

and the new heat transfer coefficient is reduced by 18%.

11.5 GRAPHICAL REPRESENTATIONS

At times, graphical representations are useful descriptions of physical situations. They provide a means to an intuitive understanding of a problem rather than a computational tool. Probably the best known example of this is the McCabe-Thiele diagram for distillation. While no one would design a distillation column nowadays using this method, the McCabe-Thiele diagram provides a means for an intuitive understanding of distillation. Another example is the reactor profiles illustrated in Chapter 14.

In this section, three graphical representations that will be used to analyze performance problems are discussed.

11.5.1   The Moody Diagram for Friction Factors

An example of a graphical representation commonly used for illustrative and for computational purposes is the Moody Diagram, which gives the friction factor as a function of Reynolds number for varying roughness factors. It is illustrated in Figure 11.1. While this diagram is often used for numerical calculations, it also provides an intuitive understanding of frictional losses. For example, it is observed that the friction factor increases as the pipe roughness increases. It is also observed that the friction factor becomes constant at high Reynolds numbers, and the dashed line represents the boundary between variable and constant friction factor.

11.5.2   The System Curve for Frictional Losses

In Table 11.1, an equation was given for frictional losses in a pipe. This equation is derived from the mechanical energy balance for constant density

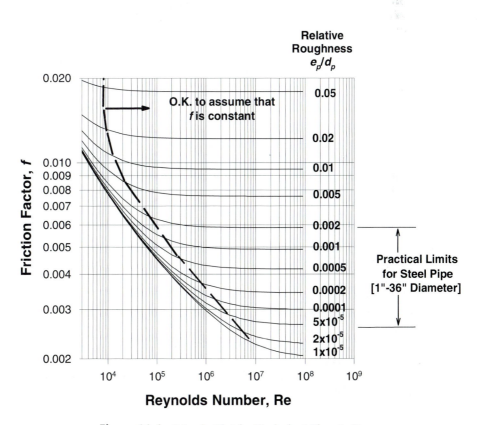

Figure 11.1   Moody Plot for Turbulent Flow in Pipes

$$\frac{\Delta P}{\rho g} + \Delta z + \Delta \frac{u^2}{2g} - \frac{W_s}{g} + \frac{F_d}{g} = 0 \tag{11.10}$$

In Eq. (11.10), Δ means out-in, and the specific energy dissipation due to drag, F_d, is a positive number. For a constant diameter length of pipe with no pump at constant elevation, $\Delta z = 0$, there is no work, the kinetic energy term is zero, and, using the expression for frictional losses, the result is

$$\Delta P = -\rho F_d = -\frac{2\rho f L_{eq} u^2}{D} \tag{11.11}$$

If fully developed turbulent flow is assumed, the friction factor, f, is constant. For process changes involving the same fluid, the density remains constant, and if the pipe is unchanged, the equivalent length and diameter are unchanged. Therefore, Eq. (11.11) describes how the pressure drop in a length of pipe changes with flowrate or velocity. This can be plotted as ΔP versus velocity, and the result is a parabola passing through the origin. This is called the "system curve," and will be shown in Chapter 12 to be a useful tool in evaluating pump performance.

If, for example, there were an elevation change over the length of pipe in question, there would be an additional term in Eq. (11.11), and the result would be a parabola with a non-zero intercept. The following examples illustrate how a system curve is obtained.

Example 11.5

Develop the system curve for flow of water at approximately 10 kg/s through 100 m of 2 in schedule 40 commercial steel pipe oriented horizontally.

The density of water will be taken as 1000 kg/m³, and the viscosity of water will be taken as 1 mPa s (0.001 kg/m s). The inside diameter of the pipe is 0.0525 m. The Reynolds number can be determined to be 2.42×10^5. For a roughness factor of 0.001, $f = 0.005$. Eq. (11.11) reduces to

$$\Delta P = -19u^2$$

with ΔP in kPa and u in m/s. This is the equation of a parabola, and it is plotted in Figure E11.5. Therefore, from either the equation or the graph, the pressure drop is known for any velocity.

Example 11.6

Repeat Example 11.5 for pipe with a 10 m vertical elevation change, with the flow from lower to higher elevation.

Here, the potential energy term from the mechanical energy balance must be included. The magnitude of this term is 10 m of water, so $\rho g \Delta z = 98$ kPa. Eq. (11.10) reduces to

$$\Delta P = -(98 + 19u^2)$$

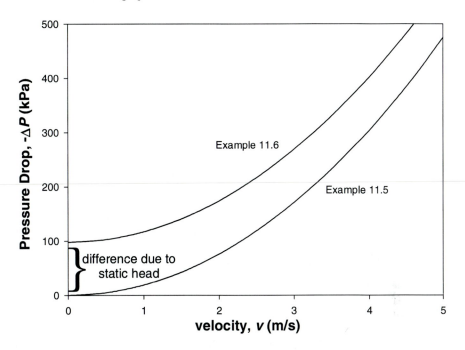

Figure E11.5   System Curves for Examples 11.5 and 11.6

with ΔP in kPa and u in m/s. This equation is also plotted in Figure E11.5. It is observed that the system curve has the same shape as that in Example 11.5. This means that the frictional component is unchanged. The difference is that the entire curve is shifted up by the constant, static pressure difference.

11.5.3   The T-Q Diagram for Heat Exchangers

Another example of a useful diagram that illustrates the behavior of a piece of equipment is the T-Q diagram for a heat exchanger. Figure 11.2 illustrates a T-Q diagram for the countercurrent heat exchanger with no phase change shown in the figure.

A T-Q diagram is a visual representation of the energy balance equation for each stream. Since, for single phase streams with constant C_p and no pressure effect on the enthalpy,

$$Q = \dot{m}C_p\Delta T \tag{11.12}$$

process streams that undergo no phase change are represented by a line with a slope of $1/(\dot{m}C_p)$. Since, for pure component streams undergoing a phase change

$$Q = \dot{m}\lambda \tag{11.13}$$

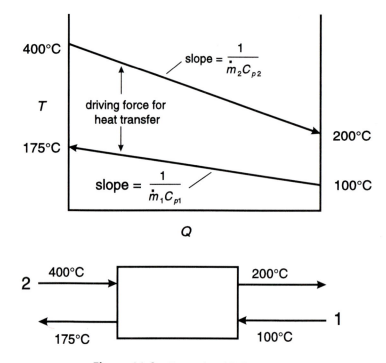

Figure 11.2   Example of T-Q Diagram

process streams that involve constant temperature phase changes (pure component vaporizers and condensers) are represented by a horizontal line.

The temperature differences between the two streams shown on the T-Q diagram provide the actual temperature driving force throughout the exchanger. The greater the temperature separation, the larger the driving force (ΔT_{lm}), and the greater the heat transferred.

The representations described above are for simple situations commonly encountered. However, more complex cases exist. If the heat capacity is not constant, then the line for heat transfer without phase change is a curve. For phase changes involving multicomponent systems, since the bubble and dew points are at different temperatures, the line representing the phase change is not horizontal. For partial condensers and vaporizers, the representation on a T-Q diagram is a curve rather than a straight line.

The T-Q diagram reveals two important truths regarding heat transfer.

1. Temperature lines cannot cross: This is an impossible situation. Temperature lines will never cross when dealing with operating equipment. If a temperature cross is encountered in doing a calculation, an error has been made.

2. Temperature lines should not approach each other too closely: As temperature lines approach each other, the area required for a heat exchanger approaches infinity. The point of closest approach is called the "pinch point." When dealing with multiple heat exchangers, the pinch point is a key to heat exchanger network integration, that is, determining the configuration for most efficient heat transfer between hot and cold streams. This concept will be discussed in Chapter 19.

The following example illustrates the construction of T-Q diagrams.

Example 11.7

Sketch a T-Q diagram for the following situations:

    a. A single-phase process stream is heated from 100°C to 200°C by condensation of saturated steam to saturated liquid at 250°C in a countercurrent heat exchanger.

    b. A single-phase process stream is heated from 120°C to 220°C by condensation of saturated steam at 250°C and subcooling of the condensate to 225°C in a countercurrent heat exchanger.

    The solution to Part a is shown in Figure E11.7a. The horizontal line at 250°C is for the steam condensing at constant temperature. The sloped line is for heating of the process stream. The arrow on the sloped line indicates the direction of flow of the process stream, since it is being heated. The arrow on the condensing steam line is opposite because of countercurrent flow.

    The solution to Part b is shown in Figure E11.7b. The difference from Part a is the subcooling zone. It is important to understand that when there is a pure-component phase change, there must be a horizontal portion of the line. It is incorrect to draw a single

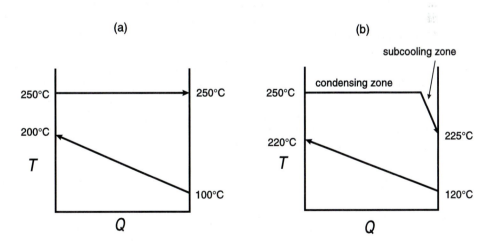

Figure E11.7   Solutions to Example 11.7

straight line between the two end temperatures. In fact, due to the amount of heat associated with phase changes relative to the heat in temperature changes, the horizontal portion, associated with the phase change, will almost always be the longer segment.

11.6 SUMMARY

In this chapter, several important tools have been introduced that are essential in analyzing equipment performance problems. Base case ratios are the most important tool. They permit comparison of two cases, usually without complex calculations, and often without the need to know physical and transport properties. Under the circumstances for which it is valid, using a limiting resistance simplifies the base case ratio. Finally, graphical representation of equipment performance is a useful tool in understanding the physical situation. All of these techniques require a minimum of calculations. However, they do provide an intuitive understanding of equipment performance that is very rarely achieved from merely doing repetitive, complex computations.

REFERENCE

1. Wales, C. E. and R. A. Stager, *Thinking with Equations,* West Virginia University, Morgantown, WV, 1990.

PROBLEMS

1. It is known that for a certain third order, elementary, gas-phase reaction (first order in A, second order in B), the rate of reaction doubles when the temperature goes from 250°C to 270°C. If the rate of reaction is 10 mole/m^{3}s at the base conditions, which are 20 atm pressure and 250°C, and equal molar flows of A and B are in the feed (no inerts), answer the following questions:
   a. Compared to the base case, by how much does the rate of reaction (at inlet conditions) change if the temperature is increased to 260°C?
   b. Compared to the base case, by how much does the rate of reaction (at inlet conditions) change if the pressure is increased by 15%?

2. A storage tank is connected to a pond (at atmospheric pressure!) by a length of 4 in pipe and a gate valve. From previous operating experience it has been found that when the tank is at a pressure of 2 atm the flow through the pipe is 20 m^3/h when the gate valve is fully open. If the pressure in the tank increases to 3 atm what will be the maximum discharge rate from the tank?

3. In a single-pass shell and tube heat exchanger, cooling water is used to condense an organic vapor. Under present operating conditions, the heat transfer coefficients are h_i = 2300 W/m^{2}K (turbulent flow of cooling water,

Re = 20,000), h_o = 950 W/m^{2}K. Fouling is negligible and the tubes are 1 in 16 BWG carbon steel. Water flows in the tubes. If the cooling water rate were to drop suddenly by 10%, estimate the change in overall heat transfer coefficient. Clearly state all assumptions made.

4. In a shell and tube heat exchanger, the overall and individual heat transfer coefficients are U = 200 W/m^{2}K, h_i = 850 W/m^{2}K, and h_o = 350 W/m^{2}K. Both the shell-side and the tube-side fluids are liquids, and no change of phase occurs. The flow of both fluids is turbulent.
   a. Estimate the new value of U if the flowrate of the shell-side fluid is increased by 50%.
   b. Estimate the new value of U if the flowrate of the tube-side fluid is increased by 50%.

5. Derive the relationship, identical to Eq. (11.9), for the situation described in Example 11.3.

6. You are considering two pipes to connect liquid leaving a pump to the entrance to a reactor. For a given mass flowrate of liquid, how much operating cost would you save (or lose) by using 2.5 in schedule 40 pipe rather than 2 in schedule 40 pipe? The pump runs on electricity, and electricity costs $0.05/kWh. The pump efficiency is 75%.

7. For laminar flow of a Newtonian liquid in a pipe, determine the effect on flowrate of the following changes:
   a. the pressure drop doubles (everything else remains constant)
   b. the pipe is changed from 1.25 in Schedule 40 to 1 in Schedule 40 (everything else remains constant)
   c. the viscosity of the fluid is reduced by 50% (everything else remains constant)
   d. the equivalent length of pipe is increased by an order of magnitude (everything else remains constant)
   e. a and b
   f. a and c
   g. b and d

8. Sketch T-Q diagrams for the following situations:
   a. the condenser on a distillation column, vapor condenses at 160°C, cooling water enters at 30°C and exits at 50°C
   b. the reboiler on the same distillation column, vapor is reboiled at 220°C, saturated steam at 4237 kPa is condensing to saturated water
   c. ethanol, initially at 30°C and atmospheric pressure is vaporized to 130°C and atmospheric pressure, saturated steam at 1135 kPa is condensed to saturated water

9. Air at approximately STP and 10 kg/s flows through 50 m of 6 in commercial steel schedule 40 pipe. Derive an expression for the system curve and sketch the system curve for this situation.

12

Performance Curves for Individual Unit Operations

As pointed out in the introduction to Section 3, the way in which a process operates will vary significantly throughout its lifetime. Plant operations do not correspond to the conditions specified in the design. This is *not* necessarily a reflection of a poor design. It is a consequence of changes in the process during the life of the plant. There are numerous reasons why a process might not be operated at design conditions. As stated previously, some examples are:

- Design/Construction: Installed equipment is often oversized. This reduces risks resulting from inaccuracies in design correlations, uncertainties in material properties, and so on.
- External Effects: Feed materials, product specifications and flowrates, environmental regulations, and costs of raw materials and utilities all are likely to change during the life of the process.
- Replacement of Equipment: New and improved equipment (or catalysts) may replace existing units in the plant.
- Changes in Equipment Performance: In general, equipment effectiveness degrades with age. For example, heat transfer surfaces foul, packed towers develop channels, catalysts lose activity, and bearings on pumps and compressors become worn. Plants are shut down periodically for maintenance to restore equipment performance.

With these factors in mind, a good design is one where operating conditions and equipment performance can be changed throughout the life of the process/plant. This is known as process flexibility. For a company or operation to remain competitive in the marketplace, it must respond to these changes. Therefore, it is essential for us to understand how equipment performs over its entire operating range and to be able to evaluate the effects of changing process conditions on the overall process performance.

Several techniques for evaluating operating systems are presented in Chapters 10 and 11. The purpose of this chapter is to apply these techniques to obtain a solution to a performance problem for a specific piece of equipment. This solution may be a single answer. However, a much more useful solution is a curve or a family of curves that represents the way an existing piece of equipment or system responds to changes in input or equipment variables. These curves are referred to as performance curves. They are the basis for predicting the behavior of existing equipment. Performance curves present a whole range of possible solutions rather than merely a single answer. In principle, we could use performance equations instead of performance curves. However, by representing equipment performance in graphical form, the performance characteristics are easier to visualize, providing a better intuitive understanding. For example, in Figure 11.1, the boundary for fully developed turbulent flow, above which the friction factor is constant, is more effectively represented by the dotted curve on the graph than by the equation of that curve.

Performance curves represent the relationship between process outputs and process inputs.

By plotting the response variable(s) as a function of the input variable(s), the sensitivity of one to the other becomes immediately obvious. Such sensitivity cannot be inferred from a single (numerical) solution. With the wide availability of spreadsheets and process simulators, the effort expended generating performance curves may be justified.

In order to construct performance and system curves, material and energy balance equations must be used along with the (design) equations relating equipment parameters. However, other constraints may also have to be considered, such as the maximum or minimum system temperature and pressure allowed for the equipment, the maximum velocity of fluid through the equipment to avoid excessive erosion, the maximum or minimum velocity to avoid flooding in packed towers and tray columns, the minimum velocity through a reactor to avoid defluidization, the maximum residence time in a reactor to avoid coking/cracking reactions or by-product formation, the minimum flow through a compressor to avoid surging, or the minimum approach temperature to avoid the condensation of acidic gases inside heat exchangers.

It is possible to construct performance curves for essentially any piece of equipment. In this chapter, we develop curves for several specific equipment types. This "case-study" approach is used to help you develop your own methods for generating performance and system curves for equipment not covered in this text.

12.1 APPLICATIONS TO HEAT TRANSFER

In this section, we analyze the steam generator shown in Figure 12.1a to illustrate the preparation and value of a performance diagram for predicting the response of a heat exchanger to changing conditions. The data shown were obtained from an actual operating system. A steam generator is similar to a shell and tube heat exchanger. Figure 12.2 contains drawings of portions of shell and tube heat exchangers.

> *Saturated steam is produced in a kettle-type vaporizer containing long vertical tubes. Heat is provided from a hot light oil stream that enters at 325°C and leaves at 300°C. The effective area for heat transfer is adjusted by changing the level of the boiling liquid in the exchanger. Figure 12.1a provides the current operating conditions, some limited thermodynamic data, data on the vaporizer, and a sketch of the equipment.*

The conditions given in Figure 12.1a designate the base case. Development of performance curves involves solving the following three equations simultaneously:

$$\text{Heat lost by light oil, } Q = \dot{m}(H_1 - H_2) = \dot{m}C_{p,\,oil}(T_{in} - T_{out}) \tag{12.1}$$

$$\text{Heat gained by water stream, } Q = \dot{m}_s(H_s - H_w) = \dot{m}_s[C_p(T_B - T_w) + \lambda_s] \tag{12.2}$$

$$\text{Heat transferred, } Q = UA\Delta T_{lm} \tag{12.3}$$

In these equations,

H_1 and H_2 refer to the specific enthalpy, kJ/kg, for the light oil, with the subscript 1 for inlet and 2 for outlet conditions.

H_s and H_w refer to the specific enthalpy, kJ/kg for the steam and water, respectively.

$\dot{m}$ and $\dot{m}_s$ refer to the stream flow rate, kg/h, for the light oil and water/steam stream, respectively.

λ_s is the latent heat of the steam.

The heat transfer equation, Eq. (12.3), contains the characteristic factors specific to the heat transfer equipment.

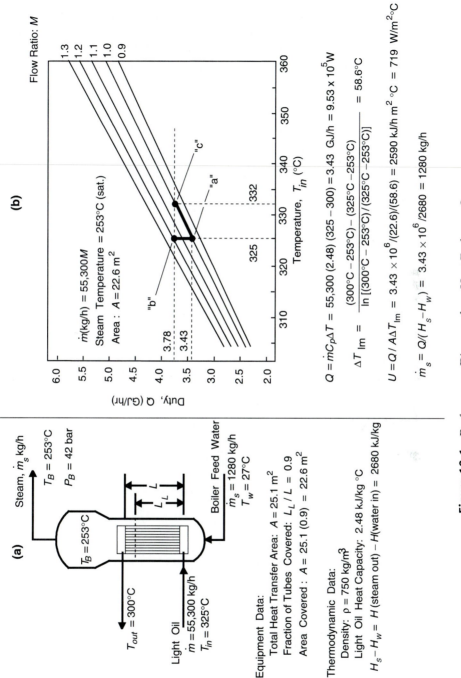

Figure 12.1 Performance Diagram for a Heat Exchange System

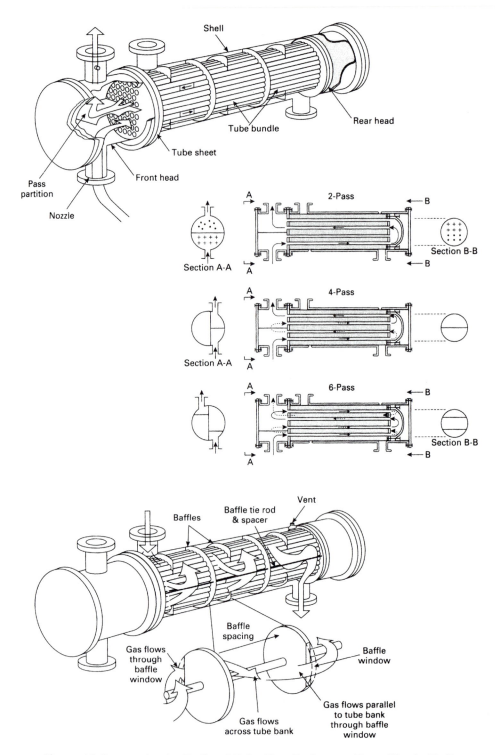

Labels in figure:

Shell

Rear head

Tube bundle

Tube sheet

Front head

Pass partition

Nozzle

A — 2-Pass — B

Section A-A

Section B-B

A — 4-Pass

Section A-A

A — 6-Pass — B

Section B-B

Baffles

Baffle tie rod & spacer

Vent

Baffle spacing

Gas flows through baffle window

Baffle window

Gas flows across tube bank

Gas flows parallel to tube bank through baffle window

Figure 12.2 Details of a Shell-and-Tube Heat Exchanger (From Woods, D. R., *Process Design and Engineering Practice*, Prentice-Hall, Englewood Cliffs, NJ, 1995.)

Figure 12.1b shows the calculations used for the base case. They include the heat transfer rate, Q, the steam generation rate, $\dot{m}_s$, the overall heat transfer coefficient, U, and the log-mean temperature difference, ΔT_{lm}.

Figure 12.1b is the performance diagram for this boiler system subject to the following constraints.

1. The temperature of the boiling water, $T_B = 253°C$, which is set by the pressure of the water in contact with the tubes.
2. The liquid level is $0.9L$ (where L = tube height).
3. Total heat transfer area, $A_T = 25.1 \text{ m}^2$.
4. The time at which the operating data were obtained.

The heat transfer rate or vaporization duty, Q, is plotted in Figure 12.1b as a function of the inlet light oil coolant temperature, T_{in}, with flowrate ratio of light oil, M, as a parameter. M is the ratio of the current oil flow rate (subscript 2) to the base case oil flow rate (subscript 1)

$$M = \dot{m}_2/\dot{m}_1 \tag{12.4}$$

where $\dot{m}_1 = 55{,}300 \text{ kg/h}$.

Before we discuss how these curves were developed, we will use the performance curves to predict the effects of changing operating conditions.

Example 12.1

An increase in steam production of 10% is needed. You are to provide the operator with the new input streams conditions for two cases, by completing the following table:

	Light Oil Flow (Mg/h)	$T_{in}(°C)$	$T_{out}(°C)$
Current Values	55.3	325	300
Case a	55.3	?	?
Case b	?	325	?

These problems can be solved by the methods developed in Chapter 11. If only a single answer is desired, this method is faster than preparation of a complete performance curve, which is described later.

If the steam production must increase by 10%, then the ratio of the new case (2) to the base case (1) from Eq. (12.2), is

$$\frac{Q_2}{Q_1} = \frac{\dot{m}_{s2}(H_s - H_w)_2}{\dot{m}_{s1}(H_s - H_w)_1} = 1.1 \tag{E12.1a}$$

Since the enthalpy difference is unchanged for the phase change of water to steam, then the ratio $Q_2/Q_1 = 1.1$. Now, the ratio of Q_2/Q_1 must be written for the remaining two equations, 12.1 and 12.3. For Case a, these are

$$\frac{Q_2}{Q_1} = 1.1 = \frac{\dot{m}_2 C_{p2}(T_{in} - T_{out})_2}{\dot{m}_1 C_{p1}(T_{in} - T_{out})_1} = \frac{(T_{in} - T_{out})_2}{25} \tag{E12.1b}$$

$$\frac{Q_2}{Q_1} = 1.1 = \frac{U_2 A_2 \Delta T_{lm2}}{U_1 A_1 \Delta T_{lm1}} = \frac{(T_{out} - T_{in})_2}{58.6 \ln\left(\dfrac{T_{out2} - 253}{T_{in2} - 253}\right)} \frac{U_2}{U_1} \qquad \text{(E12.1c)}$$

Since the mass flowrate of the oil does not change, its heat transfer coefficient remains constant. Since a boiling heat transfer coefficient does not change with flowrate, the overall heat transfer coefficient is unchanged. Therefore, $U_1 = U_2$. These assumptions were used in obtaining Eq. (E12.1c). Solving Eqs. (E12.1b) and (E 12.1c) simultaneously yields

$$T_{in2} = 332°C$$

$$T_{out2} = 304.5°C$$

For Case b, it is necessary to know more detail about the heat transfer coefficient, since the mass flowrate of the oil changes. For this system, it is known that the boiling heat transfer coefficient (h_o) is two times the oil heat transfer coefficient (h_i). Since only the oil heat transfer coefficient changes with flowrate, assuming negligible fouling and wall resistances, the overall heat transfer coefficient is expressed in terms of the oil heat transfer coefficient as

$$\frac{1}{U_1} = \frac{1}{h_{i1}} + \frac{1}{h_{o1}} = \frac{3}{2h_{i1}} \qquad \text{(E12.1d)}$$

$$\frac{1}{U_2} = \frac{1}{h_{i1} M^{0.8}} + \frac{1}{h_{o2}} = \frac{1}{h_{i1}}\left(\frac{1}{M^{0.8}} + 0.5\right) \qquad \text{(E12.1e)}$$

where $M = \dot{m}_2/\dot{m}_1$. The base case ratios now become

$$1.1 = \frac{M(325 - T_{out2})}{25} \qquad \text{(E12.1f)}$$

$$1.1 = \frac{3(T_{out2} - 325)}{2(58.6)\left(\dfrac{1}{M^{0.8}} + 0.5\right)\ln\left(\dfrac{T_{out2} - 253}{72}\right)} \qquad \text{(E12.1g)}$$

Solving these two equations simultaneously yields $T_{out2} = 301.5°C$ and $M = 1.17$, which means that the flowrate of oil must be increased by 17% to obtain a 10% increase in steam production, and the resulting outlet oil temperature rises slightly.

The same results can be obtained from the performance graph in Figure 12.1b.

New Steam Production, $\dot{m}_s = 1.1(1280) = 1410$ kg/h

New Exchanger Duty, $Q = \dot{m}_s (H_s - H_w) = 1410(2680) = 3.78 \times 10^6$ kJ/h

Case a: On Fig. 12.1b, the base case operating condition is Point "a." From Point "a," follow the constant flow line, $M = 1.0$, to a vaporizer duty, $Q = 3.78 \times 10^6$ kJ/h. This is line segment "a"–"c", and, from this, we get that the inlet oil temperature, $T_{in2} = 332°C$.
From Eq. (12.1)

$$T_{out2} = T_{in2} - Q/(\dot{m}_2 C_{p,oil}) = 332 - 3.78 \times 10^6/[55,300(2.48)] = 305°C$$

Case b: On Fig. 12.1b follow the constant temperature line for $T_{in2} = 325°C$ to the new vaporizer duty, $Q = 3.78 \times 10^6$ kJ/h. This is shown as line segment "a"–"b." From this, we get a value of $M = 1.17$.

$$\dot{m} = 1.17(55,300) = 64,700 \text{ kg/h (17\% increase in flow)}$$

From Eq. (12.1):

$$T_{out2} = T_{in2} - Q/(\dot{m}_2 C_{p,oil}) = 332 - 3.78 \times 10^6/[64,700(2.48)] = 301.5°C$$

In Example 12.1, two sets of operating conditions were calculated that would provide the required increase in steam production. The input oil flow rate, $\dot{m}$, could be increased with the temperature, T_{in}, held constant, or the oil flow rate, $\dot{m}$, could be held constant while the input oil temperature, T_{in}, is increased. These two solutions, along with an infinite number of other solutions, lie on a line of constant heat duty, $Q = 3.78$ GJ/h.

The curves for constant flow ($\dot{m} = 55,300M$) in Figure 12.1b are straight lines. Equating and rearranging Eqs. (12.1) and (12.3) produces

$$(T_{in} - T_B)/(T_{out} - T_B) = K \qquad (12.5)$$

$$\text{where, } K = \exp[UA/(\dot{m}C_{p,oil})] \qquad (12.6)$$

By solving Eq. (12.5) for T_{out} and substituting it into Eq. (12.1), we obtain

$$Q = \dot{m}C_{p,oil}(T_{in} - T_B)[1 - (1/K)] \qquad (12.7)$$

The overall heat transfer coefficient, U, varies with the light oil flow rate, $\dot{m}$. The performance for a constant value of oil flowrate is obtained once the value of K is known.

Eq. (12.7) was used to obtain the curves given in Figure 12.1b. We neglected the small amount of heat transferred above the liquid level (where the heat transfer coefficient for the gas phase will be small). In addition, it was assumed that the incoming feed water mixes rapidly with the large volume of boiling liquid water in the vaporizer. Thus, the water outside the tubes is essentially constant and equal to the saturation temperature. The simple form of Eq. (12.7) is a consequence of this constant temperature. When neither stream involves a phase change, the evaluation is more complicated. One reason is the need to consider the log-mean temperature correction factor for these types of shell and tube heat exchangers.

In reality, fouling will affect heat exchanger performance over time. Fouling is a buildup of material on tube surfaces, and it occurs to some extent in all heat exchangers. When a heat exchanger is started up, there will be no fouling; however, with time, trace impurities in the fluids deposit on the heat exchanger tubes. This is usually more significant for liquids, and among the possible impurities are inorganic salts or microorganisms. The fouling layer provides an additional resistance to heat transfer. For a heat exchanger constructed from material with high thermal conductivity operating with fluids with high heat transfer coefficients, fouling may provide a greater resistance than the convective film resistance. The influence of fouling on heat exchanger performance is the subject of problems at the end of the chapter.

Which of the two heat transfer surfaces provides the major fouling resistance in the steam generator? The water outside the tubes is designated as boiler feed water. This stream is an expensive source of water. It has been treated extensively to remove trace minerals, and hence, reduces significantly the fouling of heat transfer surfaces. This suggests that fouling on the outside of the tubes will remain low and the oil stream is the major contributor to fouling.

Example 12.2

Assume that neither the oil stream flow rate, $\dot{m}$, nor the inlet temperature, T_{in}, can be changed. What can be done to increase steam production, $\dot{m}_s$?

The heat transfer equation, $Q = UA\Delta T_{lm}$, shows that Q can be increased by:

a. Increasing U: Clean the tubes.
b. Increasing A: Increase the liquid level in the boiler.
c. Increasing ΔT_{lm}: Decrease the boiler temperature by lowering the pressure.
d. Combinations of a, b, and c.

Example 12.2 suggests ideas for alternative performance of the steam generator. Some of these are the subject of problems at the end of the chapter.

12.2   APPLICATION TO FLUID FLOW

12.2.1   Pump and System Curves

In this section, we present performance curves for a centrifugal pump and the flow of a liquid through a pipe network connecting a storage tank to a chemical reactor. Centrifugal pumps are very common in the chemical industry. Figure 12.3 shows the inner workings of a centrifugal pump. It is important to understand that the performance curves presented here are unique to centrifugal pumps. Positive displacement pumps, the other common type of pump used in the chemical industry, have a completely different performance curve and are discussed briefly in Section 12.2.3.

In situations involving the flow of fluid, the equation used to relate pressure changes and flow rate is the mechanical energy balance (or the extended Bernoulli Equation):

$$\frac{\Delta P}{\rho g} + \frac{\Delta u^2}{2g} + \Delta z = \frac{W_s}{g} - \frac{F_d}{g} \tag{12.10}$$

The terms shown on the left-hand side of Eq. (12.10) are point properties. They are independent of the path taken by the fluid. The Δ in Eq. (12.10) represents the difference between outlet and inlet conditions on the control volume. In contrast, the terms on the right-hand side of the equation are "path properties" that de-

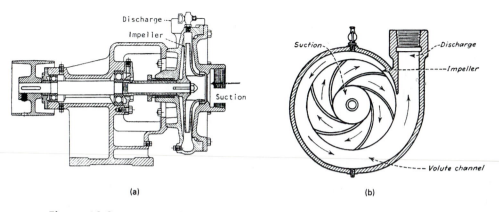

Figure 12.3   Inner Workings of a Centrifugal Pump (From Walas, S., *Chemical Process Equipment. Selection and Design,* Butterworths, Boston, 1988. Reproduced with permission.)

pend upon the path followed by the fluid. Work is defined as positive when done on the system. The terms related to the path taken by the fluid are specific to the operating system and form the basis for the performance curves.

Eq. (12.10) is written in terms of "fluid head." Each term has units of length. Fluid head is a way of expressing pressure as an equivalent static pressure of a stationary body of fluid. For example, one atmosphere of pressure is equivalent to about 34 feet of water, because the pressure difference between the top and bottom of a column of 34 feet of water is one atmosphere. Eq. (12.10), with each term defined as fluid head, is

$$\Delta h_p + \Delta h_v + \Delta h_z = h_s - h_f \qquad (12.11)$$

A centrifugal pump is shown in Figure 12.4 along with its performance diagram. For this analysis, the change in the velocity head and elevation head between points "1" and "2" are either zero or small, and Eq. (12.11) reduces to:

$$\Delta h_p = h_s - h_f \qquad (12.12)$$

Centrifugal pumps are used in a variety of applications and are available from many manufacturers. Pump impellers may have forward or backward vanes, enclosed or open configurations, operating at different speeds. Their performance characteristics depend on their mechanical design. As a result, experimentally determined performance curves are supplied routinely by the pump manufacturer.

The performance curve, or pump curve, shown in Figure 12.4 is typical of a centrifugal pump handling normal (Newtonian) liquids. It relates the fluid flowrate to the head, h_s. To understand this relationship, consider point "α" shown on the performance curve in Figure 12.2. This point lies on the curve for a 6-inch impeller. It shows that the pump delivers 0.76 m^3/min of liquid when the pressure

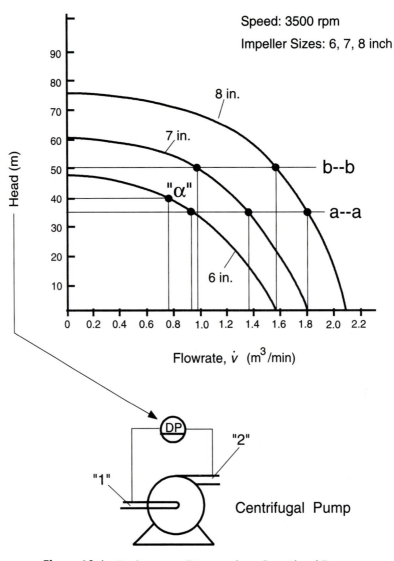

Figure 12.4   Performance Diagram for a Centrifugal Pump

differential between the suction and discharge side of the pump is equivalent to a column of liquid 40 m in height.

Figure 12.4 contains three curves, each depicting the pump performance with a different impeller diameter. The pump casing can accept any of these three impellers and each impeller has a unique pump curve. Pump curves often display efficiency and horsepower curves in addition to the curves shown in Figure 12.4.

From the pump curves in Figure 12.4, the following trends may be seen for each impeller:

1. The head produced at a given flowrate varies (increases) with the impeller diameter.
2. The pump provides low flowrates at high head and high flowrates at low head.
3. The head produced is sensitive to flowrate at high flowrates, that is, the curve drops off sharply at high flowrates.
4. The head produced is relatively insensitive to flowrate at low flowrates, that is, the curve is relatively flat at low flowrates.

The following example demonstrates how to read a pump curve.

Example 12.3

The centrifugal pump shown in Figure 12.4 is used to supply water to a storage tank. The pump inlet is at atmospheric pressure and water is pumped up to the storage tank, which is open to atmosphere, via large diameter pipes. Because the pipe diameters are large the frictional losses in the pipes and any change in fluid velocity may be safely ignored.

a. If the storage tank is located at an elevation of 35 m above the pump, predict the flow using each impeller.
b. If the storage tank is located at an elevation of 50 m above the pump, predict the flow using each impeller.

a. From Figure 12.4, $@\Delta h_p = 35$ m (see line "a—a")
   6-in. Impeller: Flow = 0.93 m^3/min
   7-in. Impeller: Flow = 1.38 m^3/min
   8-in. Impeller: Flow = 1.81 m^3/min

b. From Figure 12.4, $@\Delta h_p = 50$ m (see line "b—b")
   6-in. Impeller: Flow = 0 m^3/min
   7-in. Impeller: Flow = 0.99 m^3/min
   8-in. Impeller: Flow = 1.58 m^3/min

In Example 12.3, the flowrate into the tank is restricted to one of three discrete values (depending upon the impeller installed in the pump housing). For the system described, once the elevation of the tank and the impeller diameter are chosen, a single unique flowrate is obtained. Other flowrates cannot be obtained for this system.

Example 12.3 demonstrates the need to know the characteristic shape of the performance curve at the current operating point to predict the effect of any change in tank elevation. In Example 12.3, because the 8-in impeller is operating in a region where the flow is not greatly affected by an increase in head, the flowrate was reduced by less than 15%. In contrast, for the 6-in impeller, where a change in head has a large effect, the flowrate is reduced by 100% (no fluid flows).

> **It is essential to understand the system performance before making predictions or recommendations.**

Consider the flow situation shown in Figure 12.5. The figure illustrates a pipe network through which feed material is transported from a storage tank through a heat exchanger to a chemical reactor. From the information provided on Figure 12.5, it is possible to construct the performance curve shown, which is called the system curve.

Since the fluid is a liquid, we may assume that the change in velocity head, Δh_v, is small. In addition, the work head $\Delta h_s = 0$, and Eq. (12.11) written between points "1" and "2" reduces to:

$$\Delta h_p = -h_f - \Delta h_z \qquad (12.13)$$

The pressure head term, Δh_p, and the elevation head term, $-\Delta h_z$, are constant and independent of the flowrate. The friction term, $-h_f$, in Eq. (12.13) depends upon

1. The specific system configuration
2. The flowrate of fluid

For this system, for fully developed turbulent flow, we know the relationship needed to predict the friction term, h_f, as a function of fluid velocity. The relationship is given in Table 11.1 and in Example 11.2. This relationship in terms of fluid head is:

$$h_{f2} = h_{f1}(u_2/u_1)^2 = h_{f1}(\dot{v}_2/\dot{v}_1)^2 \qquad (12.14)$$

where, $\dot{v}$ is the volumetric flow rate, the subscript 2 refers to the new case, and the subscript 1 refers to the base case. Eq. 12.14 represents a parabola, which is the shape of the system curve in Figure 12.5. In terms of pressure, Eq. (12.14) becomes

$$\Delta P_2 = \Delta P_1(\dot{v}_2/\dot{v}_1)^2 \qquad (12.15)$$

and it is also possible to plot the system curve in terms of pressure as is done later in Figure 12.6.

The operating conditions given in Figure 12.5 serve as the base case.

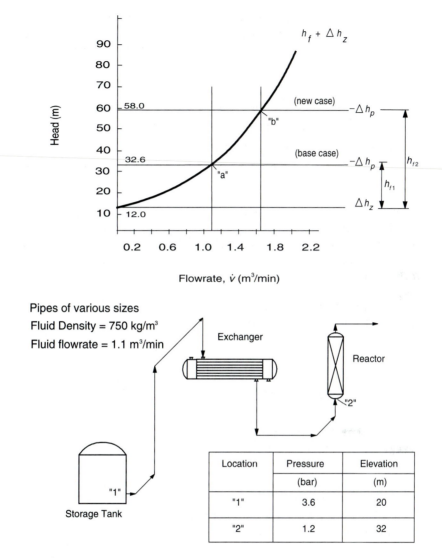

Pipes of various sizes
Fluid Density = 750 kg/m³
Fluid flowrate = 1.1 m³/min

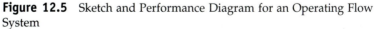

Location	Pressure	Elevation
	(bar)	(m)
"1"	3.6	20
"2"	1.2	32

Figure 12.5 Sketch and Performance Diagram for an Operating Flow System

---

Example 12.4

The fluid flowrate measured for the conditions shown in Figure 12.5 is 1.1 m³/min.

a. Using the information given on the flow diagram (see Figure 12.5) for the base case, determine the value of h_f.

b. Develop the equation for h_f as a function of the flowrate, $\dot{v}$.

a. From Eq. (12.13):

$$h_f = -\Delta h_z - \Delta h_p$$

$\Delta h_z = z_2 - z_1 = 32 \text{ m} - 20 \text{ m} = 12 \text{ m}$ (Δh_z is drawn as a horizontal line in Figure 12.5)

$$\Delta h_p = (P_2 - P_1)/\rho g = (1.2 - 3.6) \times 10^5/[750(9.81)] = -32.6 \text{ m}$$

$$h_f = -12 \text{ m} + 32.6 \text{ m} = 20.6 \text{ m}$$

b. $h_{f2} = h_{f1}(\dot{v}_2/\dot{v}_1)^2 = (20.6 \text{ m})(\dot{v}_2/1.1)^2 = 17.02 \dot{v}_2^2$

Example 12.5

The flow to the reactor in Example 12.4 is increased by 50%. The pressure of the reactor is held constant. Determine the pressure required at the exit of the storage tank.

For a 50% increase, $\dot{v}_2 = 1.5 \, \dot{v}_1 = 1.1(1.5) = 1.65 \text{ m}^3/\text{min}$.

The line of constant flow, $\dot{v} = 1.65 \text{ m}^3/\text{min}$, is shown as the vertical line through Point "b" in Figure 12.5. It intersects the performance line, $h_f + \Delta h_z = 58$ m (Point "b")

$$-\Delta h_p = 58 \text{ m} = (P_1 - P_2)/\rho g = P_1/\rho g - 1.2 \times 10^5/[750(9.81)] = P_1/\rho g - 16.3 \text{ m}$$

$$P_1 = (58 + 16.3)\rho g = 74.3(750)(9.81) = 5 \times 10^5 \text{ Pa} = 5.47 \text{ bar}$$

In this section, we considered two representative types of performance curves. For the pump, the information needed for developing a performance diagram was obtained experimentally and supplied by the manufacturer. In the flow network, a base case was established using actual operating data and the performance diagram, known as the system curve, calculated using the mechanical energy balance.

12.2.2   Regulating Flowrates

In all systems presented in this chapter, input flowrates are the primary variables that are used to change the performance of a system. In fact, in a chemical plant, process regulation is achieved most often by manipulating flowrates, which is accomplished by altering valve settings. If it is necessary to change a temperature, the flow of a heating or cooling medium is adjusted. If it is necessary to change a reflux ratio, a valve is adjusted. In this section we consider how to regulate these input flows to give desired values.

> **Process conditions are usually regulated or modified by adjusting valve setting(s) in the plant.**

Although valves are relatively simple and inexpensive pieces of equipment, they are nevertheless indispensable in any chemical plant that handles liquids and/or gases.

Figure 12.6 illustrates a fluid system containing three components.

1. a flow system, including piping and three heat exchangers
2. a pump
3. a regulating valve

The process shown in Figure 12.6 is described briefly as follows:

A liquid process stream is pumped at a rate of v̇ m³/min from a cooling pond, through a heat exchanger, through a regulating valve and two more heat exchangers connected in series, and returned to the cooling pond. The intake and discharge pipes are at the same elevation. Differential pressure gauges are installed across the pump and the regulating valve. The pump used here has the same characteristics as that shown in Figure 12.4, with a 7-inch impeller.

Conditions shown in Figure 12.6 represent a base case. The mechanical energy balance equation, Eq. (12.10), written between Points "1" and "2" in Figure 12.6 gives,

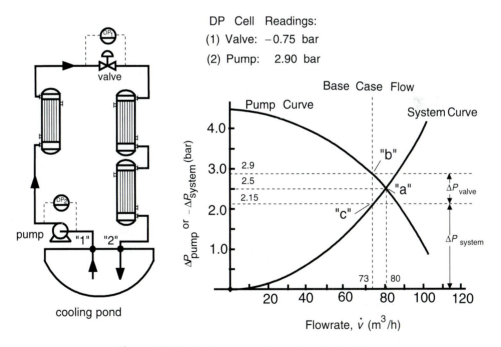

Figure 12.6   Performance Curves for Coolant System

$$-(\Delta P_{valve} + \Delta P_{pipes} + \Delta P_{exchangers}) = \Delta P_{pump} \qquad (12.16)$$

In Eq. (12.16)

ΔP_{valve} represents the pressure drop across the control valve.

ΔP_{pipe} represents the pressure drop due to friction in the piping and pipe fittings between Points "1" and "2".

$\Delta P_{exchangers}$ represents the pressure drop from due to friction in the three heat exchangers.

ΔP_{pump} represents the pressure increase produced by the pump.

The frictional drop in pressure through the pipe system and the heat exchangers is given by

$$\Delta P_{system} = \Delta P_{pipes} + \Delta P_{exchangers} \qquad (12.17)$$

ΔP_{system} is called the system pressure drop, and Equation (12.16) can be written as

$$-(\Delta P_{valve} + \Delta P_{system}) = \Delta P_{pump} \qquad (12.18)$$

The performance curve for this system, shown in Figure 12.6, consists of a plot of pressure against the liquid flowrate, as was done in Figures 12.4 and 12.5. (In this section we use pressure units instead of units of liquid head.)

The system pressure drop is estimated relative to the base case by Eq. (12.15)

$$\Delta P_{2,system} = \Delta P_{1,system}(\dot{v}_2/\dot{v}_1)^2 \qquad (12.19)$$

$$\Delta P_{2,system}(bar) = 2.15\ (\dot{v}_2/73)^2 \qquad (12.20)$$

This curve has been plotted and labeled as the system curve.

Sufficient information is available to prepare a full performance diagram. In Figure 12.6, for this system, each side of Eq. (12.18) is plotted. The pressure produced by the pump, ΔP_{pump}, depends on the volumetric flowrate, $\dot{v}$, as was shown earlier in Figure 12.4. Only the units used to express the pressure and flow rate terms have been changed.

Example 12.6

Using the pump curve and the differential pressures provided on Figure 12.6, find for the base case

   a. the system pressure drop, ΔP_{system}, and
   b. the volumetric flowrate, $\dot{v}$.

   a. For the system pressure drop, ΔP_{system}, from Eq. (12.18),

$$-\Delta P_{system} = \Delta P_{pump} + \Delta P_{valve} = 2.9\ bar - 0.75\ bar = 2.15\ bar$$

b. For the volumetric flowrate, $\dot{v}$, see Point "b" on Figure 12.6.

$$\text{for } \Delta P_{pump} = 2.9 \text{ bar}, \dot{v} = 73 \text{ m}^3/\text{h}$$

Base case conditions have been added to Figure 12.6. All pressures for the base case lie along the constant flow line, $\dot{v} = 73$ m³/h. Point "c" gives the value for the system pressure drop, ΔP_{system}, and point "b" gives the pressure increase provided by the pump. The difference between these points is the pressure drop over the regulating valve, ΔP_{valve}.

If there were no valve in the line, $\Delta P_{valve} = 0$, the system would operate at Point "a" where the pump curve and the system curve cross. We can see from Figure 12.6, as the pressure across the valve increases, we move toward the left and lower flowrates. The flow becomes zero when the valve is fully closed. Thus, by manipulating the valve setting, we can alter the flow of fluid through the system.

Example 12.7

For the base case condition shown in Figure 12.6, do the following:

a. Check the pressure drop over the valve against the guideline for control valves (Table 9.8).
b. Determine the percent increase in flow by fully opening the valve.
a. From guideline, $\Delta P \geq 0.69$ bar (see Table 9.8).
    $\Delta P = 0.75$ bar, therefore the guideline in Table 9.8 is satisfied.
b. With no valve resistance, $\dot{v} = 80$ m³/h
    the percent increase in flow = $[(80 - 73)/73]100 = 9.6\%$

The increase in flowrate is limited (less than 10% of the base case) by the pump and system, and only modest increases of flowrate are possible for this system.

12.2.3   Reciprocating or Positive Displacement Pumps

Positive displacement pumps perform differently from centrifugal pumps. They are used to achieve higher pressure increases than centrifugal pumps. Figure 12.7 is a drawing of the inner workings of a positive displacement pump. The performance characteristics are represented on Figure 12.8a. It can be observed that the flowrate through the pump is almost constant over a wide range of pressure increases. One method for regulating the flow through a positive displacement pump is illustrated in Figure 12.8b. By carefully regulating the flow of the recycle stream, the pressure rise in the pump is controlled. See Chapter 15 for more details.

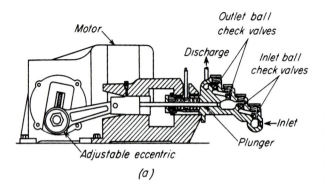

Motor

Discharge

Outlet ball check valves

Inlet ball check valves

←Inlet

Plunger

Adjustable eccentric

(a)

Figure 12.7 Inner Workings of a Positive Displacement Pump (From McCabe, W.L., J.C., Smith, and P. Harriott, *Unit Operations of Chemical Engineering* (5th ed.), McGraw-Hill, New York, 1993, Copyright © 1993 by McGraw-Hill Companies, reproduced with permission of the McGraw-Hill Companies.)

12.2.4   Net Positive Suction Head

There is a significant limitation on pump operation called Net Positive Suction Head (NPSH). Its origin is as follows. Although the effect of a pump is to raise the pressure of a liquid, frictional losses at the entrance to the pump, between the feed (suction) pipe and the internal pump mechanism, cause the liquid pressure to drop upon entering the pump. This means that a minimum pressure exists somewhere within the pump. If the feed liquid is saturated or nearly saturated, the liquid can vaporize upon entering the due to the pressure drop. This results in formation of vapor bubbles called cavitation. These bubbles rapidly collapse

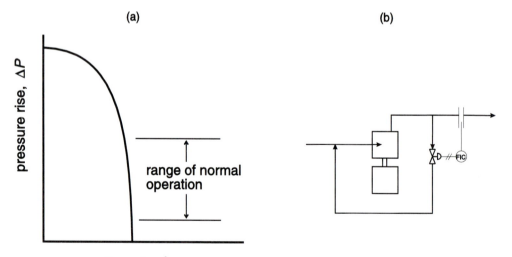

(a)

(b)

range of normal operation

pressure rise, ΔP

flowrate, $\dot{v}$

Figure 12.8 Typical Pump Curve for Positive Displacement Pump and Method for Flowrate Regulation

when exposed to the forces created by the pump mechanism. This process usually results in noisy pump operation, and, if it occurs for a period of time, will damage the pump. As a consequence, regulating valves are not normally placed in the suction line to a pump.

Pump manufacturers supply NPSH data with a pump. The required NPSH, denoted $NPSH_R$, is a function of the square of velocity, since it is a frictional loss. Figure 12.9 shows an $NPSH_R$ curve, which defines a region of acceptable pump

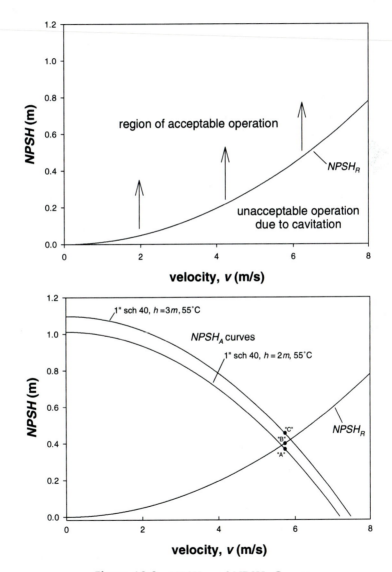

Figure 12.9 $NPSH_A$ and $NPSH_R$ Curves

operation. This is specific to a given liquid. Typical $NPSH_R$ values are in the range of 15–30 kPa (2–4 psi) for small pumps, and can reach 150 kPa (22 psi) for larger pumps. On Figure 12.9, there are also curves for $NPSH_A$, the available NPSH. The available $NPSH_A$ is defined as

$$NPSH_A = P_{inlet} - P^* \tag{12.21}$$

Eq. (12.21) means that the available NPSH ($NPSH_A$) is the difference between the inlet pressure, P_{inlet}, and P^*, which is the vapor pressure (bubble point pressure for a mixture). It is a system curve for the suction side of a pump. It is required that $NPSH_A \geq NPSH_R$ to avoid cavitation. All that remains is to calculate or know the pump inlet conditions in order to determine whether there is enough available NPSH ($NPSH_A$) to equal or exceed the required NPSH ($NPSH_R$).

For example, consider the exit from a distillation column reboiler, which is saturated liquid. If it is necessary to pump this liquid, cavitation could be a problem. A common solution to this problem is to elevate the column above the pump so that the static pressure increase minus any frictional losses between the column and the pump provides the necessary NPSH. This can be done either by elevating the column above ground level using a metal skirt or by placing the pumps in a pit below ground level, though pump pits are usually avoided due to safety concerns arising from accumulation of heavy gases in the pit.

In order to quantify NPSH, consider Figure 12.10, in which material in a storage tank is pumped downstream in a chemical process. This scenario is a very common application of the NPSH concept. From the mechanical energy balance, the pressure at the pump inlet can be calculated to be

$$P_{inlet} = P_{tank} + \rho g h - \frac{2\rho f L_{eq} u^2}{D} \tag{12.22}$$

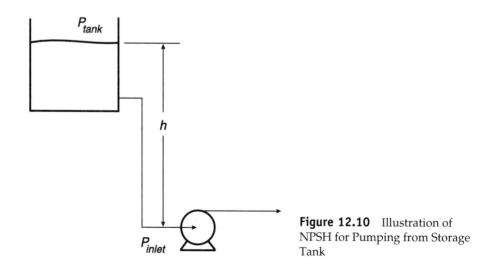

Figure 12.10   Illustration of NPSH for Pumping from Storage Tank

which means that the pump inlet pressure is the tank pressure plus the static pressure minus the frictional losses. Therefore, by substituting Eq. (12.22) into Eq. (12.21), the resulting expression for $NPSH_A$ is

$$NPSH_A = P_{tank} + \rho g h - \frac{2 \rho f L_{eq} u^2}{D} - P^* \tag{12.23}$$

which is the equation of a concave downward parabola, as illustrated in Figure 12.9.

If there is insufficient $NPSH_A$ for a particular situation, Eq. (12.23) suggests methods to increase the $NPSH_A$.

1. **Decrease the temperature of the liquid at the pump inlet.** This decreases the value of the vapor pressure, P^*, thereby increasing $NPSH_A$.
2. **Increase the static head.** This is accomplished by increasing the value of h in Eq. (12.23), thereby increasing $NPSH_A$. As was said earlier, pumps are most often found at lower elevations than the source of the material they are pumping.
3. **Increase the diameter of the suction line (feed pipe to pump).** This reduces the velocity and the frictional loss term, thereby increasing $NPSH_A$. It is standard practice to have larger diameter pipes on the suction side of a pump than on the discharge side.

The following example illustrates how to do NPSH calculations and one of the methods listed above for increasing $NPSH_A$. The other methods are illustrated in a problem at the end of the chapter.

Example 12.8

The feed pump (P-101) on Figure 1.5 pumps toluene from a feed tank (TK-101) maintained at atmospheric pressure and 55°C. The pump is located 2 m below the liquid level in the tank, and there is 6 m of equivalent pipe length between the tank and the pump. It has been suggested that 1 inch Schedule 40 commercial steel pipe be used for the suction line. Determine whether this is a suitable choice. If not, suggest methods to avoid pump cavitation.

The following data can be found for toluene: $\ln P^*(\text{bar}) = 10.97 - 4203.06/T(\text{K})$, $\mu = 4.1 \times 10^{-4}$ kg/m s, $\rho = 870$ kg/m³. For 1 inch Schedule 40 commercial steel pipe, the roughness factor is about 0.001 and the inside diameter is 0.02664 m. From Table 1.5, the flow of toluene is 10,000 kg/h. Therefore, the velocity of toluene in the pipe can be found to be 5.73 m/s. The Reynolds number is 426,000, and, from a friction factor chart, $f = 0.005$. At 55°C, the vapor pressure is found to be 0.172 bar.

From Eq. (12.23):

$$NPSH_A = 1.01325 \text{ bar} + 870(9.81)(2)(10^{-5}) \text{ bar}$$
$$- 2(870)(0.005)(6)(5.73)^2(10^{-5})/(0.02664) \text{ bar} - 0.172 \text{ bar}$$

$$NPSH_A = 0.37 \text{ bar}$$

This is shown as Point "A" on Figure 12.9. At the calculated velocity, Figure 12.9 shows that $NPSH_R$ is 0.40 bar, Point "B." Therefore, there is insufficient $NPSH_A$.

One method for increasing $NPSH_A$ is to increase the height of liquid in the tank. If the height of liquid in the tank is 3 m, with the original 1 in Schedule 40 pipe at the original temperature, $NPSH_A = 0.445$ bar. This is shown as Point "C" on Figure 12.9.

12.2.5 Compressors

The performance of centrifugal compressors is somewhat analogous to that of centrifugal pumps. There is a characteristic performance curve, supplied by the manufacturer, that defines how the outlet pressure varies with flowrate. However, compressor behavior is far more complex than that for pumps because the fluid is compressible.

Figure 12.11 shows the performance curve for a centrifugal compressor. It is immediately observed that the y-axis is the ratio of the outlet pressure to inlet pressure. This is in contrast to pump curves, which have the difference between these two values. Curves for two different rotation speeds are shown. As with pump curves, curves for power and efficiency are often included, but are not shown here. Unlike most pumps, the speed is often varied continuously to con-

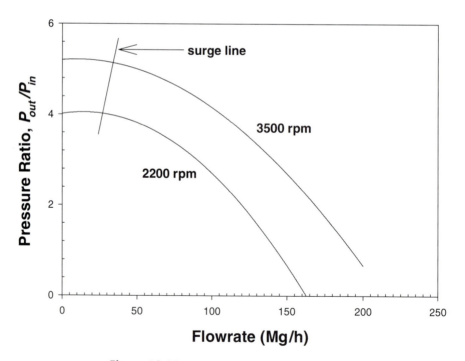

Figure 12.11   Centrifugal Compressor Curves

trol the flowrate. This is because the higher power required in a compressor makes it economical to avoid throttling the outlet as in a centrifugal pump.

Centrifugal compressor curves are read just like pump curves. At a given flowrate and rpm there is one pressure ratio. The pressure ratio decreases as flowrate increases. A unique feature of compressor behavior occurs at low flowrates. It is observed that the pressure ratio increases with decreasing flowrate, reaches a maximum and then decreases with decreasing flowrate. The locus of maxima is called the surge line. For safety reasons, compressors are operated to the right of the surge line. The surge line is significant for the following reason. Imagine that we start at a high flowrate and the flowrate is lowered continuously, causing a higher outlet pressure. At some point, the surge line is crossed, lowering the pressure ratio. This means that downstream fluid is at a higher pressure than upstream fluid, causing a backflow. These flow irregularities can severely damage the compressor mechanism, even causing the compressor to vibrate or surge (hence, the origin of the term). Severe surging has been known to cause compressors to become detached from the supports keeping them stationary and literally to fly apart, causing great damage. Therefore, the surge line is considered a limiting operating condition, below which operation is prohibited.

Positive displacement compressors also exist and are used to compress low volumes to high pressures. Centrifugal compressors are used to compress higher volumes to moderate pressures, and are often staged in order to obtain higher pressures. Figure 12.12 illustrates the inner workings of a compressor.

Problems requiring reading compressor curves are given at the end of the chapter.

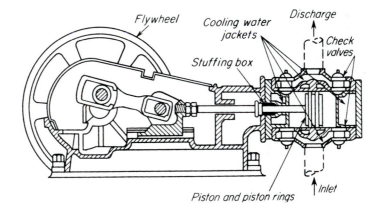

Figure 12.12 Inner Workings of a Positive Displacement Compressor (From McCabe, W.L., J.C. Smith, and P. Harriott, *Unit Operations of Chemical Engineering* (5th ed.), McGraw-Hill, New York, 1993. Copyright © 1993 by McGraw-Hill Companies, reproduced with permission of the McGraw-Hill Companies.)

12.3 APPLICATION TO SEPARATION PROBLEMS

12.3.1 Separations with Mass Separating Agents

Multistage equilibrium separations involve simultaneous solution of material balances and equilibrium relationships for each equilibrium stage. Therefore, there is no simple, closed-form relationship that describes the behavior of these systems for all situations. There are qualitative relationships that are applicable to most all situations, and these were included in Table 11.1. The key relationships are that a better separation is usually achieved by increasing the number of stages or by increasing solvent flowrate.

Similarly, continuous differential equilibrium separations, which involve simultaneous solution of material balances and mass transfer relationships, do not yield a closed-form solution either. The key qualitative relationships are that better separation is usually achieved by increasing column height or by increasing flowrate.

For the specific assumptions of dilute solutions and a linear equilibrium relationship that results in approximately constant stream flows, an analytical solution is possible for the above situations. For staged separations, the situation is shown in Figure 12.13a. The result is

$$\frac{y_{A,out} - y^*_{A,out}}{y_{A,in} - y^*_{A,out}} = \frac{1 - A}{1 - A^{N+1}} \tag{12.24}$$

where $A = (L/mG)$ is called the absorption factor, N is the number of equilibrium stages, y_A is the mole fraction of the solute in the gas phase, L and G are the molar flowrates of each stream, m is the equilibrium relationship ($y = mx$), and $y^*_{A,out} = mx_{A,in}$. Eq. (12.24), is known as the Kremser equation and is a key relation-

(a) **(b)**

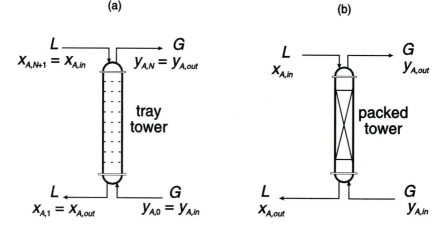

Figure 12.13 Tray and Packed Absorbers

ship for multistage, equilibrium separations obeying the assumptions listed above. It describes transport of solute, A, from the G phase to the L phase. For the reverse situation, a similar equation is derived in which the left side of Eq. (12.24) involves x_A, the mole fraction in the liquid phase, and the right side involves $S = (mG/L)$. The term $y^*_{A,out} = mx_{A,in}$ is replaced by $x^*_{A,out} = y_{A,in}/m$. If the separation involves phases that are not gas or liquid, Eq. (12.24) may still be used by defining L and G appropriately. Figure 12.14 is a plot of Eq. (12.24), and contains the performance relationship between key variables for multistage equilibrium separations following the assumptions listed above. Performance curves for a specific staged separation can be generated from the information in Figure 12.14. This is illustrated in Example 12.10. It should be noted that tray performance issues such as flooding, which are specific to a particular column design, are not predicted by the Kremser relationship.

Even though Eq. (12.24) and Figure 12.14 are only valid subject to the assumptions listed above, the qualitative performance of all staged separations can be understood from these relationships. For example, as illustrated in Figure 12.14, if a line of constant A (if $A > 1$) is followed for increasing number of stages, N, it is seen that the separation continues to improve (lower value of y-axis). For $A < 1$, a best separation is approached asymptotically. If a vertical line of constant N is followed, it is seen that as the separation improves the value of A increases. A increases if L increases or if G decreases, meaning a larger solvent to feed ratio, or if m decreases, meaning equilibrium more in favor of the L phase.

A similar relationship to the Kremser equation exists for continuous differential separations (packed beds) subject to the same assumptions. The physical situation is illustrated in Figure 12.13b. This relationship, known as the Colburn equation, is

$$\frac{y_{A,in} - y^*_{A,out}}{y_{A,out} - y^*_{A,out}} = \frac{e^{N_{tOG}[1-(1/A)]} - (1/A)}{1 - (1/A)} \tag{12.25}$$

where N_{tOG} is the number of overall gas-phase transfer units. Eq. (12.25) is for absorption, that is, transfer into the L phase. For the reverse direction of transport, the same changes are made as in the Kremser equation, with N_{tOL}, the number of overall liquid-phase transfer units replacing N_{tOG}. Eq. (12.25) is plotted in Figure 12.15. Performance curves for a specific staged separation can be generated from the information in Figure 12.15. This is the subject of a problem at the end of the chapter. Since increasing the number of transfer units increases column height, the same qualitative understanding applicable to all systems is gleaned from Figure 12.15 as was described above. For increasing N_{tOG} (which means increased column height) at constant A, a better separation is observed (for $A > 1$). For $A < 1$, a best separation is approached asymptotically. Similarly, at constant column height (constant N_{tOG}), increasing A results in better separation.

As with the Kremser equation, the Colburn equation does not describe flooding behavior of a packed bed, which is specific to a particular column design. Flooding occurs when the vapor velocity upward through the column is so

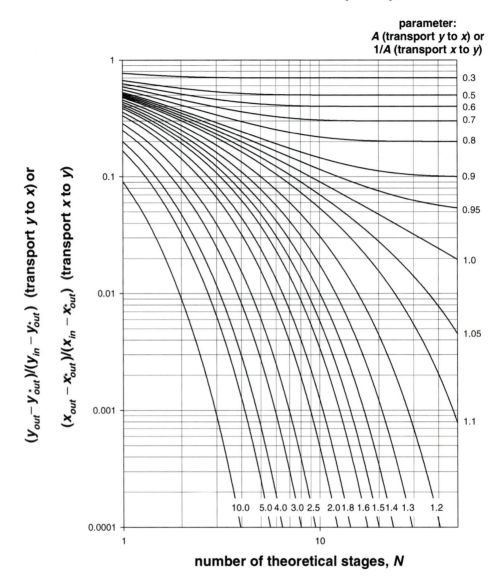

Figure 12.14 Plot of Kremser Equation, Number of Theoretical Stages for Countercurrent Operation, Henry's Law Equilibrium, and Constant A or $1/A$

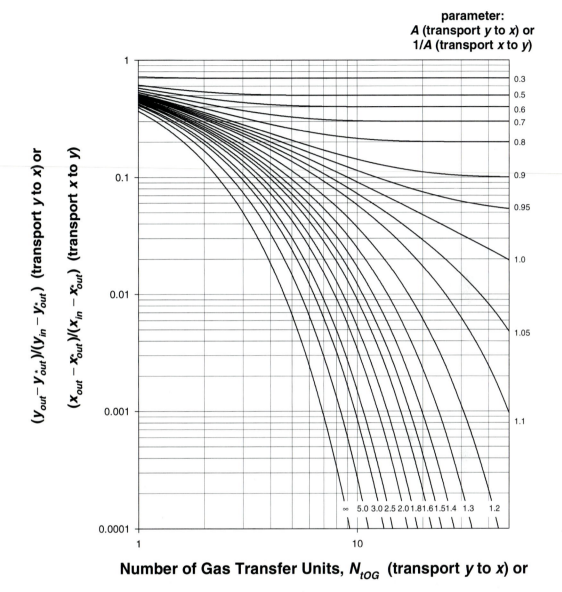

Figure 12.15   Plot of Colburn Equation, Number of Transfer Units for Countercurrent Operation, Henry's Law Equilibrium, and Constant A or $1/A$

large that liquid is prevented from flowing downward. When a column is designed, the diameter is chosen so that the vapor velocity is below the flooding limit (typically 75%–80% of the limit). However, if the vapor velocity is increased, flooding can occur. Since flooding is specific to a particular column, it cannot be illustrated on the general performance curve. Care must be taken not to use the Kremser or Colburn graphs to obtain a result that will cause flooding in a particular column or to recommend operation in the flooding zone.

The following examples illustrate the use of the Kremser and Colburn equations.

Example 12.9

A tray scrubber with 8 equilibrium stages is currently operating to reduce the acetone concentration in 40 kmol/h of air from a mole fraction of 0.02 to 0.001. The acetone is absorbed into a 20 kmol/h water stream, and it may be assumed that the water stream enters acetone free. Due to a process upset, it is necessary to increase the flow of air by 10%. Under the new operating conditions, what is the new outlet mole fraction of acetone in air?

The use of Figure 12.14 for this problem is illustrated in Figure E12.9. For the design case, $y_{A,in} = 0.02$, $y_{A,out} = 0.001$, and $x_{A,in} = 0$, which means that $y^*_{A,out} = 0$. Therefore, the y-axis is 0.05. Since $N = 8$, the design case has a value of $A = 1.2$. This is shown as Point "a" on Figure E12.9. If the flow of air (G) is increased by 10%, then the new value of $A = 1.08$. Following a vertical line (constant number of stages) from the original point to a value of $A = 1.08$ (Point "b") yields a y-axis value of 0.08. Hence, the new value of $y_{A,out} = 0.0016$.

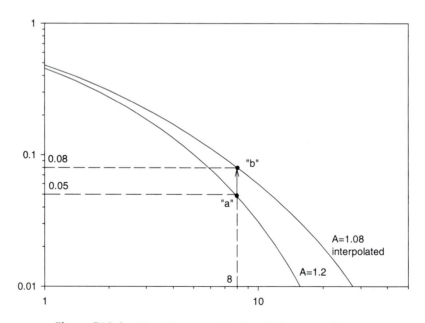

Figure E12.9   Use of Kremser Graph to Solve Example 12.9

Example 12.10

For the situation described in Example 12.9, the inlet air flowrate and the inlet mole fraction of acetone in air may vary during process operation. Prepare a set of performance curves for the liquid rate necessary to maintain the outlet acetone mole fraction for a range of inlet acetone mole fractions. There should be curves for five different values of the inlet gas rate including the original case and the following percentages of the original case: 80, 90, 110, 120. The temperature and pressure in the scrubber are assumed to remain constant.

The result is shown on Figure E12.10. The curves are generated as follows. From the original operating point, the value of m can be determined to be 0.417. By moving vertically on a line of constant number of equilibrium stages (8 in this case), values for the y-axis and A can be tabulated. The y-axis value is easily converted into $y_{A,in}$ since $y_{A,out}$ is known and constant. Then, for different values of G, values of L are obtained since m is known and constant.

12.3.2   Distillation

For distillation, there is no universal set of performance curves as for multistage, equilibrium separations involving mass separating agents. However, for a given situation, performance curves can be generated. This section demonstrates how this is done.

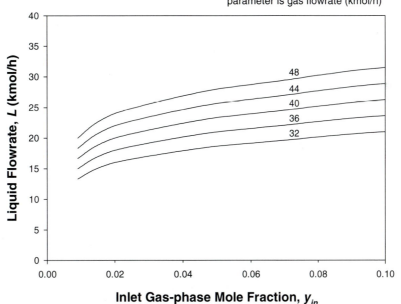

Figure E12.10   Performance Curves for Example 12.10

Figure 12.16 illustrates a multistage distillation system for separating benzene from a mixture of benzene and toluene. Figure 12.17 is a sketch of a typical distillation column. This problem is modified from Bailie and Shaeiwitz [1].

Component separation takes place within the tower. Considering the tower alone, there are three input and two output streams. Separation is achieved by the transfer of components between the two phases. For an operating plant, the size of the tower, the number and type of trays (or height of packing) in the tower, and other tower attributes are fixed. The output streams are established once the input flow streams to the tower are set.

In this section we will develop several performance diagrams for the distillation process shown in Figure 12.16. The tower is the separation unit and the two phases needed for the separation are provided by the condenser and reboiler.

In a typical distillation system, two recycle streams are returned to the tower. A condenser is added at the top of the column and a fraction of the overhead vapor, V_1, is condensed to form a liquid recycle, L_0. This provides the liquid phase needed in the tower. The remaining fraction is the overhead product, D. A vaporizer (reboiler) is added to the bottom of the column and a portion of the bottom liquid, L_N, is vaporized and recycled to the tower as stream V_{N+1}. This provides the vapor phase needed in the tower.

Figure 12.16 provides some information on the distillation process taken from an operating plant.

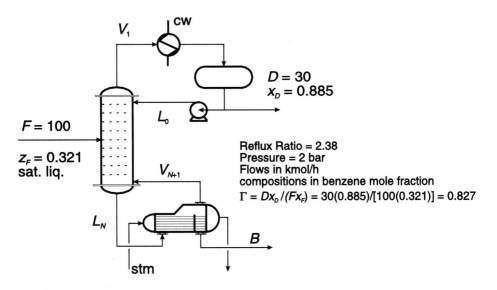

Figure 12.16   Plant Section for Distillation of Benzene from Toluene

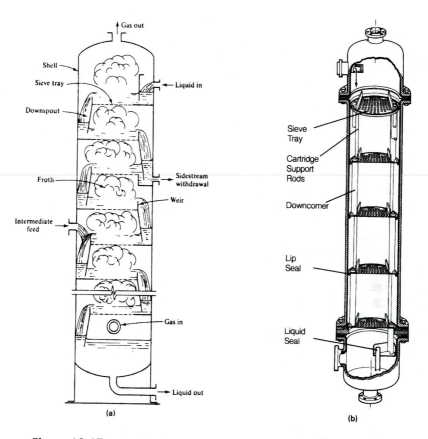

Figure 12.17 Details of Internal Construction of a Distillation Column (From Walas, S., *Chemical Process Equipment. Selection and Design*, Butterworths, Boston, 1988. Reprinted with permission.)

Product benzene is separated from a benzene-toluene mixture. The feed stream, F = 100 kmol/h, is a saturated liquid with composition $z_F = 0.321$ (mole fraction benzene) at a pressure P = 2 bar. The top product consists of a distillate stream, D = 30 kmol/h, with composition $x_D = 0.885$ and a corresponding benzene recovery, $\Gamma = 0.827$. The tower has a diameter of 0.83 m and contains 20 sieve plates with a 0.61 m plate spacing. Feed is introduced on Plate 13 and the tower operates at a reflux ratio of 2.38. No information is provided on the partial reboiler and total condenser. From a benzene balance we obtain B = 70 kmol/h and $x_B = 0.079$.

Figure 12.18 provides a McCabe-Thiele diagram for the separation given in Figure 12.16. This was constructed from the equilibrium curve and by matching the concentrations, $x_D = 0.885$, $z_F = 0.321$, and $x_B = 0.079$, the reflux ratio, $R = 2.38$, and the condition of saturated liquid feed with plant operating conditions. This

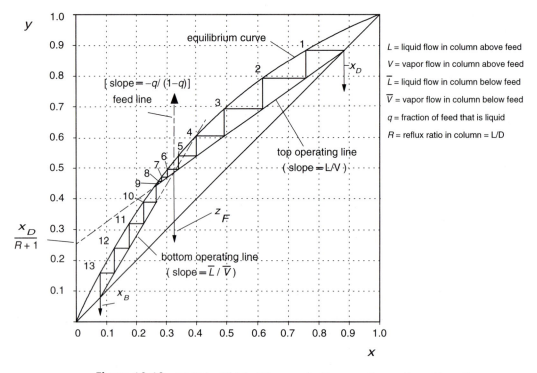

Figure 12.18   McCabe-Thiele Diagram for Benzene Separation—Base Case

McCabe-Thiele construction yielded 13 theoretical stages (12 trays plus a reboiler). This means that the average plate efficiency was 60% (100[No. of theoretical plates/No. of actual plates] = 100(12/20). The 60% plate efficiency is within the range found for columns separating benzene from toluene. The feed is added on theoretical plate number eight (13[0.60] ≈ 8). (Note: The number 13 appearing in the previous calculation refers to the actual feed location not to the number of stages.)

The McCabe-Thiele Method for evaluating theoretical stages is limited by the following assumptions and constraints:

1. It applies only to binary systems.
2. It assumes "constant molal overflow" and in most situations does not satisfy an overall energy balance.
3. It requires graphical trial-and-error solutions to solve performance problems.

While the speed of computation using the McCabe-Thiele Method is far slower than for process simulators, it nevertheless remains an important analytical tool. Its graphical representation promotes a clarity of understanding that leads to valuable insights. We illustrate this by using the McCabe-Thiele Diagram to glean some critical information required to construct a performance curve for the distillation process in Figure 12.18.

Upon studying Figure 12.18 the following factors are revealed:

1. Feed is not being introduced at the optimum location in the column. The optimal location is on Plate 10 ($6/0.60 \approx 10$) and not Plate 13 as currently used.

2. Separation steps near the feed plate are small.

For the current conditions and feed plate location, there are more than the optimum number of stages in the rectifying (top) section. This suggests that the unit may have been designed to process a lower concentration of feed material or to produce a higher concentration distillate stream than is required currently.

Increasing the slope of the top operating line, L/V, moves the operating lines away from the equilibrium line. This increases the separation of each stage (increases the "step size"). In the special case where the distillate concentration remains the same, the concentration of the more volatile material (in our case, benzene) in the bottom stream will be lowered.

From an inspection of the McCabe-Thiele diagram, the tower appears to be oversized for the present separation. It could process a lower quality of feed (lower benzene concentration), produce a higher quality product, and improve benzene recovery. These are points that may deserve further investigation and could be overlooked easily without the graphical representation provided by the McCabe-Thiele diagram.

The McCabe-Thiele analysis is difficult to utilize for developing all the information needed in a performance problem. It requires a graphical trial-and-error solution to match the number of stages to a set of values for z_F, x_D, and x_B. Fortunately, modern computer simulators can provide rigorous solutions for a large variety of separators, including distillation towers. The information needed to construct the performance diagrams for the distillation column in this chapter was obtained from a Chemcad™ simulation. We selected, as a base case, a column with 13 theoretical stages (fed on Tray 8). In our preliminary evaluation, we assume the tray efficiency remained constant. We will return to this assumption when we discuss the limitations of these performance curves.

There are three input streams entering the tower shown in Figure 12.16. The flowrate and concentration of these streams coupled with the performance of the equipment establish the tower output. The remaining streams in the distillation system can be calculated once these streams are known.

The results of a computer simulation for the base case conditions are tabulated below.

	Flowrate (kmol/h)	Mole Fractions	Concentration
Inputs			
F	100.0	z_F	0.321
Outputs from Simulation			
L_0	71.5	x_0	0.885
V_{N+1}	96.4	y_{N+1}	0.079
V_1	101.5	y_1	0.885
L_N	166.4	x_N	0.079

The value for the vapor flowrate changed from 101.5 kmol/h to 96.4 kmol/h moving from the top to the bottom of the column. Even for this "ideal" separation, the assumption of constant molal overflow is not satisfied.

The process simulator provides a rigorous solution by carrying out material balances, energy balances, and equilibrium calculations over each stage. The simulator provides tray-by-tray results that include the composition of liquid and vapor on each tray, temperature, pressure, and K-values for each component on each tray, along with transport properties of both phases. Using data from this simulation, performance diagrams may be obtained. The most important variables are the tower inputs. They cause the outputs to change. Selection of which performance diagrams to prepare depends upon the problem being considered. We illustrate this by using the following problem:

You are the engineer in charge of the toluene-benzene distillation section in a chemical plant. You have just met with your supervisor and have been informed that, in the future, the feed concentration to this unit will no longer be constant but will vary between 25 and 40% benzene. You have also been told that it is important to maintain a constant distillate flow, D, and concentration, x_D, from this distillation unit to the downstream process.

At this meeting, possibilities for replacing parts of this process, replacing the whole system, and installing a storage system to blend feed and/or product distillate, or a combination of these alternatives were discussed. The object is to maintain the distillate flow and concentration.

To make a responsible decision on this matter, it will be necessary to assess the effect that changes in feed concentration will have on the operation of the distillation process. Before making a decision, background material will be needed. You have been told to provide a performance diagram that shows the effects of changes in feed concentration on the reflux flowrate. For a preliminary assessment, you are to assume that the reboiler and condenser can meet any new demands. If they are found inadequate, they will be replaced or modified.

In addition, questions have been raised regarding the low benzene recovery in the current operation and you have been told to consider the impact of using higher benzene recoveries.

For the situation described above, the important variables are the feed concentration, x_F, and the benzene recovery, Γ. The information needed to respond to the above request can be obtained by running simulations at different feed concentrations and reflux flowrates to determine the benzene recovery for each pair of conditions. Reflux is chosen as the dependent variable here since the flow of the reflux stream is the most common way to regulate the performance of a distillation column.

Figure 12.19 gives the performance curves for the reflux, L_0. The performance diagram shows the feed concentration, z_F, on the x-axis with the benzene recovery, Γ, as a parameter. The base case is identified as Point "a" on the diagram.

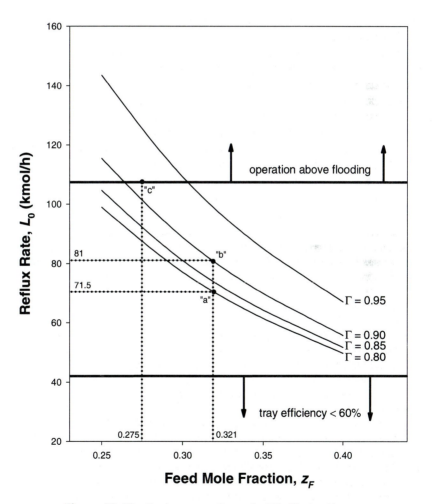

Figure 12.19   Performance Curve for Distillation Tower

The performance curves presented show the following trends:

1. For constant recovery, Γ: As z_F increases, the reflux decreases. The rate of decrease becomes less as z_F increases.
2. For a constant feed concentration, z_F: As Γ increases, the reflux increases. The rate of increase becomes greater as Γ increases.

The performance diagram given in Figure 12.19 is specific to the existing tower with fixed feed location, saturated liquid feed, and desired distillate conditions. The conditions or variables which are fixed in this problem are as follows:

$D = 30$ kmol/h

$x_D = 0.885$ mole fraction benzene

20 Trays (12 theoretical equilibrium stages plus a reboiler)

Feed plate = 13 (theoretical equilibrium Stage 8)

Diameter of Tower = 0.83 m

Tray spacing = 0.61 m

Overall Tray Efficiency = 60%

The independent input variables are the feed concentration, z_F, and the benzene recovery, Γ. One important point to note is that, like in all other performance problems, the parameters associated with the equipment are fixed. A second point to note is that with the information in Figure 12.19, it is possible to work backwards and estimate the desired reflux flowrate, for example, that would be required to obtain the desired distillate conditions if the feed concentration changed from 0.321 to 0.295. The utility of the performance curves is best illustrated with the use of the following example. Additional problems involving construction and use of performance curves for distillation are given at the end of the chapter.

Example 12.11

Using the performance diagram, Figure 12.19, find the changes to the tower input needed to achieve a recovery of 90% for the current feed concentration, $x_F = 0.321$.

New Feed Rate, F: $F = x_D D / (z_F \Gamma) = 0.885(30) / [(0.321)(0.9)] = 91.9$ kmol/h

New Reflux, L_0: From Figure 12.19 new feed rate (see Point "b") $L_0 = 81$ kmol/h

---

Example 12.11 showed that the distillation system could be operated at a recovery, $\Gamma = 0.9$. This reduces the feed that can be processed by 8.1%.

Installed equipment imposes operating limitations that restrict the range over which the performance diagrams can be used. Equipment manufacturers often provide technical information that gives the limitations of the equipment. This information may include, but is not limited to, operating temperature, pressure, installation instructions, and operating parameters such as plate efficiencies and flooding and weeping velocities. Thus this specific information adds additional constraints on the range of operations that may be considered.

From the technical information provided with the sieve trays used in the distillation tower in Figure 12.19, the following information was obtained:

1. Flooding gas velocity, $u_f = 1.07$ m/s.
2. Weeping gas velocity, $u_w = 0.35$ m/s.
3. Tray efficiency, $\epsilon = 0.60$ for $2.6 > u[\text{m/s}](\rho_v[\text{kg/m}^3])^{0.5} > 1.2$

Based on this information, two regions on Figure 12.19 are shown. These are "forbidden" or infeasible regions, in which the performance curves are not valid because of the limitations of the specific equipment or the assumptions made (e.g., $\epsilon = 0.60$). For any change to be made, the "new" condition must be checked to ensure that the new operating point lies within the feasible region for all the performance curves.

The flooding velocity limit is also shown on Figure 12.19. The limit related to weeping lies below that for tray efficiency, that is, the tray efficiency drops below 60% well before the weeping condition is reached. The constraint of operating below a tray efficiency of 60% is shown as the lower infeasible region in Figure 12.19.

Example 12.12

Find the maximum recovery possible from the distillation equipment for a feed concentration, $z_F = 0.275$.

Point "c" on Figure 12.19 provides an estimate of the recovery, $\Gamma = 0.91$.

> ### The constraints associated with the equipment must be included in any analysis of performance.

In our preliminary analysis, we assumed that the tray efficiency was 60% and the reboiler and condenser were adequate. From the manufacturer's data, it was confirmed that the efficiencies were constant over the operating range given above. The limits on the reboiler and the condenser have not been considered in the above analysis and must be checked against the performance limits for these pieces of equipment. In addition, the curves obtained assume that the feed location would not change. However, by using the optimum location, as the feed concentration changes, the performance of the tower could be improved. The use of alternative feed locations requires that feed nozzles are present at different trays. Thus, once again, the constraints of the existing equipment dictate whether this option should be considered further.

12.4 SUMMARY

In this chapter we have discovered how to construct simple performance diagrams for some common individual equipment. These performance diagrams show how a given piece of equipment responds to changes in input flows. They also can be used to predict what changes in input variables would be required in order to obtain a desired output condition. In some instances, as with valves, pumps, and compressors, performance curves are provided by equipment manufacturers. In other instances, as with the Kremser or Colburn equations, universal performance curves and equations exist subject to specific assumptions. In still other instances, as with heat exchangers or distillation columns, performance curves must be constructed by modeling equipment behavior.

In all real chemical processes, the final element that makes changes in flowrates possible is the regulating valve. Without such valves, it would be impossible for a process to adjust for unforeseen changes in operating conditions. In addition, it would be equally impossible to manipulate a process to give new desired outputs. Regulating valves are relatively simple and inexpensive pieces of equipment, yet they are absolutely essential in the day-to-day operation of a chemical process. The use of regulating valves was illustrated for a simple flow system including a pump, piping system, process equipment, and a valve.

REFERENCES

1. Bailie, R. C., and J. A. Shaeiwitz, "Performance Problems," *Chemical Engineering Education*, 28, 198 (1994).

PROBLEMS

Background Material for Problems 1–3

For the steam generator illustrated in Figure 12.1a and discussed in Section 12.1, the total resistance to heat transfer, $R_T = 1/(UA)$, is given by the following relationship:

$$R_T = R_i + R_o + R_p + R_{f,i} + R_{f,o}$$

In the above equation

R_T = total resistance to heat transfer.

R_i = convective resistance of the light oil. This resistance is a function of the flowrate, as shown in Chapter 11. Subscript 1 refers to the base case and subscript 2 refers to the new case.

$$R_{i2} = R_i(\dot{m}_2/\dot{m}_1)^{0.8}$$

R_o = convective resistance of the boiling water. This resistance is not a function of flowrate. It is a weak function of the temperature drop across the film of vapor on the outside of the tubes. For this analysis, it is assumed constant.

R_p = conductive resistance of the tube walls. This resistance is small and is neglected here.

$R_{f,i}$ and $R_{f,o}$ = fouling resistances on the inside and outside of the tubes, respectively. These two resistances are combined into a single fouling term.

$$R_f = R_{f,i} + R_{f,o}$$

For the conditions given above, the total resistance becomes,

$$R_T = R_{i1}(\dot{m}_2/\dot{m}_1)^{0.8} + R_o + R_f$$

The only resistance affected by mass flowrate, $\dot{m}_i$, is R_i.

1. Operating data taken from the plant shortly following the cleaning of the reboiler tubes show that:

$$T_{in} = 325°C,\ T_{out} = 293°C,\ \dot{m}_s = 1650\ \text{kg/h},\ T_s = 253°C$$

   a. Find the overall heat transfer coefficient.
   b. Estimate the overall resistance to heat transfer.

2. This is a continuation of Problem 1. Estimate the fouling resistance for the base case presented in Fig. 12.1a.

3. This is a continuation of Problem 1. Typical heat transfer coefficients given in Table 9.11 are used to predict the ratio of heat transfer resistances between boiling water and light oil. This ratio is estimated to be approximately 1/2. Estimate the individual resistance and heat transfer coefficient for both the oil and the boiling water.

Background Material for Problems 4–5

The performance curves presented in Figure 12.1 represent those obtained for the heat exchanger used in Problems 4–5.

4. a. Use the guideline from Tables 9.11 and 9.18 to establish "normal" limits on the amount of heat that could be exchanged.
   b. How do these limits explain the fouling of the heat exchanger observed in the operating unit?
   c. Under the constraints identified in Part a, what is the highest temperature allowed (flowrates limited to range of M values shown)?

5. a. Estimate the relative increase in heat duty that would result from increasing the liquid level from the base case to cover the tubes completely. The input streams and vapor temperature remain the same as in the base case.
    b. Explain why the duty increase is less than the increase in area.

Background Material for Problems 6–9, in Section 12.2

6. For the system presented in Example 12.3 you are asked to appraise the effects of adding a second pump and impeller set identical to the current pump in the system
    a. Prepare a performance diagram, for the two-pump system and for each impeller size, if the pumps are to be installed in series.
    b. Prepare a performance diagram, for the two-pump system and for each impeller size, if the pumps are to be installed in parallel.
    c. Estimate the pressure achieved for each pump/impeller system in Parts a and b at flow rates of 400 and 800 gallons/min.

7. Referring to Example 12.3, it has been decided that the storage tank should be kept at 1 bar and the reactor pressure at 1.2 bar. To provide the pressure head required, a pump is placed in line before the heat exchanger. For the 50% increase in flow desired in Example 12.3, answer the following:
    a. What pressure head must be produced by the pump?
    b. Which of the pump/impeller systems given in Figure 12.4 is capable of providing sufficient flow?
    c. At what location in the system would you place the pump? Explain why.

8. A centrifugal pump produces head by accelerating fluid to a velocity approaching the tip velocity of the rotating impeller. The liquid leaving the pump is at a high velocity and is forced into the discharge pipe from the pump where it comes into contact with relatively slow moving fluid in the pipe. As the fluid decelerates, the velocity head is converted into pressure head. For a given pump the following relationships can predict performance.

$$h_2/h_1 = [(N_2/N_1)(D_2/D_1)]^2 \text{ and } \dot{v}_2/\dot{v}_1 = [(N_2/N_1)(D_2/D_1)]^3$$

where:

N = speed of rotation of the impeller, rpm

h and $\dot{v}$ = head and volumetric flowrate, respectively

2 and 1 = new and base case conditions, respectively

D = impeller diameter.

The speed of rotation of the impeller can be adjusted, even when a constant speed motor is used, by the use of a pulley system as shown below:

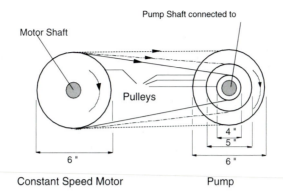

Pump Shaft connected to

Motor Shaft

Pulleys

4 "
5 "

6 "

6 "

Constant Speed Motor   Pump

The pump curve given in Figure 12.4 was obtained when the impeller was directly coupled to a constant speed motor (3500 rpm). Develop performance curves for this pump, with a 7-in diameter impeller, assuming that the pump is driven via a pulley system in which the constant speed motor (3500 rpm) has a 6-in diameter pulley and the motor impeller is connected to a series of pulleys with diameters of 4 in, 5 in, and 6 in. Construct performance diagrams for each pulley combination.

9. Naphthalene is fed to a phthalic anhydride production process. The feed is available at 200°C and 80 kPa. The flowrate is 12,800 kg/h in 1.5 in Schedule 40 pipe. A pump with NPSH characteristics as plotted in Figure 12.9 is used. Will this pump be suitable for the desired duty? If not, what modifications would be necessary in order to use the existing pump? Be quantitative.

10. Refer to Example 12.9. There are additional methods other than the two presented for increasing the NPSH in order to avoid pump cavitation. For the following situations, sketch the $NPSH_A$ curve on Figure 12.9, keeping head at 2 m keeping the flowrate constant. Identify the operating point on the $NPSH_A$ curve in order to determine whether $NPSH_A > NPSH_R$.
   a. The diameter of the suction line can be increased. Examine the effect of increasing the diameter of the suction line to 1.25 in Schedule 40.
   b. The temperature of the toluene can be decreased. Examine the effect of decreasing the temperature of the toluene to 45°C using the original pipe.

11. Refer to the compressor curves given in Figure 12.11.
   a. At a flowrate of 100 Mg/h, what is the exit pressure from the compressor operating at 3500 rpm if the inlet pressure is 3 bar?
   b. If the feed pressure is 3 bar and the compressor is operating at 2200 rpm, how can you obtain an outlet pressure of 6 bar at 100 Mg/h?
   c. It is necessary to raise the pressure from 3 bar to 48 bar at 100 Mg/h. How could this be accomplished?

12. A packed scrubber with 10 transfer units has been designed to reduce the acetone concentration in 40 kmol/h of air from a mole fraction of 0.02 to

0.001. The acetone is absorbed into a 20 kmol/h water stream, and it may be assumed that the water stream enters acetone free. If the flow of oil is decreased by 20%, what is the new outlet mole fraction of acetone in air?

13. Prepare a set of performance curves, similar to those in Example 12.10, for the packed scrubber in Problem 12.

14. Repeat Problem 13 and Example 12.10 with the following change: In each case, the gas flowrate remains constant at the original conditions. The parameter on the performance curve is now absorber temperature (assumed constant throughout the absorber). There should be 5 curves for each absorber: the original temperature, an increase of 5°C and 10°C from the original temperature, and a decrease of 5°C and 10°C from the original temperature.

15. The acetone scrubber in Problem 12 has been running well for several years. It is now observed that the outlet acetone mole fraction in air is 0.002. Suggest at least 5 reasons for this observation. Suggest at least 5 ways to compensate for this problem in the short term. Evaluate each compensation method as to its suitability.

16. It is necessary to increase the capacity of an existing distillation column by 25%. As a consequence, the amount of liquid condensed in the condenser must increase by 25%. In this condenser, cooling water is available at 30°C, and, under present operating conditions, exits the condenser at 45°C, the maximum allowable return temperature without a financial penalty assessed to your process. Condensation takes place at 75°C. You may assume that the limiting resistance is on the cooling water side. Can the existing condenser handle the scale-up without incurring a financial penalty? What is the new outlet temperature of cooling water? By what factor must the cooling water flow change?

17. In the previous problem, it may be necessary to scale up or scale down by as much as 50% from the original operating conditions. Prepare a performance curve similar to the one in Figure 12.1 for operation of the condenser over this range.

Background Material for Problems 18–22

In Section 12.3.2, it was shown that Figure 12.19 could be constructed from results of simulating the distillation column. Table P12.18 shows the results of such a simulation for the problem discussed in Section 12.3.2. The data used in Figure 12.19 are contained in this table. The data in Table P12.18 can be used to construct other performance curves, similar to Figure 12.19. In Problems 18–20, a performance curve should be constructed with the flooding and tray efficiency limits included (given on page 325), as in Figure 12.19.

18. Prepare a performance plot for vapor velocity as a function of feed composition with benzene recovery as a parameter.

Table P12.18:   Performance Data from a Process Simulator for the Benzene-Toluene Tower.
[D = 30 kmol/h, x_D = 0.885, P_{bot} = 2 bar, Overall Tray Efficiency = 60%]
[Assume ideal system for evaluation of K values and other thermodynamic data]

Feed Conc z_F	Benz Rcvry Γ	Feed F	Bottm Prod B	Reflux Ratio R	Ovhd Vapor V_1	Boil-Up V_{N+1}	Liquid Reflux L_0	Cond Duty Q_c	Rebl Duty Q_r	Bottm Temp	Vapor Vel at top u
		kmol h	kmol h		kmol h	kmol h	kmol h	GJ h	GJ h	°C	m/s
0.25	0.80	132.8	102.8	3.30	129.0	122.4	99.0	-3.93	3.87	133.4	1.01
0.30	0.80	110.6	80.6	2.57	107.1	101.7	77.1	-3.26	3.21	132.6	0.84
0.35	0.80	94.8	64.8	2.05	91.5	86.9	61.5	-2.79	2.75	131.8	0.73
0.40	0.80	83.05	53.0	1.66	79.8	75.7	49.8	-2.43	2.39	130.7	0.62
0.25	0.85	124.9	94.9	3.49	134.7	127.8	104.7	-4.10	4.04	134.1	1.05
0.30	0.85	104.1	74.1	2.70	111.0	105.4	81.0	-3.38	3.33	133.5	0.87
0.35	0.85	89.3	59.3	2.14	94.3	89.5	64.3	-2.87	2.83	132.8	0.74
0.40	0.85	78.1	48.1	1.73	81.8	77.6	51.8	-2.49	2.45	132.0	0.64
0.25	0.90	118.0	88.0	3.85	145.4	138.0	115.4	-4.42	4.36	134.9	1.14
0.30	0.90	98.3	68.3	2.96	118.8	112.7	88.8	-3.56	3.61	134.4	0.93
0.35	0.90	84.3	54.3	2.33	99.9	94.8	69.9	-3.04	3.00	133.9	0.78
0.40	0.90	73.8	43.8	1.86	85.5	81.4	55.8	-2.61	2.57	133.3	0.67
0.25	0.95	111.8	81.8	4.78	173.5	164.5	143.5	-5.26	5.20	135.6	1.36
0.30	0.95	93.2	63.2	3.64	139.2	132.1	109.2	-4.23	4.17	135.4	1.09
0.35	0.95	79.9	49.9	2.83	115.0	109.2	85.0	-3.50	3.45	135.1	0.90
0.40	0.95	69.9	39.9	2.24	97.1	92.0	67.1	-2.95	2.91	134.8	0.76

19. Prepare a performance plot for boil-up rate as a function of feed composition with benzene recovery as a parameter.

20. Prepare a performance plot for reboiler duty as a function of feed composition with benzene recovery as a parameter.

21. For the distillation problem outlined in Section 12.3.2,
    a. Determine the values of F, B, and x_B required when the concentration of the feed becomes, $z_F = 0.25$. Assume that the values of D and x_D are to remain the same (D = 30 kmol/h and x_D = 0.885) and that the recovery is to remain 90%.
    b. Report any risks associated with the solution.

c. What are the values of L_0 and V_{N+1} for this case?

22. The information given in Figure 12.18 showed that the benzene recovery column was not operating at maximum benzene recovery.
    a. Estimate the maximum value of benzene recovery for the base case $(z_F = 0.321)$.
    b. Give several reasons for operating the current column at less than maximum benzene recovery.
    c. Plot a performance curve that shows estimated values for F, L_0, and V_1 at maximum benzene recovery as a function of feed composition.

Background Information for Problems 23 and 24

In preparing the performance diagrams for the benzene-toluene system in Section 12.3.2, it was noted that under current operation, the feed stream was not introduced at the optimum location. The performance curves, Figure 12.19, were constructed for this non-optimal feed location.

23. For the same feed, distillate, and bottoms shown in Figure 12.16, determine the following:
    a. The boil-up, V_{N+1}, and reflux, L_0, required if the feed is now introduced at the optimum location.
    b. Using typical utility costs from Chapter 3, estimate the yearly savings ($/yr) obtained by changing the feed location.

24. Assume that the distillation column in Figure 12.16 is to be operated at the same value of L_0, D, and B and the same feed as the base case.
    a. Calculate x_D, x_B, and the benzene recovery, Γ, obtained by introducing the feed at the optimum location.
    b. Assuming that the benzene lost in the bottom is valued at $0.20/lb, calculate the savings resulting from locating the feed at the optimum location.

13

Performance of Multiple Unit Operations

In the previous chapter, techniques for evaluating the performance of individual unit operations were introduced. The use of performance curves was illustrated. Performance curves describe the behavior of a unit over a wide range of inputs. The techniques developed in Chapter 11 were also used to obtain results for a single set of input conditions, which is one point on a performance curve.

In a chemical process, units do not operate in isolation. If the input to one unit is changed, its output changes. However, this output is usually the input to another unit. In certain cases, what is usually considered to be a single unit is actually multiple units operating together. A distillation column is an example of this situation, since the reboiler and condenser are part of the distillation unit. Any change in the column or in either of the heat exchangers will affect the performance of the other two. Therefore, it is necessary to understand the interrelationship between performance of multiple units and gain experience in analyzing these types of problems.

Given the complexity of the performance of multiple units, performance curves will not be generated. The techniques developed in Chapters 10 through 12 will be used to solve for one point on a performance curve. As in Chapter 12, the "case-study" approach is used. Four examples of multiple units are illustrated in order to help you understand the approach to analyzing performance problems involving multiple units.

13.1 ANALYSIS OF A REACTOR WITH HEAT TRANSFER

Many industrially significant reactions are exothermic. Therefore, in a reactor, it is necessary to remove the heat generated by the heat of reaction. There are several ways to accomplish heat removal. For relatively low heats of reaction, staged adiabatic packed beds (packed with catalyst) with intercooling may be used.

Here, no heat is removed in the reaction section of the equipment. Instead, the process fluid is allowed to heat up in the reactor and heat is removed in heat exchangers between short beds of catalyst (reactor sections). For larger exothermic heats of reaction, a shell-and-tube configuration, much like a heat exchanger, is often used. Here, the reaction usually occurs in the tubes, which are packed with catalyst. As the heat of reaction increases, tube diameter is decreased in order to increase the heat transfer area. For extremely large heats of reaction, fluidized beds with internal heat transfer surfaces are often used, due to the constant temperature of the fluid bed and the relative stability of such reactors because of the large thermal mass of well-mixed solid particles.

In all three of these cases, the performance of the reactor and the heat exchanger are coupled. It is not correct to analyze one without the other. In this section, an example of the performance of a shell-and-tube type reactor (tube bundle containing catalyst immersed in pool of boiling water) with heat removal is used to illustrate the interrelationship between reactor performance and heat transfer. In Chapter 14, other examples of reactor performance are presented.

A shell-and-tube reactor, as illustrated in Figure 13.1, is used for the following reaction to produce cumene from benzene and propylene

$$C_6H_6 \; + \; C_3H_6 \rightarrow C_9H_{12}$$
$$benzene \quad propylene \quad cumene$$

The reaction kinetics are as follows

$$r = kc_b c_p \quad mol/L\,s$$

$$k = 3500 \; exp(-13.28 \; (kcal/mol)/RT)$$

Normally, the reaction occurs at 350°C and 3000 kPa (the reaction is not really isothermal, see Item 3 below). Heat evolved by the reaction is removed by producing high-pressure steam (4237 kPa, 254°C) from boiler feed water in the reactor shell. Propylene is the limiting reactant, with benzene present in excess. The feed propylene is a raw cut, and contains 5 wt % propane impurity. Recently, the propane supplier has been having difficulty meeting specifications, and the propane impurity exceeds 5 wt %. The reduced feed concentration of propylene is causing a decrease in cumene production. In order to maintain the desired cumene production rate, it has been suggested that the reaction temperature be raised to compensate.

The following assumptions will be made in order to simplify the problem:

1. There are no other impurities except for the propane.
2. The pressure drop in the reactor is negligible relative to the operating pressure.
3. The temperature profile in the reactor is flat, that is, the temperature is 350°C everywhere in the reactor. This is the most serious assumption, and it

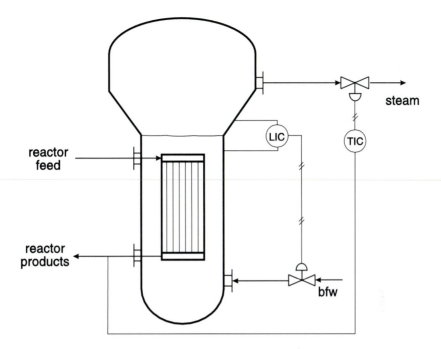

Figure 13.1 Shell-and-Tube Type Packed Bed Reactor

is clearly not correct. There will most likely be a hot spot near the reactor entrance. Figure 13.2 illustrates the temperature profiles for the reactor and for the steam. The dotted line is the anticipated temperature profile. By assuming a constant temperature, calculations are simplified, and the concept illustrated in this case study can still be understood.

Before proceeding with a detailed analysis, it is instructive to understand why a temperature increase was suggested as compensation for decreased propylene feed concentration. If the total feed rate of impure propylene is maintained constant, the feed rate of propylene decreases with increasing propane impurity. Since propylene is the limiting reactant, cumene production is related to propylene feed by the following relationship

$$F_c = F_{po} X \tag{13.1}$$

where

F_c = molar rate of cumene produced
F_{po} = molar rate of propylene feed
X = conversion of propylene

If the molar feed rate of propylene, F_{po}, is reduced, cumene production will decrease unless the conversion, X, is increased. In the reactor, the overall flow of

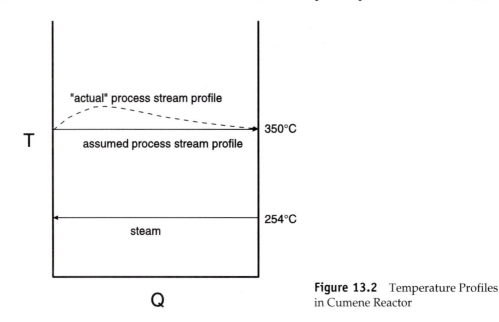

Figure 13.2   Temperature Profiles in Cumene Reactor

material is constant, since propane replaces propylene; therefore, the space time remains constant. From the reaction expression, a reduced concentration of propylene results in a reduced reaction rate. Intuitively, a reduced reaction rate at constant space time results in reduced conversion. Therefore, from Eq. (13.1), the cumene production rate is decreased because of both a decreased reactant feed rate and a decreased conversion.

One method to increase reaction rate is to increase the reactor temperature. The quantitative relationship between temperature increase and cumene production rate can be obtained by the methods learned in a typical reaction engineering class or by use of a process simulator.

Example 13.1

Analyze the heat transfer consequences of increasing the reactor temperature by 10°C and 20°C, which correspond to maintaining the desired cumene production rate at approximately 6% and 7% propane impurity, respectively.

The heat transfer performance equation must also be obeyed. Predicting trends shows that the temperature difference between the reaction side and steam side must remain constant:

$$\Delta T(\mathfrak{c}) = \frac{Q(\mathfrak{c})}{U(\mathfrak{c})A(\mathfrak{c})} \tag{E13.1a}$$

Since the cumene production rate is to remain constant, the heat load on the reactor remains constant. Since overall flows do not change, the heat transfer coefficient remains constant. Since the reactor is not modified, the heat transfer area remains constant. Therefore, ΔT must remain constant. Since the original ΔT was 96°C, the following table gives the temperatures that are required on the steam side. In order to produce higher temperature steam, the boiler feed water pressure and, therefore, steam pressure must be increased on the shell side. This is because the temperature at which boiler feed water vaporizes to saturated steam increases with pressure.

Temperature Increase (°C)	Steam Temperature (°C)	Steam Pressure (kPa)
0	254	4237
10	264	5002
20	274	5853

For a 10°C increase in reactor temperature, the pressure on the shell side must increase by almost 20%, and for a 20°C increase in reactor temperature, the pressure on the shell side must increase by about 38%.

Without considering the heat transfer analysis, it would have been easy to suggest the temperature increases be implemented. However, they could not be accomplished without raising the shell-side temperature and pressure. Because the reactor is also a heat exchanger, the performance equation for a heat exchanger must be used in the analysis. Therefore, in order to increase the reactor temperature, the steam temperature and, consequently, the steam pressure, must be increased. The unanswered question is whether the reactor can withstand the indicated pressure increases. It is unlikely that a vessel designed for a high pressure like 4237 kPa would be over-designed to handle a 20% increase in pressure, and a design for 38% increase would be extremely uncommon.

It is instructive to continue this analysis and determine if there are other ways to increase the conversion in the reactor to compensate for the reduced feed.

Example 13.2

Suggest another method for increasing the conversion in the reactor so that the cumene production rate remains constant with reduced propylene feed concentration.

The question to be answered is what can be adjusted, other than temperature that will affect the reaction rate and/or the conversion. In the reaction rate expression, the primary temperature effect is in the Arrhenius expression. For gas-phase reactions, assuming ideal gas behavior, the concentrations $c_i = P_i/RT$. The increase in temperature discussed in Example 13.2 will actually reduce the concentrations; however, the exponential increase in the reaction rate constant is the primary effect. It is also observed that increasing the pressure increases the concentration, thereby increasing the reaction rate. There is an additional effect. This can be seen from the equation of a plug flow reactor

$$\frac{\tau}{c_{Ao}} = \int_0^X \frac{dX}{-r_A}$$

Since the mass flowrate $\dot{m} = \rho A v$, and increasing the pressure increases the density, at constant mass flowrate, the velocity drops. Therefore, the residence time, τ, in the reactor increases, which results in increased cumene conversion (upper limit of integral increases), and increased cumene production. Of course, pressure limitations of the equipment in the process or in the tube side of the reactor may limit the pressure increase possible.

In this case study, the most likely solution to the problem is some combination of temperature and pressure increase. If this is insufficient, it may be necessary to try to increase the feed rates of the reactants. However, consistent with the lesson of this case study, the performance of the pumping equipment may limit possible feed rate increases.

In summary, this case study introduces the concept that performance of any unit cannot be considered in isolation. For the reactor, it seems straightforward to do an analysis involving only reaction kinetics and conclude that a temperature increase is a solution to the problem of reduced feed. However, it is now clear that since the reactor is also a heat exchanger, both reactor and heat exchanger analyses must be done simultaneously in order obtain a correct result. In order to get the correct numerical result, the correct temperature profile on the reaction side must be used. It is obtained from a process simulator or by solving the differential equations for conversion and temperature simultaneously. The correct temperature profile, as illustrated in Figure 13.2, arises because the reaction rate, $r \propto e^{-E/RT}$. Therefore, as the temperature increases, r increases by a large amount, generating even more heat, and the peak in the temperature profile may increase significantly, causing hot spots to develop. Hot spots can damage catalyst or promote undesired side reactions. Chapter 14 treats these issues in more detail.

13.2   PERFORMANCE OF A DISTILLATION COLUMN

A distillation column requires a reboiler to add the energy necessary to accomplish the separation. A condenser is also required to reject heat to the surroundings. The performance of a distillation column cannot be analyzed without consideration of the performance of both heat exchangers. This is illustrated in the following case study.

The distillation column illustrated in Figure 13.3 is used to separate benzene and toluene. It contains 35 sieve trays with the feed on tray 18. The relevant flows are given in the table below. Your assignment is to recommend changes in the tower operation to handle a 50% reduction in feed. Overhead composition must be maintained at 0.996 mole fraction benzene. Cooling water is used in the condenser, entering at 30°C and exiting at 45°C. Medium pressure steam (185°C, 1135 kPa) is used in the reboiler.

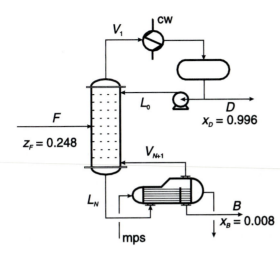

Figure 13.3 Distillation of Benzene from Toluene

The operating conditions of the tower before reduction in feed are as follows:

Input/Output	Flow (kmol/h)	Mole Fraction Benzene	Temperature (°C)
Inputs:			
Feed, F	141.3	0.248	90
Reflux, L_0	130.7	0.996	112.7
Boil-up, V_{N+1}	189.5	0.008	145.3
Outputs:			
Distillate, D	34.3	0.996	112.7
Still Bottoms, L_N	296.5	0.008	145.3
Bottoms Product, B	107.0	0.008	145.3

Among the several possible operating strategies for accomplishing the necessary scale-down are:

1. Scale-down all flows by 50%. This is only possible if the original operation is not near the lower velocity limit that initiates weeping or reduced tray efficiency. If 50% reduction is possible without weeping, lower tray efficiency, or poor heat exchanger performance, this is an attractive option.
2. Operate at the same boil-up rate. This is necessary if weeping or reduced tray efficiency is a problem. The reflux ratio must be increased in order to maintain the reflux necessary to maintain the same liquid and vapor flows

in the column. In this case, weeping, lower tray efficiency, or poor heat exchanger performance caused by reduced internal flows is not a problem.
The down side of this alternative is that a purer product will be produced
and unnecessary utilities will be used.

In the present case study, the analysis will be done by assuming that it is
possible to scale-down all flows by 50% without weeping or reduced tray efficiency.

Example 13.3

Estimate the pressure drop through the column. To what weir height does this correspond?

Interpolation of tabulated data [1] yields the following relationships for the vapor
pressures in the temperature range of interest:

$$\ln P^* \text{ (kPa)} = 15.1492 - \frac{3706.84}{T(K)} \quad \text{benzene}$$

$$\ln P^* \text{ (kPa)} = 15.3877 - \frac{4131.14}{T(K)} \quad \text{toluene}$$

It is assumed that the bottom is pure toluene and the distillate is pure benzene. This is a
good assumption for estimating top and bottom pressures given the mole fractions specified for the distillate and bottoms. Therefore, at the bottom temperature of 141.7°C, the
vapor pressure of toluene, and hence the pressure at the bottom of the column, is 227.2
kPa. At the top temperature of 104.2°C, the vapor pressure of benzene, and hence the pressure at the top of the column is 204.8 kPa.

Since there are 35 trays, the pressure drop per tray is

$$\Delta P = (227.2 - 204.8)/35 = 0.64 \text{ kPa/tray}$$

If it assumed that the weir height is the major contribution to the pressure drop on a tray,
then

$$\Delta P = \rho g h$$

where h is the weir height. Assuming an average density of 800 kg/m³, then

$$h = (640 \text{ Pa})/[(800 \text{ kg/m}^3)(9.8 \text{ m/s})] = 0.08 \text{ m} \approx 3 \text{ in}$$

This is a typical weir height and is consistent with the assumption that the weir height is
dominant.

The pressure drop is assumed to remain constant after the scale-down since the
weir height is not changed. In practice, there is an additional contribution to the
pressure drop due to gas flow through the tray orifices. This would change if column flows changed, but the pressure drop through the orifices is small and
should be a minor effect. The next two examples illustrate how the performance
of the reboiler and condenser affect the performance of the distillation column.

Example 13.4

Analyze the reboiler to determine how its performance is altered at 50% scale-down.

Figure E13.4 shows the T-Q diagram for this situation. Because the amounts of heat transferred for the base and new cases are different, the Q values must be normalized by the total heat transferred in order for these profiles to be plotted on the same scale. The solid lines are the original case (subscript 1). For the new, scaled-down case (subscript 2), a ratio of the energy balance on the reboiled stream for the two cases yields

$$\frac{Q_2}{Q_1} = \frac{\dot{m}_2 \lambda}{\dot{m}_1 \lambda} \tag{E13.4.1}$$

If it is assumed that the latent heat is unchanged for small temperature changes, then $Q_2/Q_1 = 0.5$, since at 50% scale-down, the ratio of the mass flowrates in the reboiler is 0.5. The ratio of the heat transfer equations yields

$$\frac{Q_2}{Q_1} = 0.5 = \frac{U_2 A \Delta T_2}{U_1 A \Delta T_1} \approx \frac{\Delta T_2}{43.3} \tag{E13.4.2}$$

In Eq. (E13.4.2), it is assumed that the overall heat transfer coefficient is constant. We assume here that $U \neq f(T)$. This assumption should be checked, since for boiling heat transfer coefficients with large temperature differences, the boiling heat transfer coefficient may be a strong function of temperature difference.

From Eq. (E13.4.2), it is seen that $\Delta T_2 = 21.7°C$. Therefore, for the reboiler to operate at 50% scale-down, with the steam side maintained constant, the boil-up temperature must be 163.4°C. This is shown as the dotted line in Figure E 13.4.

From the vapor pressure expression for toluene, the pressure at the bottom of the column is now 372.5 kPa. It would have to be determined whether the existing column could withstand this greatly increased pressure.

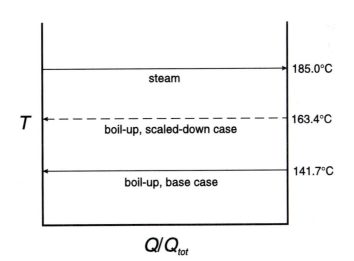

Figure E13.4   Temperature Profiles in Reboiler for Example 13.4

It is seen that the reboiler operation requires that the pressure of the boil-up stream be increased. This is not the only possible alternative. All that is necessary is that the temperature difference be 21.7°C. This could be accomplished by reducing the temperature of the steam to 163.4°C. In most chemical plants, steam is available at discrete pressures, and the steam used here is typical of medium-pressure steam. Low-pressure steam would be at too low a temperature to work in the scaled-down column. However, the pressure and temperature of medium-pressure steam could be reduced if there were a throttling valve in the steam feed line to reduce the steam pressure. The resulting steam would be superheated, so desuperheating would also be necessary. This could be accomplished by spraying water into the superheated steam. A change in a utility stream is almost always preferred to a change in a process stream.

Example 13.5

Using the results of Example 13.4, analyze the condenser for the scaled-down case.

Under the assumption that the pressure drop for the column remains at 22.4 kPa, the pressure in the condenser is now 350.0 kPa. From the vapor pressure expression for benzene, the new temperature in the condenser is 126.0°C. The remainder of the analysis is similar to Example 12.1. The T-Q diagram is illustrated in Figure E13.5. Since an organic is condensing, it is assumed that the resistance on the cooling water side is approximately equal to the resistance on the condensing side. Therefore,

$$\frac{U_2}{U_1} = \frac{\dfrac{1}{h_{o1}} + \dfrac{1}{h_{i1}}}{\dfrac{1}{h_{o2}} + \dfrac{1}{h_{i2}}} = \frac{\dfrac{2}{h_{o1}}}{\dfrac{1}{h_{o1}} + \dfrac{1}{h_{o1}M^{0.8}}} = \frac{2}{1 + \dfrac{1}{M^{0.8}}} \qquad \text{(E13.5.1)}$$

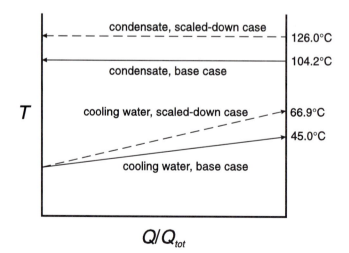

Figure E13.5   Temperature Profiles in Condenser for Example 13.5

where $M = \dot{m}_2/\dot{m}_1$

The ratio of the base cases for the energy balance on the condensing stream is

$$\frac{Q_2}{Q_1} = \frac{\dot{m}_2 \lambda}{\dot{m}_1 \lambda} \tag{E13.5.2}$$

Since the ratio of the mass flowrates is 0.5, and assuming that the latent heat is unchanged with temperature in the range of interest, $Q_2/Q_1 = 0.5$.

The ratio of the base cases for the energy balance and the heat exchanger performance equation are, respectively

$$0.5 = M\,\frac{T - 30}{15} \tag{E13.5.3}$$

$$0.5 = \frac{2(T - 30)}{(66.42)\left(1 + \dfrac{1}{M^{0.8}}\right)\ln\left(\dfrac{96}{126 - T}\right)} \tag{E13.5.4}$$

Solution of Eqs. (E13.5.3) and (E13.5.4) yields

$$M = 0.203$$

$$T = 66.9°C$$

Therefore, the cooling water rate must be reduced to 20% of the original rate and the outlet water temperature is increased by about 22°C. This would cause increased fouling problems on the cooling water side. Therefore, it may be better to use a higher reflux ratio rather than reducing the flows in the column.

It is observed that in order to operate the distillation column at 50% scale-down with process flows reduced by 50%, a higher pressure is required and the cooling water will be returned at a significantly higher temperature than before scale-down. The operating pressure of the column is determined by the performance of the reboiler and condenser.

Examples 13.4 and 13.5 reveal a process "bottleneck" at the reboiler that must be resolved in order to accomplish the scale-down. In this problem, one bottleneck is the high cooling water return temperature, which would almost certainly cause excessive fouling in a short period of time. Another potential bottleneck is tray weeping due to the greatly reduced vapor velocity. In the process of changing operating conditions (capacity) in a plant, a point will be reached where the changes cannot be increased or reduced any further. This is called a bottleneck. A bottleneck usually results when a piece of equipment (usually a single piece of equipment) cannot handle additional change. In addition to the solution presented here, there are a variety of other possible adjustments, or "debottlenecking strategies," which are presented below with a short explanation for each. Another debottlenecking problem is presented in Chapter 16.

1. Replace the heat exchangers: Eq. (E13.4.2) shows that a new heat exchanger with half of the original area allows operation at the original temperature and pressure. The heat transfer area of the existing exchanger could be reduced by plugging some of the tubes, but this modification would require a process shutdown. This involves both equipment down time or capital expense for a new exchanger. Your intuitive sense should question the need to get a new heat exchanger to process less material. You can be assured that your supervisor would question such a recommendation.

2. Keep the boil-up rate constant: This would maintain the same vapor velocity in the tower. If we are operating near the lower velocity limit that initiates weeping or lower tray efficiency, this is an attractive option. The constant boil-up increases the tower separation. This is an attractive option that results in much smaller temperature and pressure changes.

3. Introduce feed on a different plate: This must be combined with Option 4, below. The plate should be selected to decrease the separation and increase the bottoms concentration of the lower boiling fraction. This lowers the process temperature, increases the ΔT for heat transfer and the reboiler duty. This may or may not be simple to accomplish depending on whether the tower is piped to have alternate feed plates. As in Option 2, the reflux and coolant inputs will change.

4. Recycle bottoms stream and mix with feed: This must be combined with Option 3. By lowering the concentration of the feed, the concentration of the low boiler in the bottoms can be increased, the temperature lowered, and the reboiler duty increased. This represents a modification of process configuration and introduces a new (recycle) input stream into the process.

The adjustments outlined were based on

1. Replace the exchanger (Option 1)
2. Modify utility stream (Discussed after Example 13.4)
3. Modify process stream (Option 2)
4. Modify equipment (Option 3 or Option 1, if plugging of tubes to reduce area is chosen)
5. Modify process configuration (Option 4)
6. Modify operating conditions T, P (as in case study)

We have shown that there are a number a paths that remove the bottleneck that exists at the reboiler. Adjustments 1 and 2 required no changes in process inputs. All others required modifications of several input streams to maintain the output quality.

It should be noted that, at 50% scale-down, it is possible that weeping or low tray efficiency may be observed (see Section 12.3.2 and Figure 12.19). This must be considered before recommending such a scale-down. Furthermore, reduction of the cooling water flowrate in the condenser combined with increased cooling water temperatures could cause fouling problems, as was pointed out in Example 13.5. Care must taken to understand the consequences of process modifications.

During plant operations, when seeking the best operating conditions, incremental changes can be made and the effects observed. This is termed "evolutionary operation" [2]. Amoco has reported a significant increase in its annual income from evolutionary operation [3]. In the design case, changes recommended are necessarily conservative to assure that equipment purchased and installed is adequate to meet the conditions when the plant is operated.

13.3   PERFORMANCE OF A HEATING LOOP

In the previous section, we observed how the performance of a distillation column was affected by the performance of the reboiler and condenser. It was possible to analyze the reboiler, column, and condenser sequentially in order to solve the scale-down problem. In this section, we examine the performance of a reactor in which the large, exothermic heat of reaction is removed by a heat transfer fluid. It will be seen that the bottleneck for scale-up of this process is the performance of the heat removal loop. The analysis of this problem is more complex than for the distillation column since all units involved must be analyzed simultaneously in order to solve the problem.

The problem to be analyzed is illustrated in Figure 13.4. It is a part of the allyl chloride problem discussed in more detail in Appendix C.

Due to an emergency, an unscheduled shut-down at an another allyl chloride production plant operated by your company, it is necessary to increase production temporarily at your plant. Your job is to determine the maximum scale-up possible for the allyl chloride production reactor. The reactor is a fluidized bed, with a cyclone for solids recovery, operating isothermally at 510°C. The reactor is operating at two times the minimum fluidization velocity. Based on the cyclone design and the solids handling system, it is known that an increase of at least 100% of process gas flow can be accommodated. The reaction proceeds to completion as long as the reactor temperature remains at 510°C. A temperature increase destroys the catalyst and a temperature decrease results in incomplete reaction and undesired side-products. The reaction is exothermic and heat is removed by Dowtherm A™ circulating in the loop. The maximum operating temperature for Dowtherm A is 400°C, and its minimum operating pressure at this temperature is 138 psig. There are two heat exchanger units in the reactor, and they are currently operating in series. The heat transfer resistance on the reactor side is four times that on the Dowtherm A side. The inside of the reactor is always at 510°C due to the well-mixed nature of the fluidized solids. There is a spare pump. High-pressure steam is made from boiler feed water in the heat exchanger. The boiler feed water is available at 90°C, and has been pumped to 600 psig by a pump that is not shown. The pool of liquid being vaporized may be assumed to be at the vaporization temperature, 254°C. All resistance in the

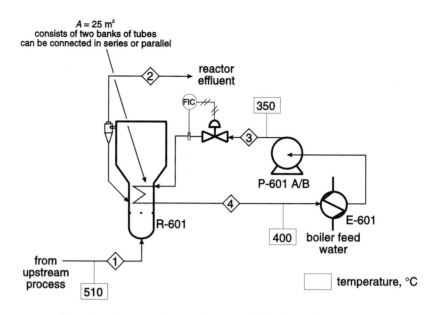

Figure 13.4   Dowtherm A Loop for Allyl Chloride Reactor

heat exchanger is assumed to be on the Dowtherm A side. At normal operating conditions, Dowtherm A circulates at 85 gal/min in the loop. The temperatures of each stream are shown on Figure 13.4. The pump curve for the pump in the loop is shown on Figure 13.5.

An energy balance on the process fluid in the reactor yields

$$Q_R = \dot{n}_{Cl}\Delta H_{rxn} \tag{13.2}$$

where $\dot{n}_{Cl}$ is the molar flowrate of chlorine, the limiting reactant, and the subscript R refers to the reactor. The heat of reaction term in Eq. (13.2) is on a unit mole basis. Since the heat of reaction is constant during scale-up, Eq. (13.2) shows that the amount of heat removed increases proportionally with the amount of scale-up. From this energy balance, there is nothing to suggest that the reactor cannot be scaled-up by a factor of two, the maximum allowed by the reactor. However, the question which must be answered is whether twice the heat can be removed from the reactor by the heat exchange loop. An energy balance on the Dowtherm A in the reactor yields

$$Q_R = \dot{m}_D C_{pD}(T_4 - T_3) \tag{13.3}$$

where the subscript D refers to the Dowtherm A. From Eq. (13.3), it is observed that the heat removal rate is proportional to both the mass flowrate and the temperature increase of the Dowtherm A. Therefore, the most obvious method is to

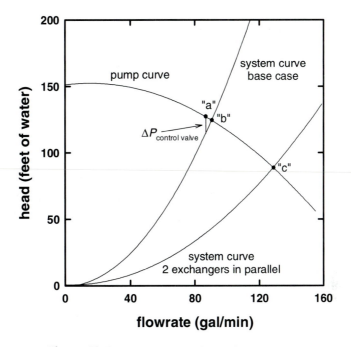

Figure 13.5   Performance of Dowtherm A Loop

increase the flowrate of Dowtherm A, so it is necessary to determine the maximum flowrate possible for Dowtherm A.

Example 13.6

For the pump curve provided in Figure 13.5, determine the maximum possible flowrate for Dowtherm A in the loop. The pump operates at only one speed, described by the curve given in the figure. The pressure drops between streams are shown in Figure E13.6.

The pump operates only at conditions shown on the pump curve. The pressure drop in the system and regulating valves must equal the pressure delivered by the pump. The base case solution gives one point on this line (Point "a"). Of this total pressure drop, 15 feet of water is from the control valve, which can be adjusted independently. The remaining pressure drop, 110 feet of water (125 − 15 feet of water), is the "system pressure drop" and is dependent on the flowrate through the system. The system pressure drop can be obtained from the equation for frictional pressure losses

$$\Delta P = \frac{2fL_{eq}v^2}{D} \tag{E13.6a}$$

Taking the ratio of Eq. (13.6a) for the scaled-up case (subscript 2) to the base case (subscript 1) for constant L, D, and the assumed constant friction factor (high Reynolds number, fully developed turbulent flow), yields

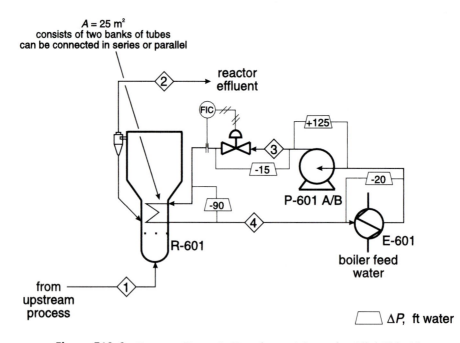

Figure E13.6   Pressure Drops in Dowtherm A Loop for Allyl Chloride Reactor

$$\frac{\Delta P_2}{110} = \frac{v_2^2}{v_1^2} = \frac{\dot{m}_2^2}{\dot{m}_1^2} = M^2 \qquad\qquad (E13.6b)$$

where M is the scale-up factor, the ratio of the scaled-up Dowtherm A flowrate to that for the base case. Eq. (13.6b) is plotted with the pump curve on Figure 13.5. The maximum flowrate occurs when the valve is wide open. When the valve is wide open, there is no (or very little) pressure drop across the valve. This is shown as Point "b," and occurs at 89 gal/min.

From Example 13.6, it is seen that the maximum scale-up is limited to about 5% by the Dowtherm A flowrate in the loop. A bottleneck has been identified. In order for additional scale-up to be possible, a method for increasing the maximum Dowtherm A flowrate must be found. Several options are possible:

1. Run only one pump but operate the reactor heat exchangers in parallel.
2. Run both pumps (the operating pump and the spare) in parallel with the reactor heat exchangers in series.
3. Run both pumps in series with the reactor heat exchangers in series.
4. Run both pumps (series or parallel) and operate the reactor heat exchangers in parallel.

Option 1 will be illustrated here. The remaining options are the subject of a problem at the end of the chapter.

Example 13.7

Determine the maximum possible Dowtherm A flowrate for one pump operating with the two reactor heat exchangers operating in parallel.

From Eq. (E13.6a), it is seen that

$$\Delta P \propto L_{eq} v^2 \qquad \text{(E13.7a)}$$

The equivalent length in the heat exchanger is reduced by half for operation in parallel. Half the flowrate goes through each heat exchanger, which means that the Dowtherm A has half the original velocity. Therefore, the pressure drop through the reactor heat exchangers drops by a factor of 8, and becomes 11.25 feet of water. The base case pressure drop is now 31.25 feet of water. Therefore, a new system curve is obtained. Its equation is

$$\Delta P = 31.25 \, M^2 \qquad \text{(E13.7b)}$$

The new system curve along with the pump curve are plotted on Figure 13.5. Point "c" represents the maximum possible flowrate of Dowtherm A, 127 gal/min. This represents a 49.4% scale-up.

Therefore, if only the original pump is used with the reactor heat exchangers in parallel, the Dowtherm A circulation rate can be increased by almost 50%. For Options 2 through 4, additional increases are possible. It should be noted that liquid velocities exceeding 10 ft/sec are not recommended due to potential erosion problems. Therefore, if the original case was designed for the maximum recommended velocity, operation at a 50% higher velocity for an extended period of time is not a good idea. We will proceed with this problem under the assumption that the original design permits 50% scale-up without exceeding the maximum recommended velocity.

Just because the flowrate of the Dowtherm A can be scaled up by 50% does not mean that the heat removal rate can be scaled up by the same factor. Analysis of the heat transfer in both heat exchangers is required. For the reactor, the energy balances were given by Eqs. (13.2) and (13.3), for the process side and the Dowtherm A side, respectively. The performance equation for the reactor heat exchanger is

$$Q_R = U_R A_R \Delta T_{lm} \qquad (13.4)$$

For the heat exchanger producing steam, there are two additional equations. One is the energy balance on the boiler feed water to steam side

$$Q_h = \dot{m}_s (C_{p,bfw} \Delta T + \lambda) \qquad (13.5)$$

where $\dot{m}_s$ is the mass flowrate of steam, ΔT is the temperature difference between the boiler feed water inlet and the vaporization temperature, and λ is the latent heat of vaporization of the steam. The second equation is the performance equation for the heat exchanger

$$Q_h = U_h A_h \Delta T_{lm} \tag{13.6}$$

The energy balance on the Dowtherm A side is identical (with the temperature difference reversed) to Eq. (13.3). It is assumed that all heat removed from the reactor is transferred to make steam ($Q_R = Q_h$).

Therefore, there are five independent equations that describe the performance of the loop. The unknowns are T_3, T_4, $Q_R = Q_h$, $\dot{m}_p$, $\dot{m}_D$, and $\dot{m}_s$. For a given value of $\dot{m}_D$, the mass flowrate of Dowtherm A, a unique solution exists.

Example 13.8

From a heat transfer analysis, determine the maximum scale-up possible for the allyl chloride reactor for the original case and for the case of one pump with the reactor heat exchangers operating in parallel.

Ratios of the scaled-up case (subscript 2) to the base case (subscript 1) will be used. Q_2/Q_1 is the ratio of the heat transfer rates, M is the ratio of the mass flowrates of Dowtherm A, and M_s is the ratio of the mass flowrates of steam. Only 3 of the 5 equations are coupled. Base case ratios of Eqs. (13.3), (13.4), and (13.6) must be solved for T_3, T_4, and Q_2/Q_1. Then M_s and M are determined from Eqs. (13.5) and (13.2), respectively. The value of Q_2/Q_1 defines the level of scale-up possible.

The base case ratios for Eqs. (13.3), (13.4), and (13.6) are

$$\frac{Q_2}{Q_1} = \frac{M(T_4 - T_3)}{50} \tag{E13.8a}$$

$$\frac{Q_2}{Q_1} = \frac{5(T_4 - T_3)}{130.44\left(4 + \dfrac{1}{M^{0.8}}\right)\ln\left(\dfrac{510 - T_3}{510 - T_4}\right)} \tag{E13.8b}$$

$$\frac{Q_2}{Q_1} = \frac{M^{0.8}(T_4 - T_3)}{119.26\ln\left(\dfrac{T_4 - 254}{T_3 - 254}\right)} \tag{E13.8c}$$

For the case of maximum scale-up with the original heat exchanger configuration, $M = 89/85 = 1.05$. The solutions are

$$T_3 = 349.8°C$$
$$T_4 = 399.1°C$$
$$Q_2/Q_1 = 1.04$$

From Eq. (13.2), since $Q_2/Q_1 = 1.04$, and since the heat of reaction is constant, the process flowrate increases by 4%. So only 4% scale-up is possible.

For one pump with the reactor heat exchangers in parallel, the mass flowrate in each reactor heat exchanger is half of the total flow. Therefore, Eq. (E 13.8b) becomes

$$\frac{Q_2}{Q_1} = \frac{5(T_4 - T_3)}{130.44\left(4 + \left(\dfrac{2}{M}\right)^{0.8}\right)\ln\left(\dfrac{510 - T_3}{510 - T_4}\right)} \tag{E13.8d}$$

For $M = 1.494$, solution of Eqs. (E13.8a), (E13.8c), and (E13.8d) yields

$$T_3 = 334.4°C$$

$$T_4 = 372.4°C$$

$$Q_2/Q_1 = 1.13$$

Therefore, 13% scale-up is possible.

Performance of the heat exchange loop is illustrated on the T-Q diagram shown in Figure 13.6. The lines for the reactor and the steam generating heat exchanger are unchanged. The line for the Dowtherm A changes slope as the mass flowrate, the reactor heat exchanger configuration, or the number and configuration of pumps changes.

This complex problem illustrates an important feature of chemical processes. Often the bottleneck to solving a problem is elsewhere in the process. In this problem, the bottleneck to reactor scale-up is not in the reactor itself, but in the heat removal loop. Nevertheless, several alternatives are available to increase the Dowtherm A flowrate and scale up the process.

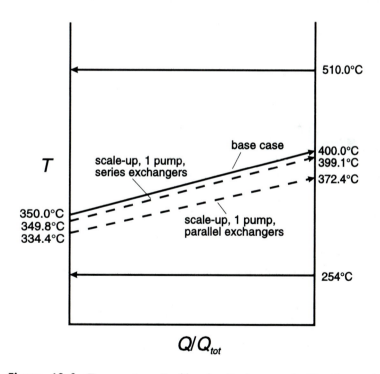

Figure 13.6 Temperature Profiles for Exchangers in Dowtherm A Loop

13.4 PERFORMANCE OF THE FEED SECTION TO A PROCESS

A very common feature of chemical processes is the mixing of reactant feeds prior to entering a reactor. When two streams mix, they are at the same pressure. The consequences of this are illustrated by the following problem.

Phthalic anhydride can be produced by reacting naphthalene and oxygen. The feed section to a phthalic anhydride process is shown in Figure 13.7. The mixed feed enters a fluidized bed reactor operating at 5 times the minimum fluidization velocity. A stream table is given in Table 13.1. It is assumed that all frictional pressure losses are associated with equipment, and that frictional losses in the piping are negligible. It is temporarily necessary to scale down production by 50%. Your job is to determine how to scale down the process and to determine the new flows and pressures.

It is necessary to have pump and compressor curves in order to do the required calculations. In this example, we will use equations for the pump curves. These equations can be obtained by fitting a polynomial to the curves provided by pump manufacturers. As discussed in Chapter 12, pump curves are usually expressed as pressure head versus volumetric flowrate. This is so that they can be

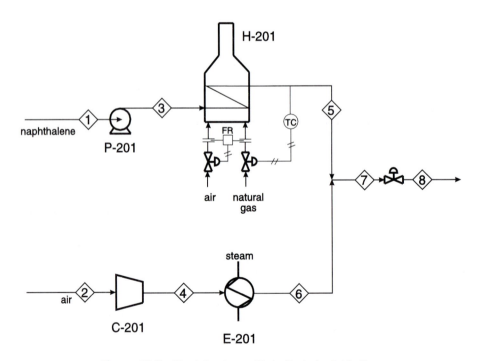

Figure 13.7   Feed Section to Phthalic Anhydride Process

Table 13.1   Partial Stream Table for Feed Section in Figure 13.7

	1	2	3	4	5	6	7	8
P (kPa)	80.00	101.33	343.00	268.00	243.00	243.00	243.00	200.00
phase	L	V	L	V	V	V	V	V
naphthalene (Mg/h)	12.82	—	12.82	—	12.82	—	12.82	12.82
air (Mg/h)	—	151.47	—	151.47	—	151.47	151.47	151.47

used for a liquid of any density. In this section, pressure head and volumetric flowrate have been converted to absolute pressure and mass flowrate using the density of the fluids involved. Pump P-201 operates at only one speed, and an equation for the pump curve is

$$\Delta P(\text{kPa}) = 500 + 4.663\dot{m} - 1.805\dot{m}^2 \quad \dot{m} \leq 16.00 \text{ Mg/h} \tag{13.7}$$

Compressor C-201 operates at only one speed, and the equation for the compressor curve is

$$\frac{P_{out}}{P_{in}} = 5.201 + 2.662 \times 10^{-3}\,\dot{m} - 1.358 \times 10^{-4}\,\dot{m}^2$$
$$+ 4.506 \times 10^{-8}\dot{m}^3 \quad \dot{m} \leq 200 \text{ Mg/h} \tag{13.8}$$

From Figure 13.7, it is seen that there is only one valve in the feed section, after the mixing point. Therefore, the only way to reduce the production of phthalic anhydride is to close the valve to the point at which the naphthalene feed is reduced by 50%.

Example 13.9

For a reduction in naphthalene feed by 50%, determine the pressures and flows of all streams after the scale-down.

Since it is known that the flowrate of naphthalene has been reduced by 50%, the new outlet pressure from P-201 can be calculated from Eq. (13.7). The feed pressure remains at 80 kPa. At a naphthalene flow of 6.41 Mg/h, Eq. (13.7) gives a pressure increase of 455.73 kPa, so $P_3 = 535.73$ kPa. Since the flowrate has decreased by a factor of two, the pressure drop in the fired heater decreases by a factor of four (see Eq. (E13.7a)). Therefore, $P_5 = 510.73$ kPa. Therefore, the pressure of Stream 6 must be 510.73 kPa. The flowrate of air can now be calculated from the compressor curve equation.

There are two unknowns in the compressor curve equation, the compressor outlet pressure and the mass flowrate. Therefore, a second equation is needed. The second equation is obtained from a base case ratio for the pressure drop across the heat exchanger. The two equations are

$$\frac{P_4}{101.33} = 5.201 + 2.662 \times 10^{-3}\dot{m}_{2,new} - 1.358 \times 10^{-4}\dot{m}_{2,new}^2 + 4.506 \times 10^{-8}\dot{m}_{2,new}^3 \quad \text{(E13.8a)}$$

$$P_4 - 510.73 = 25\left(\frac{\dot{m}_{2,new}}{151.47}\right)^2 \quad \text{(E13.8b)}$$

The solution is

$$P_4 = 512.84 \text{ kPa}$$

$$\dot{m}_2 = 43.80 \text{ Mg/h}$$

The stream table for the scaled-down case is given in Table 13.2. Though it is not precisely true, for lack of additional information, it has been assumed that the pressure of Stream 8 remains constant.

It is observed that the flowrate of air is reduced by far more than 50% in the scaled-down case. This is because of the combination of the compressor curve and the new pressure of Streams 5 and 6 after the naphthalene flowrate is scaled down by 50%. The total flowrate of Stream 8 is now 50.21 Mg/h, which is 30.6% of the original flowrate to the reactor. Given that the reactor was operating at five times minimum fluidization, the reactor is now in danger of not being fluidized adequately. Since the phthalic anhydride reaction is very exothermic, a loss of fluidization could result in poor heat transfer, which might result in a runaway reaction. The conclusion is that it is not recommended to operate at these scaled-down conditions.

The question is how can the air flowrate be scaled-down by 50%, to maintain the same ratio of naphthalene to air as in the original case. The answer is in valve placement. Because of the requirement that the pressures at the mixing point must be equal, with only one valve after the mixing point, there is only one possible flowrate of air corresponding to 50% reduction in naphthalene flowrate. Effectively, there is no control of the air flowrate. A chemical process would not

Table 13.2   Partial Stream Table for Scaled-down Feed Section in Figure 13.7

	1	2	3	4	5	6	7	8
P (kPa)	80.00	101.33	535.73	512.84	510.73	510.73	510.73	200.00
phase	L	V	L	V	V	V	V	V
naphthalene (Mg/h)	6.41	—	6.41	—	6.41	—	6.41	6.41
air (Mg/h)	—	43.80	—	43.80	—	43.80	43.80	43.80

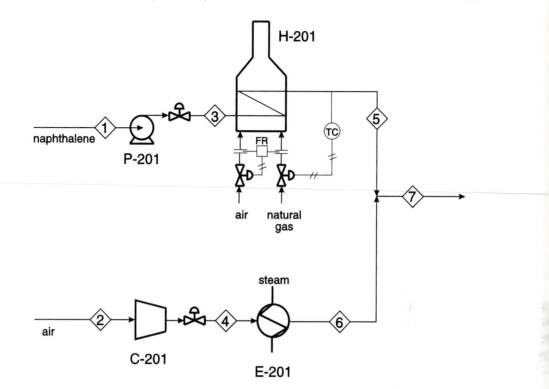

Figure 13.8   Feed Section to Phthalic Anhydride Process with Better
Valve Placement

be designed as in Figure 13.7. The most common design is illustrated in Figure 13.8. With valves in both feed streams, the flowrates of each stream can be controlled independently. Design of control systems will be discussed in more detail in Chapter 15.

13.5 SUMMARY

In this chapter, we have demonstrated that performance of existing equipment is affected by other equipment. The input/output models discussed in Chapter 10 suggested this outcome. If the input to a unit is altered, its output is altered. Since the output from one unit is the input to the next unit, the interaction observed in the case studies in this chapter are expected. It is important always to remember the interaction between equipment performance in a chemical process.

In terms of obtaining numerical solutions for performance problems involving multiple units, we have seen examples such as the reactor-heat exchanger and the distillation column, in which the adjacent units are analyzed sequentially. We have also seen examples such as the heat exchange loop and the feed section in which simultaneous solution of the relationships for multiple units was required. The exact set of calculations necessary and the difficulty of these calculations are specific to each problem encountered. However, the tools developed in Chapter 11 such as determining trends, base case ratios, and T-Q diagrams are essential to obtaining desired solutions.

REFERENCES

1. Perry, R. H., D. W. Green, and J. O. Maloney (eds.), *Perry's Chemical Engineers' Handbook* (6th ed), New York, McGraw-Hill, 1984, pp. 3-50–3-63.
2. Box, G. E. P., *Evolutionary Operation; A Statistical Method for Process Improvement*, New York, Wiley, 1969.
3. Kelley, P. E., "EVOP Technique Improves Operation of Amoco's Gas Producing Plants," *Oil and Gas Journal*, 71(44), 94 (1973).

PROBLEMS

1. It is possible to generate performance curves for the reactor in Section 13.1. One type of performance curve would have the rate of cumene production (kmol/h) on the y-axis and the percent propane impurity in the propylene on the x-axis. There would be lines for different operating temperatures all at constant pressure.

   Assume that the feed to the reactor is 108 kmol/h propylene, 8 kmol/h propane, and 203 kmol/h benzene. The pressure is kept constant at 3000 kPa. Prepare performance curves on the same graph for temperatures from 350°C to 400°C in 10°C intervals. Superimpose on this graph the maximum allowable conditions, which correspond to a steam-side pressure of 4800 kPa. A process simulator should be used to generate points on the performance curves. Other data: reactor volume = 7.89 m³, heat transfer area A = 436 m², overall heat transfer coefficient = 65 W/m²°C.

2. Repeat Problem 1 for varying reaction pressure at a constant temperature of 350°C. The pressure range is 3000 kPa to 3300 kPa, in intervals of 50 kPa.

3. For the distillation performance problem in Section 13.2, assume that scale-down occurs while maintaining constant boil-up rate. Determine the conditions in the reboiler, column, and condenser for these operating conditions. A process simulator should be used for the distillation column.

4. The benzene-toluene distillation column in Section 13.2 must temporarily handle a 25% increase in throughput while maintaining the same outlet concentrations. Determine the operating conditions required if the reflux ratio

remains constant. What other factors must be considered in order to determine whether the column can handle this amount of increased throughput?

5. Suggest alternative changes in process conditions or addition of new equipment in the distillation column in Section 13.2 that would allow 50% scale-down. Suggest as many alternatives as you can think of and discuss the advantages and disadvantages of each.

6. Consider the Dowtherm A loop described in Section 13.3. For the options involving multiple pumps either in series or parallel, determine the configuration that provides the maximum scale-up capacity. What are the temperatures of Dowtherm A, T_3 and T_4, and the maximum percent scale-up?

7. Repeat the analysis in Section 13.3 for the case when cooling water is used in the heat exchanger in place of boiler feed water. The cooling water enters at 30°C and it must exit at 45°C.

8. Repeat the analysis in Problem 6 using cooling water as described in Problem 7.

9. Suggest alternative solutions, not involving increasing the Dowtherm A flowrate, to obtain the maximum scale-up possible in the Dowtherm A loop in Section 13.3. Suggest as many alternatives as you can think of and discuss the advantages and disadvantages of each.

10. Consider the molten salt loop for removal of the heat of reaction in the production of phthalic anhydride described in Appendix C and illustrated in Figure C.5. It is necessary to scale-down phthalic anhydride production by 50%. Estimate the flow of molten salt for the scaled down case and the temperatures of molten salt entering and exiting the reactor. You may assume that sensible heat effects (energy necessary to heat reactants up to reaction temperature) are negligible.

11. Consider the situation illustrated in Figure P13.11. Due to downstream considerations, P_8 is always maintained at 200 kPa. It is known that $P_1 = 100$ kPa

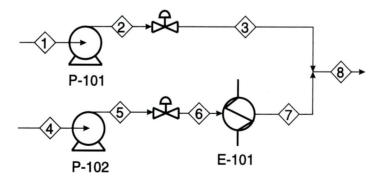

Figure P13.11   Process for Problem 11

and $P_4 = 100$ kPa, and since the feeds come from storage tanks maintained at constant pressure, they are always constant. It is also known that $P_3 = 225$ kPa, $P_5 = 375$ kPa, $P_6 = 250$ kPa, $P_7 = 225$ kPa, and $\dot{m}_1 = 15{,}500$ kg/h. The pump curves are given by the following equations, with ΔP in kPa and $\dot{m}$ in Mg/h.

$$\text{Pump 1} \quad \Delta P = 340 - 0.913\dot{m} + 0.0535\dot{m}^2 - 0.0101\dot{m}^3$$

$$\text{Pump 2} \quad \Delta P = 312 - 1.924\dot{m} + 0.0302\dot{m}^2 - 0.01124\dot{m}^3$$

   a. Calculate $\dot{m}_4$ and ΔP_{23} for this situation.
   b. Sketch the pump and system curves for this situation as illustrated in Section 12.2.1. Identify the pressure drop across the valves on the sketch.

12. Refer to Figure P13.12 and to the data given below. The compressor (C-101) exhausts to 101 kPa. Tank TK-102 is controlled to be always 101 kPa. Assume all flows are turbulent.
   a. If we close the valve, the flowrate from the compressor exhaust (Stream 7) is 100 Mg/h. What is the pressure in the storage tank (TK-101)?
   b. If the pressure of TK-101 is kept constant at the pressure calculated in part a, what is the maximum flowrate of liquid (Stream 2) when the valve is wide open (i.e., no pressure drop across the valve)?
   c. What is the flowrate of the compressor exhaust (Stream 7) when the valve is wide open?

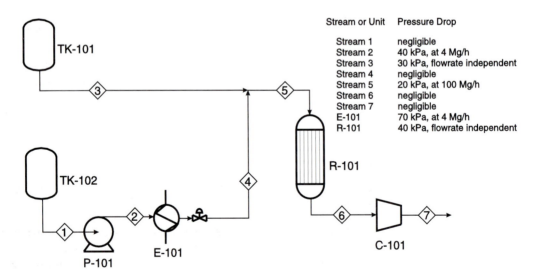

Stream or Unit	Pressure Drop
Stream 1	negligible
Stream 2	40 kPa, at 4 Mg/h
Stream 3	30 kPa, flowrate independent
Stream 4	negligible
Stream 5	20 kPa, at 100 Mg/h
Stream 6	negligible
Stream 7	negligible
E-101	70 kPa, at 4 Mg/h
R-101	40 kPa, flowrate independent

Figure P13.12 Process for Problem 12

C-101 Compressor Curve:

$$\frac{P_{out}}{P_{in}} = 4.015 + 5.264 \times 10^{-3}\dot{m} - 1.838 \times 10^{-4}\dot{m}^2$$

P-101 Pump Curve:

$$\Delta P = 500 + 4.662\dot{m} - 1.805\dot{m}^2$$

where P is in kPa and $\dot{m}$ is in Mg/h.

14

Reactor Performance

Chemical reactors are used to produce high-value chemicals from lower-value chemicals. Reactor performance depends on the complex interaction between four effects, as is illustrated in Figure 14.1. In order to understand fully the performance of a chemical reactor in the context of a chemical process, all four effects must be considered. These effects are:

1. *Reaction Kinetics and Thermodynamics*. The influence of extensive variables (pressure, temperature, concentration) on reactor performance is defined by reaction kinetics and reaction equilibrium. These extensive variables affect the reaction rate and determine the extent to which reactants can be converted into products in a given reactor or the size of reactor needed to achieve a given conversion. Additionally, catalysts are used to increase the rate of reaction. Thermodynamics sets a theoretical limit on the extent to which reactants can be converted into products and cannot be changed by catalysts.

2. *Reactor Parameters*. These include the reactor volume, space time (reactor volume/inlet volumetric flowrate), and reactor configuration. For given kinetics, thermodynamics, reactor and heat transfer configuration, and space time, the reactor volume needed to achieve a given conversion of reactants is determined. This is the design problem. For a fixed reactor volume, the conversion is affected by the temperature, pressure, space time, catalyst, and reactor and heat transfer configuration. This is the performance problem.

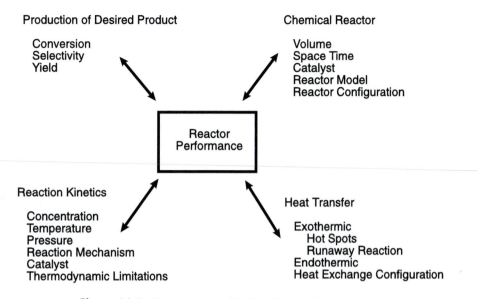

Figure 14.1   Phenomena Affecting Reactor Performance

3. *Production of Desired Product.* Conversion, selectivity, and yield are terms that quantify the amount of reactants reacted to form desired products. Reactor performance is expressed in terms of these parameters. For a fixed reactor volume, these parameters are functions of the temperature, pressure, reactor and heat transfer configuration, and space time.

4. *Heat Transfer in Reactor.* This important effect is often overlooked. Energy is released or consumed in chemical reactions. The rate of chemical reaction is highly temperature dependent. The key point to consider is the interaction between chemical kinetics and heat transfer (see Example 13.1). For exothermic reactions, the heat of reaction must be removed efficiently to avoid large temperature increases that can damage catalyst and to prevent runaway reaction. For endothermic reactions, heat must be supplied efficiently so that the reaction can proceed. The rate of heat transfer depends on reactor configuration (including the heat transfer configuration), the properties of the reacting stream, the properties of the heat transfer medium, and the temperature driving force.

In this chapter, the basic concepts needed for prediction of reactor performance are reviewed. Multiple reaction systems, reactor models, and heat transfer considerations are emphasized. Case studies are presented based on reactors found in the processes used as examples throughout the text and in the Appendices.

14.1 PRODUCTION OF DESIRED PRODUCT

Consider the following reaction scheme:

$$aA + bB \xrightarrow[rxn\ 1]{k_1} pP \xrightarrow[rxn\ 2]{k_2} uU$$

$$\tag{14.1}$$

$$\beta B \xrightarrow[rxn\ 3]{k_3} vV$$

Eq. (14.1) shows three reactions involving five species A, B, P, U, and V, with stoichiometric coefficients a, b, p, u, and v, respectively. It is assumed that P is the desired product and that U and V are unwanted by-products. The reaction scheme in Eq. (14.1) can be used to illustrate the effects commonly observed in actual reaction kinetics. These are

1. A **single reaction** produces desired product. Here $k_2 = k_3 = 0$, and only the first reaction proceeds. Acetone production (Appendix B, Figure B.3) is an example of this situation. Hydrodealkylation of toluene to benzene behaves like this because the catalyst suppresses the undesired reactions sufficiently so single reaction behavior is approached.

2. **Parallel** (competing) reactions produce desired products and unwanted by-products. Here $k_2 = 0$, and no U is formed. Species B reacts to form either P or V. In the phthalic anhydride reaction sequence (Appendix C, Figure C.5), the reaction of o-xylene to form either phthalic anhydride, maleic anhydride, or combustion products (Appendix C, Table C.10, Reactions 1, 3, and 4) is an example of a parallel reaction.

3. **Series** (sequential) reactions produce desired products and unwanted by-products. Here $k_3 = 0$, and no V is formed. Species A reacts to form desired product P, which further reacts to form unwanted by-product U. In the phthalic anhydride reaction sequence (Appendix C, Table C.10), the reaction of o-xylene to phthalic anhydride to combustion products (reactions 1 and 2) is an example of a series reaction.

4. **Series/Parallel** reactions produce desired products and unwanted by-products. Here, all three reactions in Eq. (14.1) occur to form desired product P and unwanted by-products U and V. The entire phthalic anhydride sequence (Appendix C, Table C.10) as well as the cumene production reaction (Appendix C, Table C.17) are examples of series/parallel reactions.

For reaction schemes like the one shown in Eq. (14.1), there are three key definitions used to quantify production of desired product.

The term *conversion* quantifies the amount of reactant reacted. The single-pass conversion (or reactor conversion) is

$$\text{single–pass conversion} = \frac{\text{reactant consumed in reactor}}{\text{reactant fed to reactor}} \tag{14.2}$$

and is generally reported in terms of the limiting reactant. This is different from the overall process conversion, which is defined as

$$\text{overall conversion} = X = \frac{\text{reactant consumed in process}}{\text{reactant fed to process}} \tag{14.3}$$

Most processes recover and recycle unreacted material to provide a high overall conversion.

> **High reactor conversions are neither necessary nor desirable for optimum reactor performance. At low reactor conversions, high overall conversions can be achieved with increased recycle.**

The term *selectivity* quantifies the conversion to desired product

$$\text{selectivity} = \eta = \frac{\text{rate of production of desired product}}{\text{rate of production undesired by-product(s)}} \tag{14.4}$$

A high selectivity is always desirable.

> **Competition from undesired reactions limits conversion to the desired product.**

Another term used to quantify production of the desired product is *yield*, defined as

$$\text{yield} = \frac{\text{moles of reactant reacted to produce desired product}}{\text{moles of limiting reactant reacted}} \tag{14.5}$$

The terms defined in this section will be used in examples and the case studies presented later in this chapter.

14.2   REACTION KINETICS AND THERMODYNAMICS

The kinetics of a reaction quantify the rate at which the reaction proceeds. When designing a new reactor for a given conversion, a faster reaction requires a smaller volume reactor. When analyzing an existing reactor of fixed volume, a faster reaction means increased conversion. As stated previously, thermodynamics provides limits to the conversions obtainable from a chemical reaction.

14.2.1   Reaction Kinetics

The reaction rate, r_i, is defined as

$$r_i = \frac{1}{V}\frac{dN_i}{dt} = \frac{\text{moles of } i \text{ formed}}{(\text{volume of reactor}) \,(\text{time})} \tag{14.6}$$

> **The reaction rate is an intensive property. This means that the reaction rate depends only on state variables such as temperature, concentration, and pressure, not on the total mass of material present.**

For solid catalyzed reactions, the reaction rate is often defined based on the mass of catalyst present, W

$$r_i = \frac{1}{W}\frac{dN_i}{dt}\,\rho_b = \frac{1}{V}\frac{dN_i}{dt} \tag{14.7}$$

where ρ_b is the bulk catalyst density (mass catalyst/volume reactor). The density of solid catalyst is defined as ρ_{cat} (mass catalyst particle/volume catalyst particle). So, the bulk density of the catalyst, ρ_b, is defined as

$$\rho_b = (1 - \varepsilon)\rho_{cat} \tag{14.8}$$

Here, ε is the void fraction in the reactor; so, $(1 - \varepsilon)$ is volume of catalyst/volume of reactor.

   If a reaction is an elementary step, the kinetic expression can be obtained directly from the reaction stoichiometry. For example, in Eq. (14.1), if the first reaction is an elementary step, the rate expression is

$$-r_A = k_1 c_A^\alpha c_B^\beta \tag{14.9}$$

For catalytic reactions, the rate expressions are often more complicated because the balanced equation is not an elementary step. Instead, the rate expression can be obtained by an understanding of the details of the reaction mechanism. The resulting rate expressions are often of the form

$$r_i = \frac{k_1 \prod_{i=1}^{n} c_i^{\alpha_i}}{\left[1 + \sum_{j=1}^{m} K_j c_j\right]^{\gamma}} \tag{14.10}$$

Eq. (14.10) describes a form of Langmuir-Hinshelwood kinetics. The constants (k_1 and K_j) in Eq. (14.10) are catalyst specific. The constants in Eq. (14.10) must be obtained by fitting reaction data.

In heterogeneous catalytic reacting systems, reactions take place on the surface of the catalyst. Most of this surface area is internal to the catalyst pellet/particle. The series of resistances that can govern the rate of catalytic chemical reaction are

1. Mass film diffusion of reactant from bulk fluid to external surface of catalyst
2. Mass diffusion of reactant from pore mouth to internal surface of catalyst
3. Adsorption of reactant on catalyst surface
4. Chemical reaction on catalyst surface
5. Desorption of product from catalyst surface
6. Mass diffusion of product from internal surface of catalyst to pore mouth
7. Mass diffusion from pore mouth to bulk fluid

Each step offers a resistance to chemical reaction. Reactors often operate in a region where only one or two resistances control the rate. For a good catalyst, the intrinsic rates are so high that internal diffusion resistances are usually controlling.

For solid catalyst systems, reactor performance is usually controlled by resistances to mass transfer.

The temperature dependence of the rate constants in Eqs. (14.9) and (14.10) is given by the Arrhenius equation

$$k_i = k_o e^{-\frac{E}{RT}} \tag{14.11}$$

where k_o is called the pre-exponential factor, and E is the activation energy (units of energy/mol, always positive). Eq. (14.11) reflects the significant temperature dependence of the reaction rate. For gas-phase reactions, the concentrations can be expressed or estimated from the ideal gas law, so $c_i = P_i/RT$. This is the origin of the pressure dependence of gas-phase reactions. As pressure increases, so does concentration, and so does the reaction rate. The temperature dependence of the Arrhenius equation usually dominates the opposite temperature effect on the concentration.

As temperature increases, the reaction rate always increases, and usually significantly.

14.2.2   Thermodynamic Limitations

Thermodynamics provides limits on the conversion obtainable from a chemical reaction. For an equilibrium reaction, the equilibrium conversion may not be exceeded.

Thermodynamics sets limits on possible conversions in a reacting system.

The limitations placed on conversion by thermodynamic equilibrium are best illustrated by an example.

Example 14.1

Methanol can be produced from syngas by the following reaction

$$CO + 2H_2 = CH_3OH$$

For the case when no inerts are present and for stoichiometric feed, the equilibrium expression has been determined to be

$$K = \frac{X(3 - X)^2}{4(1 - X)^3 P^2} = 4.8 \times 10^{-13} \exp(11{,}458/T) \tag{E14.1}$$

where X is the equilibrium conversion, P is the pressure in atmospheres, and T is the temperature in Kelvin. Construct a plot of equilibrium conversion vs. temperature for four different pressures: 15 atm, 30 atm, 50 atm, and 100 atm. Interpret the significance of the results.

   The plot is shown on Figure E14.1. By following any of the four curves from low to high temperature, it is observed that the equilibrium conversion decreases with increasing temperature at constant pressure. This is a consequence of LeChatelier's principle, since the methanol formation reaction is exothermic. By following a vertical line from low to high pressure, it is observed that the equilibrium conversion increases with increasing

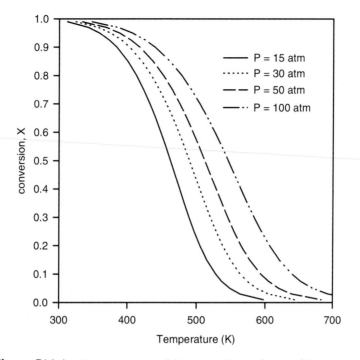

Figure E14.1 Temperature and Pressure Dependence of Conversion for Methanol from Syngas

pressure at constant temperature. This is also a consequence of LeChatelier's principle. Since there are fewer moles on the right-hand side of the reaction, increased conversion is favored at high pressures.

From thermodynamic considerations alone, it appears that this reaction should be run at low temperatures in order to achieve maximum conversion. However, as discussed in Section 14.2.1, since the rate of reaction is a strong function of temperature, this reaction is usually run at higher temperatures, with low single-pass conversion, in order to take advantage of the faster kinetics. As discussed in Section 14.1, despite a low single-pass conversion in the reactor, a large overall conversion is still achievable by recycling unreacted reactants.

14.3 THE CHEMICAL REACTOR

For continuous flow reactors, the two ideal models are the plug flow reactor and the continuous stirred tank reactor. In the plug flow reactor (PFR), which is basically a pipe within which the reaction occurs, the concentration, pressure, and temperature change from point to point. The performance equation is

$$\frac{V}{F_{Ao}} = \frac{\tau}{c_{Ao}} = \int_0^{X_A} \frac{dX_A}{-r_A} \tag{14.12}$$

In the continuous stirred tank reactor (CSTR), the reactor is assumed to be well-mixed, and all properties are assumed to be uniform within the reactor. The performance equation is

$$\frac{V}{F_{Ao}} = \frac{\tau}{c_{Ao}} = \frac{X_A}{-r_A} \tag{14.13}$$

where
V = volume of the reactor
F_{Ao} = molar flow of limiting reactant A
τ = space time (reactor volume/inlet volumetric flowrate)
c_{Ao} = inlet concentration of A
X_A = conversion of A
r_A = rate of reaction of A

Eqs. (14.12) and (14.13) must be solved for a given rate expression to get the actual relationship between the indicated parameters. However, several generalizations are possible. The reactor volume and conversion change in the same direction. As one increases, so does the other; as one decreases, so does the other. As the reaction rate increases, the volume required decreases, and vice versa. As the reaction rate increases, the conversion increases, and vice versa. It has already been discussed how temperature, pressure, and concentration affect the reaction rate. Their effect on conversion and reactor volume is derived from their effect on reaction rate.

Example 14.2

Discuss qualitatively the effect of temperature and pressure on the conversion, space time, and conversion in a chemical reactor.

Gas-Phase Reaction: From the Arrhenius equation in Eq. (14.11), the reaction rate constant increases exponentially with temperature. From the ideal gas law, the concentration decreases less drastically. From Eqs. (14.12) and (14.13), at constant pressure, if the reaction rate increases, at constant conversion, a smaller reactor volume (and space time) is required. At constant reactor volume and space time, the conversion increases.

From a typical reaction rate expression such as Eq. (14.9), as the pressure increases, so does the concentration and the reaction rate. Therefore, the same trends are observed as with temperature. However, the quantitative effect is not as great, since the concentration dependence on pressure is not exponential.

Liquid-Phase Reaction: For a liquid-phase reaction, temperature and pressure do not affect concentration. Therefore, there is no effect of pressure on reactor performance for a liquid-phase reaction. Since the temperature effect is determined by the Arrhenius equation only, the reaction rate increases exponentially with temperature.

The PFR is a hypothetical system. PFRs are elementary models for packed bed reactors. The PFR model excludes the effects of axial mixing resulting from molecular movement and fluid turbulence while assuming radial mixing is complete. This mixing takes place between adjacent fluid elements of approximately the same conversion. Reactors approximating PFRs are often built using many small diameter tubes (less than 3 inches) of significant length (20 to 30 ft) operated at high fluid velocity with small space times. This minimizes axial fluid mixing, limits radial temperature profiles, and provides needed heat transfer area. Tubes are arranged in a bundle characteristic of those found in many heat exchangers. If heat exchange is not desired in the reaction zone, a single or series of larger diameter packed beds can be used.

The CSTR is also a hypothetical system in which there is perfect mixing so that temperature, pressure, concentration, and reaction rate are constant over the reactor volume. Reactors approximating CSTRs are used for liquid-phase reactions. This represents a theoretical limit as perfect mixing can only be approached. Transit time for fluid elements varies. The exit stream is at the same temperature, pressure, and conversion as the reactor contents. Feed is mixed with the reactor contents that have a high conversion. As a result, the CSTR requires a higher volume than a PFR when operated isothermally at the same temperature and conversion for simple, elementary reactions.

Reactors are modeled using different combinations and arrangements of CSTRs and PFRs. Intermediate degrees of mixing are obtained that can be used to model various real reactor systems. Some of these models are shown in Figure 14.2.

> System A is a PFR for an exothermic reaction. The concentration and temperature vary from point to point in the reactor. This model is used to simulate the cumene reactor in Appendix C.
>
> System B is a CSTR. The temperature and concentration are constant within the reactor.
>
> System C consists of a PFR and CSTR in parallel. By increasing the ratio of CSTR feed rate to PFR feed rate, the degree of mixing is increased. By increasing the ratio of PFR feed rate to CSTR feed rate, dead space (regions that are not well-mixed) in a CSTR can be simulated.
>
> System D consists of a series of CSTRs in series. The degree of mixing is less than that in a single CSTR. While each reactor is completely mixed, there is no mixing between each section. In the limit of an infinite number of CSTRs in series, PFR behavior is approached.
>
> System E consists of an isothermal PFR with a bypass stream. In a fluidized bed, solid catalyst is circulated within the bed; however, the fluid moves through the bed essentially in plug flow. Maldistribution of fluid, due to the formation of bubbles and voids, and channeling, may cause some fluid to bypass the catalyst. This model is used in catalytic fluidized bed applica-

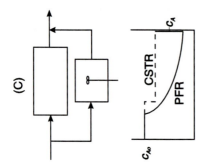

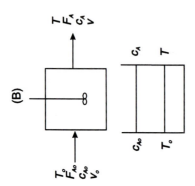

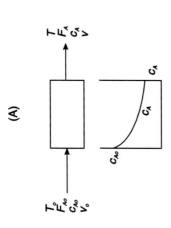

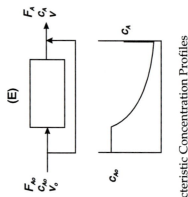

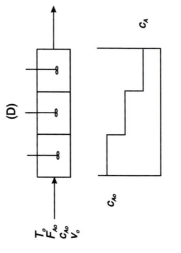

Figure 14.2 Mixing Patterns with Characteristic Concentration Profiles

tions or in any other application where the solid does not react. A portion of the reactants form an emulsion phase with little mixing (plug flow). The remaining fluid forms bubbles that move upward at a higher velocity, contact only a small portion of the solid catalyst, and do not react as much as in the emulsion phase. An important feature of the fluidized bed is that it operates isothermally as a result of the mixing of the solids. The model shown in System E is a crude approximation to what happens in a real fluidized bed, but predicts the trends found in fluidized beds. This model is used to simulate the acrylic acid reactor in Appendix B.

As the reactor model becomes more complex, it becomes more difficult to solve Eqs. (14.12) and (14.13) to predict reactor performance. Since process simulators are already programmed to do this, they are a logical choice for analyzing reactor performance, especially since real reactors must be simulated in the context of an entire process. For the chemical reactor, the required input data include identification of the dominant reactions that take place, the form of the reaction rate, and values of the kinetic constants. All of these may change with operating conditions and a given set of values apply only to a limited range of operations. They must be obtained experimentally. If the experimental data are flawed or the simulation operating conditions are outside the range of the experimental studies, the answer obtained can lead to poor prediction of reactor performance. The computer can do the calculations but is unable to recognize the reliability of the input.

The following example summarizes the concepts presented thus far in this chapter.

Example 14.3

Consider the reaction scheme given in Eq. (14.1) where P is the desired product, with both U and V as undesired by-products. Assume that Eq. (14.1) represents elementary steps, that $a = b = p = u = v = 1$, that $\beta = 2$, and that the activation energies for the reactions are as follows: $E_1 > E_2 > E_3$.

   a. For the case where $k_2 = 0$, what conditions maximize the selectivity for P?
   b. For the case where $k_3 = 0$, what conditions maximize the selectivity for P?

   a. For this case, from Eq. (14.4), the selectivity is written as

$$\eta = \frac{r_P}{r_V} = \frac{k_1 c_A c_B}{k_3 c_B^2} = \frac{k_1 c_A}{k_3 c_B} \tag{E14.3a}$$

There are several ways to maximize the selectivity. Increasing c_A/c_B increases the selectivity. This means that excess A is needed and that B is the limiting reactant. Many reactions are operated with one reactant in excess. The reason is usually to improve selectivity as shown here. Since pressure affects all concentrations equally, it is seen that pressure does not affect the selectivity here. Temperature has its most significant effect on the rate constant. Since the activation energy for rxn 1 is larger than that for rxn 3, k_1 is more strongly affected by temperature changes than is k_3. Therefore, increasing the temperature increases the selectivity. In summary, higher temperatures and excess A maximize the selectivity for P.

b. For this case, the selectivity is written as

$$\eta = \frac{r_P}{r_U} = \frac{k_1 c_A c_B - k_2 c_P}{k_2 c_P} = \frac{k_1 c_A c_B}{k_2 c_P} - 1 \tag{E14.3b}$$

Since pressure appears to the second power in the numerator and to the first power in the denominator (it is in each concentration term), increasing the pressure increases the selectivity. Since the activation energy for rxn 1 is larger, increasing the temperature increases the selectivity. Increasing both reactant concentrations increases the selectivity, but increasing the concentration of component P decreases the selectivity. The question is how can the concentration of component P be kept to a minimum. The answer is to run the reaction at low conversions (small space time, small reactor volumes). Quantitatively, the selectivity in Eq. (14.3b) is maximized. Intuitively, since this is a series reaction with the desired product intermediate in the series, a low conversion maximizes the intermediate product and minimizes the undesired product. This can be illustrated by the concentration profiles obtained by assuming that these reactions take place in a PFR, as shown in Figure E14.3. It is seen that the ratio of P/U is at a maximum at low reactor volumes, which corresponds to low conversion. Therefore, increasing the temperature and pressure both increase the selectivity for component P. Running at low conversion probably does more to increase the selectivity than can be accomplished by manipulating temperature and pressure alone. However, there is a trade-off between selectivity and overall profitability, since low conversion per pass means very large recycles and larger equipment.

14.4   HEAT TRANSFER IN THE CHEMICAL REACTOR

Chemical reactions are either exothermic (release heat) or endothermic (absorb heat). Therefore, heat transfer in reactor systems is extremely important. For exothermic reactions, heat may have to be removed so that the temperature in the reactor does not increase above safe limits. For endothermic reactions, heat may have to be added so that the reaction will occur at an acceptable rate.

> **Reactor performance is often limited by the ability to add or remove heat.**

Most industrially significant chemical reactions are exothermic. Highly exothermic reactions, especially in a reactor behaving like a PFR, can be dangerous if the rate of heat removal is not sufficient. For a rate expression such as Eq. (14.9), the concentration of reactants is largest at the entrance to the reactor. The reaction at the entrance generates heat if the reaction is exothermic. The concomitant in-

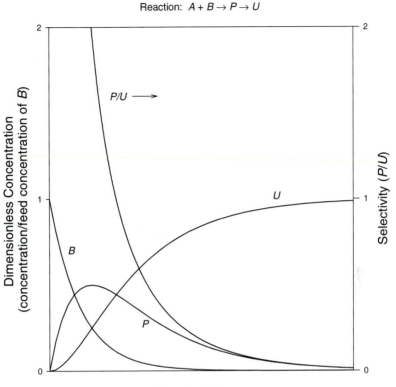

Figure E14.3   Concentration Profiles in PFR for Series Reaction

crease in temperature increases the reaction rate even further, which increases the temperature further, and so forth. Therefore, it is possible to have large temperature increases, called hot spots. The large temperature increase can be offset by removing heat from the reactor. This can cause radial temperature gradients, with severe hot spots near the center of the tube. Hot spots can damage and deactivate catalyst particles. In the extreme case, if the rate of heat removal is not sufficient to offset the rate of heat production by reaction, the temperature may increase rapidly causing damage to the reactor and its contents or an explosion. This is called runaway reaction. Reactor temperature profiles illustrating the situation leading to runaway reaction are illustrated in Figure 14.3b.

> **Beware of exothermic reactions! Runaway reaction is possible if there is insufficient heat removal.**

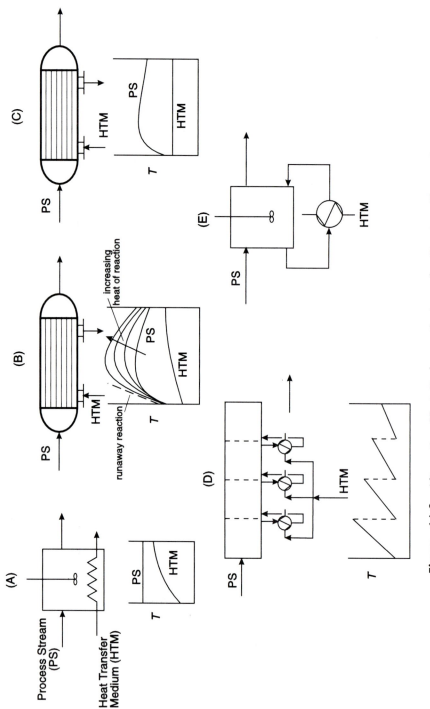

Figure 14.3 Alternative Heat Exchange Systems for Reactors Shown for Exothermic Reaction: A-C: Internal Heat Exchanger D-E: External Heat Exchanger

For endothermic reactions, heat must be added to the reactor. Cold zones are possible based upon the same logic as for the hot spots for exothermic reactions. However, if there is insufficient heat addition to an endothermic reaction, the reaction will be quenched. Therefore, endothermic reactions are inherently safe.

There are many methods used to avoid hot spots in a reactor with an exothermic reaction. Some of these are:

1. Run heat transfer medium cocurrently: The heat transfer medium enters at its lowest temperature. By matching this low temperature with the highest temperature zone in the reactor, the temperature gradient is largest at the reactor entrance, providing more heat removal.

2. Use inert solid: By randomly packing the bed with inert solid (no catalytic activity), the heat released per unit volume of reactor is reduced, thereby minimizing hot spots. This results in a larger reactor.

3. Use catalyst gradients in reactor: This is similar to 2, above. Here, a larger fraction of inert solid is used where hot spots are anticipated. This minimizes hot spots, but also results in a larger reactor.

4. Use a fluidized bed: Due to the mixing behavior in a fluidized bed, isothermal operation is approached.

Both internal and external heat exchangers can be used to exchange heat with a reactor. In the internal systems (Figure 14.3a–c), the heat transfer area is in contact with the reacting fluid. Quite often, this is like a shell-and-tube heat exchanger, with the catalyst most often placed in the tubes. In the external systems (Figure 14.3d–e), reaction fluid is withdrawn from the reactor, sent to a heat exchanger, and then returned to the reactor.

a. CSTR with single-phase heat transfer medium: The temperature profile is similar to that for the condensation of a single component vapor. The volume taken up by the heat exchanger increases the reactor volume. The mixing intensity of the reacting fluid may be reduced and the potential for local non-isothermal behavior close to the heat exchanger surfaces exists.

b. PFR with single-phase heat transfer medium: Temperature on the process side goes through a maximum.

c. PFR heat transfer media boils: Temperature on the utility side remains constant and goes though a maximum on the process side.

d. Adiabatic PFR with heat removal between stages: This configuration is usually used for exothermic reactions that do not have extremely large heats of reaction. The reaction proceeds adiabatically with a temperature rise, and fluid is removed periodically to be cooled.

e. External heat exchanger in CSTR: Fluid in the reactor is circulated through an external heat exchanger.

In order to solve a problem involving a chemical reactor with heat transfer, the following equations must be solved:

1. Reactor performance equation (such as Eqs. (14.12) and (14.13))
2. Energy balance on reaction side
3. Energy balance on heat transfer medium side
4. Heat transfer performance equation ($Q = UA\Delta T_{lm}$)
5. Pressure drop in packed bed (Ergun equation)

Therefore, as the heat transfer configuration becomes more complex, it becomes very difficult to solve the necessary equations to predict reactor performance. Since process simulators are already programmed to do this, they are a logical choice for analyzing reactor performance with heat transfer.

For the chemical reactor with heat transfer, in addition to the input discussed earlier, the required input data must include the heat transfer coefficient, the heat transfer area, the length and diameter of tubes, and the number of tubes. The output obtained is the temperature, pressure, and concentration of each component at each point in the reactor.

Heterogeneous catalytic reactions are even more difficult to simulate. The major resistance to reaction can change from chemical reaction to external diffusion or to pore diffusion within a single reactor. Stream temperatures are not necessarily the temperatures of the catalyst surface where the reaction takes place. Large temperature gradients may exist in the radial direction and within the catalyst particles. Although they do not consider these important factors, process simulators can often be used to obtain approximate solutions.

14.5 REACTOR SYSTEM CASE STUDIES

This section presents case studies involving reactors from process flow diagrams presented throughout this text. The process changes presented illustrate typical problems a chemical engineer working in production may face on a routine basis.

All studies presented contain three parts:

Part A: A statement of the problem.

Part B: A solution to the specific problem, which includes discussion of the strategy used in the solution.

Part C: A discussion of the significance of the solution to a wider range of applications. Figures of process response to important process variables are developed and used to reveal trends that yield insight into the problem. These performance curves provide better understanding of reactor behavior. They are also used as a guide to develop alternative problem solutions.

The availability of computers and computer software permit process simulation. Once the process is simulated, variables are easily varied to discover important performance characteristics that can be plotted over a wide range of operating conditions.

14.5.1   Replacement of Catalytic Reactor in Benzene Process

Problem.   The catalyst in our benzene reactor has lost activity and must be replaced. The benzene process flow diagram presented in Chapter 1 (Figure 1.5) is used to produce benzene. The reaction is carried out in an adiabatic PFR. Because of the cost of periodic catalyst replacement, it has been suggested that the catalytic reactor system could be replaced with a non-catalytic adiabatic PFR system that might be less costly. The new system is to match the input/output conditions of the current system so that the remainder of the plant will operate with little, if any, modification. You are asked to investigate the merits of this suggestion by identifying major cost items. Table 14.1 provides reactor information based on experimental studies on the toluene/benzene reaction.

Figure 14.4a provides a flow diagram of the current reactor taken from Figure 1.5. A-A identifies the boundaries of the reactor system to be replaced. Flowrates and stream temperature and pressure at these boundaries are given.

The dominant chemical reaction at plant operating conditions is

$$C_7H_8 \quad + \quad H_2 \quad = \quad C_6H_6 \quad + \quad CH_4$$
$$\text{\textit{toluene}} \qquad \text{\textit{hydrogen}} \quad \text{\textit{benzene}} \qquad \text{\textit{methane}}$$

Table 14.1   Findings from Experimental Studies on Dealkylation of Toluene [1]

1.	Reaction Rate (mol/liter reactor s)	$-r_{tol} = 3.0 \times 10^{10} \exp(-25{,}164/T) \, c_{tol} c_H^{0.5}$
2.	Temperature Range	700–950°C
3.	Carbon Formation	None observed < 850 °C with ratio of H_2/Tol > 1.5
4.	Selectivity	Benzene/Unwanted > 19/1 with toluene conversion < 0.75
5.	H_2/Toluene	>2
6.	By-product yield W = mass of all by-products/mass feed	W(Mass % of feed) = $1.0 \times 10^8 \, (\theta/b) \exp(-25{,}667/T)$ where θ is the residence time (in seconds) and b is the H_2/toluene ratio
7.	Observed components in output	hydrogen, methane, toluene, various diphenyls

T in K, c in mol/liter

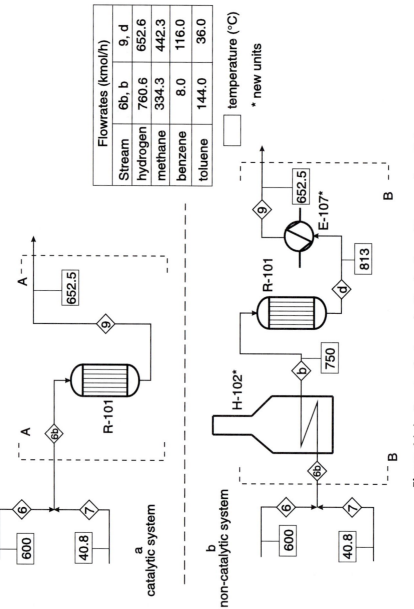

Flowrates (kmol/h)		
Stream	6b, b	9, d
hydrogen	760.6	652.6
methane	334.3	442.3
benzene	8.0	116.0
toluene	144.0	36.0

☐ temperature (°C)

* new units

Figure 14.4 Reactor Systems for Production of Benzene from Toluene

378

Reactor conversion, $X_{toluene} = 1 - (36.1/144.0) = 0.749$. The amount of by-product formed by side reaction is negligible; however, the small amounts of carbon formed can deposit on the catalyst surface.

Solution.   The reactor volume calculated for a non-catalytic adiabatic PFR using the kinetics expression given in Table 14.1 that met the conditions shown in Figure 14.4a is 257 m^3 (7800 ft^3). This is six times larger than the current catalytic reactor (41.5 m^3, see Table 1.7).

Intuitively, the reactor volume can be reduced by increasing the pressure and/or temperature in the new reactor system. Increasing the pressure increases the concentrations in the rate expression and increasing the temperature increases the rate constant. This requires a gas compressor and/or a heat exchanger (additional fired heater) be added in the new reactor system.

Installing an additional fired heater that increases the feed temperature to 750°C was also analyzed. The calculated reactor size is about 5 m^3 and the outlet temperature is 814°C.

An evaluation of the preliminary findings for the non-catalytic reactor configuration shows:

1. Using the same inlet conditions as for the catalytic reactor (inlet temperature 588°C):
   a. Extrapolation of the data in Table 14.1 is uncertain. This is because the reaction rate expression is extrapolated outside the region of experimental data (see item 1 of Table 14.1).
   b. The predicted reactor size is much larger than current reactor.
2. At increased temperature (inlet temperature 750°C):
   a. The reaction rate is known from experimental data.
   b. The reaction volume is much smaller than the current reactor.
   c. The temperature, H$_2$/toluene ratio, and conversion are all in a range that gives high benzene selectivity.
3. Increasing reactor feed pressure: This is considered an expensive option due to gas compression costs, and was not evaluated.

The non-catalytic replacement system is shown in Figure 14.4b as region B-B. It is a more complicated system. Higher feed temperature requires the addition of another fired heater before the reactor and a gas cooler following the reactor. (Heat integration might be another alternative.) Major capital costs are required for the reactor, fired heater, and cooler. Materials of construction for the reactor are more expensive at the higher temperature (At the exit temperature of 814°C, a reactor vessel would glow red). Fuel for the fired heater increases operating cost. The economic optimum reactor temperature that considered these items could be determined if needed.

Unless the cost of catalyst is extremely large, replacement of the catalytic re-actor with an non-catalytic system does not appear to be an attractive option.

Discussion.   It is observed that the low-temperature reactor (at the temper-ature of the catalytic reaction) was very large, a result obtained by assuming that it is valid to extrapolate the data in Table 14.1 outside the stated temperature range. Assuming the data to be accurate in the low temperature range, we now investigate reasons why a large reactor was obtained.

Several important factors considered in selecting a chemical reactor are re-vealed in this case study. Normally the advantages of a catalytic over a non-catalytic reactor are:

1. reduction in operating temperature and pressure
2. improvement in product selectivity

In this case study, the catalyst reduced the temperature and pressure require-ments, as expected. However, high selectivity for the desired product could be obtained from both the catalytic and non-catalytic reactors.

Carbon is a common by-product formed during organic reactions carried out in high temperature reactors. It is most significant in catalytic reactors where carbon forms on the catalyst surface and reduces activity. High hydrogen concen-trations discourage carbon formation. This helps explain the high H_2/toluene ratio of 5.1 used in the catalytic system compared to the lower value of 2.0 re-quired for the non-catalytic reactor.

Equilibrium constants for the reactions involving the components found in the laboratory studies on the benzene reaction are published [2]. The primary by-product at high temperatures is diphenyl, formed by the following reaction.

$$2C_6H_6 \quad = \quad C_{12}H_{10} \quad + \quad H_2$$

     benzene   *diphenyl*   *hydrogen*

At equilibrium, excess H_2 increases the toluene conversion, decreases diphenyl conversion, and improves benzene selectivity. This trend is assumed to hold for non-equilibrium conditions, which suggests that high excess H_2 be used.

Increasing the reaction temperature decreases the equilibrium constant of the benzene reaction because the main reaction is exothermic and increases the equilibrium constant of the diphenyl reaction because the side reaction is en-dothermic. As a result, the selectivity decreases and the raw material costs in-crease.

In our system, any diphenyl present in the reactor effluent, Stream 9, would be largely removed in the separation section and recycled with the toluene to the reactor. The concentration of diphenyl in the reactor feed, Stream b, would be in-creased. When sufficient diphenyl is introduced in Stream b to provide the equilibrium concentration, little diphenyl is formed in the reactor. This retains a high benzene selectivity at higher reactor temperatures.

For a new design, we might want to consider scenarios to increase the reaction rate for the non-catalytic reaction. Figure 14.5 shows the reaction rate, temperature, and toluene mole fraction plotted against the reactor volume for the original feed temperature. This figure shows low reaction rates exist at the feed end of the reactor with rapidly falling reaction rates at the reactor exit. The reactor temperature increases over the volume of the reactor. Figure 14.5 identifies steps to take that would improve reactor performance. Increased performance can be obtained by

    a. Elimination of the low reaction rate region at the beginning of the reactor. The low reaction rates are due to the low reaction rate constant, k, at low temperature. Increasing the feed temperature would increase reaction rate in this section.

    b. Avoid operating in the region at the end of the reactor where the reaction rate falls rapidly. The falling reaction rate results from the lowering of toluene concentrations at high conversions. This region can be avoided by operating at lower reactor conversion (and higher recycle rates of unreacted feed).

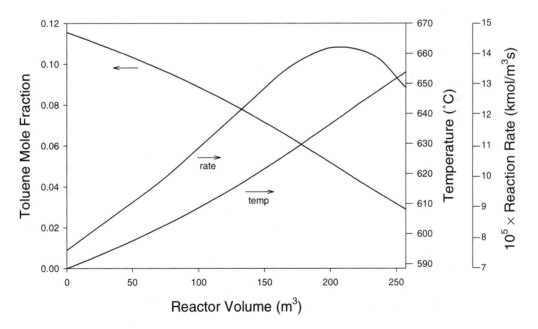

Figure 14.5   Behavior of Benzene PFR at Original Feed Temperature (585°C)

This case study demonstrates that a dramatic decrease in the reactor volume required for a given conversion results from an increase in feed temperature.

One could consider replacing the current reactor with an adiabatic CSTR. The reaction rate throughout the reactor would be constant at reactor exit conditions. This reaction rate is shown in Figure 14.5. The average reaction rate for the PFR is less than that of the CSTR and the volume for a CSTR is less than that of the PFR. The elimination of the low temperature region in the PFR overcomes the effect of lower toluene concentration in the CSTR.

The conclusion of this case study is that replacing the reactor with a non-catalytic PFR is not an attractive option. This may not be the conclusion if a new plant were to be built. In a new plant, the excess hydrogen could be substantially reduced. This would increase the toluene concentration and reduce the amount of recycle. Lower reactor conversions could be considered. Higher temperatures, higher toluene concentrations, and lower conversions would all contribute to a smaller reactor.

14.5.2   Replacement of Cumene Catalyst

Problem.   Your company produces cumene. At your plant, the shell-and-tube packed bed reactor shown in Appendix C, Figure C.8 is used. At other plants owned by your company, proprietary fluidized bed reactors providing low bypass are used. The remainder of the process is identical at all locations. Your supplier has informed you that they have developed a new cumene catalyst that will improve fluidized bed reactor performance. The new catalyst is supported on the same inert material as the old catalyst. Therefore, the fluidizing properties are identical to that of the old catalyst. You are assigned the task of verifying claims of improved performance and identifying changes in operating conditions that would maximize the improvement this new catalyst could provide.

Figure 14.6 provides the operating conditions for this base case. In addition to the general process information provided with the process flow diagram (see cumene flow diagram in Appendix C, Figure C.8), the following specific background information is provided for your plant.

In the first stage of a plant start-up, the plant is operated without the recycle stream, Stream 11. Benzene normally provided by this recycle stream is added with the feed stream. In the initial plant start-up, the plant operated for a substantial period without the recycle. During this time, the plant was operated at a range of temperatures, pressures, and excess benzene rates. These data were used to determine effective kinetics that are presented in Table 14.2. The active reactor volume is 7.88 m^3. It was found that the fluidized bed could be modeled accurately over a narrow range of superficial velocities as a PFR with 5% bypass.

The information from the open system (no recycle) was used to establish a base case. At each plant start-up following a catalyst change, the plant is run at base case conditions. If the same reactor performance is not obtained, the catalyst

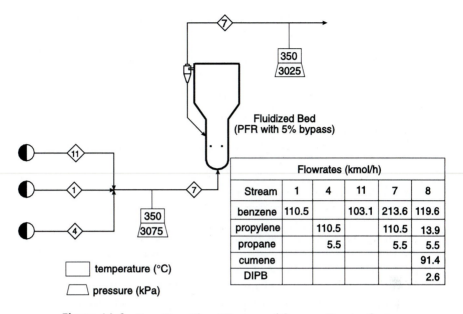

Figure 14.6   Base Case Flow Diagram of Cumene Reactor System

and flow pattern in the reactor are examined. Problems of reactor performance are resolved before continuing plant start-up.

Solution.   The dominant reactions involved within the cumene plant are

$$\underset{\text{propylene}}{C_3H_6} + \underset{\text{benzene}}{C_6H_6} \xrightarrow{k_1} \underset{\text{cumene}}{C_9H_{12}} \qquad \text{Reaction 1}$$

$$\underset{\text{propylene}}{C_3H_6} + \underset{\text{cumene}}{C_9H_{12}} \xrightarrow{k_2} \underset{\text{p-diisopropyl benzene}}{C_{12}H_{18}} \qquad \text{Reaction 2}$$

The kinetics for the catalyst are provided in Table 14.2.

In this analysis, to assure low bypass ratio (5%), the superficial gas velocity was held constant at the base case conditions. For higher reactor temperatures, the molar flow rate to the reactor was decreased slightly so as to maintain the same superficial velocity. This was done by retaining the base case flows of Stream 11 and Stream 4, and lowering the excess benzene Stream 1. The fluidized bed was assumed to behave as an isothermal PFR with a 5% bypass stream. This is shown in Figure 14.7. Some of the simulation results are shown in Figures 14.8 and 14.9. Figure 14.8 presents the production of cumene obtained for the current and new catalyst for temperatures ranging from 350°C–410°C. The new catalyst is more sensitive to a change in temperature than the current catalyst. At the base case temperature, the new catalyst produces only 81 kmol/h, point "a," compared to 91 kmol/h, point "b," for the current catalyst. At a reactor temperature

Table 14.2 Kinetic Constants Obtained for Hypothetical Cumene Catalysts

Catalyst	Current	New
Reaction Rate (mol/liter s)		
Reaction 1 (cumene formation)	$r_1 = k_1 c_p c_b$	$r_1 = k_1 c_p c_b$
Reaction 2 (DIPB formation)	$r_2 = k_2 c_p c_c$	$r_2 = k_2 c_p c_c$
Rate Constant (liter/mol s)		
Reaction 1	$k_1 = 3{,}500\exp(-6{,}680/T)$	$k_1 = 2.8 \times 10^7 \exp(-12{,}530/T)$
Reaction 2	$k_2 = 290\exp(-6{,}680/T)$	$k_2 = 2.32 \times 10^9 \exp(-17{,}650/T)$

T has the units of K, liters refers to liters of reactor

of 361°C, the new catalyst produces the same amount of cumene, point "c," as the old catalyst operating at the base case temperature. At 367°C, both systems produce the same amount of cumene, point "d." The maximum production rate of 101 kmol/h is obtained for the new catalyst at about 390°C, point "e." The current catalyst provides a maximum cumene production of 96 kmol/h, point "f."

Figure 14.9 shows the trends in product selectivity (moles of cumene produced per mole of DIPB produced) for both catalysts. At 350°C, the selectivity of the new catalyst is 160, point "a," compared to 35 for the current catalyst, point "b." At a temperature of 390°C, where the cumene reaches maximum production

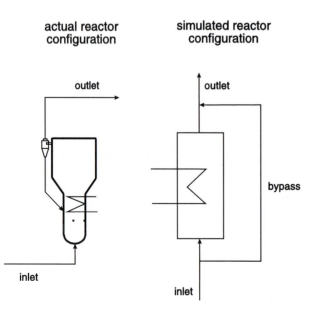

Figure 14.7 Fluidized Bed Reactor Showing Configuration (5% bypass) Used for Simulation

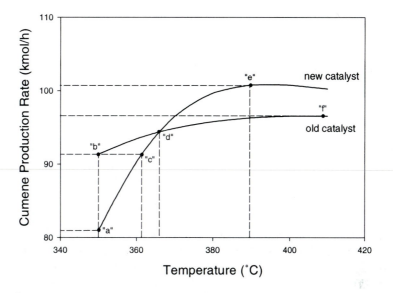

Figure 14.8   Cumene Production Rates at Constant Reactor Volume
and Superficial Velocity

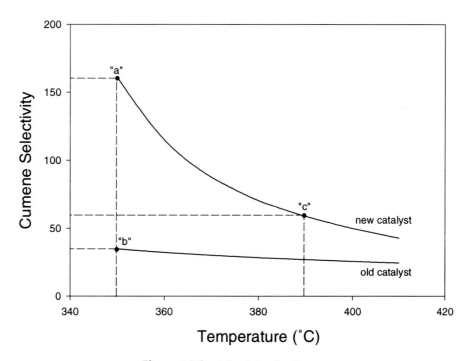

Figure 14.9   Selectivity for Cumene

for the new catalyst, the selectivity is 59, point "c," well above that for the current catalyst.

In conclusion, the new catalyst is expected to provide higher product selectivity and higher cumene production at increased reactor temperature.

Discussion.   It was found in the simulation of the dominant reactions in the production of cumene using the new catalyst that high temperatures increase the formation of the undesirable product, DIPB. This is expected based on the relative activation energies of the two reactions. The higher activation energy for the DIPB reaction leads to a more rapid increase in the rate of reaction of DIPB with temperature compared to cumene. Therefore, it is important to avoid hot spots in the reactor to avoid DIPB production. From a reaction point of view, isothermal reactors represent a desirable goal. This goal is obtained with a CSTR, but is not easy to obtain in a PFR with a high heat of reaction. The fluidized bed system can approach isothermal behavior because of rapid mixing of the solid particles making up the bed. For highly exothermic reactions, the fluidized bed is inherently safe, hot spots do not develop, heat can be removed from the bed.

However, fluid bypassing takes place in a fluidized bed. This adversely affects performance. This is because the bypassed fluid fraction does not come into contact with the catalyst so a portion of the feed does not react. In the case study given above, the bypass fraction had a low value for a fluidized bed, 5%. As a result the highest conversion that can be achieved is 95%.

If the products from the reactor are easily separated, the unreacted reactants can be recovered and recycled to achieve high overall conversion. From a kinetic point of view, the formation of DIPB in the product could be effectively eliminated by operating at low reactor conversions. This can be seen from the reaction rate expression that shows the formation of DIPB is directly proportional to the concentration of cumene. If cumene concentration is kept low, the amount of DIPB formed is small.

Separation of the reactor products containing propane/propylene presents a problem. Propylene and propane have similar volatilities and are difficult to separate. If propylene and propane are not separated, there are three options to be evaluated:

1. Propylene-propane mix is discarded as waste. Fuel value may be recovered.
2. Total recycle of propylene-propane mix. This is not possible. Since the propane does not react, it must be purged from system.
3. Partial recycle of propylene-propane mix. The propane in the feed would increase. This decreases the concentrations of the reactants and reduces the production of cumene. The superficial velocity in the reactor would increase. This reduces residence time and the fraction bypassed in the fluidized bed.

In the case study, the constant superficial velocity was maintained at higher temperatures by decreasing the excess benzene. The total concentration of gas in the reactor was reduced. This tends to decrease the reaction rate. In addition, the reaction equations show the lower benzene concentration reduces the cumene reaction rate and has no impact on the DIPB reaction. As a result of maintaining a constant required superficial velocity by lowering the excess benzene, the selectivity is reduced.

An alternative approach to maintain the superficial velocity at increased temperature and retain the excess benzene flow is to increase the system pressure. This should provide higher reaction rates at a given temperature, resulting in higher conversions, and would not effect the selectivity.

The conclusions given regarding the catalyst replacement assumed that the reaction took place in a region where chemical reaction controlled. When chemical reaction controls, the effect of temperature is large. Figure 14.8 indicates little effect of temperature on cumene production using the current catalyst, but temperature had major impact on cumene production for the new catalyst. The low sensitivity in the base case suggests that the reaction is not chemical reaction controlled. Often when dealing with catalytic reactions, one finds that due to high catalyst activity, the reaction may be diffusion controlled. The activation energy for diffusion is significantly lower than that for chemical reactions. Therefore, the effective activation energy for a diffusion controlled reaction is also significantly lower. This means that intrinsic reaction kinetics supplied by manufacturers may not be directly applicable in process simulation. The reader is referred to standard reaction engineering texts.

> **For catalytic reactions, always consider mass transfer effects. If present, they reduce the temperature sensitivity of the reaction.**

14.5.3   Increasing Acetone Production

Problem. You are responsible for the operation of the acetone production unit in your plant. The acetone produced is used internally elsewhere in your plant. A need to increase internal production by as much as 50% is anticipated. If this full increase cannot be met from your acetone unit you will have to purchase additional acetone. You have been requested to evaluate how much increase in capacity can be obtained using our current chemical reactor.

Acetone is produced by the endothermic reaction

$$(CH_3)_2CHOH \rightarrow (CH_3)_2CO + H_2$$

$$\text{isopropyl alcohol} \qquad \text{acetone} \qquad \text{hydrogen}$$

There are no significant side reactions. The reaction is endothermic and takes place in the gas phase on a solid catalyst. The reaction kinetics are given by

$$-r_{IPA} = kc_{IPA} \quad \text{kmol/m}^3 \text{ reactor s}$$

where

$$k = [3.156 \times 10^5 \text{ m}^3 \text{ gas}/(\text{m}^3 \text{ reactor s})]\exp[-8702/T]$$

where T is in Kelvin. The process flowsheet for our plant is given in Appendix B, Figure B.3. Figure 14.10 shows the reactor portion of this process along with reactor specifications obtained from Table B.5. The process uses a PFR with the heat transfer medium, HTM, on the shell side. Because we are concerned about operation of the downstream separation section, the IPA reaction conversion is held

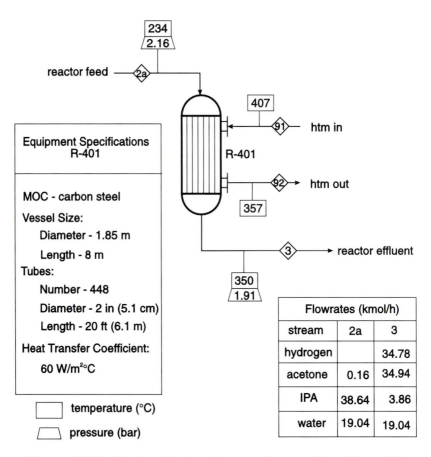

Figure 14.10   Reaction Section of Acetone Process (From Figure B.3 and Table B.5)

constant at 90% to ensure a constant feed composition to the separation section even though the flowrate of this stream may change.

Solution.   The conversion of an endothermic reaction is most often limited by the amount of heat that can be provided to the reaction. Current operating conditions define the base case. Figure 14.11a shows the temperature profile in the reactor for the reacting stream and HTM for the base case. There is a small temperature driving force for heat transfer over most of the reactor. Only in a small region near the reactor entrance is there a large temperature difference needed to provide a high heat flux. To increase production, it is necessary to increase the overall heat flux into the reactor.

Figure 14.11a reveals four options that will increase the average heat flux:

1. Increase the temperature of the HTM.
2. Decrease the temperature of the reaction stream.
3. Operate with the HTM introduced countercurrent to the reacting stream.
4. Heat the feed stream to a higher temperature.

An evaluation of these options follows:

Option 1: *Increase HTM temperature*. Figure 14.12 presents results obtained for increased acetone production resulting from a change in HTM inlet temperature (flow rate of HTM held constant). Increasing the HTM by 50°C increased acetone production by 48%. This would require a 48% increase in heat duty and a 50°C increase in HTM temperature from the fired heater, H-401.

Option 2: *Decrease reactor temperature*. Lower reactor temperatures decreases the reaction rate (lower reaction rate constant, k) reducing acetone production. This option has no merit and is rejected.

Option 3: *Countercurrent flow*. Figure 14.11b shows the temperature profile obtained by introducing the HTM countercurrent to the reaction stream. Compared to the base case (cocurrent flow), it provides a larger, more constant temperature driving force (and heat flux). This change provides a 13% increase in acetone production. The heat duty of the fired heater, H-401, must be increased by 13% to provide this increase in duty.

Option 4: *Heat feed stream to higher temperature*. This provides energy in the form of sensible heat that adds to the energy transferred from the HTM. For an increase in feed temperature of 100°C, the acetone production increases by 14%. This would require an additional fired heater for the reactor feed stream.

This investigation showed there are a number of options that can effect a significant increase in acetone production. A 50% increase in capacity can be obtained from the reactor. All options require an increase in fired heater capacity.

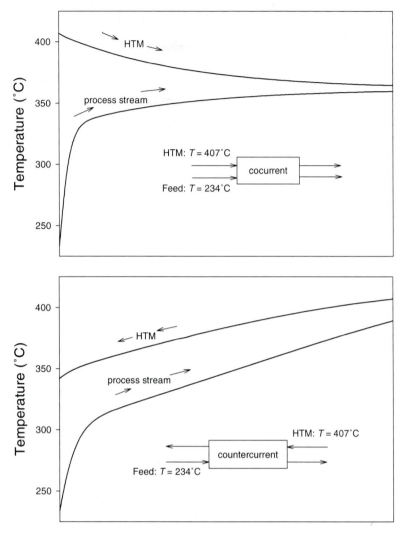

Position in Heat Exchanger

Figure 14.11   Temperature Profiles in Acetone Reactors for Cocurrent and Countercurrent Flow (IPA conversion of 90%)

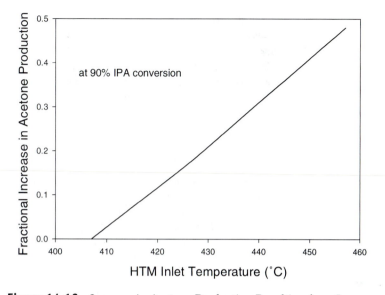

Figure 14.12   Increase in Acetone Production Resulting from Increase in HTM Inlet Temperature (HTM flow constant)

Discussion.   It has been shown that there are a number of options that provide increased acetone production. Except for Option 2, the options provided can be combined. The heat transfer configuration, Option 3, can be combined with an increase in HTM temperature, Option 1. In addition, the reactor feed temperature, Option 4, can be increased. In most situations, there is more than one option to consider to make a desired change in performance.

All options to increase acetone production require an increase in capacity of the fired heater (or addition of a new fired heater). Therefore, the fired heater is likely to represent a bottleneck to increased operations. Fired heaters are expensive, and it is not likely that the current heater was oversized enough to provide a 50% increase in heat duty. Until this bottleneck is removed, the additional acetone production required cannot be obtained.

Once the fired heater bottleneck is resolved, the separation system must be able to process the large increase in throughput. It is unlikely that the separation system can process the 50% increase. Therefore, if the fired heater bottleneck is removed, the separation system becomes the bottleneck.

Once a decision is made to eliminate these two bottlenecks (fired-heater and separation system), the analysis of reactor behavior could be modified to consider a number of secondary effects. The case study did not consider the effect of increasing reactor flowrates on the heat transfer coefficient (gas phase resistance dominates). The impact of the increased pressure drop over the reactor was not examined. These options were not necessary to confirm that the reactor is capable of providing a 50% increase in acetone production. The option of increasing the HTM flowrate was not appraised.

Endothermic reactions are inherently safe systems in terms of undergoing an uncontrolled reaction that is characteristic of an exothermic reactions. If the heat source is withdrawn, the temperature reactor drops and the rate of reaction decreases.

14.6   SUMMARY

In this chapter, chemical reactor performance was analyzed differently from the way it is usually done in reaction engineering classes. Key qualitative trends were emphasized. This was done in the context of Figure 14.1, which illustrated the interrelationship between product production, reaction kinetics, reactor behavior, and heat transfer limitations. The case studies presented reinforce the interrelationship between these factors.

REFERENCES

1. Zimmerman, C. C. and R. York, "Thermal Demethylation of Toluene," *Ind. Eng. Chem. Proc. Des. Dev.*, **3**, 254 (1964).
2. Rase, H. F., *Chemical Reactor Design for Process Plants, Vol. 2: Case Studies and Design Data*, Wiley-Interscience, New York, 1977, p. 38.

PROBLEMS

1. For the hypothetical, endothermic, gas-phase reaction

$$A = B + C$$

   sketch a plot similar to Figure E14.1. Discuss the relationship between equilibrium conversion, temperature, and pressure.

2. For the situation in Example 14.3, if all of the reactions take place, what conditions maximize the selectivity for P?

3. Examine the reaction network and the reaction kinetics for phthalic anhydride production from o-xylene in Appendix C, Table C.10. What conditions will maximize the selectivity for phthalic anhydride? Sketch concentration profiles for the case when this reaction is run in a packed bed modeled as a PFR.

4. Examine the reaction network and the reaction kinetics for cumene production in Appendix C, Table C.17. What conditions will maximize the selectiv-

ity for cumene? Sketch concentration profiles for the case when this reaction is run in a packed bed modeled as a PFR.

5. Sketch the T-Q diagram for an exothermic reaction in a PFR for
   a. countercurrent flow of HTM (heat transfer medium)
   b. cocurrent flow of HTM

   Which configuration minimizes hot spot formation? Why? What are the consequences on reactor size of choosing the configuration that minimizes hot spot formation? Explain your answer by examining the trends (See Chapter 11).

For Problems 6–20, it is recommended that a process simulator be used.

Problems 6–10 investigate the performance of the reactor section for a non-catalytic process for the hydrodealkylation of toluene to produce benzene. Reactor R-101 in Figure 14.4b operated at the flow conditions shown represents a base case (inlet pressure is 25 bar). Kinetics are given in Table 14.1. Neglect reactor pressure drop. Reactor volume for base case is 4.76 m³.

6. Investigate the effect of increasing reactor volume on toluene conversion. Plot the results.

7. Investigate the effect of a variation of hydrogen feed rate (all other parameters held constant at base case values).
   a. Plot toluene conversion versus percent excess oxygen.
   b. Determine the maximum conversion possible.

8. Investigate a two-reactor system to replace the single adiabatic PFR in the base case. The two-reactor system consists of an adiabatic CSTR followed by an adiabatic PFR. For the base case toluene conversion (0.75):
   a. Determine the size of the CSTR and PFR that provides the minimum total reactor volume.
   b. Determine the fraction of the total conversion obtained in the CSTR, from part a, above.

9. Investigate the effect of increasing the feed rate on reactor performance.
   a. Prepare a performance curve of the toluene conversion versus feed rate.
   b. Prepare a performance curve of benzene generated versus feed rate.

10. Investigate the effect of feed temperature on reactor performance. Prepare a performance curve for toluene conversion versus feed temperature.

Problems 11–15 investigate the performance of the cumene reactor (fluidized bed), described in Section 14.5.2, that results from replacing the catalyst. The replacement catalyst kinetics are given in Table 14.2. The base case selected is the currently operating reactor (with an active volume of 7.49 m³). Process output for the base case is given in the flow table in Figure 14.6.

11. a. Derive an equation for the ratio

$$\eta = \text{rate of cumene formation/rate of DIPB formation}$$

Rewrite the equation in the form $\eta = A/B - 1$.

b. Present the equation for η given above in terms of the following variables: T, P, k_o, E_{act}, and mole fractions.

c. Using the equation from Part b, determine the trend resulting from a change in the following variables. After each item, provide ↑ (increases), ↓ (decreases), ¢ (constant) or ? (cannot be determined).

   Increasing system temperature
   Increasing system pressure
   Increasing benzene mole fraction benzene
   Increasing propylene mole fraction
   Increasing propane mole fraction
   Increasing cumene mole fraction
   Increasing DIPB mole fraction

12. Replace the current catalyst with the new catalyst (maintain same feed rate and conditions) and
   a. Provide a flow table (similar to that in Figure 14.6).
   b. Compare cumene production with the new to the old catalyst.
   c. Compare DIPB production with the new to the old catalyst.

13. Evaluate the effectiveness of the new catalyst at 350°C that results from increasing the reactor volume.
   a. Plot cumene production versus reactor volume:
   b. Plot selectivity (defined as cumene generated/DIPB generated) versus volume.

14. In Figures 14.8 and 14.9, the volumetric flowrate to the reactor was held constant at increased temperatures by reducing the hydrogen in the feed. This altered the feed concentration. Investigate the alternative of reducing the total feed rate (at the same composition) to maintain constant volumetric flowrate.
   a. Plot the cumene production rate versus temperature and compare to Figure 14.8.
   b. Plot the selectivity (cumene generated/DIPB generated) versus temperature and compare to Figure 14.9.

Problems 15–20 investigate the performance of the acetone reactor system. The reaction kinetics are presented in Section 14.5.3. The reactor and internal heat exchanger are fixed. For the base case, the active volume of the reactor is 5.1 m³ at the operating conditions given in Figure 14.10. Unless otherwise stated, the heat transfer coefficient and reactor pressure drop are to be assumed constant. The feed temperature, pressure, and component concentrations remain con-

stant. For these calculations, assume the utility fluid to be n-heptadecane (at pressures over 30 bar). At the base case, the utility flow is 58 kmol/h.

15. Reconfigure the heating system to operate in countercurrent fashion. Retain the same utility stream input and IPA conversion (90%).
    a. Determine the change in process feed rate.
    b. Determine the percent change in process feed rate.

16. Increase the utility feed rate by 50% and maintain the base case process feed stream.
    a. Determine the conversion obtained.
    b. Determine the percent change in the acetone produced.

17. Increase the utility feed rate by 50%. Retain the IPA conversion of 0.9.
    a. Determine the process feed rate.
    b. Determine the percent change in acetone produced.

18. Increase the utility feed temperature by 50°C. Retain the IPA conversion at 0.9.
    a. Determine the process feed rate.
    b. Determine the percent change in acetone production.

19. Prepare a performance curve for the change in utility flowrate necessary to maintain an IPA conversion of 0.9 at increased process feed rates.

20. Assume the individual film coefficients change according to the relationship

$$h_i \propto \text{velocity}^{0.8}$$

    a. Estimate the effect of process flow changes on the overall heat transfer coefficient.
    b. Determine the process flowrate that provides an IPA conversion of 0.9 resulting from an increase in utility temperature of 50°C.
    c. Does the assumption of constant overall heat transfer coefficient result in a higher or lower estimate of the process feed rate? If you worked Problem 19, you have this value for comparison.

CHAPTER

15

Regulating Process Conditions

It is important, over the life of a plant, for operating conditions to be regulated in order to obtain stable operations and to produce quality products efficiently and economically. In this chapter, we consider regulation; a subset of this is process control. Regulation establishes the strategy by which the process can be controlled. Regulation also involves the dynamic response (transient behavior) of the process to changes in operating variables. The latter will not be considered here since it is covered in the typical undergraduate process control course.

The regulation of process operations involves an understanding of two facts:

In most situations, processes are regulated, either directly or indirectly, by the manipulation of the flowrates of utility and process streams.

Changes in flowrates are achieved by opening or closing valves.

In order to decouple the effects of changes in process units, adjustments are most often made to the flow of the utility streams. Utility streams are generally supplied via large pipes called "headers." Changes in flowrates from these headers have little effect on the other utility flows and, hence, they can be changed independently of each other. The one notable exception to this concept is when the flowrate of a process stream is to be controlled. Clearly, in this case, the process stream must be controlled or regulated directly by a valve placed in the process line.

In this chapter, we consider that utility streams are available in unlimited amounts at discrete temperatures and pressures. This is consistent with normal plant operations, and a typical (although not exhaustive) list of utilities is given in Table 3.4.

This chapter looks at several aspects of regulation that are important for the successful control of processes. The following topics are covered:

The characteristics of regulating valves

The regulation of flowrates and pressures

The measurement of process variables

Some common control strategies

Exchange of heat and work between process streams and utility streams

Case studies of a reactor and a distillation column

Before we look at these topics, however, we start out with a simple regulation problem from the overhead section of a distillation column.

15.1   A SIMPLE REGULATION PROBLEM

Consider the flow of liquid from an overhead condenser on a distillation column to a reflux drum as illustrated in Figure 15.1a. This is a section of the process flow diagram for the DME process shown in Figure B.1 and discussed in Appendix B. The liquid condensate flows from the heat exchanger, E-205, to the reflux drum, V-201. From the drum, the liquid flows to a pump, P-202, from which a portion (Stream 16) is returned to the distillation column, T-201, and the remainder (Stream 10) is sent to product storage. Let us assume that the amount returned to the column as reflux is set by a control valve, shown in the diagram. The amount of reflux is fixed in order to maintain the correct internal flows in the column and hence the product purity. Consider what happens if there is an upset and there is an increase in the amount of liquid being sent to the reflux drum. If no additional control strategy is employed, then the level of liquid in the reflux drum will start to increase, and at some point the drum may flood, causing liquid to back up into the overhead condenser causing the condenser to malfunction. Clearly, this is an undesirable situation. The question is how do we control the situation? The answer is illustrated in Figure 15.1b. A control valve is placed in the product line, and a level indicator and controller are placed on the reflux drum. As the level in the drum starts to increase, a signal is sent from the controller to open the control valve. This allows more flow through the valve and causes the level in the drum to drop. Although this may seem like a simple example, it illustrates an important principle of process control, namely, that a major objective of any control scheme is to maintain a steady-state material balance.

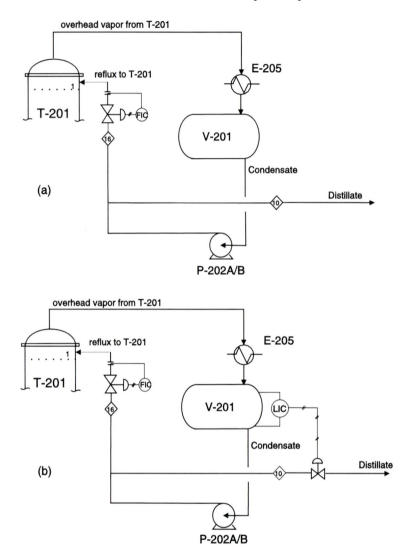

Figure 15.1 Basic Material Balance Control for DME Overhead Product

When controlling a process, it is important to ensure that the steady-state material balance is maintained, that is, to avoid the accumulation (positive or negative) of material in the process.

In the following sections, we will look at the role and functions of valves in the control of processes and the types of control strategies that are commonly found in chemical processes.

15.2   THE CHARACTERISTICS OF REGULATING VALVES

Figure 15.2a is a crude representation of a regulating or control valve. Fluid enters on the left of the valve. It flows under the valve seat where it changes direction and flows upward between the valve seat and the disk. It again changes direction and leaves the valve on the right. The disk is connected to a valve stem

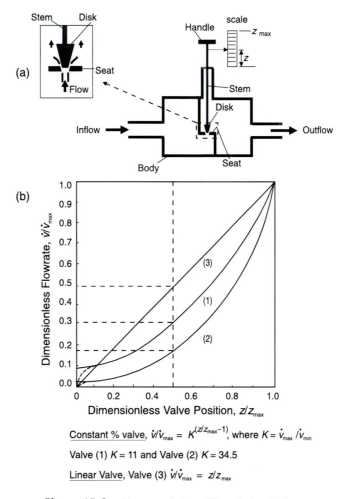

Constant % valve, $\dot{v}/\dot{v}_{max} = K^{(z/z_{max}-1)}$, where $K = \dot{v}_{max}/\dot{v}_{min}$

Valve (1) $K = 11$ and Valve (2) $K = 34.5$

Linear Valve, Valve (3) $\dot{v}/\dot{v}_{max} = z/z_{max}$

Figure 15.2   Characteristics of Regulating Valves

that can be adjusted (in the vertical direction) by turning the valve handle. The position of the disk is given on the linear scale. The enlarged section shows the critical region involving the disk and the valve seat. The liquid flows through the annular space between the disk and the seat. As the disk is lowered the area of this annulus decreases. The relationship between flowrate and valve position depends upon the shape of both the valve disk and the valve seat. It should be noted that the direction of flow through the valve could also be from the right to the left. In this case, the flow of the fluid pushes down on the disk and pushes it towards the seat. This configuration is preferred if the fluid is at a relatively high pressure or there is a large pressure drop across the valve.

Figure 15.2b shows the pressure drop versus flowrate characteristics of regulating valves. To make a specific change in pressure drop across the valve, the valve must be opened or closed, as appropriate. In order to evaluate the desired valve position, it is necessary to know the performance diagram or characteristic for the valve.

Figure 15.3 is a mechanical drawing of the cross section of a globe valve. Globe valves are often used for regulation purposes. There are two general valve types used for regulation.

Linear Valves.   In a linear valve, the flowrate is proportional to the valve position. This is given by the relationship, $\dot{v} = kz$, where $\dot{v}$ is the volumetric flow and z is the vertical position of the disk or valve stem.

Constant or Equal Percentage Valves.   For an equal percentage valve, an equal change in the valve position causes an equal percentage change in flowrate. The valve equation for the constant percentage valve is given by, $\dot{v}/\dot{v}_{max} = K^{(z/z_{max}-1)}$. We note that a true constant percentage valve does not completely stop the flow. However, in reality, the valve does stop the flow when fully closed, and the constant percentage characteristic does not apply to very low flowrates, as indicated by the dashed lines at the lower end of the curves in Figure 15.2b.

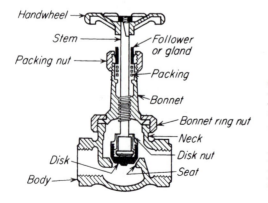

Figure 15.3   Cross Section of a Globe Valve (From McCabe, W. L., J. C. Smith, and P. Harriott, *Unit Operations of Chemical Engineering* (5th ed.), McGraw-Hill, New York, 1993, copyright © 1993 by McGraw-Hill Companies, reproduced with permission of the McGraw-Hill Companies).

Both of these valve characteristics are described by the equations shown on Figure 15.2b. Note that these equations only strictly apply for a constant pressure drop across the regulating valve.

Example 15.1

A valve on a coolant stream to a heat exchanger is operating with the valve stem position at 50% of full scale. You are asked to find what the maximum flowrate would be if the valve were opened all of the way (at the same pressure drop). What is an appropriate response given the information in Figure 15.2b?

The flow would increase by 100% or more. The exact increase depends upon the design of the valve.

From Figure 15.2b we see that a linear valve (Curve 3 on Figure 15.2b) has the smallest increase, from $v/v_{max} = 0.5$ to $v/v_{max} = 1$ (100% increase). For constant percent valve 1 (Curve 1), the increase is from $v/v_{max} = 0.3$ to $v/v_{max} = 1.0$ (233%), and for constant percent valve 2 (Curve 2), the increase is from $v/v_{max} = 0.17$ to $v/v_{max} = 1.0$ (488%). The flow increases in the range of 100 to 488%. The problem of predicting the flow from the valve position is even more complex because there usually exists some form of hysteresis. Thus, the pressure drop over the valve changes not only with flowrate of fluid through the valve but also with the direction of change, that is, whether the flow was last increased or decreased.

Most valves installed are constant percentage valves, and there is no generally accepted standard design. Such valves offer fast response at high flowrates and fine control at low flowrates. The prediction of the disk position needed to give the required flowrate would be difficult using the performance diagram in Figure 15.2. In practice, the flowrate is controlled by observing a measured flowrate while changing the valve stem position. The valve position continues to be adjusted until the desired flowrate is achieved. This approach forms the basis of the "feedback control system" used for many automatic flow control schemes and discussed further in Section 15.5.

In practice, automatic control systems change valve positions to obtain a desired flowrate. The valve position is modified by installing a servomotor in place of the valve handle or by installing a pneumatic diaphragm on the valve stem.

15.3   REGULATING FLOWRATES AND PRESSURES

The rate equation describing the flowrate of a stream is given by

$$\text{Flowrate} = \frac{\text{Driving force for flow}}{\text{Resistance to flow}} \tag{15.1}$$

The driving force for flow is proportional to ΔP (pressure head), and the resistance to flow is proportional to friction. The resistance to flow can be varied by opening or closing valves placed in the flow path.

Figure 15.4 shows a pipe installed between a high pressure header, containing a liquid, and a low pressure header. The pressure difference, ΔP, between these two headers is the driving force for flow. Two valves are shown in the line, and when these valves are fully open they offer little resistance to flow. The resistance in the transport line is due to frictional losses in the pipe. The flowrate is at

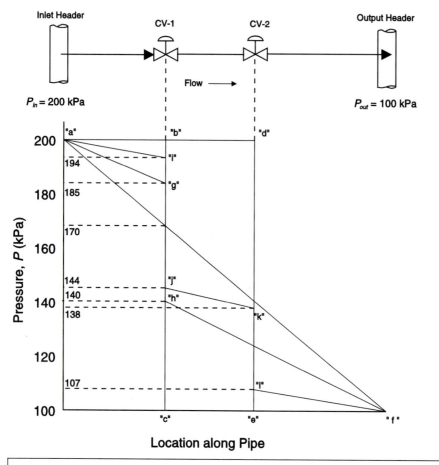

Profile	CV-1	CV-2	Flow (kmol/h)	ΔP (friction) (kPa)	P (before CV-1) (kPa)	ΔP over CV-1 (kPa)	ΔP over CV-2 (kPa)
1: a–f	open	open	100	100	170	0	0
2: a–b–c–f	closed	open	0	0	200	100	0
3: a–d–e–f	open	closed	0	0	200	0	100
4: a–g–h–f	partially open	open	74	55	185	45	0
5: a–i–j–k–l–f	partially open	partially open	44	19	194	50	31

Figure 15.4   Pressure Profiles in a Pipe Containing Two Valves

a maximum value when the valves are fully open. When either valve in the line is fully closed, the resistance to flow is infinite, and the flowrate is zero. The valves may be adjusted to provide flows between these limits.

Plotted below the diagram in Figure 15.4 are pressure profiles for various valve settings. The pressure (in kPa) is plotted on the y-axis, and we note that for the example shown $P_{in} = 200$ kPa and $P_{out} = 100$ kPa. The x-axis indicates the relative location of the valves in the process. In addition, Figure 15.4 includes a table that provides information on the flowrate, pressure drop due to pipe friction, and the pressure before valve CV-1. The resistance in the pipe is proportional to the square of the flowrate, which is the case for fully developed turbulent flow (see Chapter 11).

Profile 1 (a–f) shows the pressure profile with both valves fully open. It gives the maximum flowrate, 100 kmol/h, possible for this system. Profile 2 (a–b–c–f) is for the case when CV-1 is fully closed and CV-2 is fully open. For this case, the flow is zero, and all the pressure drop occurs over CV-1. The pressure upstream of CV-1 is P_{in} (200 kPa), and the downstream pressure is P_{out} (100 kPa). Profile 3 (a–d–e–f) is the case for when CV-2 is fully closed and CV-1 is fully open. The pressure upstream of CV-2 is P_{in} (200 kPa), the downstream pressure is also equal to P_{in} (200 kPa), and the flowrate is zero.

For Profile 4 (a–g–h–f), valve CV-1 is partially open, providing a pressure drop of 45 kPa, and valve CV-2 is fully open. The pressure drop across CV-1 (ΔP_{g-h}) can be varied by changing the valve position of CV-1. The greater the pressure drop across CV-1, the lower the pressure drop available to overcome friction and the lower the flowrate. In Profile 4, the flow is reduced to 74% of the maximum flow.

For every setting of CV-1 (with CV-2 fully open), a unique value for pressure and flowrate is obtained.

Either pressure or flowrate, but not both simultaneously, can be regulated by altering the setting of a single valve.

Two valves are required to regulate simultaneously both the pressure and flowrate of a stream. The total system resistance (pipe and valves) determines the flowrate. The ratio of valve resistances establishes the pressure profile through the process. This is shown in Profile 5 (a–i–j–k–l–f), where 50% of the available pressure drop is taken over CV-1 and 31% is taken over CV-2. The resistance ratio, for the two valves, is $31/50 = 1.61$, the flow is 44% of the maximum flow, and the pressure upstream of CV-1 is 194 kPa. To illustrate this concept further, we consider the following example:

Example 15.2

Consider the flow diagram in Figure 15.4. At design conditions, we have 70% of the total available pressure drop across the two control valves, and the flowrate of fluid at these

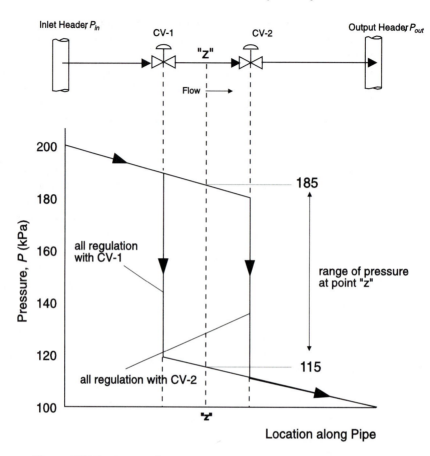

Figure E15.2   Range of Pressure at Point "z" for Design Flow

conditions is given as $100 [30/100]^{0.5} = 54.8$ kmol/h. If we consider a point, "z," midway between the two valves, over what range can the pressure at point "z" be varied at the design flowrate?

To solve this problem we consider the two extreme cases: (1) CV-1 regulates the flow and CV-2 is fully open, and (2) CV-2 regulates the flow and CV-1 is fully open. Both these situations are illustrated in Figure E15.2. From the diagram, we can see that the pressure at point "z" may vary between 185 kPa (CV-1 fully open) and 115 kPa (CV-2 fully open). Note that all possible combinations of partially open valves give pressures at point "z" between these two limits.

15.4   THE MEASUREMENT OF PROCESS VARIABLES

The process variables that are most commonly measured and used to regulate process performance are:

- Temperature: Several instruments are available that provide continuous measurement of temperature, for example, thermocouples, thermometers, thermopiles, and resistance thermometric devices.

- Pressure: A variety of sensors are available to measure the pressure of a process stream. Many sensors use the deflection of a diaphragm, in contact with the process fluid, to infer the pressure of the stream. In addition, the direct measurement of process pressure by gauges, for example, Bourdon gauge, is still commonly used.

- Flowrate: Fluid flowrates, until recently, were most often measured using an orifice or venturi to generate a differential pressure. This differential pressure measurement was then used to infer the flowrate. More recently, several other instruments have gained acceptance in the area of flowrate measurement. These devices include vortex shedding, magnetic, ultrasonic, and turbine flowmeters.

- Liquid Level: Liquid levels are commonly used in the regulation of chemical processes. There are many types of level sensors available and these vary from simple float operated valves to more sophisticated load cells and optical devices.

- Composition and Physical Properties: Many composition measurements are obtained indirectly. Physical properties such as temperature, viscosity, vapor pressure, electric conductivity, density, and refractive index are measured and used to infer the composition of a stream, in place of a direct measurement. A number of other measurement techniques have become commonplace for the on-line analysis of composition. These include gas chromatography and mass and infrared spectrometry. These instruments are very accurate, but are expensive and often fail to provide the continuous measurements that are required for rapid regulation.

15.5   COMMON CONTROL STRATEGIES USED IN CHEMICAL PROCESSES

There are many strategies used to control process variables in an operating plant. For more details, see Anderson [1] and Shinskey [2]. In this section, we consider only feedback, feedforward, and cascade systems.

15.5.1   Feedback Control/Regulation

The process variable to be controlled is measured and compared to its desired value (set-point value). If the process variable is not at the set-point value, then appropriate control action is taken.

Advantage.   The cause of the change in the output variable need not be identified for corrective action to be taken. Corrections continue until the set point value is achieved.

Disadvantage.   No action is taken until after an error has propagated through the process and the error in the process variable has been measured. If there are large process lag times, then significant control problems may occur.

Example 15.3

Identify all the feedback control loops in the process flow diagram for the production of DME, Figure B.1, and explain the control action of each. Note: There are other control valves on the utility streams, but these are only shown on the P&ID and not considered here.

All the control loops associated with the control valves shown on the PFD exhibit feedback control strategies. These are shown individually in Figure E15.3. Each control action is explained below:

Figure E15.3a: In this figure, the object of the controller is to control the flowrate of Stream 4, which is sent through two heat exchangers and then into reactor R-201. Because P-201 is a positive displacement pump, the flow is controlled by varying the amount of liquid bypassed from the exit to the inlet of the pump. (See Section 12.2.3) The signal to adjust the setting of the valve in the bypass line is obtained from an orifice placed in Stream 3. Thus, if the flow sensed by the pressure cell (not shown) across the orifices below the set-point value, a signal is sent to the bypass valve to close the valve slightly. This has the effect of reducing the amount of liquid bypassed around the pump which increases the flow of Stream 2 and hence increases the flow of Stream 3. If the flow of Stream 3 lies above the set-point, then the opposite control action is initiated.

Figure E15.3b: The flowrate of the bottom product in both distillation columns is adjusted so as to maintain a constant level of liquid in the bottom of the tower. The control strategy for the bottoms of both T-201 and T-202 are identical. Assume that the level in the bottom of the column drops below its set-point. A signal from the level sensor would be sent to close slightly the valve on the bottom product stream. This would reduce the flow of bottom product and result in an increase in liquid inventory at the bottom of the column, causing the liquid level to rise. If the liquid level were to increase above the set-point value, then the opposite control action would be initiated.

Figure E15.3c: The control strategy at the top of both columns (T-201 and T-202) is illustrated in Figure E15.3c. The control action was explained previously in Example 15.1, where the reflux stream is held constant by a control valve, and the liquid level in the reflux drum is held constant by adjusting the flow of the overhead product.

It should be noted that at present there is no product quality (purity) control on either column. This is addressed under cascade control below.

15.5.2 Feedforward Control/Regulation

Process input variables are measured and used to provide the appropriate control action.

Advantages.   Changes in the output process variable are predicted and adjustments are made before any deviation from the desired output takes place. This is useful, especially when there are large lag times in the process.

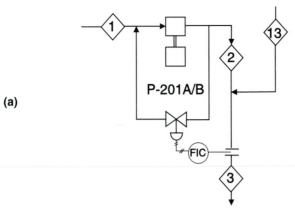

(a)

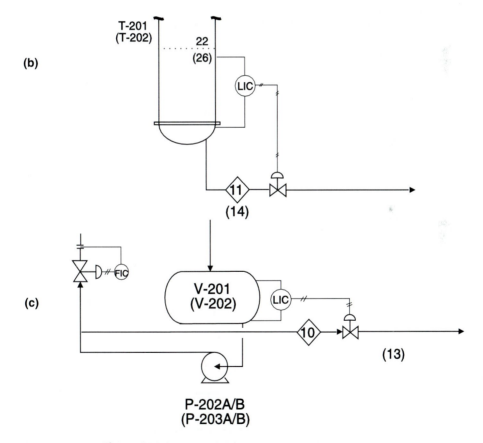

(b)

(c)

Figure E15.3   Feedback Control Loops on DME Flowsheet

Disadvantages. It is necessary to identify all factors likely to cause a change in the output variable and to describe the process by a model. The regulator must perform the calculations needed to predict the response of the output variable. The output variable being regulated is not used directly in the control algorithm. If the control algorithm is not accurate and/or the cause of the deviation is not identified, then the process output variable will not be at the desired value. The accuracy and effectiveness of the control scheme is directly linked to the accuracy of the model used to describe the process.

Example 15.4

Consider the process illustrated in Figure E15.4. The object is to cool, in a heat exchanger, a process stream (Stream 1) to a desired temperature (Stream 2) using cooling water supplied from a utility header. In Figure E15.4a, a feedback control scheme is illustrated. The control strategy is to measure the temperature of the process stream leaving the heat exchanger (Stream 2) and to adjust the flow of cooling water to obtain the desired temperature. Thus, if the temperature of Stream 2 were to be greater than the desired set-point value, then the control action would be to increase the flow of cooling water and vice versa.

A feedforward control scheme for this process is illustrated in Figure E15.4b. Here, both the flowrate and temperature of the input process stream and the inlet cooling water temperature are measured. A calculation is then made that predicts the flowrate of cooling water needed to satisfy the energy balance and the exchanger performance equations as given in Chapter 12. Using the subscript c and p to refer to the cooling water and process streams, respectively, we get:

$$Q = \dot{m}_c C_{p,c}(T_b - T_a) \tag{E15.4a}$$

$$Q = \dot{m}_p C_{p,p}(T_1 - T_2) \tag{E15.4b}$$

$$Q = UA \frac{(T_1 - T_b) - (T_2 - T_a)}{\ln\dfrac{(T_1 - T_b)}{(T_2 - T_a)}} \tag{E15.4c}$$

From Eqs. (E15.4a)–(E15.4c), we can categorize the variables as unknown, measured, known, and specified according to the following table:

Unknown	Measured	Known	Specified
$Q, \dot{m}_c, T_b$	$\dot{m}_p, T_a, T_1$	$C_{p,c}, C_{p,p}, U, A$	T_2

Therefore, for a given value of the exit process temperature (T_2), the three equations above can be solved to give the unknowns in the table. Thus, by measuring the inlet process stream flow and temperature and the inlet cooling water temperature, the model, given by Eqs. (E15.4a)–(E15.4c) can be solved to yield the desired value of the cooling

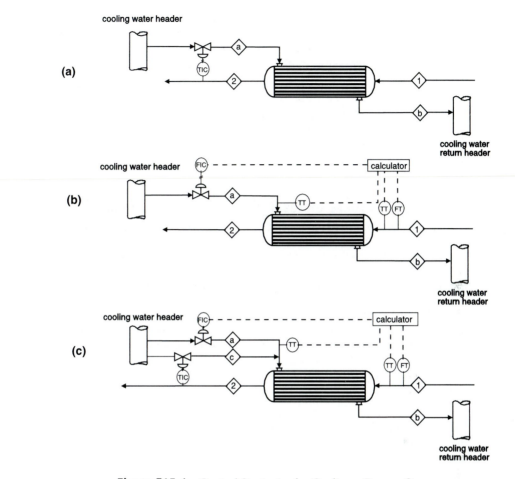

Figure E15.4   Control Strategies for Cooling a Process Stream

water flowrate. The valve on the cooling water inlet can then be adjusted to give this value. There are some assumptions associated with this model that will affect the performance of this control scheme. The first assumption is that both streams flow countercurrently, and hence the exchanger effectiveness factor, F, is assumed to be equal to one. If the value of F is not close to unity, then the model given above will not be accurate. However, the model may be modified by including another equation to calculate F based on the process variables. A more serious assumption is that the overall heat transfer coefficient, U, has a constant value. In real processes, this is seldom the case, either because of flowrate changes or due to fouling of heat exchange surfaces occurring throughout the life of the process. As the heat transfer surfaces foul or flowrates change, the value of U changes and the predictions of the above model will deviate increasingly from the true value of T_2. This illustrates one of the major disadvantages of feedforward control strategies, in that the efficacy of the control strategy is only as good as the predictions of the equations used to model the process.

An alternative scheme to pure feedforward control is to use a combination of feedback and feedforward control strategies. This is illustrated in Figure E15.4c and discussed below.

In feedforward control, the process must be modeled using material and energy balances and the performance equations for the equipment.

The effectiveness of feedforward control is strongly influenced by the accuracy of the process model used.

15.5.3   Combination Feedback/Feedforward Control

By correctly combining these two control strategies, the advantages of both methods can be realized. The feedforward strategy uses measured input variable(s) to predict changes necessary to regulate the output. The feedback regulator then measures the regulated output variable and makes additional changes to assure the process output is at the set point.

Advantages.   The advantages of both the feedback and feedforward regulation are achieved. The feedforward regulator makes the major corrections before any deviation in process output process takes place. The feedback regulator assures that the final set-point value is achieved.

Disadvantages.   The control algorithm and hardware to achieve this form of control are somewhat more complicated than either individual strategy.

Example 15.5

We return to Example 15.4 and look at the operation of the combined control strategy illustrated in Figure E15.4c. This system is essentially a combination of Figures E15.4a and E15.4b. The feedforward scheme of Figure E15.4b has been retained but is used to regulate only a fraction of the cooling water required, say 80%. In addition, a feedback control loop has been added. This feedback loop consists of a second parallel cooling water feed stream that will be adjusted so as to maintain the desired set-point value for the temperature of Stream 2. The amount of cooling water regulated by the feedback loop is the remaining 20% required to satisfy the regulation. This combined system will respond quickly to changes in process flowrate or inlet temperatures via the feedforward control loop. In addition, a second cooling water stream is adjusted, based on the feedback control loop, to control exactly the process outlet temperature.

15.5.4   Cascade Regulation

The cascade system uses controllers connected in series. The first control element in the system measures a process variable and alters the set point of the next control loop.

Advantages.   Reduces lags and allows finer control.

Disadvantages.   More complicated than single loop designs.

Example 15.6

Consider the control of the top product purity in distillation column T-201 in the DME process. As shown in Example 15.3, the material balance control is achieved by a flow controller on the reflux stream and a level controller on the top product stream. From previous operating experience, it has been found that the top product purity can be measured accurately by monitoring the refractive index (RI) of the liquid product from the top of the column, Stream 10. Using a cascade control system, indicate how the control scheme at the top of T-201 should be modified to include the regulation of the top product composition.

The solution to the problem is illustrated in Figure E15.6. An additional control loop has been added to the top of the column that consists of a refractive index measuring element that sends a signal to the flow controller placed on the reflux stream (Stream 16). The purpose of this signal is to change the set-point of the flow controller. Consider a situation where the purity of the top product has fallen below its desired value. This change may be due to a number of reasons; for example, the flowrate or purity of the feed to the column may have changed. However, the reason for the change is not important at this stage. The decrease in purity will be detected by the change in refractive index of the liquid product, Stream 10, and a signal will be sent from the RI sensor to increase the reflux flowrate of Stream 16. The flow of Stream 16 will increase, which will result in an improved separation in the column and an increase in purity of the top product. For an increase in product purity, the reverse control action will occur.

These control strategies do not represent a comprehensive list of possibilities. The range of strategies is very large. This is due, in part, to the flexibility that software has added to the control field by allowing sophisticated process models to be used in simulating processes, as well as the use of more complicated control algorithms.

No mention has been given to the types of controller (proportional, differential, integral, etc.) and strategies to tune these controllers. Such information is

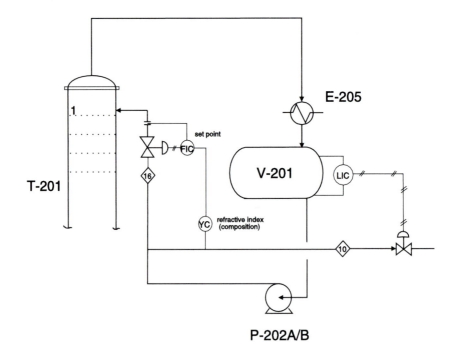

Figure E15.6   Cascade Control for DME Overhead Product Purity

covered in traditional process control texts, such as Stephanopolous [3], Coughanowr [4], Smith and Corripio [5], Seborg et al. [6], and Marlin [7].

15.6   EXCHANGING HEAT AND WORK BETWEEN PROCESS AND UTILITY STREAMS

In order to obtain the correct temperatures and pressures for process streams, it is often necessary to exchange energy, in the forms of heat and work, with utility streams. In order to increase the pressure of a process stream, work is done on the process stream. The conversion of energy (usually electrical) into pressure head is provided by pumps (for liquid streams) or compressors (for gas streams) that are inserted directly into the process stream. For a process gas stream undergoing a significant reduction in pressure, the recovery of work using a gas turbine may have a significant impact (favorable) on process economics. In order to change the temperature of a process stream, heat is usually transferred to or from a process by exchange with a heat transfer medium, for example, steam and cooling water. As shown in Chapter 19, it may also be beneficial for heat to be exchanged between process streams, using heat integration techniques.

15.6.1   Increasing the Pressure of a Process Stream and Regulating Its Flowrate

In the preceding examples, streams were available at constant pressure. This is normally true for utility streams. However, for a process stream, the pressure may change throughout the process. When higher pressures are needed, pumps or compressors are installed directly into a process stream. These units convert electrical energy into the required pressure head for the stream.

It should be noted that when a pump or compressor is required in a process, it is necessary to specify the type of fluid and fluid conditions, the design flowrate of fluid, the inlet (suction) and outlet (discharge) pressures at the design conditions, and an overdesign (safety) factor. The pump or compressor conforming to these specifications must be able to operate at and somewhat above the design conditions. However, in order to regulate the flow of the process streams, as will be required throughout the life of the process, it is necessary to implement a control or regulation system. Some typical methods of regulating the flow of process streams are outlined below and shown in Figure 15.5.

Figure 15.5a shows a centrifugal pump system that increases the pressure of a process stream. The position of the control valve, CV-1, is at the discharge side of the pump. The flowrate of Stream 2 is changed in order to regulate the flow of the stream passing through the pump. The operation of the pump and the interaction of the pump curve, system curve, and the control valve were discussed in the presentation of pump performance curves in Chapter 12. In reviewing this material, it is important to remember that the maximum flowrate of the fluid is determined by the intersection of the pump and system curves. Regulation of flowrate is only possible for flows less than this maximum value, that is, to the left of the intersection of the pump and system curves, point "a" in Figure 12.6.

Figure 15.5b shows a positive displacement pump increasing the pressure of a process stream. The positive displacement pump can be considered to be a constant flow, variable head device, that is, it will deliver the same flowrate of fluid at a wide range of discharge pressures. Thus we may write that:

$$\dot{v}_2 + \dot{v}_3 = \dot{v}_1 = \text{a constant} \tag{15.2}$$

For this reason, in order to regulate the flow of the process stream, Stream 2, a portion of the output stream from the pump must be recycled. In Figure 15.5b this is accomplished by returning Stream 3 to the suction side of the pump. By altering the position of valve CV-2, the recycle stream flowrate is altered, and since the flow of liquid through the pump is almost constant, this also provides flow regulation for the main process stream, Stream 2. It should be noted that one should never throttle the output stream of a positive displacement pump since the pump curve is almost vertical and the flow will change little with throttling, although the discharge pressure will increase drastically (see Section 12.2.3, Figure 12.8).

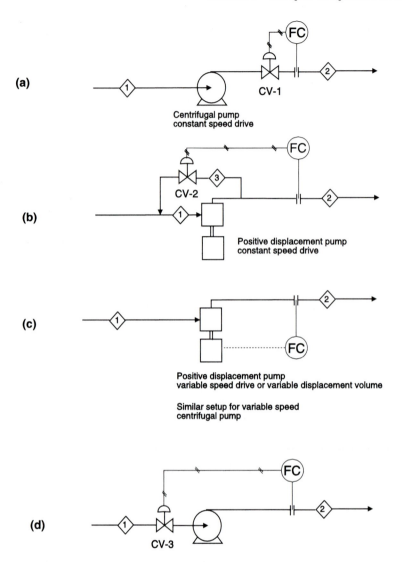

Figure 15.5   Flowrate Feedback Control Schemes for Pumping Liquids

Figure 15.5c uses a variable speed drive or variable displacement volume, for a positive displacement pump, to regulate the flow of the process stream. For a centrifugal pump, the impeller speed is regulated to provide the required flow, at the desired pressure, for the process stream. The advantage of using this type of control strategy is that there is no wasted energy due to throttling in a control valve. However, variable speed controls and motors are expensive and less efficient. Therefore, this type of control scheme is usually only cost effective for gas blowers, compressors, and large liquid pumps, due to the large savings in utilities required to offset the large capital investment for the variable speed drives and controls.

Figure 15.5d regulates the input flow with a valve, CV-3, installed in the suction line to the pump. The position of CV-3 is altered to provide the desired flowrate of fluid. This may work well for gases but is seldom used for liquid pumps. For liquid streams, the reduction in pressure at the pump inlet increases the possibility of cavitation, that is, $NPSH_A$ is reduced drastically. The causes of cavitation and NPSH calculations were covered in Section 12.2.4. On the other hand, for the case of gas compression using centrifugal machines, the throttling of the inlet stream is essential for start-up purposes. Thus, this method is often the preferred control strategy for flow regulation in centrifugal blowers and compressors.

All of the control systems shown in Figure 15.5 for the regulation of liquid process streams can be applied to compressor systems to regulate the flow of gaseous process streams.

15.6.2   Exchanging Heat between Process Streams and Utilities

The amount of heat added to or removed from a process stream is usually altered by changing the flow or pressure of utility streams. The primary utility streams in a chemical plant are:

1. Cooling water is used to remove heat from a process stream. The heat transferred to the coolant adds to the sensible heat (enthalpy) of the coolant stream (temperature increases).
2. Air can be used to remove heat from a process stream and is often used in product coolers and overhead condensers. The heat transferred to the coolant adds to the sensible heat (enthalpy) of the coolant stream (temperature increases).
3. Steam condensation is used to add heat to a process stream. The heat transferred from the steam decreases the enthalpy of the steam (steam condenses).
4. Boiler feed water (bfw) is used to remove heat from a process stream to make steam that can be used elsewhere in the process. The heat transferred to the bfw increases the enthalpy of the bfw stream (bfw vaporizes).

When using cooling water or air, in which there is no change of phase of the utility stream, regulation is generally performed by changing the flowrate of the cooling stream. For example, see Figure E15.4 and Example 15.4. In this section, we concentrate on systems in which the utility stream undergoes a change of phase. The focus is on boiler feed water and steam. However, the results apply equally well to other heat transfer media that undergo a phase change. It is further assumed that utility streams are taken from and returned to headers and that the process streams are a single phase. The control of reboilers and condensers in a distillation column is covered in the case study on distillation given in the following section.

There are many ways to regulate systems using steam. We do not try to present an exhaustive list of possible schemes but concentrate on several of the more common techniques. Figures 15.6 and 15.7 show several systems used to control the transfer of heat between a utility stream and a process stream. The systems on the left-hand side of the figures add heat to the process stream by condensing steam. Systems on the right-hand side remove heat from the process stream by vaporizing bfw. In either case, there are two phases present in equilibrium on the utility side of the exchanger. The temperature and pressure are related by the vapor pressure relationship for water. If steam from the header is significantly superheated, it may be passed through a desuperheater prior to being fed to the heat exchanger. The desuperheater adds just enough condensate in order to saturate the steam. The reason for using a desuperheater is that highly superheated steam acts like a gas with a correspondingly low film heat transfer coefficient. Thus, a significant amount of heat exchanger area may be taken up in desuperheating the steam. In such situations, it is more cost-effective to saturate the steam before entering the heat exchanger, reducing significantly the heat exchange area required.

Figures 15.6a and 15.6b show a process heater (steam condenser) and a process cooler (bfw vaporizer) without control valves. The heater shown in Figure 15.6a includes a steam trap (shown as a box with a T in the middle) on the

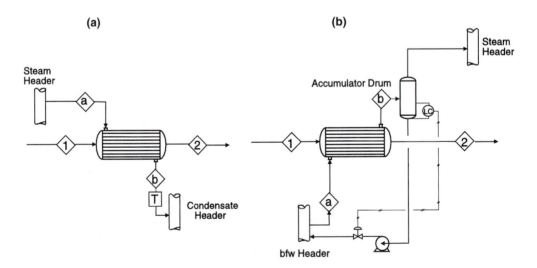

Figure 15.6   Unregulated Heat Exchanger Using Utility Streams with a Phase Change (a) Process Heater, (b) Process Cooler

utility line. This steam trap separates the condensate formed from the vapor. The condensate is collected by the trap in the bottom of the exchanger and discharged intermittently to the condensate header. The steam flowrate, Stream a, is established by the rate at which heat is exchanged. This is different from normal utility flow regulation. As the steam condenses, there is a large decrease in the specific volume of the utility stream. Thus, steam is continually "sucked" from the header into the heat exchanger. For this system, no regulation of process temperature is attempted.

The process cooler shown in Figure 15.6b vaporizes bfw. In this case, the flow of bfw is more than sufficient to provide for the vaporization taking place. The exit stream is a two-phase mixture of steam and bfw. The bfw is separated from the vapor in an accumulation drum (phase separator) and recycled back to the inlet of the exchanger. Again, for this system, no regulation of process temperature is attempted.

Although no control schemes are implemented in the systems in Figures 15.6a and 15.6b, to some extent, both systems are self regulating. The reason for this self-regulating property is that the major resistance to heat transfer is on the process side (film heat transfer coefficients for condensing steam and vaporizing bfw are very high). Consider an increase in the flowrate of the process stream in Figure 15.6a. As the flow increases, so does the process side heat transfer coefficient, and this increase is proportional to the (process flowrate)$^{0.8-0.6}$, see Table 11.1 for an explanation of the range of exponents. Since the major resistance to heat transfer is on the process side, this suggests that the overall heat transfer coefficient changes by the same amount. Thus, for a 10% increase in process flowrate, the overall heat transfer coefficient increases by 6–8%. Since the change in the overall heat transfer coefficient does not quite match the change in process flowrate, the exit temperature of the process stream, Stream 2, must drop slightly. Although the temperature of the exit stream drops, it does not drop by much due to the partial regulation provided by the increase in heat transfer coefficient.

In discussing the operation of this heat exchanger, no mention was made of the change in flowrate of the steam that would be required to provide the increase in heat exchanger duty caused by the increased process flowrate. As the exchanger duty increases, more steam will condense and this increased demand for steam does have an effect on the operation. It should be noted that steam is supplied from the header to the exchanger in the quantity that is required, that is, it is determined by the energy balance. As the steam flowrate increases, the frictional pressure losses in the supply pipe from the header to the exchanger increases. This causes the pressure on the steam side of the heat exchanger to decrease. Thus, the utility side of the heat exchanger essentially "floats" on the header pressure, with an appropriate pressure loss to account for flow through the supply line. As the pressure of the steam side of the exchanger drops, this causes the temperature at which the steam condenses to drop slightly, thus reducing the temperature driving force for heat exchange. However, compared to the change in heat transfer coefficient, this change is minor.

In order to regulate the temperature of the process stream leaving the heat exchanger, it is necessary to control the heat duty. In order to do this, we must be able to regulate a variable on the right hand side of the heat exchanger performance equation

$$Q = UA\Delta T_{lm} \tag{15.3}$$

This can be achieved by doing one or more of the following:

1. Regulate the temperature driving force (ΔT_{lm}) between the process fluid and the utility.
2. Adjust the overall heat transfer coefficient (U) for the heat exchanger.
3. Change the area (A) for heat exchange.

We consider two alternatives for regulating the temperature of the process exit stream.

Regulate the Temperature Driving Force between the Process Fluid and the Utility.   The systems shown in Figures 15.7a and 15.7b include a valve on the utility line. This valve is used to change the pressure of the utility stream in the exchanger. Since there are two phases present, the pressure establishes the temperature at which the phase change takes place. Reducing pressure reduces temperature according to the vapor pressure-temperature relationship for water, for example, from Antoine's equation.

To increase the heat duty for the process heater, Figure 15.7a, the temperature driving force is increased by increasing the pressure. This is achieved by opening CV-1 to decrease the frictional resistance to flow. Since CV-1 is located on the input side of the heat exchanger, opening the valve increases the pressure of the steam in the exchanger and the temperature at which condensation occurs will also increase. To increase the process cooler heat duty, Figure 15.7b, the temperature driving force is increased by opening CV-2, which decreases the resistance to flow. CV-2 is located on the discharge side of the utility, that is, between the exchanger and the steam header. In this case, the exchanger once again floats on the steam header pressure. As CV-2 is opened the pressure on the utility side of the exchanger decreases, this reduces the temperature at which the bfw boils and increases the temperature driving force for heat transfer.

Adjust the Overall Heat Transfer Coefficient for the Heat Exchanger by Adjusting the Area Exposed to Each Phase.   The systems shown in Figures 15.7c and 15.7d operate such that the interface between the liquid and vapor on the utility side of the exchanger covers some of the tubes. For this situation, a fraction of the heat exchange area is immersed in the liquid region with the remaining fraction in the vapor phase region. The heat transfer coefficients and the

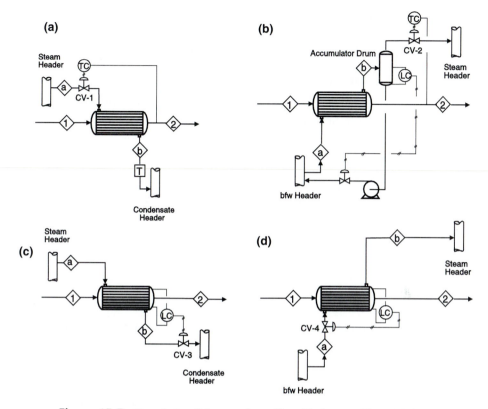

Figure 15.7   Regulation Schemes for a Heat Exchanger Using Utility
Streams with a Phase Change

temperature driving forces for these two regions differ significantly. Thus, by adjusting the level of the liquid-vapor interface in the exchanger it is possible to regulate the duty.

For the process heater, Figure 15.7c, decreasing the level increases the heat duty. The reason for this is that in the vapor-phase region the utility-side film coefficient and temperature driving force are large since steam is condensing. For the liquid-phase region, both the film coefficient and temperature driving force are low since the condensate is being subcooled. Therefore, as the liquid level decreases, more heat transfer area is exposed to condensing steam, the overall average value of UA increases, and the duty increases.

For the process cooler, Figure 15.7d, increasing the liquid level increases the heat duty. The reason for this is that in the vapor-phase region, the utility side film coefficient and temperature driving force are low since steam is being superheated. For the liquid-phase region, both the film coefficient and temperature driving force are high since the bfw is boiling. Therefore, as the liquid level increases more heat transfer area is exposed to boiling water, the overall average value of UA increases, and the duty increases.

15.6.3   Exchanging Heat between Process Streams

The integration of heat between process streams is often economically advantageous, and a strategy to do this is explained in Chapter 19. When heat integration is implemented in a process, it is likely that heat exchange will occur between two process streams. The regulation of the flow of these process streams cannot be used to control the heat transfer effectively, since the flow of the two streams will be directly coupled. One method to control the exit temperature in such an exchanger is to bypass some portion of one or both of the process streams around the exchanger. By altering the amount of the stream which bypasses the equipment, temperature regulation is possible. This concept is considered further in Problem 15.4.

15.7   CASE STUDIES

In this section we look in detail at two operations that require more complex control schemes. Again, we emphasize that these are not the only ways to control these units, but are typical schemes that might be employed. Many alternatives exist, and in the course of a career in process engineering, you will encounter a wide variety of regulation systems.

15.7.1   The Cumene Reactor, R-701

Consider the reactor system for the production of cumene shown in Figure C.8 and redrawn in Figure 15.8. The reactor feed, Stream 7, comes from a fired heater, where it is heated to the temperature required in the catalytic reactor. The reaction takes place in a bank of parallel catalyst-filled tubes and is mildly exothermic. In order to regulate the temperature of the reacting mixture and catalyst, the tubes are immersed in boiler feed water. The heat is removed by vaporizing the water to make high pressure steam, which is then sent to the high-pressure steam header.

The first thing we need to ask ourselves is, what exactly are we trying to control in the reactor and why? This is not a simple question to answer since many things must be considered. Some of the important considerations for exothermic catalytic reactions are discussed in Chapter 14. Here, we assume that the exit temperature from the reactor, Stream 8, is the variable that must be controlled. If the reactor is designed such that the temperature increases monotonically from the inlet to the outlet, then the exit stream is the hottest point in the reactor. More commonly, there will be a temperature bump or warm (hot) spot somewhere within the reactor, and the exit temperature will be cooler than at the

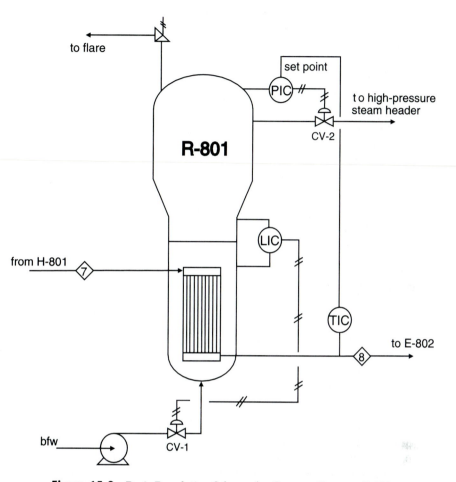

Figure 15.8   Basic Regulation Scheme for Cumene Reactor, R-801

temperature bump. In this case, we could use a series of in-bed thermocouples to measure the temperature profile within the catalyst tubes and use the maximum temperature as our controlled variable. In either case, it is important to control the reactor temperature, since the reaction rate and catalyst activity are both affected strongly by temperature.

The first part of the control scheme is a feedback material balance control on the boiler feed water. Control valve CV-1 is adjusted using a signal from the level controller mounted on the side of the reactor. This scheme assures that the bank of catalyst tubes is always totally immersed in the bfw. Regulation of the bfw level below the top of the tubes, as was discussed in the previous section, is inadvisable for this reactor, since the poor heat transfer coefficient for the portion of the tubes above the liquid level might cause a hot spot to occur at the entrance of the reactor. The second part of the control scheme uses cascade control to regulate the reactor exit temperature. The temperature at which the bfw boils in the reactor is regulated by the setting of CV-2. The logic is the same as was described for the situation in Figure 15.7b, namely that the pressure of the steam in the reactor is set by the steam header pressure (fixed) and the pressure drop across CV-2 (adjustable). The temperature of Stream 8 is monitored and used to adjust the set point on CV-2 as required.

Consider a situation in which the reactor exit temperature is seen to drop slowly. This situation may be caused by a number of reasons, for example, the catalyst slowly loses its activity and has the undesirable effect of reducing the conversion in the reactor with the possibility of a reduction in cumene production. With the control scheme shown in Figure 15.8, the system would respond in the following manner. As the temperature of Stream 8 drops, the set-point on CV-2 would be adjusted to close the valve and increase the pressure of the steam in the reactor. This would have the effect of increasing the temperature at which the bfw boiled and would reduce the temperature driving force for heat transfer. The net effect would be to reduce the amount of heat removed from the reactor causing the exit stream temperature (and reactor temperature profile) to increase and also causing the amount of reaction to increase. If the reactor exit temperature were to increase for some reason, then the control strategy would be the opposite of that described above. The increase in temperature of Stream 8 would be sensed by the temperature controller, which would then adjust downward the set-point pressure for the reactor. This would cause CV-2 to open reducing the pressure and temperature of the steam in the reactor. This, in turn, would increase the temperature driving force for heat transfer and increase the amount of heat removed from the reactor causing the exit stream temperature to decrease and also causing the amount of reaction to decrease. Clearly, for exothermic reactions, the control of temperature is imperative to safe reactor operation. For such equipment, several safety features would be incorporated into the overall control scheme. These safety features are extremely important and are treated separately in Chapter 21.

15.7.2   A Basic Control System for a Binary Distillation Column

The purpose of a control system for a distillation column is to provide stable operation and to produce products with the desired purity. Figure 15.9 shows a control scheme typically used to regulate a binary distillation system. Flowrates, pressure, stream composition, and liquid levels are regulated. The five control variables to be regulated are shown (column pressure, composition of distillate, composition of bottom product, liquid level in the bottom of the column, and the liquid level in the reflux drum).

Shown in Figure 15.9 are five feedback control loops and one cascade control loop used to regulate the five control variables:

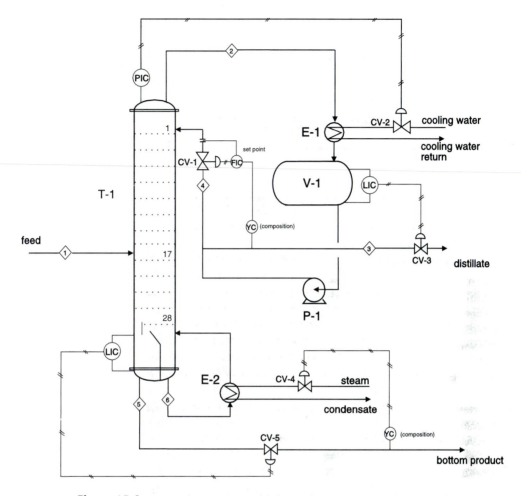

Figure 15.9   Typical Basic Control Scheme for a Binary Distillation Column

Valve CV-1 regulates the liquid reflux, Stream 4, in order to maintain the desired distillate composition. In addition, a composition measurement is made on the overhead product stream, Stream 3, and this is used to change the set point of CV-1.

Valve CV-2 regulates the condenser heat duty (by changing the flowrate of cooling water) in order to maintain a constant column pressure.

Valve CV-3 regulates the distillate flowrate, Stream 3, in order to maintain a constant liquid level in the reflux drum, V-1.

Valve CV-4 regulates the reboiler heat duty (by changing the pressure of the stream flowing into E-2) in order to maintain a constant distillate bottoms composition.

Valve CV-5 regulates the bottom product flowrate, Stream 5, in order to maintain a constant liquid level in the bottom of the column.

Consider the changes that result from a decrease in the concentration of the light component in the feed, assuming the feed flowrate remains unchanged, and the regulation that the control system provides. It is observed for this system that as a lower amount of light material is fed to the column, the purity of the distillate and amount of overhead vapor begins to drop. The concentration detector located on Stream 3 will detect this change and will send a signal to change the set point of CV-1. Valve CV-1 will begin to open in order to increase the reflux rate. Simultaneously, the level of liquid in V-1 will start to drop. The level control will sense this change and start to close CV-3 thus reducing the flow of overhead product, Stream 3. In addition, as the amount of overhead vapor, Stream 2, flowing to E-1 also drops, the column pressure will start to decrease slightly and the cooling water flowrate, regulated via CV-2, will be reduced. At the bottom of the column, the heavier material will start to accumulate. For this system, it is observed that this will result in an increase in the liquid level in the bottom of the column and an increase in the purity of the bottom product. The level controller at the bottom of the column will sense the increase in liquid level and open valve CV-5, increasing the bottom product flowrate. Finally, as the composition detector on Stream 5 senses the increase in purity of the bottom product, it will send a signal to CV-4 to close, reducing the boil-up.

The action of the control scheme to the change in feed composition essentially regulated five streams in the following manner:

1. Stream 3$\downarrow$
2. Stream 5$\uparrow$
3. Stream 4$\uparrow$
4. Cooling water to condenser (E-1)$\downarrow$
5. Steam to reboiler (E-2)$\downarrow$

Actions 1, 2, 4, and 5 are consistent with satisfying the material and energy balances for the distillation system. Action 3 is required to adjust the purity of the products.

The regulation scheme shown in Figure 15.9 is only one of many possible systems that can be used to regulate a binary distillation column. If there is a sudden change in a process variable, this system may become unstable and careful tuning of the controllers is necessary to avoid such problems. It should also be pointed out that the material balance for the column is automatically satisfied by using the liquid levels in the column and reflux drum to control the product flowrates. The same is not true for the energy balance. For example, if valve CV-2 opens rapidly while valve CV-4 opens slowly, then a disparity in the energy balance will occur. Less vapor is produced in the reboiler than is condensed in the condenser and the pressure drops. Fortunately, the distillation column tends to be self-regulating in response to pressure. If pressure decreases, the temperature decreases because of saturation conditions in the column. This increases the driving force for heat transfer in the reboiler and decreases the driving force for heat transfer in the condenser. This increases the boil-up and decreases the condensation of overhead vapor. Therefore, this results in an increase in the system pressure that tends to correct for the disparity in the energy balance.

For distillation columns with side products, the control strategy is more complicated. In general, for each additional variable that is to be regulated or controlled, an additional control loop must be added.

It is worth pointing out that most major column upsets arise from a change in the conditions of the feed stream to the column. These changes are most often caused by process upsets occurring upstream of the distillation column. The effects of sudden changes in column feed conditions can be reduced significantly by installing a surge tank upstream of the column. This tank acts as a buffer or capacitor by storing and mixing feed of differing composition. The overall effect is to dampen the amplitude of concentration fluctuations and reduce the impact of sudden changes on the column operation. The use of surge tanks is not only restricted to distillation columns. In fact, any place where a significant inventory of liquid is stored acts to dampen changes in feed conditions to the downstream units.

A surge tank reduces the effects of sudden changes in feed conditions.

15.7.3   A More Sophisticated Control System for a Binary Distillation Column

Figure 15.10 shows a control system for a binary distillation process, adapted from Skrokov [8]. The distillation unit is similar to the one shown in Figure 15.9. Valves CV-1, CV-2, CV-3, CV-4, and CV-5 are in the same location and perform the same function in both systems.

However, for the system shown in Figure 15.10, the concentrations and flows of the input stream, exit streams, and reflux stream are measured and sent to the monitor/analyzer. This unit performs energy and material balances and uses performance relationships to evaluate new set-points for the recycle, distillate and bottom streams. It is a feedforward system that predicts and makes changes in operations based on the predictions of a detailed process model.

Modern control capabilities involving sophisticated computer hardware and software also require a high degree of chemical engineering expertise in order to develop a system that optimizes the performance of process units.

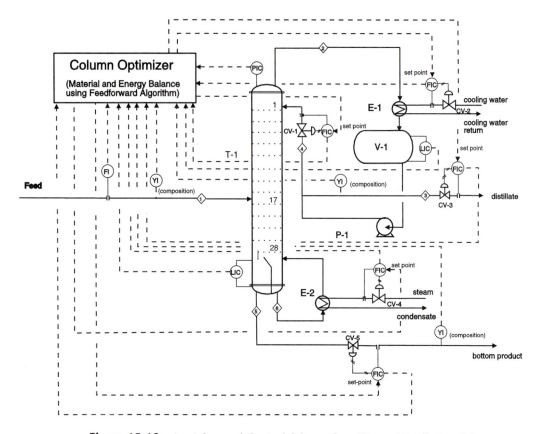

Figure 15.10   An Advanced Control Scheme for a Binary Distillation Column. (Adapted from Skrokov, M.R., *Mini- and Microcomputer Control in Industrial Processes: Handbook of Systems and Applications Strategies,* reproduced by permission of Van Nostrand Reinhold, New York, 1980.)

15.8   SUMMARY

In controlling any chemical process, the final control element is almost always a valve that regulates the flow of a process or, more commonly, a utility stream. The basic construction and operation of a control valve were reviewed. The way in which valves regulate flow and the pressure profiles in a pipe system were illustrated with several examples.

In this chapter, we reviewed the basic regulation systems that are commonly used in simple process control schemes. These systems were feedback, feedforward, cascade, and combinations of feedback and feedforward control. The logic behind each of these control strategies was explained and examples illustrating this logic were given. Methods of controlling the temperature of process streams using heat transfer media that undergo phase changes were discussed. Several strategies for controlling the temperature of process streams exchanging energy with these heat transfer media were given. Finally, two case studies were presented in which several control schemes were used to regulate variables in an exothermic reactor and a simple binary distillation column.

REFERENCES

1. Anderson, N. A., *Instrumentation for Process Measurement and Control*, 3rd ed., Chilton Co., Radnor, PA, 1980.
2. Shinskey, F. G., *Process Control Systems*, 3rd ed., McGraw-Hill, New York, 1988.
3. Stephanopolous, G., *Chemical Process Control: An Introduction to Theory and Practice*, Prentice Hall, Englewood Cliffs, NJ, 1984.
4. Coughanowr, D. R., *Process Systems Analysis and Control*, 2nd ed., McGraw-Hill, New York, 1991.
5. Smith, C. A., and A. B. Corripio, *Principles and Practice of Automatic Process Control*, Wiley, New York, 1985.
6. Seborg, D. E., T. . Edgar, and D. A. Mellichamp, *Process Dynamics and Control*, Wiley, New York, 1989.
7. Marlin, T. E., *Process Control: Designing Process and Control Systems for Dynamic Performance*, McGraw-Hill, New York, 1995.
8. Skrokov, M. R., *Mini- and Microcomputer Control in Industrial Processes: Handbook of Systems and Application Strategies,* Van Nostrand Reinhold, New York, 1980.

PROBLEMS

1. Consider an alternative reactor configuration for the cumene reactor, R-801, discussed in Section 15.7.1. The alternative is shown in Figure P15.1 and consists of a series of adiabatic catalytic packed beds through which the reactants pass. As the reactants and products pass through each bed, they are cooled by a cold shot of inert diluent, in this case, propane, before entering the next bed. The catalyst must be protected from excessive temperatures that cause it to sinter and to lose activity rapidly. For each of the control strategies given below, sketch the control loops that must be added to the system and discuss how the regulation is achieved, that is, as x increases, valve y opens, causing . . .

   a. Use the exit temperature from a given bed to adjust the amount of diluent fed to that bed.

   b. Use the exit temperature from the previous bed to adjust the amount of diluent fed to the next bed.

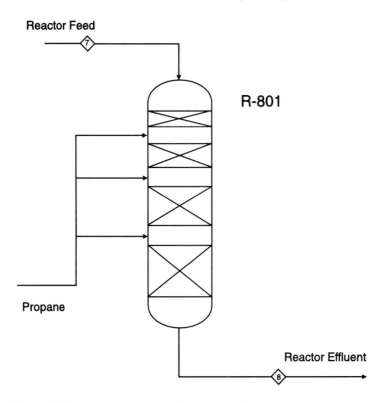

Figure P15.1   Alternative Configuration for Cumene Reactor from Problem 15.1

   c. To what types of control strategy do Parts a and b above conform? Which scheme do you think is better for this type of reactor? Explain your reasoning.

   d. Devise a control strategy that incorporates both the ideas used in Parts a and b above.

2. Consider the benzene purification column, T-101, in Figure 1.5 for the toluene hydrodealkylation process. The column feed contains non-condensables, mainly methane and a small amount of hydrogen, that must be vented from the system. The vent is taken off the top of the reflux drum, V-102. Sketch a control system in which the valve on this vent line is used to control the pressure of the column.

   Explain what happens when:

   a. The column pressure begins to drop.

   b. The amount of non-condensables fed to the column suddenly increases.

3. It has been proposed that the feed toluene to the process shown in Figure 1.5 be vaporized prior to mixing with the hydrogen. It is believed that this

scheme will allow more flexibility in the operation of the plant. You have been asked to devise a conceptual design for this vaporizer. The source of heat is from the condensation of high pressure steam. The system that you design should be capable of controlling the amount of toluene that is vaporized and the pressure at which the toluene leaves the vaporizer. Sketch a diagram showing the major pieces of equipment and major control loops.

4. In the DME process shown in Figure B.1, the reactor feed exchanges heat with the reactor effluent stream in exchanger E-202. From the diagram, it is evident that the temperature of the stream entering the reactor is controlled by regulating the amount of reactor effluent that is bypassed around the heat exchanger. For this system, answer the following questions:
   a. Explain how this system works. For example, explain what happens if the temperature of the feed into E-202 (Stream 4) were to increase or decrease.
   b. How would this system respond to fouling in the heat exchanger or a loss of activity of catalyst in the reactor?
   c. What type of control strategy is used in this example—feedback, feedforward, etc. ?
   d. Design a control system that would regulate the exit temperature of the reactor (Stream 6) rather than the inlet stream temperature.

5. For the Benzene Column (T-101) in Figure 1.5, do the following:
   a. Implement a control scheme that will regulate the purity of the benzene product (Stream 15). You may assume that the purity of this stream can be evaluated by an on-line refractive index monitor placed on Stream 15.
   b. What is the type of control system that you have designed in Part a called?

6. Consider the feed section of the phthalic anhydride process in Figure C.5. A single control valve is used to control the flow of the combined vaporized naphthalene and air streams, Stream 7. In this process it is important to regulate both the flowrate of Stream 7 and the relative amounts of air and naphthalene in Stream 7. Does the current control scheme allow this type of regulation to occur? If not, devise a control scheme that will allow both these variables to be controlled independently.

7. Describe which form of control strategy (feedforward, feedback, cascade, or a combination of these) best describes how a responsible person drives an automobile.
   What would be the consequence of using a feedback control strategy alone?

8. Figure P15.8 shows a CSTR, R-901, carrying out a liquid-phase exothermic reaction. The feed, Stream 1, comes from an upstream unit, and its flowrate is known to vary. The heat of reaction is removed by circulating a portion of the contents of the reactor through an external heat exchanger. Cooling water is used as the cooling utility in E-901.

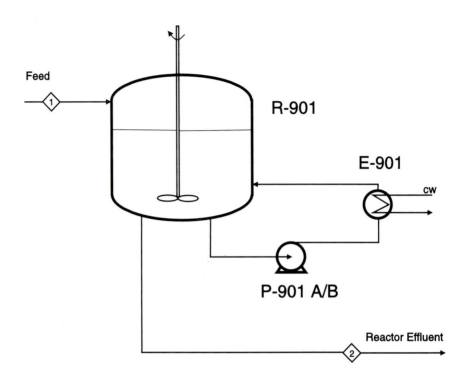

Figure P15.8   CSTR Configuration for Problem 15.8

For this reactor scheme, devise a system to implement the following control actions:

a. Regulate the temperature of the reactor.

b. Regulate the inventory of the reactor, that is, keep a constant liquid level.

Explain how your control system would compensate for a change in the flowrate of Stream 1.

What additional control scheme would be needed to maintain a constant conversion in the reactor, assuming that constant conversion can be achieved by ensuring constant residence time in the reactor?

9. Figure P15.9 shows a CSTR carrying out a liquid-phase exothermic reaction. The heat released partially vaporizes the contents of the reactor. This vapor is condensed in an external condenser, and the condensate is returned to the reactor. The reactor operates at the boiling point (T and P) of the mixture in the CSTR. The vapor formed consists of reactants and products.

For this system, devise a system to regulate the following variables:

a. temperature in reactor

b. residence time in the reactor

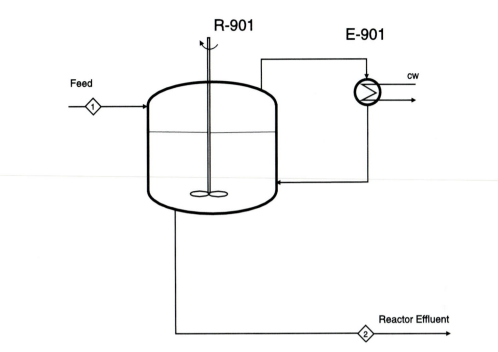

Figure P15.9   CSTR Configuration for Problem 15.9

Explain how your system would respond to the following changes:
c. increase in flowrate of feed
d. fouling of the tubes (on cw side) of the condenser.

10. Consider the feed section of the toluene process given in Figure 1.5. The flow of fresh toluene, Stream 1, is regulated by monitoring the level in the Toluene Storage Tank, TK-101. If the level is seen to drop, then the valve on Stream 1 opens and vice versa. This is an example of a feedback material balance control loop. For this system, implement a feedforward control scheme in which the flow of the recycle, Stream 11, is measured and used to control the flow of Stream 1. Do you foresee any potential problems with this control strategy?

16

Process Troubleshooting and Debottlenecking

Imagine that you are responsible for a chemical process unit. The pressure in a chemical reactor begins to increase. You are concerned about material failure and explosion. What do you do? For a case such as this with potential catastrophic consequences, it may be necessary to shut the process down. However, process shutdown and start-up are very costly, and if a safe alternative were available, you would certainly want to consider it as an option. In another scenario, what would you do if it had been observed that the purity of product from your unit had been decreasing continuously for several days and customers had begun to complain of poor product quality and have threatened to cancel lucrative contracts?

The situations described above may be classified as process troubleshooting problems. Once a plant is built and operating, it is anticipated that it will operate for a number of years (10–30 years). During this time, there will be times when the plant displays unusual behavior. This unusual behavior may represent a problem or a symptom of a problem that has not yet become apparent. The procedure for identifying the root cause of unusual behavior is part of troubleshooting. The other part is to provide guidance into what action should be taken to correct the problem, or, in the case of a symptom that has not yet resulted in a problem, to prevent a problem from developing. Problems that affect process performance represent financial losses and potential safety hazards, so these problems must be quickly identified and resolved. The key to smooth plant operations is preventive action based on correct diagnosis of early symptoms. Troubleshooting problems associated with process start-up are beyond the scope of this text.

Even during a period of successful operation, the process does not operate at a steady state. Distillation units operate differently in summer than in winter (between night and day) as a result of internal reflux changes resulting from heat losses or gains from the tower. Feed materials fluctuate, the temperature of cooling water changes, catalyst decays, heat exchangers foul, and so on. The control system responds to these changes and alters utility flows to maintain process streams close to normal operating conditions.

The key to solving troubleshooting problems is to make use of the information regarding the process taken for periods of successful operation. Based on operating experience, we learn the range over which changes can take place without a significant effect on the performance of the process. Consequently, there is no single base case to represent process behavior as there was in Chapter 12. When comparing current operations to normal operations, we check to determine if current operation lies within the range of normal operations. The range of normal process operation provides the base case.

Three steps can be identified to troubleshoot a process:

1. Treat the symptoms: In this situation, the observed problem is addressed without investigation of the root cause. If the reactor pressure is increasing, find a way to relieve the pressure. This is a short-term solution. Since the root cause of the pressure increase has not been identified and addressed, the pressure may increase again. However, the immediate problem (an explosion) has been avoided, and there is now time to seek the root cause of the problem.

2. Identify the cause of the problem: Eventually, the cause of the problem should be diagnosed. This is particularly true if the problem recurs or if it is safety-related. Since this may take time, the symptoms must continue to be treated.

3. Fix the problem: Ultimately, the problem should be fixed.

Process troubleshooting involves solving open-ended problems for which there are likely to be several possible solutions. It is necessary for the engineer faced with such a problem to consider many identifiable solutions. Failure to consider a sufficient number of possible solutions may result in missing the actual solution.

Now, consider a different situation. It is necessary to determine how much scale up is possible for the process for which you are responsible. You determine that one process unit can only be scaled up by 10%, whereas all other process units can be scaled up by at least 15%. The process unit that can only be scaled up by 10% is called a bottleneck. Elimination of this bottleneck is called debottlenecking and involves determining how to remove obstacles limiting process changes.

To put troubleshooting and debottlenecking problems in context, consider the input/output model shown in Figure 16.1. This model was first introduced in Chapter 10 and was used to define the design and performance problems. The input/output model can also be used to define troubleshooting and debottlenecking problems. Troubleshooting problems involve identification and correction of the change in inputs and/or the process responsible for observed changes

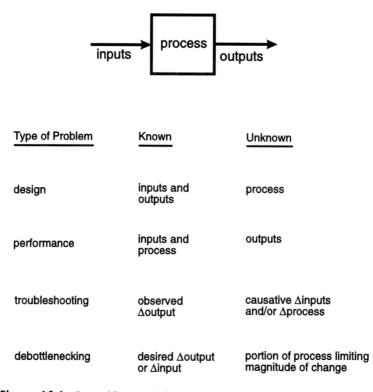

Figure 16.1   Input/Output Relationships for Various Types of Problems

in outputs. Debottlenecking problems involve identification and modification of the portion of a process limiting the ability to change inputs or outputs.

In this chapter, a recommended methodology for attacking process troubleshooting problems is presented. Then, five case studies of increasing complexity are presented that will serve to enhance the reader's skills in attacking troubleshooting problems. Finally, an example of a debottlenecking problem is presented.

16.1   RECOMMENDED METHODOLOGY

16.1.1   Elements of Problem-Solving Strategies

Three elements of successful problem-solving strategies are recommended here for attacking process troubleshooting problems (including debottlenecking problems). This is not meant to be an exhaustive list of strategies; it is simply some of the strategies that can be used. The three strategies discussed here are:

1. Brainstorming
2. Using known or observed data plus your understanding of equipment behavior
3. Considering the unexpected

Each is considered below separately. They can also be combined into a general methodology for solving open-ended problems.

Brainstorming involves generating an extensive list of possible ideas. This need not be a formal process; it can be done informally and rapidly. In fact, the natural response to a troubleshooting situation is to think immediately and rapidly of several possible causes. If the product is not meeting specifications, think of all of the reasons that can be considered as potential causes, no matter how remote the possibility. Then, brainstorm possible solutions. The main rule of brainstorming is that there are no bad ideas. The goal is to generate as many ideas as is possible. After brainstorming, only then is it time to evaluate all items critically to generate the most likely causes and solutions. While brainstorming is usually a group activity, it often must be done individually. For a large, long-term problem, time can be taken to brainstorm in a group. For an everyday problem, especially an emergency situation, one should train oneself to brainstorm automatically. Brainstorming is a component of most problem-solving strategies.

> **When brainstorming a troubleshooting problem, consider all ideas, no matter how unusual they may seem.**

When troubleshooting a chemical process, an **understanding of equipment behavior** should be used to narrow the list of possibilities. For example, in a staged separation using a mass separating agent, the Kremser equation (see Section 12.3.1) quantifies the relationship between process variables. The most important of these relationships for a variety of unit operations were summarized in Table 11.1. Use of these relationships will be illustrated in the case studies presented later in this chapter.

When there is a problem with process operation, the cause of the problem must be identified. The problem may or may not be located at the unit where poor operation is observed. For example, the output from the separator may be off-spec due to lower quality product exiting the reactor rather than due to poor separator operation. In troubleshooting plant problems, a vast amount of data exists that can be used to help identify the problem. These data would likely include current and historical operating conditions. In addition, the plant P&IDs would show where to look for additional current operating data. Any methodology for troubleshooting must consider this information, and any solution must be consistent with the operating data. Often, solving the problem is facilitated by selection of the data that will lead to identification of the problem.

While a knowledge of equipment behavior is essential to solving troubleshooting problems, it is important that all alternatives, no matter how **unexpected**, be considered. If the pressure drop in a tray tower is increasing, a knowledge of equipment function suggests that there may be loading or flooding, so an increased liquid or vapor flowrate is a possible cause. However, was the possibility that someone left a toolbox in the downcomer during a recent maintenance shut-down also considered? The lesson here is to expect the unexpected! Consider all possibilities no matter how remote they may seem.

Given the open-ended nature of troubleshooting problems, their solution may best be attacked by creative problem solving strategies. One such strategy, presented by Fogler and LeBlanc [1], is discussed here in the context of process troubleshooting. (Another similar strategy is presented in Chapter 20.) Their problem-solving strategy involves five steps:

1. Define
2. Generate
3. Decide
4. Implement
5. Evaluate

First, the correct problem must be defined. If the problem is incorrectly defined, it is likely that an incorrect solution will be found. If the product is not meeting specifications, this is the problem. However, if after further investigation, it is found that the stream leaving the reactor is not at design conditions while the reactor feed is at design conditions, it may be necessary to redefine the problem to be incorrect reactor performance.

Once the problem is defined, ideas must be generated. This is identical to brainstorming. It is important to generate as many ideas as possible. It is poor problem-solving strategy to focus on one possible solution or to assume that there is only one possible solution. For process troubleshooting, ideas may need to be generated both for the cause of the problem and for remedies to the problem.

Once ideas have been generated, the next step is to decide how to proceed. This is where knowledge of equipment can be used to select the most likely item(s) from the brainstorming list to implement first. The next step is to implement the chosen solution.

Once the chosen solution method is implemented, it is necessary to evaluate the chosen solution. Is it working? If not, why not? Should another solution be implemented? If several solutions have been attempted, none of which appear to be solving the problem, this may be the time to think about whether there is an "unexpected" solution that can solve the problem.

16.1.2 Application to Troubleshooting Problems

A trouble shooting strategy is given in Table 16.1. It involves five steps. This sequence of steps is shown to parallel the problem solving strategy of Fogler and LeBlanc discussed in the previous section.

Phase 1. Screen the whole process for the most common causes of problems or symptoms of problems in the process. This might involve brainstorming done informally as part of your thought process.

It is important to **Define** the correct problem, or to determine if one even exists. One common situation is false identification of a problem or symptom. An instrument could have been read incorrectly or could be broken, the analytical analysis (on line or in lab) may not be correct, reagents may have been prepared incorrectly, and so on. In this situation, there is no problem with the process, it is a false indication of a problem. Never accept the initial problem identification without verification.

Once it has been determined that a problem really exists, the suspects should be screened. This is the **Generate** step. A common cause of problems results from changes in process inputs. The adage "garbage in, garbage out" is universal and applies to chemical processes. Component flows into the process must be verified and compared to those for normal operations. If a problem appears as a result of process inputs, go to Phase 3. If a problem is known to involve a single unit (which may only be true for academic problems), consider possible unit mal-

Table 16.1 Strategy for Troubleshooting Existing Plants

Phase 1	Check out primary suspects.
	a. Verify the identified problem or symptom.
	b. Check input to the process.
	c. If only one unit is involved, check operating conditions of unit.
	d. Check for fully open or closed control valves.
Phase 2	Identify the unit operation producing the problem or symptom.
	a. System size is reduced systematically until the unit operation that is the source of the problem or symptom is determined.
	b. Inputs to each system are checked.
Phase 3	Perform a detailed analysis of unit operation uncovered in Phase 2 to determine and to verify the root cause of the problem or symptom.
Phase 4	Report your diagnosis of the root cause of the problem and recommend action to remove the problem or symptom.
Phase 5	Report significant observations uncovered during the analysis that may be important to your organization.

functions (some are listed under Phase 2, below). In this case, you should also go directly to Phase 3.

A third common cause results from limitations of the control systems on the utility streams used to maintain the temperature and pressure of process streams. If any control valves are found to be fully open or fully closed, there is a high probability that the desired control is not being achieved. All of the utility control valves are not normally shown on the PFD, and it is necessary to review the P&ID. All control valves should be checked. This information is used as input in later phases to identify the cause of the problem observed.

Phase 2.   Locate unit operation that is producing problem or symptom. This is part of the **Decide** step. The process is divided into subsections. If there is no obvious choice for selecting subsections, the process sections identified in the block flow diagram represent a reasonable starting point. Each system analyzed contains the stream identified as having a problem or symptom. Analyze system inputs. This identifies the subsection containing the cause of the problem. The subsection size is reduced and the inputs are again analyzed. This is continued until the unit operation producing the problem is identified. Then, the operation of the identified unit should be checked. Ask key questions about each unit. Is there evidence of heat exchanger fouling? Is the reflux ratio on the distillation column within its normal range? Are the temperature and pressure of the reactor at normal conditions?

Phase 3.   Determine the root cause of the problem or symptom by a detailed analysis of the unit operation identified in Phases 1 or 2. This is the remainder of the **Decide** step. Normal operation is used as a base case. This is the first place where the utility flows are analyzed. This involves using heuristics (Chapter 9), operating conditions of special concern (Chapter 8), calculation tools, and other material presented in Chapters 10–12.

Phase 4.   Present all available evidence that establishes the root cause identified in Phase 3 as valid. Also, present any evidence that may not support the argument. Recommend action to be taken to correct the problem or treat the symptom. This is the **Implement** step.

Phase 5.   Present an evaluation of any significant observations that resulted from your analysis that could impact this or other processes within your company. This is the **Evaluate** step. If you identify process improvements or potential for future problems, you are acting professionally. This is especially true if the suggestions are related to the environment or safety of personnel.

In summary, the cause of a change in output originates with a change in input and/or a change in process operation. Some possibilities are illustrated in Figure 16.2. This list is not meant to be exhaustive. It illustrates some possible causes for changes in output. The strategy discussed above is illustrated in the problems which follow.

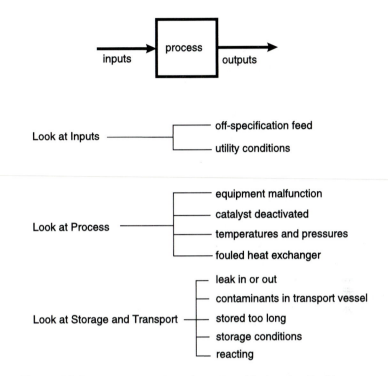

Figure 16.2   Suggestions for Solving Troubleshooting Problems

16.2 TROUBLESHOOTING INDIVIDUAL UNITS

The first two case studies presented involve troubleshooting individual pieces of equipment. While real troubleshooting situations usually involve an entire process, the necessary skills can be developed on simpler problems such as the two presented in this section.

16.2.1 Troubleshooting a Packed Bed Absorber

The first troubleshooting problem involves a packed bed absorber. The absorber has been designed to remove a contaminant from an air purge stream, and has been operating for some time as designed. Then, it is observed that the outlet air contains more contaminant than it should. A similar problem involving a tray absorber is given at the end of chapter.

A packed absorber (Figure 16.3) has been designed to reduce the acetone concentration in 40 kmol/h of air from a mole fraction of 0.02 to 0.001. Acetone is absorbed into pure water at 20 kmol/h. Acetone is recovered from the effluent liquid, and the water, which is as-

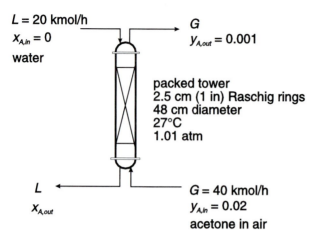

L = 20 kmol/h

$X_{A,in} = 0$

water

G

$y_{A,out} = 0.001$

packed tower
2.5 cm (1 in) Raschig rings
48 cm diameter
27°C
1.01 atm

L

$X_{A,out}$

G = 40 kmol/h

$y_{A,in} = 0.02$

acetone in air

Figure 16.3   Packed Absorber for Troubleshooting Case Study

sumed pure, is recycled to the absorption unit. After a period of successful operation, it is observed that the exit acetone mole fraction in air is now 0.002.

The column is packed with 2.5 cm (1 in) Raschig rings, and has a 48 cm diameter, which was obtained by designing for 75% of flooding. The column is assumed to operate isothermally at 27°C and the nominal pressure is 1 atm. Raoult's law is assumed, and the partition coefficient for acetone, $m = y/x = P^*/P$, where $\ln P^* = 10.92 - 3598/T(K)$ has been determined from tabulated data [2]. At 27°C and 1.01 atm, $m = 0.337$.

This problem, which involves dilute solutions, can be analyzed using the Colburn graph, Figure 12.15. On this graph, the interrelationship between the number of transfer units, N_{toG}, the absorption factor, A ($A = L/mG$), and the mole fraction is defined. The base case point can be located. The y-axis is at a value of 0.05, and A = 1.48. This gives $N_{toG} = 6.2$. This is shown as point "a" on Figure 16.4.

The first step is to verify that the acetone concentration has indeed increased. This might involve having an operator or technician make flow and concentration measurements on the effluent stream in question. We will assume that the increased concentration at the normal flowrate has been verified. The next step in Phase 1 is to check process inputs and process operation (since only one unit is known to be involved) for potential causes for the observed change in output concentration. Changes in input are evaluated first. If the inputs are not found to be the cause of the problem, process operation is then investigated.

Example 16.1

Generate a list of causes for the observed change in absorber output. First, examine input changes, then examine process operation changes.

The knowledge of packed bed absorber performance illustrated on the Colburn graph is a good starting point. Any or all of these parameters could be different from design conditions.

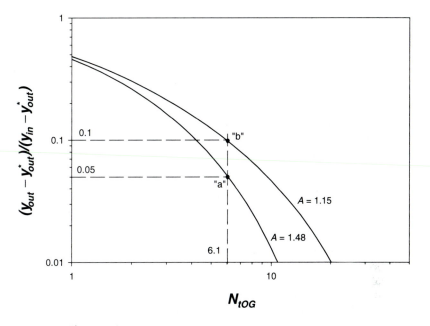

Figure 16.4   Use of Colburn Graph to Solve Case Study

Potential input problems include:

1. Increased flowrate of gas to be treated: A disturbance in A is a possible cause. Intuitively, if it is the gas rate, G, that has been disturbed, an increased gas rate would be the cause of an increased acetone concentration in the exit gas stream. This would result in a decrease in A.

2. Decreased flowrate of water: This would also result in a decrease in A.

3. The water does not enter acetone free: If the entering water contains acetone, intuitively, less acetone can be absorbed.

4. There is more acetone in the feed: If the mole fraction of acetone in the feed were increased, the same fractional removal of acetone from air would result in a higher acetone mole fraction in the exit air stream.

These four items would be checked first to determine if they are the cause of the faulty absorber performance. If they were not found to be the cause, the process operation would be investigated. Potential process operation problems include:

5. A decrease in column pressure: Intuitively, a decrease in column pressure favors the vapor phase. Mathematically, since $A = L/mG$, and $m = P^*/P$, a decrease in P increases m, causing a decrease in A.

6. An increase in column temperature: Intuitively, an increase in column temperature favors the vapor phase. Mathematically, since P^* increases with temperature, so does m, causing a decrease in A.

7. There is channeling in the packed bed: If there were channeling in the packed bed, all of the available (and designed for) mass transfer area would not be used. This could cause faulty absorber performance.

Example 16.1 shows seven possible causes ("suspects") for the observed change in output concentration. To reiterate, the process inputs would be checked. If they were not found to be the problem, then process operation would be investigated. The next step would be to check for fully open or closed control valves. In this problem, it is assumed that none are involved. Phase 2 is also not applicable since there is only one unit operation involved in this problem.

Phase 3 involves a detailed analysis of the "suspects." In Example 16.1, Items 1, 2, 5, and 6 can be represented on the Colburn graph. For these items, diagnosis is that the absorption factor has decreased, which moves the operating point for the column vertically at constant $N_{toG} = 6.2$ to point "b." The new absorption factor is 1.15. The problem could be in any (or all) of the parameters of the absorption factor. L could have decreased to 17,391 mol/h or G could have increased to 46,000 mol/h. Alternatively, the value of m could have changed to 0.388, meaning that the temperature of the column increased to about 30°C (30.5°C), or that the column pressure decreased to 0.87 atm. However, since the air stream discharges to the atmosphere, a decrease in column pressure below 1 atm is not possible.

There is an alternative diagnosis, however, which is Item 3 in Example 16.1. The operating point can remain fixed in the original position (point "a"), but the outlet acetone concentration in air increased due to the presence of acetone in the water fed to the column. This makes the second term in the numerator and denominator of the y-axis non-zero. Solution for the inlet acetone concentration in water yields a mole fraction of 0.00312.

Thus far, five possible causes for the observed increase in outlet acetone concentration in air have been identified. There are certainly additional possible causes, some of which are associated with equipment operation such as liquid distribution, channeling, and fouling, which could also contribute the observed performance decrease. One of these is identified as Item 6 in Example 16.1.

At this point, there is not enough information given to complete Phase 3 to identify the root cause of the problem. The next step would be to measure the input flows and concentrations, column temperature, column pressure, and column pressure drop (a measure of channeling). This should allow identification and verification of the root cause of the problem.

If the root cause of the reduced performance of the absorber is understood, then possible methods of compensation are straight forward (Phase 4). Here, it is assumed that compensation cannot be achieved by altering the cause of the disturbance, that is, if the cause is an increased gas rate, then the gas rate cannot be lowered. However, in the context of a chemical process, the possibility of reversing the disturbance should be investigated.

If the gas rate is too high, the liquid rate can be increased to compensate. However, flooding could be a problem, especially if both gas and liquid rates are increased. The Colburn graph does not account for flooding, which is specific to a given packed column.

If the liquid rate is too low, it is unlikely that the gas rate can be decreased without scaling down the entire process. A better choice might be to decrease the temperature of the absorber to 22°C, increase the pressure in the absorber to 1.15 atm, or a combination of changes in temperature and pressure, in order to make the absorption equilibrium more favorable, and bring the absorption factor back to a value of 1.48.

If a temperature increase is the problem and altering flowrates is not desirable due to flooding considerations, one possible compensation is to alter the pressure. Another might be to decrease the temperature of the water used to remove the acetone from the air. Increasing the liquid rate moves the column towards flooding, but a small increase should not be a serious problem.

Finally, if the cause of the disturbance is acetone in the water, compensation can be accomplished by decreasing the temperature, increasing the liquid rate, increasing the pressure. For this situation, the stripping column used to remove scrubbed acetone from the water stream should also be investigated, since the cause of the faulty absorber performance may lie in an adjacent piece of equipment.

In general, adjusting the temperature or pressure is probably the best method of compensation, since flooding is not an issue. Of course, there can be multiple causes of the disturbance and compensation can be achieved by adjusting two variables by smaller amounts rather than by adjusting only one variable.

As an example of Phase 5, suppose that the root cause was determined to be an increase in acetone content of the scrubber liquor (water). It is necessary for you to notify anyone else using the same water supply of the acetone contamination. Similarly, suppose that channeling in the absorber was identified as the problem. It is necessary to report the operating conditions that caused the channeling to other parts of the company using the same packing material.

To review, this relatively simple problem illustrates how there can be multiple possible causes and multiple solutions to a troubleshooting problem. The strategy presented here for solving troubleshooting problems was illustrated.

16.2.2   Troubleshooting the Cumene Process Feed Section

This problem deals with the feed section to the cumene process in Appendix C, Figure C.8. This problem is actually part of the problem presented in Appendix C, and the process flowsheet along with calculations associated with the pumps can be found there. The problem is restated here.

A problem has recently arisen regarding the feed pumps to the cumene process. A maintenance check showed that P-802 (propylene feed pump) needed a new bearing and a new one was installed. Premature bearing failure in this pump, often associated with cavitation, has occurred several times. The latest problem occurred during a recent warm spell, when the ambient temperature reached 110°F. The same maintenance check showed that P-801 (benzene feed pump) was fine. The ambient temperature has now returned to an average of 70°F, and both pumps seem to be working fine. Suggest a diagnosis and a method for compensation for the problem with P-802. In the process description in Appendix C, it is stated that propylene is stored in a tank as a saturated vapor/liquid mixture with liquid drawn from the tank as feed, and that liquid benzene is stored in a tank (most likely with an inert vapor blanket such as nitrogen) at atmospheric pressure.

In this case, the problem is immediately verified; the bearing had to be replaced. A check of the pump input might identify a contaminant that could have caused the bearing to deteriorate. We will assume that no such contaminant was found. Since the unit operation producing the problem has already been identified, we skip to Phase 3 to identify the root cause of the problem.

An intuitive understanding of pump operation suggests that the primary suspect is cavitation of P-802. This is consistent with the need for replacement of a bearing since cavitation can damage the internals of a pump. This may also be consistent with the recent warm spell, since the available net positive suction head, $NPSH_A$, often decreases at higher temperatures due to increasing vapor pressure. Recalling the discussion in Section 12.2.4,

$$NPSH_A = P_{inlet} - P^* \tag{16.1}$$

Since the vapor pressure P^* increases with increasing temperature, it seems logical that the available $NPSH$ decreased during the warm spell causing cavitation that damaged the pump bearing.

Example 16.2

Calculate the $NPSH_A$ for the propylene feed pump under normal ambient conditions (70°F) and for the warm spell (110°F).

Eq. (16.1) can be rewritten as in Section 12.2.4, with an expression for P_{inlet} substituted

$$NPSH_A = P_{tank} + \Delta P_{static} + \Delta P_{friction} - P^* = P_{tank} + \rho g h - \frac{2 \rho f L_{eq} u^2}{D} - P^* \tag{E16.1}$$

For the case of propylene being stored as a vapor/liquid mixture, the pressure in the tank is equal to the vapor pressure. As a consequence, $NPSH_A$ does not change with temperature since $P_{tank} = P^*$. Therefore, the $NPSH_A = 7$ feet of liquid at the level alarm low (LAL—the tank level at which an alarm goes off warning of too low of a liquid level in the tank) in the tank at all temperatures, as per the calculations shown in Appendix C.

The lesson learned from Example 16.2 is that while a belief in the understanding of how equipment works can help you focus in on a solution to a troubleshooting problem, the tendency to focus on the first solution or on only one solution can lead to an erroneous solution.

> **Resist the temptation to focus on only one solution or the first solution that comes to mind.**

The reason why the pump malfunctioned is still not clear. If cavitation were the reason, the $NPSH_A$ expression should be investigated further. It is now known, for this situation, that

$$NPSH_A = \Delta P_{static} + \Delta P_{friction} \tag{16.2}$$

so we are seeking reasons why this expression may have decreased in value. There may be reasons for the pump malfunction not associated with the $NPSH_A$. Therefore, five additional possibilities that should be analyzed (Phase 3) for causing the pump malfunction are:

1. The static pressure has decreased. Perhaps the level in the tank has fallen below the low alarm level and the alarm has failed. Therefore, these items should be checked.

2. The frictional pressure has increased. Perhaps the flowrate from the tank has increased, causing an increase in the frictional losses. The flowrate monitors and the settings on the control valve after the pump should be checked. This scenario would have additional consequences, since an increased propylene flowrate would have an effect on reactor conversion and possibly on product production rate and purity. Therefore, the flowrate and purity of the reactor effluent should also be checked.

3. The bearing in the pump wore out with age. Just because pump cavitation often causes damage to pump internals such as bearings does not mean that it was the cause of this pump's malfunction. Given the analysis above, it is possible, indeed likely, that the pump malfunction observed here was simply caused by a worn bearing.

4. Mechanical problems in the pump. The bearing could be wearing out prematurely due to poor shaft alignment within the pump.

5. Manufacturing defect within the pump. If the pump shaft or other internals were defective (incorrect size, for example), the bearing might wear out prematurely.

The recommended action (Phase 4) will depend upon which (if any) of the five causes of bearing failure listed above is identified to be the problem. If problems are found with the pump (Items 4 and/or 5), it would be appropriate for you to notify others in your company using the same pump or similar pumps from the same manufacturer of the problem you have had with your pump.

16.3 TROUBLESHOOTING MULTIPLE UNITS

In this section, two troubleshooting case studies are presented that involve multiple unit operations. One of the lessons of these case studies is that the symptoms of a problem are not necessarily observed at the source of the problem. In these two case studies, only one solution is discussed. Generation of alternative possibilities is the subject of problems at the end of the chapter.

16.3.1 Troubleshooting Off-Specification Acrylic Acid Product

This problem concerns an acrylic acid production process similar to the one in Appendix B, Figure B.2. The process flow diagram, stream flows, and equipment specifications are presented there.

At another acrylic acid plant owned by your company, process shutdown for modifications and improvements has recently been completed, and the process has been started up once again. Customers have begun to complain that the acrylic acid product does not appear to be meeting specifications. They have observed that the acrylic acid has a yellowish color, which is different from the clear liquid they had previously received. Their tests also found that the viscosity of the acrylic acid has increased.

The following process modifications were completed during the recent shutdown:

1. A new catalyst that is supposed to minimize side-reactions was installed in the reactor. The reactor specifications were not changed.

2. A new solvent is now being used in the extraction unit. It is less expensive than the previous solvent, and the performance of the extraction unit is supposed to be unchanged. No modifications were made to the extraction equipment.

3. As a cost-cutting measure, refrigerated water (entering at 10°C) has been replaced by cooling water (entering at 30°C) in the acrylic acid purification column. The column has 25 actual trays with 2.25 in weirs, a total condenser, and a partial reboiler.

The following operating restrictions are also known. For the new catalyst, the operating range is between 250°C and 350°C and between 1 bar and 5 bar. Once the acrylic acid has been produced and condensed into the liquid phase, the temperature is to be maintained below 90°C to avoid polymerization of the acrylic acid.

The first step is to verify the problem. Let us assume that you had a technician take a sample of acrylic acid product and the yellowish color and increased viscosity were both verified. This means that the problem is not in the shipping and storage steps, but is within the plant. Therefore, the problem is yours, not the customer's. Further, let us assume that all process inputs have been checked and found to be within normal operating conditions, and that a check has found no control valves to be fully open or closed.

There are several possible causes for the off-specification acrylic acid product. The most likely causes are due to changes that may have occurred during shutdown. Perhaps the new catalyst is not performing as designed. Perhaps the new solvent is contaminating the product. You would want to check the reactor effluent and the extractor effluent streams for contamination. Suppose that you have done this and have found everything to be within normal conditions. Therefore, the most likely cause involves the temperature in the acrylic acid distillation column. In this column, acrylic acid and acetic acid are separated. It will be assumed that this is a simple binary distillation.

There must have been a reason for using refrigerated water in the distillation column prior to the recent shutdown. This can be understood using an analysis similar to the one in Section 13.2. From tabulated data [2], the following vapor pressure expressions can be obtained for acrylic acid and acetic acid:

$$\text{acrylic acid} \quad \ln P^* \text{ (mm Hg)} = 19.776 - \frac{5450.06}{T(\text{K})} \qquad (16.3)$$

$$\text{acetic acid} \quad \ln P^* \text{ (mm Hg)} = 18.829 - \frac{4786.41}{T(\text{K})} \qquad (16.4)$$

Since acrylic acid is the heavier component, the bottom of the column must remain below 90°C to avoid undesired acrylic acid polymerization. From Eq. (16.3), assuming that pure, saturated acrylic acid leaves the bottom of the column, the pressure at the bottom of the column is 118 mm Hg, which is 5.26 ft of liquid assuming that liquid acrylic acid and liquid acetic acid have the same density as liquid water. The pressure drop per tray will be approximated by the height of liquid on the tray. The height of liquid on each tray will be approximated by the weir height. Since there are 25 trays, the pressure drop in the column is

$$25 \text{ trays}[(2.25/12)\text{ft liquid/tray}] = 4.69 \text{ ft liquid}$$

Therefore, the pressure at the top of the column is 5.26 − 4.69 = 0.57 ft liquid = 12.8 mm Hg. From Eq. (16.4), assuming pure, saturated acetic acid at the top of the column, $T = 21°C$. Since cooling water enters at 30°C, it is not possible to condense acetic acid at 21°C with cooling water. This is why refrigerated water was used in the original design.

The above discussion suggests that the switch to cooling water could be one reason for the off-specification acrylic acid product. With cooling water available at 30°C, it is not possible for the top of the column to be below 30°C, which places a lower bound on the top pressure of the column. Since the pressure drop in the column is fixed by the weir height on the trays, there is also a lower bound on bottom pressure in the column, which places a lower bound on the bottom temperature, which is above 90°C, thereby promoting polymerization. Apparently, this column has no control room pressure reading (perhaps the instrument is out of order). Therefore, an operator would be sent to measure the pressure at the top of the column.

There is one simple remedy. Refrigerated water should be used in the condenser. If this were not desirable for the long-term, one could consider the incremental economics of modifying the trays to have lower weirs, which could only happen at the next plant shutdown. It is also necessary for you to report this problem to any other plants within your company making acrylic acid so that they can avoid (or correct) problems arising from the use of cooling water instead of refrigerated water.

It is observed that rough calculations involving certain reasonable, simplifying assumptions were used to obtain an approximate result very quickly. It is neither necessary nor desirable to do detailed calculations when screening alternatives in a troubleshooting problem.

> **When screening alternatives, rough calculations using reasonable approximations are more useful than detailed simulations.**

The approximations that were made to facilitate a rapid calculation for the distillation column were that the top and bottom products are pure and that the height of liquid on the trays equaled the weir height. While these are not exactly true, detailed calculations would show that these are good approximations.

16.3.2   Troubleshooting Steam Release in Cumene Reactor

This problem involves the reactor in a cumene process similar to the one in Appendix C, Figure C.8. The process flow diagram is presented in Appendix C.

Our company has been testing a new cumene catalyst at a facility producing the identical amount of cumene to the one in Appendix C. The new catalyst completely suppresses the undesired DIPB formation reaction. However, the reaction rate for the desired reaction is lower. Therefore, when the new catalyst is used in the existing reactor, a single-pass conversion of only 50% is obtained. This new catalyst, which is less expensive than the previous catalyst, is known to have a higher initial activity that decays rapidly to constant activity. During the initial activity period, a 33% increase in cumene production has been observed. The operating parameters of the benzene distillation column have been altered, the recycle benzene stream has been increased, the DIPB column has been taken off-line, and the plant has been producing cumene successfully.

The reactor for cumene production, which is shown in Figure 16.5, is of the shell-and-tube design, with catalyst in the tubes and boiler feed water vaporized to form high-pressure steam in the shell. The pipe to the steam header is 2 in schedule 40, and contains

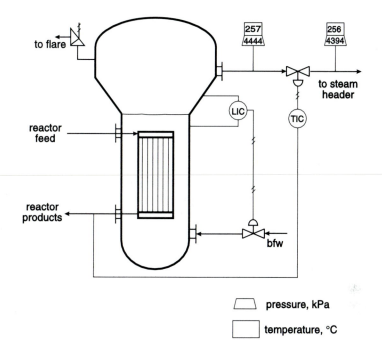

Figure 16.5   Cumene Reactor in Case Study

32.5 m of equivalent pipe length after the regulating valve. Under normal operating conditions, the pressure drop across the regulating valve is 50 kPa. There is a pressure-relief valve on the shell side of the reactor rated at 4500 kPa. The tubes are completely submerged in the boiler feed water. As part of the test of the new catalyst, the plant has been shut down briefly every three months for the past year, and the catalyst has been replaced so that the spent catalyst could be studied.

After start-up subsequent to each shutdown, the pressure-relief valve on the shell side of the reactor has opened periodically for a few days. Since only steam was released, and it is not in a part of the plant where anyone could be harmed, the problem was attributed to start-up transients and was ignored. A recent safety audit has resulted in the suggestion that the cause of this problem be identified and corrected.

The problem is verified by the observation of a steam release. Let us assume that you have checked all process inputs, the operation of all other units, and control valves and you have decided that the problem must be at the reactor. A check of the reactor feed and the boiler feed water input reveals normal conditions. Therefore, the problem must be with reactor operation, which points to the higher activity catalyst.

First, it is necessary to calculate some operating parameters for the design conditions. Under normal operating conditions, from Appendix C, the heat gen-

erated in the reactor is 9,800 MJ/h. The amount of steam formed can be calculated

$$Q = \dot{m}_{stm}[C_p(T_{stm} - T_{bfw}) + \lambda_{stm}] \tag{16.5}$$

Solving for the mass flow of steam yields

$$\dot{m}_{stm} = \frac{9{,}800{,}000 \text{ kJ/h}}{[4.35 \text{ kJ/kg}°\text{C}(256 - 90°\text{C}) + 1700 \text{ kJ/kg}]} = 4046 \text{ kg/h}$$

Here, the heat capacity of water is taken at an average temperature of 173°C, and
the temperature of steam formed in the shell corresponds to the pressure in the
shell calculated below. Since high-pressure steam is made, the pressure of steam
downstream of the reactor is that of high-pressure steam, 4237 kPa. Under design
conditions, the pressure of steam in the reactor exceeds this value to account for
the pressure drop in the pipe leading to the steam header and across the regulating valve. The density of the steam varies with the pressure. It will be assumed
that the average steam conditions between the reactor shell and the steam header
are 255°C and 4300 kPa. At these conditions, the density of steam is about
17.6 kg/m³. For 2 in schedule 40 pipe ($D = 0.05250$ m), at the given mass flow and
density of steam, the velocity in the pipe is 28.3 m/s, so the Reynolds number is

$$\text{Re} = \frac{(0.0525 \text{ m})(28.3 \text{ m/s})(17.6 \text{ kg/m}^3)}{1.88 \times 10^{-5} \text{ kg/m s}} = 1.39 \times 10^6$$

For commercial steel pipe, $e/d = 4.6 \times 10^{-5}$ m/0.0525 m = 0.0009. From Figure 11.1,
the friction factor $f = 0.009$. Therefore, the pressure drop in the pipe leading to the
steam header is

$$\Delta P = \frac{2f\rho L_{eq}u^2}{D} = \frac{2(0.009)(17.6 \text{ kg/m}^3)(32.5 \text{ m})(28.3 \text{ m/s})^2}{0.0525 \text{ m}} = 157 \text{ kPa}$$

The pressure in the reactor shell, under normal conditions, is the pressure at the
steam header plus the pressure drop in the pipe plus the pressure drop across the
valve, which is

$$P_{rxr} = 4237 + 157 + 50 = 4444 \text{ kPa}$$

The situation immediately after catalyst replacement can be analyzed using base
case ratios. If the activity of the new catalyst is higher than the old catalyst, then
the reaction rate is increased. Since the only reaction involved is for cumene production, 33% more steam is produced. Therefore, the steam velocity in the pipe
leading to the steam header is increased by 33%. Since the flow is fully turbulent,
the friction factor remains constant, and, assuming unchanged density,

$$\frac{\Delta P_2}{\Delta P_1} = \left(\frac{v_2}{v_1}\right)^2 = 1.33^2 = 1.77 \tag{16.6}$$

where subscript 2 refers to new conditions and subscript 1 refers to design conditions. Therefore,

$$\Delta P_2 = 157 \text{ kPa}(1.77) = 278 \text{ kPa}$$

If it is assumed that the control system has responded by opening the regulating valve completely, so there is no pressure drop across the valve, then the pressure in the reactor shell is

$$P_{rxr} = 4237 + 278 = 4515 \text{ kPa}$$

which exceeds the rating for the pressure-relief valve, causing it to open and release steam.

Now that the problem has been identified, the next question is how to compensate for this problem. It is likely that necessary changes can only be made during the next shutdown, which is not a problem since there are no releases for most of an operating cycle. One alteration would be to reset the pressure-relief valve to a higher pressure. Caution is warranted here because a 4500 kPa limit was originally chosen for a reason. The pressure rating for the materials of construction and reactor design should be checked very carefully to determine their limits.

Another simple solution would be to replace the line leading to the steam header with larger diameter pipe. The next larger size, 2.5 in schedule 40, has a diameter $D = 0.06271$ m. Using a relationship from Table 11.1, a base case ratio at the original flowrate is

$$\frac{\Delta P_2}{\Delta P_1} = \left(\frac{D_1}{D_2}\right)^5 = \left(\frac{0.05250}{0.06271}\right)^5 = 0.41$$

so the pressure drop, under normal operating conditions with the new pipe is

$$\Delta P_2 = 157 \text{ kPa}(0.41) = 64.4 \text{ kPa}$$

and the pressure drop with the new pipe with increased catalyst activity is

$$\Delta P_2 = 64.4 \text{ kPa}(1.77) = 114 \text{ kPa}$$

and the pressure in the reactor is

$$P_{rxr} = 4237 + 114 = 4351 \text{ kPa}$$

With the larger diameter pipe, it is possible for there to be a pressure change across the regulating valve without exceeding the cut-off pressure of the relief valve.

Finally, you should report these results and explanations clearly so that proper modifications can be made in similar plants prior to switching to the new catalyst.

16.4   A PROCESS TROUBLESHOOTING PROBLEM

In this section, a troubleshooting problem involving an entire process is presented [3]. It is based on the production of cumene problem presented in Appendix C, Figure C.8. The process flowsheet and stream flow table are included in Appendix C.

Lately, Unit 800 has not been operating within standard conditions. We have recently switched suppliers of propylene; however, our contract guarantees that the new propylene feed will contain less than 5 wt% propane.

 Upon examining present operating conditions, we have made the following observations:

1. Production of cumene has dropped by about 8%, and the reflux in T-801 was increased by approximately 8% in order to maintain 99 wt% purity. The flows of benzene (Stream 1) and propylene (Stream 2) have remained the same. Pressure in the storage tanks (not shown on flowsheet) has not changed appreciably when measured at the same ambient temperature.

2. The amount of fuel gas being produced has increased significantly and is estimated to be 78% greater than before. Additionally, it has been observed that the pressure control valve on the fuel gas line (Stream 9) leading from V-801 is now fully open, while previously it was controlling the flow.

3. The benzene recycle Stream 11 has increased by about 5% and the temperature of Stream 3 into P-201 has increased by about 3°C.

4. Production of steam in the reactor has fallen by about 6%.

5. Catalyst in the reactor was changed 6 months ago, and previous operating history (over last 10 years) indicates that no significant drop in catalyst activity should have occurred over this time period.

6. p-diisopropyl benzene (p-DIPB) production, Stream 14, has dropped by about 20%.

Suggest possible causes and potential remedies for the observed problems.

The reactions are as follows:

$$\underset{\text{propylene}}{C_3H_6} \quad + \quad \underset{\text{benzene}}{C_6H_6} \quad \rightarrow \quad \underset{\text{cumene}}{C_9H_{12}} \tag{16.7}$$

$$\underset{\text{propylene}}{C_3H_6} \quad + \quad \underset{\text{cumene}}{C_9H_{12}} \quad \rightarrow \quad \underset{\text{p-diisopropyl benzene}}{C_{12}H_{18}} \tag{16.8}$$

Assume that all of the above symptoms have been verified. The next step would be to check process inputs, which might reveal off-specification feed. This might immediately identify the problem. A check of the control valves verifies that only the valve in Stream 9 is fully open, as stated in Observation 2. If there were no problems with the feed, the next step would be to check the individual units to determine which one(s) were not operating within normal limits. As part of

Phase 2, let us analyze the six observations above to determine what they suggest.

An analysis of the six observations above suggests the following:

1. Observation 1 suggests that either less cumene is being produced in the reactor or that significant cumene is being lost as fuel gas. If it is assumed that the feeds are unchanged, these are the only possibilities.

2. Observation 2 suggests that the fuel gas rate has increased significantly. Components of fuel gas could be cumene, unreacted propylene and benzene, propane, and p-DIPB.

3. Observation 3 suggests that additional benzene is being processed in the distillation column. The temperature increase could be due to the increased concentration of benzene relative to propane and propylene at the top of T-801.

4. Observation 4 suggests that less cumene is being formed in the reactor. This is the opposite situation as in the case study presented in the previous section.

5. Observation 5 suggests that catalyst deactivation should not be a problem. This does not necessarily guarantee that catalyst deactivation is not a problem.

6. Observation 6 suggests that the selectivity for the desired reaction, Eq. (16.7), has increased.

Among the possible causes of some of these observations are:

1. The propylene feed contains propane impurity in excess of 5 wt%. Even though the new supplier claims that the propylene meets specifications, it is possible that the propylene is off-specification. This would be verified in Phase 1. If the propylene feed contains excess propane that was not detected, it is likely that the feed rate would be unchanged (This could also be verified.). Therefore, the concentration of propylene in the reactor feed would decrease, thereby decreasing the reaction rate. Conversion in the reactor would then decrease. An examination of Figure 16.6, the approximate reactor profiles, shows that a decrease in reactor conversion would have a larger percentage effect on the p-DIPB, since its concentration is lower. Examination of the kinetics, which are based on the assumption that Eqs. (16.7) and (16.8) are elementary steps, also supports this diagnosis. If the concentration of propylene is decreased, the rates of Eqs. (16.7) and (16.8) both decrease. This decreases the concentration of cumene in the reactor, which causes the rate of Eq. (16.8) to decrease even further, reducing the p-DIPB concentration more than the cumene concentration. If this reactor scenario were true, then there would be additional propane leaving the reactor. This would increase the flow of fuel gas, which is mostly propane and

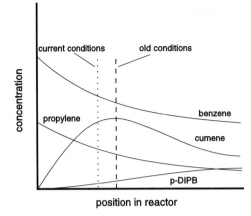

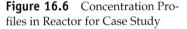

Figure 16.6 Concentration Profiles in Reactor for Case Study

propylene, and increase the recycle, which is mostly benzene. It is seen that this scenario is consistent with all 6 observations.

2. The catalyst is defective and has begun to deactivate. This could be checked by analyzing samples of the reactor input and reactor output. If the catalyst is defective and has begun to deactivate, all reaction rates would decrease. This would result in less cumene and p-DIPB and more unreacted benzene and propylene. This scenario is consistent with 5 of the 6 observations. Since the possibility of bad catalyst must be acknowledged, this scenario is also possible.

3. Reactor temperature has decreased. This could be checked in the control room, but the possibility of a faulty thermocouple or indicator should also be considered if the "correct" temperature was observed. A decrease in reactor temperature would result in decreased reaction rates. Qualitatively, the results are similar to those for deactivated catalyst in 2, above.

4. Reactor pressure has decreased. This could be checked in the control room, but the possibility of a faulty pressure transducer or indicator should also be considered if a "correct" pressure was observed. A decrease in reactor pressure would result in a decrease in all gas-phase concentrations and a decrease in all reaction rates. Qualitatively, the results are similar to 2 and 3, above.

5. The flow controllers on Streams 1 and/or 2 have failed. Suppose that the benzene flow were increased or the propylene flow were decreased. The benzene concentration would increase, which would increase the excess benzene and increase the selectivity for the desired reaction. This is consistent with the last observation above. If this were not observed in the control room, the possibility of faulty instrumentation should be considered.

6. The temperature and/or pressure in V-801 is not at specification. This could be checked, but a "correct" reading could be due to faulty instrumentation.

If the temperature and/or pressure in the flash vessel were incorrect, the desired separation would not be accomplished. If the temperature were too high and/or the pressure were too low, additional fuel gas would be produced causing less cumene to leave as product and less benzene in the recycle. If the opposite were true, additional feed to the distillation column would be produced, which could cause flooding in the column and compromise the desired separation.

7. A combination of any or all of the above items. There is no guarantee that the problem has only one cause. Therefore, combinations of the above possibilities should be considered.

An examination of Items 1–7 above suggests that taking a sample of the fuel gas would identify the root cause of the problem. If there were too much propane in the fuel gas, the problem would be with the feed. If there were too much benzene in the fuel gas, the problem would be reduced conversion in the reactor. If there were too much or too little cumene in the fuel gas, the problem would be that the temperature and/or pressure of the flash were incorrect.

The next question is how to remedy the problem. It is not always necessary to identify the cause of a problem in order to begin to compensate. However, it is necessary to identify the problem to ensure that it does not recur. It is also necessary to be certain that what appears to be a remedy is not a problem itself. For example, if the propylene feed contains excess propane or if the catalyst was deactivated, it would still be necessary to find a way to compensate temporarily until new feed and/or catalyst could be obtained. Also, detailed quantitative solutions may not be necessary as long as the qualitative trends are understood.

In order to suggest remedies, it will be assumed that the cause of the observed process upset is in the process feed and in the reactor (1 and 2, above). Possible remedies include, but are not limited to,

1. Increase the temperature in the reactor: Intuitively, this seems like a reasonable possibility. However, Chapters 13 and 14 should be consulted for the limitations associated with increasing reactor temperature. Increasing the reactor feed temperature can be accomplished by increasing the air and natural gas flows to the fired heater, but the reactor operation is limited by heat transfer. Increasing the reaction temperature increases the reaction rate exponentially, so a large temperature increase should not be needed. However, increasing the inlet temperature does not guarantee that the reactor operation will change appreciably. This situation was illustrated in Example 13.1, and is an example of heat transfer limitations of reactor performance discussed in Section 14.4. One consequence of increasing reactor temperature is that the temperature of the boiling water used to remove the heat of reaction must increase. This requires an increased steam pressure. It must

be determined whether the materials of construction of the reactor can withstand the required pressure increase. The temperature limitations of the catalyst support and of catalyst activity must also be considered. Increasing the temperature high enough to damage the catalyst is an example of the remedy causing another problem.

2. Increase the pressure on the process side in the reactor: Increasing the pressure in the reactor can be accomplished by closing the valve after the reactor. The increase in pressure increases the reaction rate by increasing the concentration. The effect is not as significant as for temperature since temperature increases the reaction rate exponentially.

3. Increase the flow of propylene feed: Increasing the propylene feed can only be accomplished by using the spare pump (P-202B) in series or in parallel with the operating pump (P-202A), due to limitations in pumping capacity shown on the pump and system curve plot, which is shown in Appendix C. In principle, the flow of propylene can be increased enough so that the specified cumene production rate is achieved in the reactor. However, it must be determined how each piece of equipment will perform when subjected to the increased capacity and concentration changes caused by this remedy. The required calculations are performance problems like those discussed in Chapters 12 and 13.

There are several lessons to be learned from this process troubleshooting problem. As was discussed earlier in this chapter, it is important not to focus on one possible solution to the exclusion of others. It is important to consider as many alternatives as possible. There were several possible causes presented here for the observed process upsets. Without detailed measurements and/or simulations, which take more time to perform, the other possibilities could not be ruled out. They would cause the same qualitative trends, just different quantitative values for the upset parameters.

It is also important to observe that a problem in the reactor was manifested in process locations far removed from the reactor: the fuel gas, the benzene recycle, the cumene production rate, and the p-DIPB production rate.

The cause of an observed process upset may be located in a different part of the process.

Finally, if the problem were identified to be impure propylene feed or defective catalyst, it would be necessary for you to report this problem to other plants in your company using the same propylene feed supplier or the same catalyst.

16.5 DEBOTTLENECKING PROBLEMS

In the course of operating a chemical process, it may become necessary to modify operating conditions. Possible changes include scale-up, scale-down, handling a new feed composition, and so on. A bottleneck is defined as the part of the process that limits the desired change. Troubleshooting the bottleneck is called debottlenecking. Debottlenecking is a sequential process. A bottleneck is identified and removed. Then, the next bottleneck is identified and removed, and so on. The removal of these bottlenecks does not usually involve a large capital investment.

The performance of a heating loop problem discussed in Section 13.3 was actually a debottlenecking problem. In this problem, described in detail in Appendix C, it is necessary to scale up production of our allyl chloride facility (See Appendix C, Figure C.4) due to an unscheduled shutdown at a similar facility owned by our company. The problem is to determine the maximum level of scale-up possible for our allyl chloride facility. Therefore, it is necessary to determine the maximum scale-up possible for the reactor. (Note: Some conditions in the problem in the appendix are different from those in Section 13.2.)

For the problem in Section 13.2, it was determined that the pump in the Dowtherm A loop could only handle 5% increased flow, which only allowed 4% increased reactor operation. Within the reactor portion of the allyl chloride process, the pump is identified as the bottleneck. From the perspective of the entire process, assuming that all other units can be scaled up by more than 5%, the Dowtherm A loop is the bottleneck. In Section 13.2, it was shown that by operating the two reactor heat exchange coils in parallel, that the reactor could be scaled up by 13%. Altering configuration of the reactor heat exchange coils is the act of debottlenecking the process. For this problem, it can be shown that additional scale-up in the reactor and Dowtherm A loop are possible by operating the pump and the spare either in series or in parallel (Problem 13.6). Suppose that the maximum scale-up were now found to be 25% (not the correct answer to Problem 13.6). What if one of the distillation columns downstream could only handle 20% increased throughput before flooding? Then, the distillation column would become the new bottleneck. You would now focus on methods for debottlenecking the distillation column. If this were possible, then another unit would become the bottleneck, and so forth. Therefore, debottlenecking is a progressive problem in which bottlenecks are removed from the process one at a time. Eventually, the maximum possible change, where a bottleneck cannot be removed, will be reached. At this point, a decision would have to be made whether a significant capital investment should be made in order to increase further the maximum scale-up. When significant process modifications involving new equipment are required, the procedure is called retrofitting.

16.6 SUMMARY

In summary, process troubleshooting problems and debottlenecking problems, such as the ones described here, are very realistic problems in terms of what the process engineer will experience. Unlike comprehensive design problems, the troubleshooting problems rely on simple, approximate calculations along with an intuitive understanding of a chemical process rather than repetitive, complex calculations. In order to solve troubleshooting and debottlenecking problems, it important to develop both an intuitive feel for chemical processes plus the ability to do approximate calculations to complement your ability to do repetitive, detailed calculations.

REFERENCES

1. Fogler, H. S., and S. E. LeBlanc, *Strategies for Creative Problem Solving*, Prentice-Hall, Englewood Cliffs, NJ, 1995.
2. Perry, R. H., D. W. Green, and J. O. Maloney (eds.), *Perry's Chemical Engineers' Handbook* (6th ed), McGraw-Hill, New York, 1984, p. 3–50.
3. Shaeiwitz, J. A., and R. Turton, "A Process Troubleshooting Problem," *1996 Annual ASEE Conference Proceedings*, Session 3213.

PROBLEMS

1. For the absorber problem in Section 16.2.1, it is necessary to adjust process operation temporarily to handle a 20% increase in gas to be treated.
   a. Can this be accomplished by increasing the liquid rate by 20%?
   b. Suggest at least two additional methods for handling the increase. Be quantitative.
2. A five equilibrium-stage tray absorber is used for the acetone separation described in Section 16.2.1. It is now observed that the outlet mole fraction of acetone in air is 0.002.
   a. Suggest at least six individual causes for the faulty absorber performance.
   b. For the causes listed in Part a that are represented on the Kremser graph, determine the exact value of the parameter (i.e., what flowrate would cause the observed outlet mole fraction?).

    c. For each cause listed in Part b, suggest at least three compensation methods. Be quantitative.

3. For the situation in Problem 2, how would you handle the following temporary process upsets? Be quantitative. Suggest at least three alternatives for each situation.
   a. The gas rate must increase by 10%.
   b. The feed mole fraction of acetone must increase to 0.025.
   c. The outlet mole fraction of acetone in air must be reduced to 0.00075.

4. Suggest additional alternatives for the off-specification acrylic acid in the case study in Section 16.3.1. Analyze the alternatives quantitatively.

5. Our acrylic acid facility has been designed to operate successfully using cooling water in the distillation column condenser. However, after a recent warm spell in which the temperature exceeded 100°F for a week, our customers complained that our acrylic acid product had the same yellowish color and increased viscosity observed in our other plant, as described in Section 16.3.1. Suggest possible causes and remedies for this situation.

6. Suggest additional alternatives for the steam release in the cumene reactor in the case study in Section 16.3.2. Analyze the alternatives quantitatively.

7. During the start-up of a chemical plant, one of the final steps before introducing the process chemicals is a steam-out procedure. Essentially, this step involves filling all the equipment with low pressure steam and leaving it for a period of time in order to clean the equipment. During one such steam-out, in a plant in Wisconsin, a vessel was accidentally isolated from other equipment and left to stand overnight. Upon inspection, the following morning, it was found that this vessel had ruptured.

   The company responsible for the design of the plant claimed that the design was not at fault and cited a similar situation that occurred at a plant in Southern California in which no damage to the vessel was seen to occur.

   From the above information can you explain what happened? What would you suggest be done in the future in order to ensure that this problem does not reoccur?

8. During the hydrogenation of a certain plant-derived oil, the fresh feed is pumped from a vessel to the process unit. The process is illustrated in Figure P16.8a.

   Because the oil is very viscous, it is heated with steam that passes through a heating coil located in the vessel. The present system uses 50 psig saturated steam to heat the oil. Due to unforeseen circumstances, the 50 psig steam supply will be down for maintenance for about a week. It has been suggested that a temporary connection from the high pressure steam line (600 psig, saturated) be made via a regulator (to reduce the pressure to 50 psig) to supply steam to the steam coil as shown in Figure P16.8b.

(a)

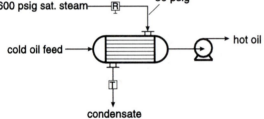

(b)

Figure P16.8   Heat Exchanger Configurations for Problem 16.8

Do you foresee any problems with the recommendation regarding the high-pressure steam? If so, what recommendations do you suggest?

9. During the start-up and operation of a new plant (see Figure P16.9) the pressure relief/safety valve on top of the steam drum (V-101) of a waste heat boiler (E-101) has opened and low pressure steam is escaping to the atmosphere through the open valve.

Upon questioning the operators from another unit, you discover that similar incidents have occurred during the initial operation of other process units. However, it appears that the situation remedies itself after a few months.

What could be causing this phenomenon to occur? If the situation does not remedy itself, what permanent solution (if any) would you suggest to fix the problem?

10. For the cumene troubleshooting problem in Section 16.4, it has been determined that off-specification propylene will have to be used for the next several months. To assist in handling this feed, prepare the following performance graphs to assist in determining what reactor temperature and pressure are required to compensate for increased propane impurity. Determine the maximum possible propane impurity that can be handled in the

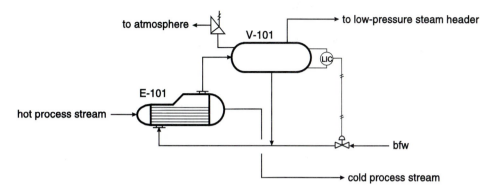

Figure P16.9   PFD for Problem 16.9

given temperature or pressure range without loss of cumene production capability.

  a. A plot of cumene production rate vs. propane impurity (from 5 wt% to 10 wt%) at pressures from 2700 kPa to 3300 kPa at 400°C and the original propylene feed rate.

  b. A plot of cumene production rate vs. propane impurity (from 5 wt% to 10 wt%) at temperatures from 300°C to 400°C at 3300 kPa and the original propylene feed rate.

11. In the cumene problem in Section 16.4, assume that one possible remedy involves increased benzene recycle with a corresponding decrease in fresh benzene feed. What are some potential consequences on the benzene feed pump, P-201? Support your answer with calculations.

12. You are in charge of a process to manufacture a polymer in which acetone is used as a solvent. The last step is a drying oven in which residual acetone is removed from the polymer into an air stream. The air is fed to the absorber in Example 16.1. All flows, mole fractions, and physical property data in Example 16.1 are assumed to hold.

    It is necessary to scale up polymer production as much as possible. Upstream, there is a spare reactor, dryer, and peripherals since the plant has been operating below capacity for many years. However, the acetone scrubber was designed and installed recently for the current production capacity. Therefore, it is anticipated that the acetone scrubber will be the process bottleneck. What limit does the acetone scrubber place on scale-up? How would you debottleneck this situation?

13. The phthalic acid production scale down project (Appendix C, Project 3) involves determining a method for scale down of phthalic acid production by 50%. The feed section of this process was discussed in Section 13.4.

a. Identify potential bottlenecks to the 50% scale down.

b. Quantify the answer to Part a. That is, determine the primary bottleneck to 50% scale-down.

c. Debottleneck your answer to Part b.

d. Repeat Parts b and c until 50% scale-down or a maximum possible scale-down is achieved.

14. For the cumene problem in Section 16.4, it has been determined that the feed is off-specification. Until we can contract with a new propylene supplier we will have to use the propylene from the current supplier. Tests have shown that we can expect this propylene to have between 5 wt% and 10 wt% propane impurity. It has been decided to try to maintain the design cumene production rate by increasing the propylene feed rate so that a constant, design amount of propylene enters the reactor. Identify the bottlenecks to the proposed process change. Debottleneck this situation.

4

Synthesis and Optimization of a Process Flow Diagram

In this section, the problem of how to create, simulate, and optimize a process and how to develop a PFD is addressed. In order to create a process flow diagram, a considerable amount of information needs to be gathered. This includes reaction kinetics, thermodynamic property data, the required purity for products and by-products, the types of separations to be used in the process, the reactor type, the range of conditions for the reaction, and many others. Once this information has been gathered, it must be synthesized into a working process. In order to accommodate the synthesis of information, the chemical engineer relies on solving material balances, energy balances, and equilibrium relationships using a process simulator. The basic data required to perform a simulation of a process are covered, and other aspects of using a process simulator are discussed. Once the PFD has been simulated, the optimization of the process can proceed. In general, process optimization involves both parametric and topological changes and both these aspects are discussed.

This material is treated in the following chapters:

Chapter 17: Synthesis of the PFD from the Generic BFD
The information required to obtain a base case process flow diagram is discussed and categorized into the six basic elements of the generic block flow process diagram. The need to obtain reaction kinetics, thermodynamic data, and alternative separation methods is discussed in the context of building a base case process.

Chapter 18: Synthesis of a Process using a Simulator and Simulator Troubleshooting

The structure of a typical process simulator and the basic process information required to simulate a process are discussed. The various types of equipment that can be simulated, and the differences between alternative modules used to simulate similar process equipment are reviewed. The importance of choosing the correct thermodynamic package for physical property estimation is emphasized, and strategies to eliminate errors and solve simulation problems are presented.

Chapter 19: Process Optimization

Basic definitions used to describe optimization problems are presented. The need to look at both topological changes in the flowsheet (rearrangement of equipment) and parametric changes (varying temperature, pressure, etc.) are emphasized. Strategies for both types of optimization are included. Special attention is paid to the integration of heat within a process. A variety of examples are given to illustrate these principles that show how optimization is applied to chemical processes.

CHAPTER

17

Synthesis of the PFD from the Generic BFD

The evolutionary procedure to create a full PFD (as presented in Chapter 1) from the Generic Block Flow Process Diagram (GBFD) (Chapter 6) is described in this chapter. This full PFD truly defines the process in a chemical engineering sense and is the starting point for chemical and other engineers to design the machines, structures, and electrical and electronic components needed to make the chemical engineer's vision a reality.

This crucial step in the design of the chemical plant involves all of the sub-areas of chemical engineering: reaction engineering, thermodynamics, process control, unit operations and transport, and material and energy balances. Each is applied to put details into the six general sections of the GBFD (Reactor Feed Preparation, Reactor, Separator Feed Preparation, Separator, Recycle, and Environmental Control).

In this synthesis, the broader context of the project (e.g., environmental concerns, customer expectations, return on investment) is integrated with the important details such as the type of heat-transfer medium or the number of stages in a column. It is crucial to consider as many alternatives as possible in the early stages to try to avoid becoming trapped in a suboptimal design.

> **It is a common human trait to resist change more strongly as more effort is expended on a task, design, or product. We describe this as not wanting to abandon our "investment" in the activity.**

17.1   INFORMATION NEEDS AND SOURCES

Before the detailed synthesis of the PFD can be completed, one needs basic physical property and kinetics information. We assume here that the very basic "chemistry" of the desired reaction is known, that is, what main feed materials go to what main product. Before PFD synthesis begins, the marketing engineers or others have identified a market need for a specific product, and the chemists have identified at least one way to produce the chemical in the laboratory. Even the marketing and chemistry information, however, will need to be refined. Flowsheet synthesis will uncover the need for more detailed data on the reaction rate, temperature and pressure effects, and market values of products of different purities.

17.1.1   Interactions with Other Engineers and Scientists

Teams of engineers work on the development of the process. For example, the marketing department will find the customer for the product of the plant, and product specifications will be identified. Many chemical engineers are employed as marketing engineers, and they will understand that product purity affects product price, often dramatically so. However, the details of this interplay can only be determined by the process design engineers as the PFD is being developed; only through discussions and negotiations with customers can the marketing engineer determine the relationship between product purity and the product value (i.e., maximum selling price) to the customer.

Similarly, there may be more than one chemical pathway to the product. Pathways of greatest interest to the chemical engineer are not necessarily those of greatest interest to the chemist. The abilities to use impure feed materials and to avoid the production of by-products reduce costs but may not be of interest to a chemist. The costs of small-lot, high-purity laboratory reagents may not even qualitatively correlate to those of multiple tank-car, industrial-grade raw materials. Isothermal operation of small laboratory reactors is common but essentially impossible to achieve on a large scale. It is more economical per unit volume to maintain high pressures on the plant scale than it is in the lab. Batch operations are common in laboratory work but less so in the plant. Thus, the chemical process design engineer must be in touch with the chemist to make sure that expensive constraints or conditions suggested by laboratory studies are truly needed.

17.1.2   Reaction Kinetics Data

Before reactor design can begin, the kinetics of the main reaction must be known. However, a knowledge of the kinetics of unwanted side reactions is also crucial to the development of PFD structure or topology (number and position of recycle streams; types, numbers, and locations of separators). Knowledge of detailed reaction pathways, elementary reactions, and unstable reaction intermediates is not required. Rather, the chemical process design engineer needs to know the rate of reaction (main and by-product reactions) as a function of temperature, pressure, and composition. The greater the range of these independent variables, the better the design can be.

For some common homogeneous reactions, kinetics are available [1,2,3]. However, most commercial reactions involve catalysts. The competitive advantage of the company is often the result of a unique catalyst. Thus, kinetics data for catalyzed reactions are not as readily available in the open literature but should be available within the company files or must be obtained from experiment. One source of kinetics data for catalytic reactions is the patent literature. The goal of someone writing a patent application, however, is to present as little data as possible about the invention while obtaining the broadest possible protection. This is why patent information is often cryptic. However, this information is often sufficient to develop a base case PFD. The key data to obtain from the patent are the inlet composition, temperature, pressure, outlet composition, and space time. If the data are for varying compositions, one can develop crude kinetics rate expressions. If the data are for more than one temperature, an activation energy can be determined. These data reduction procedures are described in undergraduate textbooks on reaction engineering [4,5].

Without kinetics data, a preliminary PFD and cost analysis can still be done [6]. In this type of analysis, the differing process configurations and costs for different assumed reaction rates provide estimates of the value of a potential catalyst. If doubling the reaction rate reduces the cost of manufacture by $1 million per year, for example, the value of catalysis research to increase the reaction rate (all other things being equal) is clear. As a guideline, the economic breakpoint is often a catalyst productivity to desired product of ~0.10 kg product per kg catalyst per hour [7]. Another guide is that activation energies are usually between 40 to 200 kJ/mol.

17.1.3   Physical Property Data

In addition to kinetics data, physical property data are required for material and energy balances, as well as for sizing of heat exchangers, pumps and compressors, and separation units. These data are, in general, easier to obtain and, when necessary, easier than kinetics data to estimate.

For the material and energy balances, pure-component heat capacity and density data are needed. These are among the most widely measured data and are available on process simulators for well over a thousand substances. (See Chapter 18 for details of process simulators.) There are also reasonably accurate group-contribution techniques for use when no data are available [8]. The enthalpies of mixtures require an accurate equation of state for gases and nonionic liquids. The equations of state available on process simulators are accurate enough for these systems. However, additional heat of solution data are needed for electrolyte solutions, and these data may not be as readily available. For these systems, care should be taken to use accurate experimental data, as estimation techniques are not as well defined.

The design of heat exchangers and the determination of pressure drops across units require thermal conductivity and viscosity data. These data are usually available (often in the databanks of process simulators) and, if unavailable, can be estimated by group contribution techniques [8].

The most crucial and least available physical property data are for phase equilibrium. Most separators are based on equilibrium stages; thus, these data are usually needed for a process design. For vapor-liquid equilibrium (such as for distillation), either (1) a single equation of state for both phases or (2) a combination of vapor-phase equation of state, pure-component vapor pressure, and liquid-state activity coefficient model is required. The choice of thermodynamics package for process simulators is explained in Section 18.4. The key experimentally determined mixture parameters for either equations of state or activity-coefficient models are called BIPs (binary interaction parameters), and they have great effect on the design of separation units. Any estimation of them (for example, assuming them to be zero!) can lead to severely flawed designs. The solubilities of "non-condensables" in the liquid phase are also essential, but difficult to estimate.

17.2   REACTOR SECTION

For a process with a reactor, often the synthesis of the PFD begins with the reactor section of the GBFD. (See Chapter 14.) A base case reactor configuration is chosen according to the procedures described in reaction engineering textbooks. This configuration (e.g., plug flow, CSTR, adiabatic, isothermal) is used at some base conditions (temperature, pressure, feed composition) and some preliminary base specification (e.g., 60% conversion) to calculate the outlet composition, pressure, and temperature. The goal at this stage is to develop a feasible PFD for the process. Optimization of the PFD can begin only after a suitable base case is developed. If there are obvious choices that improve the process (such as using a fluidized bed instead of a packed bed reactor), these choices are made at this stage; however, these choices should be revisited later.

To enable later optimization, the general effects of varying the feed conditions should be investigated at this point with the trend prediction approach of Chapter 11. A list of possible reactor configurations should also be developed. These choices often have dramatic effects on the other parts of the GBFD. The earlier that these effects are understood, the better will be the final design.

At this stage, the utility needs of the reactor should be considered. If heating or cooling is required, an entire additional system may be required to be designed. The choice of heating or cooling medium must be made based on strategies described in Chapter 19, the heuristics presented in Chapter 9, and the costs of these utilities.

The trade-offs of different catalysts, parallel vs. series reactors, and conversion vs. selectivity should be considered, even though the optimization of these choices occurs after the base case is developed. Again, early identification of alternatives improves later detailed optimization.

Once the base case reactor configuration is chosen, the duties of the Reactor Feed Preparation and Separator Feed Preparation units are partially determined.

For the reactor, important questions to be considered include:

1. *In what phase does the reaction take place (liquid, vapor, mixed, etc.)?* The answer will affect the reactor feed section. For example, it will determine whether a vaporizer or fired heater is required upstream of the reactor when the feed to the plant is liquid.

2. *What are the required temperature and pressure ranges for the reactor?* If the pressure is higher than the feed pressure, pumps or compressors are needed in the Reactor Feed Preparation section. If the required reactor feed temperature is greater than approximately 250°C, a fired heater is probably necessary.

3. *Is the reaction kinetically or equilibrium controlled?* The answer affects both the maximum single-pass conversion and the reactor configuration. The majority of gas- and liquid-phase reactions in the CPI are kinetically controlled. The most notable exceptions are the formation of methanol from synthesis gas, synthesis of ammonia from nitrogen and hydrogen, and the production of hydrogen via the water-gas shift reaction.

4. *Does the reaction require a solid catalyst or is it homogeneous?* This difference dramatically affects the reactor configuration.

5. *Is the main reaction exothermic or endothermic, and is a large amount of heat exchange required?* Again, the reactor configuration is more greatly affected by the heat transfer requirements. For mildly exothermic or endothermic gas-phase reactions, multiple packed beds of catalyst or shell-and-tube reactors (catalyst in tubes) are common. For highly exothermic gas-phase reactions, heat transfer is the dominant concern, and fluidized beds or shell-and-tube reactors with catalyst dilution (with inert particles) are used. For liquid-phase reactions, temperature control can be achieved by pumping the reacting mixture through external heat exchangers (for example, in Figure B.4). For some highly exothermic reactions, part of the reacting mixture is vaporized to help regulate the temperature. The vapor is subsequently condensed and returned to the reactor. External jackets and internal heat-transfer tubes, plates, or coils may also be provided for temperature control of liquid-phase reactions. (See Chapters 14 and 15.)

6. *What side reactions occur, and what is the selectivity of the desired reaction?* The formation of unwanted by-products may significantly complicate the separation sequence. If these by-products are formed in large quantities and are to be purified for sale, this is especially important. For high-selectivity reactions, it may be more economical to dispose of by-products as waste or to burn them

(if they have high heating values), which simplifies the separation section. However, with the heightening of environmental concerns, more emphasis is being placed on producing salable by-products, or none at all.

7. *What is the approximate single-pass conversion?* The final single-pass conversion is determined from detailed parametric optimizations (Chapter 19); however, the range of feasible single-pass conversions affects the structure of the separations section. If extremely high single-pass conversions are possible (e.g., above ~98%), it may not be economical to separate and recycle the small amounts of unreacted feed materials. In this case, the feed materials become the impurities in the product, up to the allowable concentration.

8. *For gas-phase oxidations, should the reactor feed be outside the explosive limits?* For example, there are many reactions that involve the partial oxidation of hydrocarbons (see acrylic acid production in Appendix B and phthalic anhydride production in Appendix C). Air or oxygen is fed to a reactor along with hydrocarbons at high temperature. The potential for explosion from rapid, uncontrolled oxidation (ignition) is possible whenever the mixture is within its explosive limits. (Note that the explosive limits widen significantly with increase in temperature.) An inherently safe design would require operation outside these limits. Often, steam is added both as a diluent and to provide thermal ballast for highly exothermic reactions, for example in the acrylic acid reactor (Figure B.2).

17.3   SEPARATOR SECTION

After the reactor section, the separator section should be studied. The composition of the separator feed is that of the reactor effluent, and the goal of the separator section is to produce a product of acceptable purity, a recycle stream of unreacted feed materials, and a stream or streams of by-products. The ideal separator used in the GBFD represents a process target, but it generally represents a process of infinite cost. Therefore, one step is to "de-tune" the separation to a reasonable level. However, before doing that, one must decide what the by-product streams will be. There may be salable by-products, in which case a purity specification is required from the marketing department. For many organic chemical plants, one by-product stream is a mixture of combustible gases or liquids that are then used as fuel. There may also be a waste stream (often a dilute aqueous stream) to be treated downstream; however, this is an increasingly less desirable process feature.

Prior to enactment of stringent environmental regulations, it was generally less expensive to treat waste streams with so-called "end-of-pipe" operations. That is, one produced waste, concentrated the waste, and disposed of the waste in an acceptable manner. As the regulations have stiffened on what is an "acceptable manner" of disposal, the strategy of *pollution prevention* has grown. Some details are given in Chapter 21, but the overall strategy is to minimize wastes at their source or to turn them into salable products.

The separation section then generally accepts one stream (from the pre-separation unit) and produces product, by-product, and (sometimes) waste streams. In the development of the PFD, one must consider the most inclusive or flexible topology so that choices can be made in the optimization step. Thus, each of the types of streams should be included in the base case.

Next, the minimum number of simple separation units must be determined. Although there are single units that produce multiple output streams (such as a petroleum refining pipe still with many sidedraws), most units accept a single inlet stream and produce two outlet streams. For such simple separators, we need at least $(N-1)$ units, where N is the number of outlet streams (products, by-products, and waste). There are two types of questions to answer concerning these units in the separation section: (1) What types of units should be used? and (2) How should the units be sequenced?

There are general guidelines concerning choice of separation unit. One such set of rules is given in Table 17.1 for the most common choices of separation units on process simulators.

For a base case, it is essential that the separation technique chosen be reasonable, but it is not necessary that it be the best.

Table 17.1   Guidelines for Choosing Separation Units

- Use distillation as a first choice for separation of fluids when purity of both products is required.
- Use gas absorption to remove one trace component from a gas stream.
- Consider pressure-swing adsorption to purify gas streams, especially when one of the components has a cryogenic boiling point.
- Consider membranes to separate gases of cryogenic boiling point and relatively small flowrates.
- Choose an alternative to distillation if the boiling points are very close or if the heats of vaporization are very high.
- Consider extraction as a choice to purify a liquid from another liquid.
- Use crystallization to separate two solids or to purify a solid from a liquid solution.
- Use evaporation to concentrate a solution of a solid in a liquid.
- Use centrifugation to concentrate a solid from a slurry.
- Use filtration to remove a solid in almost dry form from a slurry.
- Use screening to separate solids of different particle size.
- Use float/sink to separate solids of different density from a mixture of pure particles.
- Consider reverse osmosis to purify a liquid from a solution of dissolved solids.
- Use leaching to remove a solid from a solid mixture.

For sequencing of the separation units, there is another set of guidelines, given below. In the base case, it is often helpful to consider the same type of separator for each unit. During optimization, one can compare different separator types for the different duties. Again, some separators can do multiple separations in one unit, but these can be found during optimization. Some rules for sequencing are given in Table 17.2. Additional heuristics for separation unit sequencing are given in Table 9.13 and in reference [9].

As with all sets of heuristics, these can be mutually contradictory. However, in the initial topology of the separation section, the main goal is to follow as many of these heuristics as possible.

Beyond these general guidelines, beware of azeotropes and multiple phases in equilibrium (especially when water and organics are present). Special techniques are available to deal with these problems. On the other hand, if a single-stage flash will do the separation, do not use a column with reflux.

For the separations section, other important questions to be considered include:

1. *What are the product specifications for all the products?* Product specifications are developed to satisfy customers who will use these products in their own processes. The most common specification is a minimum concentration of the main constituent, such as 99.5 wt%. Maximum impurity levels for specific contaminants may also be specified as well as requirements for specific physical properties such as color, odor, and specific gravity. A single separation technique may not be sufficient to meet all the required product specifications.

Example 17.1

In the production of benzene via the hydrodealkylation of toluene, it is necessary to produce a benzene product stream that contains >99.5 wt% benzene that is water white in color (i.e., absolutely clear). If the feed toluene to the process contains a small amount of color, determine a preliminary separation scheme to produce the desired benzene product.

Table 17.2   Guidelines for Sequencing Separation Units

• Remove the largest product stream first. This makes all of the subsequent separation units smaller.
• For distillation, remove the product with the highest heat of vaporization first, if possible. This reduces the heating/cooling duties of subsequent units.
• Do not recombine separated streams. (This may seem obvious, but it is often disobeyed.)
• Do the easy separations first.
• Do not waste raw materials, and do not overpurify streams based on their uses.
• Remove hazardous or corrosive materials first.

As a guide, we may look at the Toluene hydrodealkylation process shown in Figure 1.5. Since the volatilities of toluene and benzene are significantly different, the main purification step (the separation of benzene from toluene) can be accomplished using distillation, which is consistent with Figure 1.5. However, it has been found that the compound causing the discoloration of the toluene is equally soluble in benzene and toluene, causing the benzene product to be discolored. It is further found by laboratory testing that the benzene product can be decolorized by passing it through a bed of activated carbon. Thus, a second separation step, consisting of an activated carbon absorber, will be added to the process to decolorize the benzene product.

2. *Are any of the products heat sensitive?* If any of the products or by-products are heat sensitive (i.e., they decompose or polymerize at elevated temperatures), the conditions used in the separations section may have to be adjusted.

Example 17.2

It is known that acrylic acid starts to polymerize at 90°C when it is in a concentrated form. Acrylic acid must be separated from acetic acid to produce the required purity product, and the volatilities of both acids are significantly different. This points to distillation as the separation method. The normal boiling points of acrylic acid and acetic acid are 140°C and 118°C, respectively. How should the separation be accomplished to avoid degradation of the acrylic acid product?

The distillation column must be run under vacuum to avoid the problem of acrylic acid degradation. The pressure should be set so that the bottom temperature of the column is below 90°C. From Figure B.2 and Table B.4, we see that a column pressure of 0.16 bar at the bottom can accomplish the desired separation without exceeding 90°C.

3. *Are any of the products, by-products, or impurities hazardous?* Since separation between components is never perfect, small quantities of toxic or hazardous components may be present in product, fuel, or waste streams. Additional purification or subsequent processing of these streams may be required, depending on their end use.

17.4   REACTOR FEED PREPARATION AND SEPARATOR FEED PREPARATION SECTIONS

The purpose of these sections is to match the temperature and pressure desired on the inlet streams to the reactor section and to the separation section. If the reactor operates at high temperature (a common occurrence as this increases reaction rate), the Reactor Feed Preparation and Separator Feed Preparation sections are often combined in a single process-process heat exchanger. Such heat-integration can be built into the base case flowsheet, but, if not, it should be caught during the heat-integration step of optimization (see Chapter 19).

Pressure may also need to be increased for the reactor (or, infrequently, decreased), and this requires a pump for liquids or a compressor for gases. Pumps are preferable to compressors, when there is a choice, because the operating, capital, and maintenance costs are all lower for pumps. If pressure is reduced between the reactor and the separator sections (or anywhere else in the process), an expander can be considered (for gases), but it is often not economical both because of its high cost and also because it reduces the controllability of the process. A valve allows control at a modest cost, but energy is not recovered.

In these temperature and pressure matching sections, the lowest cost utility should always be used. For heating the feed to an exothermic reaction, heat integration can be used with the reactor effluent. For low-temperature heating, low-pressure steam or another low-temperature utility is used. For safety reasons, exothermic reactions (when reactor runaway is possible) should be run with the reactor feed coming in at a temperature high enough to ensure a significant reaction rate. This avoids the buildup of large inventories of unreacted feed materials, which can happen if cold material enters the reactor and quenches the reaction. When sufficient heat is later provided, the entire contents of the reactor could react very rapidly, a process called ignition.

When possible, consider operation between 1 and 10 bar. High pressures increase pumping, compression, and capital costs, while low pressures tend to increase the size and cost of vessels. The temperature of the feed to the separation unit (at least for the base case) is usually set between the boiling points of the top and bottom product for distillation, or based on similar considerations for the other separation options.

17.5   RECYCLE SECTION

This section is relatively straightforward. The stream or streams of unreacted raw materials are sent back to the reactor to reduce feed costs, to reduce impurities in the product, or to improve the operation of the process. If the conditions of the recycle stream(s) are close to those of the raw material feed, the recycle stream should mix with the raw materials prior to the Reactor Feed Preparation section. Otherwise, any heating/cooling or pressure increase/decrease should be done separately. Thus, the recycle stream is combined with the raw material streams when they are all at a similar temperature.

17.6   ENVIRONMENTAL CONTROL SECTION

As stated previously and in Chapter 21, this section should be eliminated or minimized. However, especially if the contaminant has little or no value if concentrated, there will be relatively dilute waste streams generated and sent to the Environmental Control section. Here, they are concentrated (by the separations techniques discussed above) and then disposed of (by incineration, neutralization, oxidation, burial, or other means). The keys are to concentrate the waste and to make it benign.

17.7   MAJOR PROCESS CONTROL LOOPS

During the initial synthesis of the PFD, the major control loops are developed. These control loops affect more than just one unit of a process. For example, the level control in the condensate tank of a distillation column is necessary for plant operation. On the other hand, reactor temperature controller that changes the flowrate of molten salt through the cooling tubes of a reactor is a major loop and should be shown on the PFD.

It is through the early development of major control loops that significant design improvements can be made. In the high-temperature exothermic reactor, for example, failure to consider the control loop might lead one to propose an integrated heat exchanger network that is difficult or impossible to control.

Beyond the importance of the control loops in maintaining steady-state material balance control, assurance of product purity, and safety, they provide focal points for the optimization that will follow the initial PFD synthesis. As described in Chapter 19, the controlled variables are the variables over which we have a choice. We find the best values of these variables through optimization. These loops also provide early clues to the flexibility of the process operation. For example, if the feed to the reactor is cut in half, less heat needs to be removed. Therefore, there must be an increase in the temperature of the cooling medium, which occurs when the coolant flowrate is reduced.

17.8   FLOW SUMMARY TABLE

The format for the flow summary table is given in Chapter 1. Each of the conditions (temperature, pressure, flowrate, composition, and phase) should be estimated early in the flowsheet development. All are needed to get preliminary costs, for example. Even estimates based on perfect separations can provide sufficient data to estimate the cost of a recycle versus the value of burning the impure, unreacted feed material as a fuel.

Completeness of the estimates, not their accuracy, is important at this stage. For example, an early determination of phase (solid vs. liquid vs. gas) is needed to help choose a separation scheme or reaction type (see Table 17.1).

17.9   MAJOR EQUIPMENT SUMMARY TABLE

Chapter 1 explains the requirements of a major equipment summary table. In the context of initial PFD development, the process of creating the table forces the process design engineer to question the size (and cost) of various units for which there may well be other options. If, early on, one must specify a large compressor, for example, the process can be changed to a lower pressure or it can be modified to use liquid pumping followed by vaporization rather than vaporization followed by compression. The early consideration of materials of construction provides the clue that normal temperatures and pressures usually result in less expensive materials.

17.10 SUMMARY

The inclusion of enough detail and the freedom to look at the big picture without the burden of excessive detail are the keys to successful PFD synthesis. One must remain fully aware of the broadest goals of the project while looking for early changes in the structure of the flowsheet that can make significant improvements.

The beginning of the process is the Generic Block Flow Process Diagram. While the flow summary table at this stage is based on crude assumptions and the equipment summary table is far from the final equipment specifications, they help keep the chemical engineer cognizant of key choices that need to be made.

REFERENCES

1. *Tables of Chemical Kinetics: Homogeneous Reactions*, National Bureau of Standards, Circular 510, 1951; Supplement 1, 1956; Supplement 2, 1960; Supplement 3, 1961.
2. Kirk, R. E., and D. F. Othmer, *Encyclopedia of Chemical Technology*, 4th ed., Wiley, New York, 1991.
3. McKetta, J. J., *Encyclopedia of Chemical Processes and Design*, Marcel Dekker, New York, 1977.
4. Fogler, H. S., *Elements of Chemical Reaction Engineering*, 2nd ed. Prentice-Hall, Englewood Cliffs, NJ, 1992.
5. Levenspiel, O., *Chemical Reaction Engineering*, 2nd ed., Wiley, New York, 1972.
6. Viola, J. L., "Estimate Capital Costs via a New, Shortcut Method," in *Modern Cost Engineering: Methods and Data*, volume II, edited by J. Matley, McGraw-Hill, New York, 1984, p. 69 (originally published in *Chemical Engineering*, April 6, 1981).
7. J. B. Cropley, Union Carbide Technical Center, South Charleston, WV, personal communication, 1995.
8. Reid, R. C, J. M. Prausnitz, and B. E. Poling, *The Properties of Gases and Liquids*, 4th edition, McGraw-Hill, New York, 1987.
9. Rudd, D. F., G. J. Powers, and J. J. Siirola, *Process Synthesis*, Prentice-Hall, Englewood Cliffs, NJ, 1973.

PROBLEMS

1. Choose one of the cases from Appendix B. Identify:
   a. All required physical property data.
   b. Sources for all data needed but not provided.

2. Search the patent literature for kinetics information for one of the processes in Appendix B. Convert the data provided to a form suitable for use on a process simulator.

3. Develop five heuristics for reactor design.

4. For design of an exothermic reactor with cooling, one needs to choose an approach temperature (the nominal temperature difference between the reaction zone and the cooling medium). One engineer claims that the temperature difference across the tube walls should be as small as possible. Another claims that a large temperature is better for heat transfer.
   a. Describe the advantages and disadvantages of these two choices from the point of view of capital costs.
   b. Describe the advantages and disadvantages of these two choices from the point of view of operating costs.
   c. Describe the advantages and disadvantages of these two choices from the point of view of operability of the process.

5. Are there any safety considerations in the choice of heat-transfer driving force (ΔT) in Problem 4 above? Explain.

6. From the flowsheets in Appendices B and C, identify four examples of each of the following types of recycles described in this chapter:
   a. Recycles that reduce feed costs.
   b. Recycles that reduce impurities in products.

7. Recycles are also used for the following purposes. Identify four examples of each.
   a. Heat removal
   b. Flowrate control
   c. Improvement of separation
   d. Reduction of utility requirements

8. From your previous courses and experience, identify four process recycle streams that serve different purposes than those identified in 7 a–d above.

9. What information is needed about the available utilities before a choice between them can be made for a specific heating duty?

10. Take a flowsheet for one of the processes in Appendix B and identify the Reactor Feed Preparation, Reactor, Separator Feed Preparation, Separator, Recycle, and Environmental Control sections. Is there any ambiguity concerning the demarcation of these sections? Explain.

18

Synthesis of a Process Using a Simulator and Simulator Troubleshooting

The advancement in computer-aided process simulation over the past generation has been nothing short of spectacular. Until the late 1970s, it was rare for a graduating chemical engineer to have any experience in using a chemical process simulator. Most material and energy balances were still done by hand by teams of engineers. The rigorous simulation of multistaged separation equipment and complicated reactors was generally unheard of, and the design of such equipment was achieved by a combination of simplified analyses, short-cut methods, and years of experience. In the present day, however, companies now expect their junior engineers to be conversant with a wide variety of computer programs, especially a process simulator.

To some extent, the knowledge base required to simulate successfully a chemical process will depend on the simulator used. Currently there are several process simulators on the market, for example, CHEMCAD™, ASPEN PLUS™, HYSIM, PRO/II™. Many of these companies advertise their product in the trade magazines, for example, *Chemical Engineering*, *Chemical Engineering Progress*, *Hydrocarbon Processing*, or *The Chemical Engineer*. The availability of such powerful software is a great asset to the experienced process engineer, but such sophisticated tools can be potentially dangerous in the hands of the neophyte engineer. The bottom line in doing any process simulation is that you, the engineer, are still responsible for analyzing the results from the computer. The purpose of this chapter is not to act as a primer for one or all of these products. Rather, the general approach to setting up processes is emphasized, and we aim to highlight some of the more common problems that process simulator users encounter and to offer solutions to these problems.

18.1   THE STRUCTURE OF A PROCESS SIMULATOR

The six main features of all process simulators are illustrated in the left-hand column of Figure 18.1. These elements are:

1. Component Database—This contains the constants required to calculate the physical properties from the thermodynamic models.
2. Thermodynamic Model Solver—A variety of options for vapor-liquid (VLE) and liquid-liquid (LLE) equilibrium, enthalpy calculations, and other thermodynamic property estimations are available.
3. Flowsheet Builder—This part of the simulator keeps track of the flow of streams and equipment in the process being simulated. Often this information can be both input and displayed graphically.

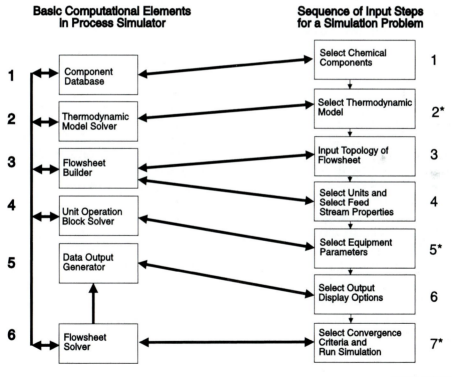

Figure 18.1 Relationship between Basic Computational Elements and Required Input to Solve a Process Simulation Problem

4. Unit Operation Block Solver—Computational blocks or modules are available that allow energy and material balances and some design calculations to be performed for a wide variety of process equipment.

5. Data Output Generator—This part of the program serves to customize the results of the simulation in terms of an output report. Often, graphical displays of tower profiles, heating curves, and a variety of other useful process data can be produced.

6. Flowsheet Solver—This portion of the simulator controls the sequence of the calculations and the overall convergence of the simulation.

There are several other elements commonly found in process simulators that are not shown in Figure 18.1. For example, there are file control options, the option to use different engineering units, and possibly some additional features associated with regressing data for thermodynamic models, and so on. The availability of these other options are dependent on the simulator used and will not be discussed further.

Also shown on the right-hand side of the diagram in Figure 18.1, are the seven general steps to setting up a process simulation problem. The general sequence of events that a user must follow in order to set up a problem on a simulator are as follows:

1. Select all of the chemical components that are required in the process from the component database.

2. Select the thermodynamic models required for the simulation. These may be different for different pieces of equipment. For example, to simulate correctly a liquid-liquid extractor, it is necessary to use a thermodynamic model that can predict liquid-phase activity coefficients and the existence of two liquid phases. However, for a pump in the same process, a less sophisticated model could be used.

3. Select the topology of the flowsheet to be simulated by specifying the input and output streams for each piece of equipment.

4. Select the properties (temperature, pressure, flowrate, vapor fraction, and composition) of the feed streams to the process.

5. Select the equipment specifications (parameters) for each piece of equipment in the process.

6. Select the way in which the results are to be displayed.

7. Select the convergence method and run the simulation.

The interaction between the elements and steps and the general flow of information is shown by the lines on the diagram. Of the seven input steps given above, steps 2, 5, and 7 are the cause of most problems associated with running process simulations. These areas will be covered in more detail in the following sections. However, before these topics are covered, it is worth looking at the basic solution algorithms used in process simulators.

There are basically three types of solution algorithm for process simulators [1]: sequential modular, equation solving (simultaneous non-modular), and simultaneous modular.

In the sequential modular approach, the equations describing the performance of equipment units are grouped together and solved in modules, that is, the process is solved equipment piece by equipment piece. In the equation solving, or simultaneous non-modular, technique, all the relationships for the process are written out together and then the resulting matrix of nonlinear simultaneous equations is solved to yield the solution. This technique is very efficient in terms of computation time, but requires a lot of time to set up and is unwieldy. The final technique is the simultaneous modular approach, which combines the modularizing of the equations relating to specific equipment with the efficient solution algorithms for the simultaneous equation solving technique.

Of these three types, the sequential modular algorithm is by far the most widely used. In the sequential modular method, each piece of equipment is solved in sequence, starting with the first then followed by the second, and so on. We assume that all the input information required to solve each piece of equipment has been provided (see Section 18.2.5). Therefore, the output from a given piece of equipment, along with specific information on the equipment, becomes the input to the next piece of equipment in the process. Clearly, for a process without recycle streams, this method requires only one flowsheet iteration to produce a converged solution. The term flowsheet iteration means that each piece of equipment is only solved once. However, there may be many iterations for any one given piece of equipment. This concept is illustrated in Figure 18.2.

The solution sequence for flowsheets containing recycle streams is more complicated, as shown in Figure 18.3. In Figure 18.3a we see that the first equip-

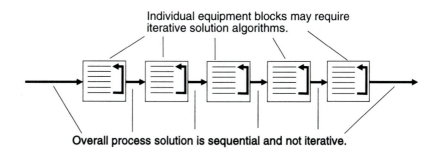

Individual equipment blocks may require iterative solution algorithms.

Overall process solution is sequential and not iterative.

Figure 18.2 Solution Sequence Using Sequential Modular Simulator for a Process Containing No Recycles

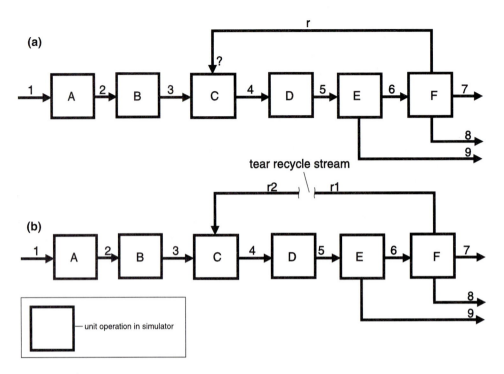

Figure 18.3   The Use of Tear Streams to Solve Problems with Recycles Using the Sequential Modular Algorithm

ment in the recycle loop (C) has an unknown feed stream (r). Thus, before Equipment C can be solved, some estimate of Stream r must be made. This leads to the concept of tear streams. A tear stream, as the name suggests, is a stream that is torn or broken. If we now consider the flowsheet in Figure 18.3b, with the recycle stream torn, we can see that, provided information is supplied about Stream r2, the input to Equipment C, we can solve the flowsheet all the way around to Stream r1 using the sequential modular algorithm. We then compare Streams r1 and r2. If they agree within some specified tolerance, then we have a converged solution. If they do not agree, then Stream r2 is modified and the process simulation is repeated until convergence is obtained. The splitting or tearing of recycle streams allows the sequential modular technique to handle recycles. The convergence criterion and the method by which Stream r2 is modified can be varied, and multivariable successive substitution, Wegstein, and Newton-Raphson techniques [2,3] are all commonly used for the recycle loop convergence. Usually, the simulator will identify the recycle loops and automatically pick streams to tear and a method of convergence. The tearing of streams and method of convergence can also be controlled by the user, but this is not recommended for the novice.

18.2   INFORMATION REQUIRED TO COMPLETE A PROCESS SIMULATION—INPUT DATA

Referring back to Figure 18.1, we consider each input block separately. The input data for the blocks without asterisks (1, 3, 4, and 6) are quite straightforward and require little explanation. The remaining blocks (2, 5, and 7) are often the source of problems, and these are treated in more detail.

18.2.1   Selection of Chemical Components

Usually, the first step in setting up a simulation of a chemical process is to select which chemical components are going to be used. The simulator will have a databank of many components (over a thousand chemical compounds are commonly included in these databanks). It is important to remember that all components—inerts, reactants, products, byproducts, utilities, and waste chemicals—should be identified. If the chemicals that you need are not available in the databank, then there are usually several ways that you can add *components (user added components)* to your simulation. How to input data for *user added components* is simulator specific, and the simulator user manual should be consulted.

18.2.2   Selection of Thermodynamic Model(s)

This is a very important part of any simulation. If the wrong thermodynamic package or model is used, then the simulated results will not be accurate. The choice of what thermodynamic models to use is often overlooked by the novice, and can cause many simulation problems down the road. The two most important aspects of thermodynamic model selection are the choice of a model to predict enthalpy and a model to predict phase equilibria. Several of the popular simulators have *expert systems* to help users select the appropriate model for their system. The expert system determines the range of temperatures and pressures that the simulation is to take place over (usually with additional user input), and, with data on the components to be used, makes an informed guess of the thermodynamic models that will be best for the process being simulated. The user should be aware that the model chosen by the expert system may not be best for a given piece of equipment.

Due to the importance of thermodynamic model selection and the many problems that the wrong selection leads to, a separate section (Section 18.4) is dedicated to this subject. An example of how the wrong thermodynamic package can cause serious errors is given in Example 18.1.

Example 18.1

Consider the HCl absorber (T-602) in the separation section of the allyl chloride process, Figure C.3. This equipment is shown in Figure E18.1. The function of the absorber is to contact countercurrently Stream 10a, containing mainly propylene and hydrogen chloride,

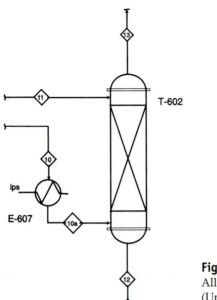

Figure E18.1   HCl Absorber in Allyl Chloride Separation Section (Unit 600)

with water, Stream 11. The HCl is highly soluble in water and is almost completely absorbed to form 32 wt% hydrochloric acid, Stream 12. The gas leaving the top of the absorber, Stream 13, is almost pure propylene, which is cleaned and then recycled.

The table below shows the results for the two outlet streams from the absorber, Streams 12 and 13, for two simulations each using a different thermodynamic model for the vapor-liquid equilibrium calculations. The first two columns in the table show the results using the SRK (Soave [4], Redlich and Kwong [5]) model, which is the preferred model for many common components. The third and fourth columns show the results

Phase Component Flows (kmol/h)	Using SRK Model		Using PPAQ* Model	
	Stream 12 Liquid	Stream 13 Vapor	Stream 12 Liquid	Stream 13 Vapor
propylene	0.05	57.48	—	57.53
allyl chloride	0.01	—	0.01	—
hydrogen chloride	0.91	18.78	19.11	0.58
water	81.37	0.63	81.88	0.12
Total	82.34	76.89	101.00	58.23

*This is a model used in the CHEMCAD™ simulator especially for HCl-water and similar systems.

using a model that is specially designed to deal with ionic type compounds (HCl) that dissolve in water and then dissociate. The difference in results is remarkable. The HCl-water system is highly non-ideal and, even though the absorption of an acid gas into aqueous solutions is quite common, the SRK model is not capable of modeling correctly the phase behavior of this system. With the SRK model, virtually all the HCl leaves the absorber as a gas. Clearly, if the simulation were done using only the SRK model, then the results would be drastically in error. This result is especially disturbing since SRK is the default thermodynamics package in many simulators.

More details of model selection are given in Section 18.4.

The importance of thermodynamic model selection and its impact on the validity of the results of a simulation are discussed at length by Horwitz and Nocera [6]. The following statement is from this paper.

> **"You absolutely must have confidence in the thermodynamics that you have chosen to represent your chemicals and unit operations. This is your responsibility, not that of the software simulation package. If you relinquish your responsibility to the simulation package, be prepared for dire consequences."**

18.2.3   Input the Topology of the Flowsheet

The most reliable way to input the topology of the process flow diagram is to make a sketch on paper and have this in front of you when you construct the flowsheet on the simulator. Contrary to the rules given in Chapter 1 on the construction of PFDs, every time a stream splits or several streams combine, a simulator equipment module (splitter or mixer) must be included. These "phantom" units were introduced in Chapter 7, and are useful in tracing streams in a PFD as well as being required for the simulator. Most simulators allow the flowsheet topology to be input both graphically and by keyboard. Certain conventions in the numbering of equipment and streams are used by the simulator to keep track of the topology and connectivity of the streams. When using the graphical interface, the streams and equipment are usually numbered sequentially in the order they are added. These can be altered by the user if required.

18.2.4   Select Feed Stream Properties

As discussed in Section 18.1, the sequential modular approach to simulation requires that all feed streams be specified (composition, flowrate, vapor fraction, temperature, and pressure). In addition, estimates of recycle streams should also be made. Although feed properties are usually well-defined, some confusion may exist regarding the number and type of variables that must be specified to define completely the feed stream. In general, feed streams will contain n components and consist of one or two phases. For such feeds, a total of $n + 2$ specifications completely defines the stream. This is a consequence of the phase rule. Giving the flowrate (kmol/h, kg/s, etc.) of each component in the feed stream takes care of n of these specifications. The remaining two specifications should also be independent. For example, if the stream is one phase, then giving the temperature and pressure of the stream completely defines the feed. Temperature and pressure also completely define a multicomponent stream having two phases. However, if the feed is a single component and contains two phases, then temperature and pressure are not independent. In this case, the vapor fraction and either the temperature or the pressure must be specified. Vapor fraction can also be used to specify a two-phase multicomponent system, but if used, only temperature or pressure can be used to specify completely the feed. To avoid confusion, it is recommended that vapor fraction (vf) be specified only for saturated vapor ($vf = 1$), saturated liquid ($vf = 0$) and two-phase, single-component ($0 < vf < 1$) streams. All other streams should be specified using the temperature and pressure.

Use the vapor fraction (vf) to define feed streams only for saturated vapor ($vf = 1$), saturated liquid ($vf = 0$), and two-phase, single-component ($0 < vf < 1$) streams.

By giving the temperature, pressure, and vapor fraction for a feed, we overspecify the stream and errors will result.

18.2.5   Select Equipment Parameters

It is worth pointing out that process simulators, with a few exceptions, are structured to solve process material and energy balances, reaction kinetics, reaction equilibrium relationships, phase equilibrium relationships, and equipment performance relationships for equipment in which sufficient process design variables have been specified. For example, consider the design of a liquid-liquid extractor to remove 98% of a component in a feed stream using a given solvent. In general, a process simulator will not be able to solve this design problem directly, that is, it cannot determine the number of equilibrium stages required for this separation. However, if the problem is made into a simulation problem, then it can be solved by a trial-and-error technique. Thus, by specifying the number of stages in the extractor, case studies in which the number of stages are varied can be performed and this information used to determine the correct number of stages required to obtain the desired recovery of 98%. In other cases, such as a plug-flow reactor module, the simulator can solve the design problem directly, that is, calculate the amount of catalyst required to carry out the desired reaction. Therefore, before one starts a process simulation it is important to know what equipment parameters must be specified in order for the process to be simulated.

There are essentially two levels at which we can carry out a process simulation. The first level, Level 1, is one in which the minimum data are supplied in order for the material and energy balances to be obtained. The second level, Level 2, is one in which we use the simulator to do as many of the design calculations as possible. The second level requires more input data than the first. An example of the differences between the two levels is illustrated in Figure 18.4. In this figure, we see a heat exchanger in which a process stream is being cooled using cooling water. At the first level, Figure 18.4a, the only information that is specified is the desired outlet

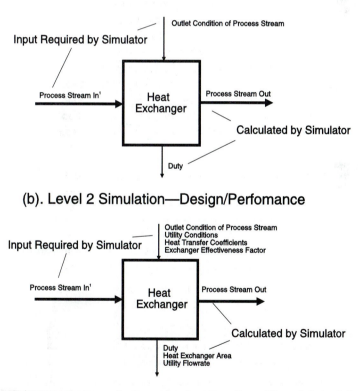

(a). Level 1 Simulation—Basic

Input Required by Simulator

Outlet Condition of Process Stream

Process Stream In'

Heat Exchanger

Process Stream Out

Calculated by Simulator

Duty

(b). Level 2 Simulation—Design/Perfomance

Input Required by Simulator

Outlet Condition of Process Stream
Utility Conditions
Heat Transfer Coefficients
Exchanger Effectiveness Factor

Process Stream In'

Heat Exchanger

Process Stream Out

Calculated by Simulator

Duty
Heat Exchanger Area
Utility Flowrate

' This information may come from the preceding unit operation and thus would be supplied automatically by the simulator.

Figure 18.4   Information Required for Different Levels of Simulation

condition of the process stream, for example, pressure and temperature or vapor fraction, if the stream is to leave the exchanger as a two-phase mixture. However, this is enough information for the simulator to calculate the duty of the exchanger and the properties of the process stream leaving the equipment. At the second level, Figure 18.4b, additional data are provided: the inlet and desired outlet temperature for the utility stream, the fact that the utility stream is water, the overall heat transfer coefficient and the heat exchanger configuration or effectiveness factor, F. Using this information, the simulator calculates the heat exchanger duty, the required cooling water flowrate, and the required heat transfer area.

When attempting to do a simulation on a process for the first time, it is recommended that you provide the minimum data required for a Level 1 simulation. When you have obtained a satisfactory, converged solution, you can go back and provide more data to obtain desired design parameters, that is, a Level 2 solution.

When first simulating a process, input only the data required to perform the material and energy balances for the process.

The structure of the process simulator will determine the exact requirements for the input data, and such information will be available in the user manual for the software. However, for Level 1 simulations, we present a brief list of typical information that may help a novice user prepare the input data for a process simulation.

Pumps, Compressors, and Power Recovery Turbines (Expanders). For pumps, the desired pressure of the fluid leaving the pump *or* the desired pressure increase of the fluid as it flows through the pump is all that is required.

For compressors and turbines, the desired pressure of the fluid leaving the device *or* the desired pressure increase of the fluid as it flows through the equipment are required. In addition, the mode of compression or expansion—adiabatic, isothermal, or polytropic—is required.

Heat Exchangers. For exchangers with a single process stream exchanging energy with a utility stream, all that is required is the condition of the exit process stream. This can be the exit pressure and temperature (single-phase exit condition) or the exit pressure and vapor fraction (two-phase exit condition).

For exchangers with two or more process streams exchanging energy (as might be the case when heat integration is being considered), the exit conditions (pressure and temperature or vapor fraction) for both streams are required. The user must be aware of the possibility of temperature crosses in heat exchange equipment. The simulator may or may not warn the user that a temperature cross has occurred but will continue to simulate the rest of the process. The results from such a simulation will not be valid, and the temperature cross must be remedied before a correct solution can be obtained. Therefore, it is recommended that the user check the temperature profiles for all heat exchangers after the simulation.

Fired Heaters (Furnaces). The same requirements for heat exchangers with a single process fluid apply to fired heaters.

Mixers and Splitters. Mixers and splitters used in process simulators are usually no more than simple tees in pipes. Unless special units must be provided, for example, when the fluids to be mixed are very viscous and in-line mixers might be used, the capital investment of these units can be assumed to be zero.

Mixers represent points where two or more process streams come together. The only required information is an outlet pressure or pressure drop at the mixing point. Usually, the pressure drop associated with the mixing of streams is small and the pressure drop can be assumed to be equal to zero with little error. If feed streams enter the mixer at different pressures, then the outlet stream is assumed to be at the lowest pressure of the feed streams. This assumption causes little error in the material and energy balance. However, the correct analysis of the pressure profiles in a system where several streams mix is given in Section 13.4.

Splitters represent points at which a process stream splits into two or more streams with different flowrates but identical compositions. The required information is the outlet pressure or pressure drop across the device and the relative flows of the output streams. Usually, there is little pressure drop across a splitter, and all streams leaving the unit are at the same pressure as the single feed stream.

Valves. Either the outlet pressure or pressure drop is required.

Reactors. The way in which reactors are specified depends on a combination of the input information required and the reactor category. Generally there are three categories of reactor: Stoichiometric Reactor, Kinetic (Plug Flow or CSTR) Reactor, and Equilibrium Reactor. All these reactor configurations require input concerning the thermal mode of operation—adiabatic, isothermal, amount of heat removed or added. Additional information is also required. Each reactor type is considered separately below.

Stoichiometric Reactor: This is the simplest reactor type that can be simulated. The required input data are the number and stoichiometry of the reactions, the temperature and pressure, and the conversion of the limiting reactant. Reactor configuration (plug flow, CSTR) is not required since no estimate of reactor volume is made. Only basic material and energy balances are performed.

Kinetic (Plug Flow and CSTR) Reactor: This reactor type is used to simulate reactions for which kinetics expressions are known. The number and stoichiometry of the reactions are required input data. Kinetics constants (Arrhenius rate constants and Langmuir-Hinshelwood constants, if used) and the form of the rate equation (simple first-order, second-order, Langmuir-Hinshelwood kinetics, etc.) are also required. Reactor configuration (plug flow, CSTR) is required. Options may be available to simulate cooling or heating of reactants in shell and tube reactor configurations in order to generate temperature profiles in the reactor.

Equilibrium Reactor: As the name implies, this reactor type is used to simulate reactions that obtain or approach equilibrium conversion. The number and stoichiometry of the reactions and the fractional approach to equilibrium are the required input data. In addition, equilibrium constants as a function of temperature may be required for each reaction or may be calculated directly from information in the database. In this mode, the user has control over which reactions should be considered in the analysis.

Minimum Gibbs Free Energy Reactor is another common form of the equilibrium reactor. In the Gibbs reactor, the outlet stream composition is calculated by a free energy minimization technique. Usually data are available from the simulator's databank to do these calculations. The only input data required are the list of components that one anticipates in the output from the reactor. In this mode the equilibrium conversion that would occur for an infinite residence time is calculated.

As a general rule, one should initially use the least complicated reactor module that will allow the heat and material balance to be established. The reactor module can always be substituted later with a more sophisticated one that allows the desired design calculations to be performed. It should also be noted that a common error made in setting up a reactor module is the use of the wrong component as the limiting reactant when a desired conversion is specified. This is especially true when several simultaneous reactions occur and the limiting component may not be obvious just from the amounts of components in the feed.

Flash Units. In simulators, the term *flash* refers to the module that performs a single-stage vapor-liquid equilibrium calculation. Material, energy, and phase equilibrium equations are solved for a variety of input parameter specifications. In order to specify completely the condition of the two output streams (liquid and vapor), two parameters must be input. Many combinations are possible, for example, temperature and pressure, temperature and heat load, or pressure and mole ratio of vapor to liquid in exit streams. Often, the flash module is a combination of two pieces of physical equipment, that is, a phase separator and a heat exchanger. These should appear as separate equipment on the PFD.

Distillation Columns.   Usually, both rigorous (plate-by-plate calculations) and short-cut (Fenske and Underwood relationships using key components) methods are available. In preliminary simulations, it is advisable to use short-cut methods. The advantage of the short-cut methods is that they allow a design calculation (which estimates the number of theoretical plates required for the separation) to be performed. For preliminary design calculations, this is a very useful option and can be used as a starting point for using the more rigorous algorithms, which require that the number of theoretical plates be specified. It should be noted that, in both methods, the calculations for the duties of the reboiler and condenser are carried out in the column modules and are presented in the output for the column. Detailed design of these heat exchangers (area calculations) often cannot be carried out during the column simulation.

Short-Cut Module: The required input for the design mode consists of identification of the key components to be separated, specification of the fractional recoveries of each key component in the overhead product, the column pressure and pressure drop, and the ratio of actual to minimum reflux ratio to be used in the column. The simulator will estimate the number of theoretical plates required, the exit stream conditions (bottom and overhead products), optimum feed location, and the reboiler and condenser duties.

If the short-cut method is used in the performance mode, the number of plates must also be specified, but the R/R_{min} is calculated.

Rigorous Module: The number of theoretical plates must be specified, along with the condenser and reboiler type, column pressure and pressure drop, feed tray location(s), side product locations (if side stream products are desired). In addition, the total number of specifications given must be equal to the number of products (top, bottom, and side streams) produced. These product specifications are often a source of problems, and we illustrate this in Example 18.2 below.

Several rigorous modules may be available in a given simulator. Differences between the modules are the different solution algorithms used and the size and complexity of the problems that can be handled. Tray-to-tray calculations can be handled for several hundred stages in most simulators. In addition, these modules can be used to simulate accurately other equilibrium staged devices, e.g., absorbers and strippers.

Example 18.2

Consider the benzene recovery column in the toluene hydrodealkylation process shown in Figure 1.5. This column is redrawn in Figure E18.2. The purpose of the column is to separate the benzene product from unreacted toluene, which is recycled to the front end of the process. The desired purity of the benzene product is 99.6 mol%. The feed and the top and bottom product streams are presented below in the table, which is taken from Table 1.5.

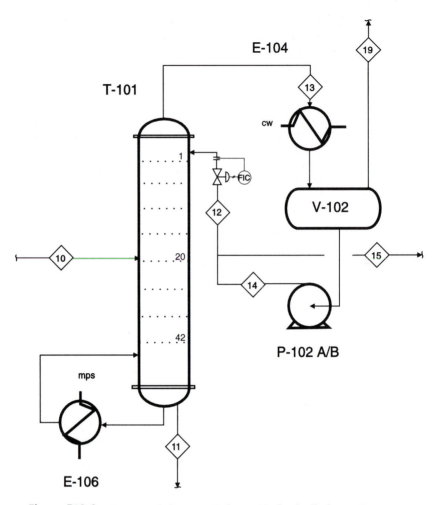

Figure E18.2   Benzene Column in Toluene Hydrodealkylation Process
(from Figure 1.5)

Component	Stream 10	Stream 15	Stream 19	Stream 11
Hydrogen	0.02	—	0.02	—
Methane	0.88	—	0.88	—
Benzene	106.3	105.2	—	1.1
Toluene	35.0	0.4	—	34.6

There are many ways to specify the parameters needed by the rigorous column algorithm used to simulate this tower. We give two examples:

1. We identify the key components for the main separation as benzene and toluene. We specify the composition to be 99.6 mole% benzene and that the recovery of toluene in the bottom product is 0.98.

2. We specify the top composition to be 99.6 mol% benzene and that the recovery of benzene in the bottom product is 0.01.

The first specification violates the material balance while the second specification does not. Looking at the first specification we see that if 98% of the toluene in the feed is recovered in the bottom product, then 2% or 0.7 kmol/h must leave with the top product. Even if the recovery of benzene in the top product were 100%, this would yield a top composition of 106.3 kmol/h benzene and 0.7 kmol/h of toluene. This corresponds to a mole fraction of 0.993. Therefore, the desired mole fraction of 0.996 can never be reached. Thus, by specifying the recovery of toluene in the bottom product, we automatically violate the specification for the benzene purity.

Looking at the second specification, we see that we can achieve both specifications and not violate the material balance. The top product contains 99% of the feed benzene (105.2 kmol/h) and 0.4 kmol/h toluene, which gives a top composition of 99.6 mol% benzene. The bottom product contains 1.0% of the feed benzene (1.1 kmol/h) and 34.6 kmol/h of toluene.

> **When giving the top and bottom specifications for a distillation column, make sure that the specifications do not violate the material balance.**

If problems continue to exist, one way to ensure that the simulation will run is to specify the top reflux rate and the boil-up rate (reboiler duty). Although this strategy will not guarantee the desired purities, it will allow a base case to be established. With subsequent manipulation of the reflux and boil-up rates the desired purities can be obtained.

Absorbers and Strippers.   Usually these units are simulated using the rigorous distillation module given above. The main differences in simulating these types of equipment are that condensers and reboilers are not normally used. In addition, there are two feeds to the unit, one feed enters at the top and the other at the bottom.

Liquid-Liquid Extractors.   A rigorous tray-by-tray module is used to simulate this multistaged equipment. It is imperative that the thermodynamic model for this unit be capable of predicting the presence of two liquid phases, each with appropriate liquid-phase activity coefficients.

18.2.6   Selection of Output Display Options

Several options will be available to display the results of a simulation. Often, a report file can be generated and customized to include a wide variety of stream and equipment information. In addition, a simulation flowsheet (not a PFD), T-Q diagrams for heat exchangers, vapor and liquid flows, temperature and composition profiles (tray-by-tray) for multistaged equipment, and a wide variety of phase diagrams for streams can be generated. The user manual should be consulted for the specific options available for the simulator you use.

18.2.7   Selection of Convergence Criteria and Running a Simulation

For equipment requiring iterative solutions, there will be user-selectable convergence and tolerance criteria in the equipment module. There will also be convergence criteria for the whole flowsheet simulation, which may be adjusted by the user.

 The two most important criteria are number of iterations and tolerance. These criteria will often have default values set in the simulator. Unless specific problems arise, these default values should be used in your simulations.

If the simulation has not converged, the results do not represent a valid solution and should not be used.

When convergence is not achieved, three common causes are:

1. The problem has been ill posed. This normally means that an equipment specification has been given incorrectly. For example, see the first specification in Example 18.2 for the rigorous column module.
2. The tolerance for the solution has been set too tightly, and convergence cannot be obtained to the desired accuracy no matter how many solution iterations are performed.
3. The number of iterations is not sufficient for convergence. This occurs most often when the flowsheet has many recycle streams. Rerunning the flowsheet simulation with the results from the previous run may give a converged solution. If convergence is still not obtained, then one way to address this problem is to remove as many recycle streams as possible. The simulation is then run, and the recycle streams are added back, one by one, using the results from the previous simulation as the starting point for the new one. This method is discussed in more detail in Section 18.3.

Of the three reasons above, the first one is by far the most common.

> **The most common reason for the failure of a simulation to con-verge is the use of incorrect or impossible equipment specifica-tions.**

18.3   HANDLING RECYCLE STREAMS

Recycle streams are very important and common in process flowsheets. Computationally, they can be difficult to handle and are often the cause for unconverged flowsheet simulations. There are ways in which we can minimize the problems caused by recycle streams, and when we attempt to simulate a flowsheet for the first time, it is wise to consider carefully any simplifications that may help the convergence of the simulation. Let us first consider the simulation of the DME flowsheet illustrated in Figure B.1, Appendix B. This flowsheet is shown schematically in Figure 18.5a. The DME process is simple, no by-products are formed, the separations are relatively easy, and the methanol can be purified easily prior to being returned to the front end of the process. In attempting to simulate this process for the first time, it is evident that two recycle streams are present. The first is the unreacted methanol that is recycled to the front of the process, upstream of the reactor. The second recycle loop is due to the heat integration scheme that is used to preheat the reactor feed using the reactor effluent stream. The best way to simulate this flowsheet is to eliminate the recycle streams as shown in Figure 18.5b. In this figure, two separate heat exchangers have been substituted for the heat integration scheme. These exchangers allow the streams to achieve the same changes in temperature while eliminating the interaction between the two streams. The methanol recycle is eliminated in Figure 18.5b by producing a methanol pseudo output stream. The simulation of the flowsheet given in Figure 18.5b is straightforward; it contains no recycle streams and will converge in a single flowsheet iteration. Troubleshooting of the simulation, if input errors are present, is very easy since the flowsheet converges very quickly. Once a converged solution has been obtained, the recycle streams can be added back. For example, we would first add the methanol recycle stream back into the simulation. We know the composition of this stream from the previous simulation, and this will be a very good estimate for the recycle stream composition. Once the simulation has been run successfully with the methanol recycle stream, the heat integration around the reactor can be added back and the simulation run again. Although this method may seem unwieldy, it does provide a reliable method for obtaining a converged simulation.

For the DME flowsheet in Figure 18.5, the unreacted methanol that was recycled was almost pure feed material. This means that the estimate of the recycle stream composition, obtained from the once-through simulation using Figure 18.5b, was very good. When the recycle stream contains significant amounts of

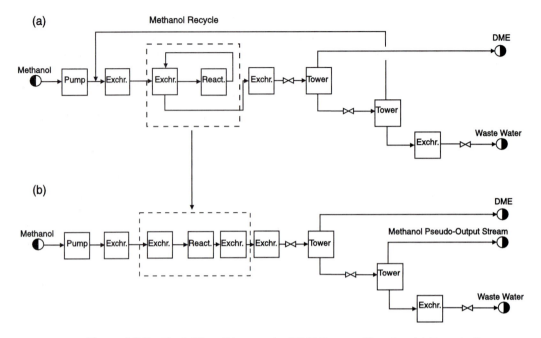

Figure 18.5   Block Flow Diagram for DME Process Showing (a) Recycle Structure, (b) Elimination of Recycles

by-products, as is the case with the hydrogen recycle stream in Figure 1.5 (Streams 5 and 7), the estimate of the composition using a once-through simulation will be significantly different from the actual recycle stream composition. For such cases, when purification of the recycle stream does not occur, it is best to keep this recycle stream in the flowsheet and eliminate all other recycle streams for the first simulation. Once a converged solution is reached, the other recycle streams can be added back one at a time.

Often, we will want to run a series of case studies using a base case simulation as a starting point. This is especially true when we are performing a parametric optimization on the process (see Chapter 19). When performing such case studies, it is wise to make small changes in input parameters in order to obtain a converged simulation. For example, assume that a converged simulation for a reactor module at 350°C has been obtained, and we want to run a case study at 400°C. When the equipment temperature in the reactor module is changed and the simulation is rerun, we may find that the simulation does not converge. If this is the case, for example, we should go back to the base case run, change the reactor temperature by 25°C and see if this converges. If it does, then we can change the input by another 25°C to give the desired conditions, and so on. The use of small increments or steps when simulating changes in flowsheets often produces a converged simulation when a single large change in input will not.

Often when simulating a process, it is the flowrate of products (not feeds) that is known, for example, production of 60,000 tonne/yr of chemical X, with a purity of 99.9 wt%. Assume that a converged solution has been found in which all the product specifications have been met except that the flowrate of primary product is not at the desired value. For this case, it is a simple matter to multiply all the feeds to the process by a factor to obtain the desired flowrate of product, that is, just scale the solution up or down by a constant factor and rerun the simulation to get the correct equipment specifications.

For more advanced simulation applications, such as optimization or simulating existing plants, it may be necessary or useful to use controller modules in the simulation to obtain a desired result. For example, in a recycle loop it might be required that the ratio of two components entering a reactor be set at some fixed value. A controller module could be used to adjust the purge flowrate from the recycle stream to obtain this ratio. The use of controller modules introduces additional recycle loops. The way in which specifications for controllers are given can cause additional convergence problems, and this topic is covered in detail by Schad [7].

18.4 THERMODYNAMIC MODELS

The results of any process simulation are only as good as the input data.

> **Everything from the energy balance to the volumetric flowrates to the separation in the equilibrium-stage units is dependent on accurate thermodynamic data.**

No other choice made in the simulation has as great an effect on the results. If reaction kinetics information is missing, the simulator will not calculate the conversion from a given reactor volume. The user is reminded that such a calculation is not possible, and only equilibrium reactor modules and those with specified conversion can be used.

> **Unfortunately, process simulators have default thermodynamics packages, which will—without warning—blindly miscalculate the entire flowsheet.**

This problem arises because only a few, readily available data are required to estimate the parameters in simple thermodynamic models. If the critical temperature and critical pressure are known for each pure component, the parameters for simple, cubic equations of state can be estimated. Even if these critical properties are unknown, they in turn can be estimated from one vapor pressure and one liquid density. Group-contribution models require even less information: merely the chemical structure of the molecule. In thermodynamics, as elsewhere, *you only get what you pay for—or less*!

Compounding this problem is the development and implementation of "expert systems" to help choose the thermodynamic model. To date, these features offer the promise of false hope. Human thermodynamics experts do not recommend them.

A safe choice of thermodynamic model requires knowledge of the system, the calculational options of the simulator, and the margin of error. In this section, we offer guidance on choosing and using a thermodynamic model.

18.4.1   Physical Properties

Physical properties such as density, viscosity, thermal conductivity, and heat capacity are generally not a serious problem in simulation. The group-contribution methods are reasonably good, and experimental data for heat capacity are included in the simulator databanks for over a thousand substances. Although these correlations have random and systematic errors of several percent, this is close enough for most purposes. (However, clearly, they are not sufficient when you are paying for a fluid crossing a boundary based on volumetric flowrate.)

The main problems for simulation are (1) enthalpy and (2) phase equilibrium.

18.4.2   Enthalpy

Although the pure-component heat capacities are calculated with acceptable accuracy, the enthalpies of phase changes often are not. Care should be taken in choosing the enthalpy model for a simulation. If the enthalpy of vaporization is an important part of a calculation, simple equations of state should not be used. In fact, the "latent heat" or "ideal" options would be better. If the substance is above or near its critical temperature, equations of state must be used, but the user must beware, especially if polar substances such as water are present.

Example 18.3

A gas stream at 3000°F of the following concentration is to be cooled by evaporation of 500 kg/h of water entering at 70°F. Assume atmospheric pressure.

H_2	22.72 kg/h
N_2	272.24
CO	268.40
HCl	26.84

Perform a simulation to determine the final temperature of the cooled gas stream with the default thermodynamic model and with the ideal model.

The default on most process simulators is an equation of state, either PR or SRK. These models give an outlet temperature of 480°F. The ideal model gives an outlet temperature of 348°F.

18.4.3   Phase Equilibria

Extreme care must be exercised in choosing a model for phase equilibria (sometimes called the fugacity coefficient or K-factor model). Any available phase equilibrium data for the system should be used to regress the parameters in the model, or at least the deviation between the model predictions and the experimental data should be studied.

There are two general types of fugacity models: equations of state and liquid-state activity-coefficient models. An equation of state is an algebraic equation for the pressure of a mixture as a function of the composition, volume, and temperature. Through standard thermodynamic relationships, the fugacity, enthalpy, and so on, for the mixture can be determined. These properties can be calculated for any density; therefore, both liquid and vapor properties, as well as supercritical phenomena, can be determined.

Activity-coefficient models, however, can only be used to calculate liquid-state fugacities and enthalpies of mixing. These models provide algebraic equations for the activity coefficient (γ_i) as a function of composition and temperature. Since the activity coefficient is merely a correction factor for the ideal-solution model (essentially Raoult's Law), it cannot be used for supercritical or "non-condensable" components. Modifications of these models for these types of systems have been developed, but they are not recommended for the process simulator user without constant consultation with a thermodynamics expert.

Equations of state are recommended for simple systems (non-polar, small molecules) and in regions (especially supercritical conditions for any component in a mixture) where activity-coefficient models are inappropriate. For complex liquid mixtures, activity coefficient models are preferred, but only if *all* of the BIPs are available.

Equations of State.   The default fugacity model is normally either the SRK (Soave-Redlich-Kwong) or the PR (Peng-Robinson) equation. They (like most popular equations of state) normally use three pure-component parameters per substance and one binary-interaction parameter per binary pair. Although they give qualitatively correct results even in the supercritical region, they are known to be poor predictors of enthalpy changes, and (except for light hydrocarbons) they are not quantitatively accurate for phase equilibrium.

The predicted phase equilibrium is a strong function of the binary-interaction parameters (BIPs). Process simulators have regression options to determine these parameters from experimental phase equilibrium data. The fit gives a first-order approximation for the accuracy of the equation of state. This information should always be considered in estimating the accuracy of the simulation. Additional simulations should be run with perturbed model parameters to get a feel for the uncertainty, and the user should realize that even this approach gives an optimistic approximation of the error introduced by the model. If BIPs are provided in the simulator, and the user has no evidence that one equation of state is better than another, one should perform a separate, complete simulation for each of these equations of state. The difference between the simulations is a crude measure of the uncertainty introduced into the simulation by the uncertainty in the models. Again, the inferred uncertainty will be on the low side.

If BIPs are available for only a subset of the binary pairs, extreme caution must be exercised. Assuming the unknown BIPs to be zero is dangerous. There are group-contribution models for determining BIPs for equations of state, but, again, the user should beware.

Liquid-State Activity Coefficient Models.   The uncertainty of the simulation should be estimated through resimulations at different, seemingly equally accurate parameter sets.

> **For most chemical systems below the critical region, a liquid-state activity coefficient model is the better choice.**

The most popular liquid-state activity coefficient models are Wilson, NRTL, UNIQUAC, and UNIFAC. The first three require phase equilibrium data for parameter regression (2 BIPs per binary pair). The UNIFAC model is a group-contribution model for determining the BIPs for the UNIQUAC model. Only chemical structure data are needed, but the calculations are not very accurate. When determining the numbers of groups within a molecule, one should always start with the largest group. This strategy minimizes the assumptions (and, therefore, the errors) in the model. Group-contribution models should be used with extreme care.

For many systems, a model such as UNIFAC may be the only option. If so, even a very crude uncertainty estimate can be difficult. If just one phase equilibrium datum for the system can be found, its deviation from the model prediction is at least some estimate of the uncertainty (as long as that datum was not used to regress parameters).

Using Scarce Data to Calibrate a Thermodynamic Model.   Any experimental data on phase equilibria can be used to perform a crude calibration or verification of the model. It need not be the type of data that would be taken in the lab. If one knows the recovery in a column for one set of conditions, for example, and if only one BIP is unknown, one can easily find the only value of the BIP that will reproduce that datum. Such data are sometimes found in patents.

More Difficult Systems.   The above discussions pertain to "easy" systems: (1) small, nonpolar or slightly polar molecules for equations of state and (2) non-electrolyte, non-polymeric substances considerably below their critical temperatures for liquid-state activity-coefficient models. Most simulators have some models for electrolytes and for polymers, but these are likely to be even more uncertain than for the easy systems. Again, the key is to find some data, even plant operating data, to verify and to calibrate the models. If the overall recovery from a multistage separation is known, for example, one can simulate the column, using the best known thermodynamic model, and the deviation between the plant datum and the simulator result is a crude (optimistic) estimate of the uncertainty.

Since most thermodynamic options are semitheoretical models for small, nonpolar molecules, the more difficult systems require another degree of freedom in the model. The most common such modification is to make the parameters temperature dependent. This requires additional data, but there is some theoretical justification for using effective model parameters that vary with temperature.

18.4.4   Using Thermodynamic Models

In summary, one should use the assumed best thermodynamic model, based on the rough guidelines above. However, one should always resimulate with either another model that is assumed to be equally good or with different model parameters. The appropriate perturbation to apply to the parameters is available from the experimental data regression or from comparison of calculated results with experimental or plant data. Such data should always be sought for the conditions closest to those in the simulation. If the application is liquid-liquid extraction, for example, liquid-liquid equilibrium data (rather than vapor-liquid equilibrium data) should be used in the parameter regression, even though the same activity-coefficient models are used for both liquid-liquid and vapor-liquid equilibria.

> **The availability of BIPs in the databank of a process simulator must never be interpreted as an indication that the model is of acceptable accuracy.**

These parameter values are merely the "best" for some specific objective function, for some specific set of data. The model may not be able to correlate even these data very well with this optimum parameter set. And one should treat a decision to use a BIP equal to zero as equivalent to using an arbitrary value of the BIP. The decision to use zero is, in fact, a decision to use a specific value based on little or no data.

Example 18.4

Consider the DME Tower, T-201, in the DME process (Figure B.1). Simulate this unit using a shortcut model with different liquid-state activity coefficient models to determine the required number of stages for the reflux ratio and the recoveries shown in Appendix B. Include no corrections for heat of mixing. Then perform the rigorous column simulation to check the distillate purity. Compare the results.

The base case (Figure B.1 and Tables B.1 and B.2) uses the UNIFAC model with vapor-phase fugacity correction. The specifications are:

- Column pressures. Top: 10.3 bar; Bottom: 10.5 bar
- Key components. Light: DME; Heavy: methanol
- Light key recovery in distillate: 98.93%
- Heavy key recovery in bottoms: 99.08%
- Reflux ratio: 0.3631
- 70% plate efficiency

An initial CHEMCAD simulation confirms the value given in Appendix B of 22 actual stages.

Table E.18.4 shows the results obtained from six simulations, all with the same input specifications but with different thermodynamic options. The number of actual stages calculated ranges from 15 to 24. Without further information about the ability of the various models to correlate experimental vapor-liquid equilibrium data, a precise solution to the problem is not possible. However, the differences in the results obtained indicate that the

Table E.18.4   Comparison of Simulations for DME Column, T-201, for Various Thermodynamic Options

Thermodynamic Option	Number of Actual Stages Obtained from Shortcut Method	Methanol in Distillate (mol%) from Rigorous Simulation with 22 Stages
UNIFAC w/RK correction	22	3.26
UNIFAC	15	5.90
NRTL w/RK correction	23	3.89
NRTL	16	6.55
UNIQUAC w/RK correction	24	3.58
UNIQUAC	16	5.15

choice of thermodynamic model is a crucial one. Of special concern here is the choice of vapor-phase correction of fugacities with the RK equation.

To determine which model is best, one finds data on the various binaries in the mixture. These data consist of measurements of temperature, pressure, and composition of a liquid in equilibrium with its vapor. Sometimes, the concentration of the vapor is also measured. From the temperature and the liquid-phase composition, the pressure and vapor-phase compositions are calculated with the thermodynamic model. The sum of the squared deviations between the experimental and the calculated pressure and between the experimental and the calculated vapor compositions is the objective function. The decision variables are the adjustable parameters in the thermodynamic model. Through the procedures of Chapter 19, the objective function is minimized and the optimum set of parameters are found.

Of great concern is the purity of the overhead product. This stream is the DME product stream from the process, and the PFD in Appendix B (Figure B.1) shows its concentration as 0.46 mol% methanol. The third column in Table E.18.4 shows the results of rigorous, tray-by-tray simulations of this column. In each case, the number of stages, feed location, column pressures, reflux ratio, and reboiler duty were set at those shown in Appendix B. Again, only the thermodynamic option was varied.

The variations between simulations for different thermodynamic models are significant, but the greater deviation comes from the type of calculation done. In the shortcut calculation, many assumptions are made, including the constancy of relative volatilities. For the rigorous calculation, the full power of the thermodynamic package is used in the phase-equilibrium and energy-balance calculations. In this example, the distillate purity specification would be very far off from that calculated with the short-cut method.

18.5 CASE STUDY—TOLUENE HYDRODEALKYLATION PROCESS

The purpose of this section is to present the input information necessary to make a basic simulation of the toluene hydrodealkylation process presented in Chapter 1. The required input data necessary to obtain a Level 1 simulation is presented in Table 18.1. The corresponding simulator flowsheet is given in Figure 18.6. In Table 18.1, the equipment numbers given in the third column correspond to those used in Figure 18.6. In the first column, the equipment numbers on the toluene hydrodealkylation PFD, Figure 1.5, are given. It should be noted that there is not a one-to-one correspondence between the actual equipment and the simulation modules. For example, 3 splitters and 6 mixers are required in the simulation, but these are not identified in the PFD. In addition, several pieces of equipment associated with the benzene purification tower are simulated by a single simulation unit. The numbering of the streams in Figure 18.6 corresponds to that given in Figure 1.5, except when additional stream numbers are required for the simulation. In order to avoid confusion, these extra streams are assigned numbers greater than 90.

Table 18.1 Required Input Data for a Level 1 Simulation of Toluene Hydrodealkylation Process

Equipment Number	Simulator Equipment	Simulator Equip. No	Input Streams		Output Streams		Required Input
TK-101	mixer	m-1	1	11	90	—	Pressure drop = 0 bar
P-101	pump	p-1	90	—	2	—	Outlet pressure = 27.0 bar
E-101	hexch	e-1	92	—	4	—	Outlet stream vapor fraction = 1.0
H-101	heater	h-1	4	—	6	—	Outlet temperature = 600°C
R-101	stoic react	r-1	93	—	9	—	Conversion of toluene = 0.75
E-102	flash	f-1	9	—	8	94	Temperature = 38°C Pressure = 23.9 bar
V-101	flash	f-1	9	—	8	94	No input required since vessel is associated with flash operation.
V-103	flash	f-2	94	—	17	18	Temperature = 38°C Pressure = 2.8
E-103	hexch	e-2	18	—	10		Outlet temperature = 90°C
T-101	Shortcut tower	t-1	10	—	19	11	Recovery of benzene in top product = 0.99 Recovery of toluene in top product = 0.01 R/R_{min} = 1.5 Column pressure drop = 0.3 bar
E-104	Shortcut tower	t-1	10	—	19	11	Included in tower simulation
E-106	Shortcut tower	t-1	10	—	19	11	Included in tower simulation
V-102	Shortcut tower	t-1	10	—	19	11	Not required in simulation
P-102	Shortcut tower	t-1	10	—	19	11	Not required in simulation
E-105	hexch	e-3	95	—	15	—	Outlet temperature = 38°C
C-101	compr	c-1	97	—	98	—	Outlet pressure = 25.5 bar
	mixer	m-2	3	5	91	—	Pressure drop = 0 bar
	mixer	m-3	2	91	92	—	Pressure drop = 0 bar
	mixer	m-4	6	7	93	—	Pressure drop = 0 bar
	mixer	m-5	17	96	99	—	Pressure drop = 0 bar
	mixer	m-6	99	100	16	—	Pressure drop = 0 bar
	splitter	s-1	8	—	97	96	Pressure drop = 0 bar
	splitter	s-2	98	—	5	7	Pressure drop = 0 bar
	splitter	s-3	19	—	100	95	Pressure drop = 0 bar

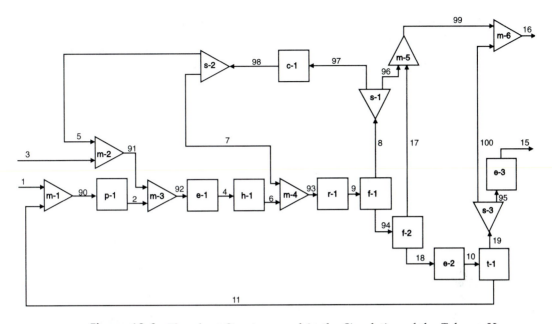

Figure 18.6   Flowsheet Structure used in the Simulation of the Toluene Hydrodealkylation Process

In Table 18.2, the specifications for the feed streams are given. For this process, there are only two feed streams, Streams 1 and 3, corresponding to toluene and hydrogen, respectively. In addition, estimates of all the recycle streams should be given prior to beginning the simulation, and these are given in

Table 18.2   Feed Stream Properties and Estimates of Recycle Streams

	Stream 1	Stream 3	Stream 11	Stream 5	Stream 7
Temperature (°C)	25.0	25.0	150.0	50.0	50.0
Pressure (bar)	1.9	25.5	2.8	25.5	25.5
Hydrogen (kmol/h)	—	286.0	—	200.0	20.0
Methane (kmol/h)	—	15.0	—	200.0	20.0
Benzene (kmol/h)	—	—	—	—	—
Toluene (kmol/h)	108.7	—	30.0	—	—

Table 18.2. However, these estimates need not be very accurate. Usually, any estimate is better than no estimate at all.

The data given in Tables 18.1 and 18.2 are sufficient to reproduce the material and energy balances for the toluene hydrodealkylation process. The use of these data to reproduce the flow table in Table 1.5 is left as an example problem at the end of the chapter. As mentioned in Section 18.3, some difficulty may arise when trying to simulate this flowsheet because of the three recycle streams. If you encounter problems in obtaining a converged solution, you should try to eliminate as many recycle streams as possible, run the simulation, and then add the recycle streams back into the problem one at a time. The thermodynamic models for this simulation should be chosen using the guidelines in Section 18.4 or using the expert system in the simulator that you use. The results given in Chapter 1 for this process were obtained using the SRK models for enthalpy and phase equilibria.

18.6   SUMMARY

In this chapter, we reviewed the general components of a process simulator and the seven types of input required to simulate a process successfully. We covered in detail each of the seven required inputs: selection of chemical components, selection of thermodynamic models, selection of process topology, selection of feed stream properties, selection of equipment parameters, selection of output options, and selection of convergence criteria.

Special attention was paid to the role of recycle streams in obtaining converged solutions, and methods to help convergence were discussed. The selection of thermodynamic models and their importance were discussed in depth. Finally, a case study for the toluene hydrodealkylation process given in Chapter 1 was given and the required data to complete a process simulation was presented.

REFERENCES

1. Westerberg, A. W., H. P. Hutchinson, R. L. Motard, and P. Winter, *Process Flowsheeting* (Chapter 2), Cambridge University Press, Cambridge, 1979.
2. Franks, R. G. E., *Modeling and Simulation in Chemical Engineering* (Chapter 2), Wiley, New York, 1972.
3. Carnahan, B., H. A. Luther, and J. O. Wilkes, *Applied Numerical Methods* (Chapter 5), Wiley, New York, 1969.
4. Soave, G., "Equilibrium Constants from a Modified Redlich-Kwong Equation of State," *Chem. Eng. Sci.*, **27,** 1197 (1972).

5. Redlich, O., and J. N. S. Kwong, "On the Thermodynamics of Solutions: V: An Equation of State. Fugacities of Gaseous Solutions," *Chem. Rev.*, **44**, 233 (1949).

6. Horwitz, B. A., and A. J. Nocera, "Are You 'Scotamized' by your Simulation Software?," *CEP*, **92** (9), 68 (1996).

7. Schad, R. C., "Don't Let Recycle Streams Stymie Your Simulations," *CEP*, **90** (12), 68 (1994).

PROBLEMS

1. For the toluene hydrodealkylation process, using the data given in Tables 18.1 and 18.2, simulate the process and compare the results with those given in Chapter 1, Table 1.5. Remember that the number of actual plates is given in Table 1.7 and an efficiency of 60% was assumed.

2. For the DME flowsheet given in Appendix B, Figure B.1 list the minimum input information required to obtain mass and energy balances for this process. Using the process simulator available to you, simulate the DME process and compare your results to those given in Table B.2. Assume a tray efficiency of 70% for both distillation columns.

3. For the IPA to acetone process flowsheet given in Appendix B, Figure B.3, list the minimum input information required to obtain mass and energy balances for this process. Using the process simulator available to you, simulate the DME process and compare your results to those given in Table B.6. Assume a tray efficiency of 70% for columns T-402 & T-403.

4. Using the results from Problem 1 above, and Tables 1.5 and 1.7, compare the results for the simulation of the benzene recovery column, T-101, using a shortcut method and a rigorous method. One way to do this comparison is to use the number of theoretical plates from the shortcut method as an input to the rigorous method. The rigorous method is used to simulate the same separation as the shortcut method, that is, same overhead purity and recovery. The difference in the methods is then reflected by the difference between the reflux required for both methods. Comment on the difference for this near ideal system. Remember that there is no need to simulate the whole flowsheet for this problem just the use the input to the column from Table 1.5.

5. In Problem 1 above, you should have simulated the reactor as a stoichiometric reactor with 75% per pass conversion. In order to estimate the volume of the reactor, it is necessary to have kinetics expressions. For the catalytic hydrodealkylation of toluene, assume that the reaction is kinetically controlled with the following kinetics:

$$-r_{tol} = kc_{tol}c_{hyd}^{0.5} \frac{\text{kmol}}{\text{m}^3 \text{ reactor s}}$$

where

$$k = 2.833 \times 10^7 e^{-\frac{17814}{T(K)}} \frac{\text{m}^{1.5}}{\text{kmol}^{0.5}\text{s}}$$

With these kinetics, simulate the reactor in Figure 1.5 as a two-stage, packed bed adiabatic reactor with a "cold shot" (Stream 7) injected at the inlet to the second bed. The maximum temperature in the reactor should not exceed 655°C, and this will occur at the exit of both beds. Design the system for this maximum outlet temperature for both packed beds. Compare your results with the total volume of the catalyst given in Table 1.7.

19

Process Optimization

Optimization is the process of improving an existing situation, device, or system such as a chemical process. In this chapter, we present techniques and strategies to:

- Set up an optimization problem
- Quantify the value of a potential improvement
- Identify quickly the potential for improvement
- Identify the constraints, barriers, and bottlenecks to improvement
- Choose an appropriate procedure to find the best change
- Evaluate the result of the optimization

We start with some basic definitions of terms and then investigate several techniques and strategies to perform the optimization of a process.

19.1 BACKGROUND INFORMATION ON OPTIMIZATION

In optimization, various terms are used to simplify discussions and explanations. These are defined below.

Decision variables are those independent variables over which the engineer has some control. These can be continuous variables such as temperature or discrete (integer) variables such as number of stages in a column. Decision variables are also called *design variables*.

An *objective function* is a mathematical function that, for the best values of the decision variables, reaches a minimum (or a maximum). Thus, the objective function is the measure of value or goodness for the optimization problem. If it is a profit, one searches for its maximum. If it is a cost, one searches for its minimum. There may be more than one objective function for a given optimization problem.

Constraints are limitations on the values of decision variables. These may be linear or nonlinear, and they may involve more than one decision variable. When a constraint is written as an equality involving two or more decision variables, it is called an *equality constraint*. For example, a reaction may require a specific oxygen concentration in the combined feed to the reactor. The mole balance on the oxygen in the reactor feed is an equality constraint. When a constraint is written as an inequality involving one or more decision variables, it is called an *inequality constraint*. For example, the catalyst may operate effectively only below 400°C, or below 20 MPa. An equality constraint effectively reduces the *dimensionality* (the number of truly independent decision variables) of the optimization problem. Inequality constraints reduce (and often bound) the search space of the decision variables.

A *global optimum* is a point at which the objective function is the best for all allowable values of the decision variables. There is no better acceptable solution. A *local optimum* is a point from which no small, allowable change in decision variables in any direction will improve the objective function.

Certain classes of optimization problems are given names. If the objective function is linear in all decision variables and all constraints are linear, the optimization method is called *linear programming*. Linear programming problems are inherently easier than other problems and are generally solved with specialized algorithms. All other optimization problems are called *nonlinear programming*. If the objective function is second order in the decision variables and the constraints are linear, the nonlinear optimization method is called *quadratic programming*. For optimization problems involving both discrete and continuous decision variables, the adjective *mixed-integer* is used. Although these designations are used in the optimization literature, we deal in this chapter mainly with the general class of problems know as MINLP, *mixed-integer nonlinear programming*.

The use of linear and quadratic programming is limited to a relatively small class of problems. Some examples in which these methods are used include the optimal blending of gasoline and diesel products and the optimal use of manufacturing machinery. Unfortunately, many of the constraints that we run into in chemical processes are not linear, and the variables are often a mixture of continuous and integer. A simple example of a chemical engineering problem is the evaluation of the optimal heat exchanger to use in order to heat a stream from 30°C to 160°C. This simple problem includes continuous variables, such as the area of the heat exchanger and the temperature of the process stream, and integer variables, such as whether to use low-, medium-, or high-pressure steam as the heating medium. Moreover, there are constraints such as the materials of construction that depend, nonlinearly, on factors such as the pressure, temperature, and composition of the process and utility streams. Clearly, most chemical process problems are quite involved, and we must be careful to consider all the constraints when evaluating them.

19.1.1 Common Misconceptions

> **A common misconception is that optimization is a complex, esoteric, mathematical exercise. In fact, optimization is usually a dynamic, creative activity involving brainstorming, exploring alternatives, and asking "what if .. ?"**

Although some problems are simple enough to be posed in closed form and to be solved analytically, most real problems are more interesting.

For example, Figure 19.1a shows a classical curve of annualized pumping cost vs. pipe diameter. The annualized pumping cost includes annualized equipment (pump plus pipe) cost and power (operating) cost. These two components of cost have been calculated based on smooth cost functions and are shown as

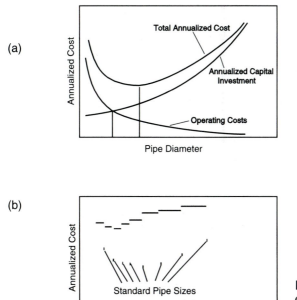

Figure 19.1 Optimization using Continuous and Discrete Functions of Pipe Diameter

separate curves on Figure 19.1a. The combined curve represents the total annualized cost. Using these results, one can analytically determine the pipe diameter at which the slope of the combined cost curve is zero and the second derivative is positive. It is important to note that, even with all the assumptions used in this analysis, the minimum annualized cost does not occur at the pipe diameter where the two component curves cross.

> **The point of intersection of the curves of annual operating cost and of annualized capital cost is not the optimum.**

In reality, the cost function (if one were to bother to calculate it) would look something like Figure 19.1b, because only certain pipe diameters and pump sizes are standard equipment. Other sizes could be produced, but only at much higher costs. Thus, only a few cost evaluations are needed—those at the standard pipe sizes.

Another common misconception is that the optimum will usually be found at a point where the first derivatives of the cost function are zero. Even when the cost function is continuous and smooth, this is seldom the case and should never be assumed. Nearly all problems of any reasonable complexity have optima along at least one constraint. Figure 19.2 shows such a case. Again, both annualized capital costs and operating costs are included. Although there is a point of zero slope (point A), the best design (minimum annual cost) shown is at point B.

One must not assume that the best solution has been found when it merely has been bracketed. Such an assumption not only leads to a false conclusion, but

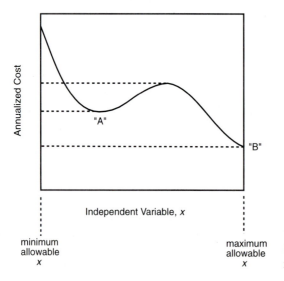

Figure 19.2   Location of the Optimum Value of a Variable

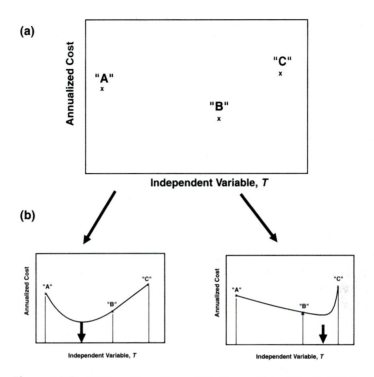

Figure 19.3   Different Locations of Optimum Using Identical Data

it ignores useful information for estimating the true optimum solution. For example, Figure 19.3a shows three points that have been calculated for the optimization of the temperature of a flash unit. If the objective function is continuous and smooth, and if there is indeed only one minimum, the optimum lies between points A and C. However, there is no reason to believe that Point B is the optimum, and there are insufficient data to determine whether the optimum lies to the left or to the right of Point B. However, by approximating the objective function simply by a quadratic, for the three points bracketing the optimum, we can estimate the optimum, as shown on the left-hand side of Figure 19.3b. However, the true curve may look like the right hand side of Figure 19.3b. To be certain that the optimum has been located correctly, we continue to evaluate more points between A and C, using successive quadratic approximations based on the best point achieved plus the closest point on either side of it.

19.1.2   Estimating Problem Difficulty

A key step in any problem-solving strategy is to estimate the effort required to reach the solution. Not only can this estimate provide motivation, but it can also help to redirect resources if the effort (time, money, personnel) required for a "complete" solution is more than the effort available. The first step is to decide

Table 19.1   Characteristics of Easy and Difficult Optimization Problems

Easy Problems	Difficult Problems
Few decision variables	Many decision variables
Independent (uncorrelated) decision variables	Correlated decision variables
Discrete decision variables	Mixed discrete and continuous decision variables
Topological optimization first	Parametric optimization first
Single process units	Multiple, interrelated process units
Separate constraints for each decision variable	Constraints involving several decision variables
Constraints are obvious	Constraints are not obvious or become obvious only after the optimization has begun
Single objective	Multiple objectives
Objective function easy to quantify	Objective function difficult to quantify
Linear objective function	Highly nonlinear objective function
Smooth objective function	Kinked objective function
No local optima	Many deep, local optima

whether the problem at hand is an "easy" problem or a "difficult" problem. Table 19.1 lists some characteristics of inherently easy and inherently difficult problems.

For a typical flowsheet, such as the DME (dimethyl ether) PFD in Figure B.1 (Appendix B), there are many decision variables. The temperature and pressure of each unit can be varied. The sizes of each piece of equipment involve decision variables (usually several per unit). The reflux in tower T-201 and the purity of the distillate from T-202 are decision variables. There are many more. Clearly, the simultaneous optimization of all of these decision variables is a difficult problem. However, some sub-problems are relatively easy. If Stream 4 (the exit from the methanol preheater) must be at 154°C, for example, the choice of which heat source to use (lps, mps, or hps) is easy. There is only a single decision variable, there are only three discrete choices, and the choice has no direct impact on the rest of the process. The problem becomes more difficult if the temperature of Stream 4 is not constrained.

19.1.3   Top-down/Bottom-up Strategies

For any real process of any complexity, the true "global" optimum (at which any change in any decision variable would degrade the system) will not be found. If it could be found with an acceptable level of effort, one would concentrate on minute changes in decision variables. Rather than become bogged down in such details, experts at optimization tend to look alternately at the big picture and at the details. The overview encourages one to make bold changes in process configuration or variable values, but the closer study is needed to confirm whether the changes are true improvements.

The strategy of looking at the big picture first, followed by the detailed study, is called *top-down*. Looking at the DME flowsheet, such a strategy might lead to one questioning the need for E-201 (Methanol Preheater) and E-203 (DME Cooler). A detailed study of the flowsheet would confirm whether these two units should be eliminated in favor of a larger E-202 (Reactor Cooler). On the other hand, a detailed study of incremental changes in the heat duties of these two heat exchangers would have led to the same solution. Such a strategy is called *bottom-up*. Before investing too much time on detailed calculations, check the big picture. When the big picture is not clear or leads to conflicting alternatives, run the detailed calculations. The key is flexibility.

Details of the top-down and bottom-up approaches are covered in Sections 19.3 and 19.4 under topological and parametric optimization.

19.1.4   Communication of Optimization Results

As will be discussed in Chapter 22, the best strategy for obtaining results is not the best strategy for presenting those results to others. While the goal of optimization is to find the best solution (and the sensitivity of that solution as explained in Section 19.6) efficiently, the goal of the presentation is to convince the audience clearly that the solution is the best. Thus, one must explain the ranges of decision variables that were searched, show that the solution is (most likely) not merely a local minimum, and show the degradation in the objective function from moving away from the solution. For example, as is discussed in Section 19.4, it is more efficient to change more than one decision variable at a time when searching for the optimum; however, it is better to communicate the validity of the optimum with families of curves in which any single curve involves the variation of only one decision variable (Figure 19.14).

19.2   STRATEGIES

In this section, the DME process from Appendix B is used as an example. Stream and unit numbers refer to those in Figure B.1.

19.2.1   Base Case

A base case is the starting point for optimization. It may be a very simple conceptual flowsheet, it may be a detailed design, or it may be an actual plant whose operation one wishes to improve.

> **Since the goal of optimization is to *improve* the process, it is essential that one start from a defined process, that is, a base case.**

The choice of a base case is straightforward: Choose the best available case. For example, if you have already determined (through prior analysis) that heat integration greatly improves your process, the base case should include the heat integration. The level of detail of the base case is also a crucial decision. At a minimum, the analysis must be detailed enough to provide the calculation of the objective function. If the objective function includes both capital costs and operating costs, the base case analysis must include sizing and costing of equipment as well as the material and energy balances and utility costs. The analysis must be detailed enough to show the effect of all important decision variables on the objective function. In our example, since the temperature of R-201 affects unit E-202, the base case analysis must include this effect.

Although the ability to calculate the objective function and how it changes with variations in all decision variables is essential, the base case certainly need not be a completed design. For example, details of T-201 such as tray spacing, weir height, and so on, are not important at this point in the problem, and a "shortcut" rather than a "rigorous" tower simulation will suffice.

However, it is essential to include enough detail in the base case. Capital costs for individual major pieces of equipment are needed, as are utility costs broken down by type and unit. Creating a calculational model that lumps all costs (even if correctly annualized by the methods shown in Chapter 5) is not very helpful because specific modifications to the flowsheet are masked. For example, a lumped EAOC would not show whether the cost for the reactor, R-201, was relatively small (leading to a conclusion to increase its size) or if the steam costs for E-201 were especially high (leading to a search for an alternative heat source). Table 19.2 provides a list of useful information that should be available for the base case. If the optimization problem is for an existing plant, the original capital cost is not relevant. However, methods for the estimation of capital costs for new equipment should be available, as should some estimate of the value that any removed equipment has when used in other parts of the total plant.

The scope of the base case (and, therefore, of the optimization) must be chosen. Usually, the battery limits for the analysis are defined by functional groupings, company organization, location of large surge or storage facilities, or convenience. However, if the scope is too small, important effects will be missed. For example, if the scope of optimization is R-201, the effect of reactor product composition on tower T-201 is missed. On the other hand, if the scope is too broad, one can become overwhelmed by the interactions. One could include the effect on the internal cost of medium-pressure steam of changes in steam usage, for exam-

Table 19.2   Data Required for Base Case (in Addition to PFD and Flow Tables)

Capital	Operating	Material
Installed cost, each equipment	Utility flowrates (each type)	Total cost for each raw material
Installed cost, each category	Utility targets (19.3)	Value of purged or wasted material
Estimated credit for equipment used elsewhere in plant (for existing processes)	Utility costs on $/GJ basis	
	Estimated uncertainties	Total product value Estimated uncertainties
Estimated uncertainties	Other operating costs	

ple. Such an expansion of scope clearly would not be warranted. Typically, column reflux ratio can be optimized separately, but reactor conditions (T, P, conversion), recycle rates, and separator recoveries should be optimized together.

19.2.2   Objective Functions

The optimization can only begin after the objective function is selected. It must be chosen such that the extreme maximum (or minimum) is the most desired condition. For example, we typically look at minimizing the EAOC rather than the installed capital cost or at maximizing net profit rather than gross sales. If the objective function is poorly chosen or imprecise, the solution will be worthless.

The most common objective functions have units of dollars. The recurring costs can be discounted to obtain a net present value (NPV), or the capital costs can be annualized to obtain an equivalent annual operating cost (EAOC), as discussed in Chapter 5. Usually, this value is for the process unit. However, if a smaller scope is defined for part of the optimization (e.g., T-201 reflux), the EAOC or NPV used should not include capital and recurring cost contributions that do not change significantly over the ranges of the varied decision variables. For example, optimization of the reflux for T-201 does not significantly affect the costs for the rest of the flowsheet (other than the ancillary units E-204, E-205, V-201, P-202). Thus, the raw material costs, reactor costs, other capital and utility costs should not be included. If they were, the total variation in the objective function might be in the fourth or fifth digit. The total EAOC or NPV is not accurate to this level, but the changes caused by the tower optimization are.

This focus on the changes to the costs (or savings) rather than on the total costs is called "incremental analysis." We continue to modify the design or operation of a process plant as long as the improvement gained (the "return") for an incremental investment is greater than our "benchmark" rate. In Chapter 5, the details of this form of investment analysis are covered. In optimization, the implementation of incremental analysis requires careful choice of the objective function, as described above.

Some objective functions are not directly based on economics. For example, one may be asked to maximize the production of DME from an existing plant or to minimize the contaminant concentration in the waste water stream. However, the objective function should be quantitative. Furthermore, a rational basis for any objective function (monetary or nonmonetary) should be developed. For example, if the goal is to maximize profit (rather than revenues), maximizing the production of DME may not be desirable. Similarly, if the goal is to cause the least harm to the environment, minimization of the contaminant concentration (rather than the total contaminant flowrate) may not be the best approach.

19.2.3   Analysis of the Base Costs

The first analysis of the base costs should be to determine targets of an idealized process. The value for the objective function should be determined based on the assumption of equilibrium conversion, no equipment or utility costs, and perfect separations. The profit for such a case is called the gross profit margin or, simply, the margin (see Chapter 5.)

Next, a form of *Pareto* [1] analysis should be used. This analysis is based on the observation that, for most problems, a large fraction of the objective function (e.g., 80%) is due to a small fraction of the contributing factors (e.g., 20%). For example, Table 19.3 shows the contribution to the EAOC for the different types of equipment and for the operating costs and raw materials for the DME process given in Appendix B. Only a few of these categories account for the bulk of the total cost. For our case, as with many chemical processes, the raw material costs swamp all other costs and are treated separately. In this case, the overriding goal is to convert as much of the raw materials into product as possible. From this

Table 19.3   Ranking of Contributions to the EAOC

Category	Contribution to EAOC ($/yr)
Raw materials	11,215,000
Raw material (@ 100% conversion) = $ 11,185,000	
Target savings = $ 30,000	
Medium pressure steam	695,000
Towers and vessels	210,000
Heat exchangers	170,000
Pumps (including electricity)	160,000
Reactor	70,000
Cooling water	31,000
Waste water treatment	1,000
Total	**12,552,000**

standpoint, the waste of methanol is very low and the net savings that can be made from better use of the methanol amounts to only \$30,300/yr. The calculations for this analysis are covered in the next section. However, assuming that we are already buying methanol from the cheapest source, we should not concentrate on reducing the cost of raw materials, since this will not be productive in reducing the cost of producing DME. The single largest operating cost is the medium pressure steam. A greater improvement in the objective function can be made by lowering steam costs than can be obtained by lowering any other cost category.

Basically, this first analysis of the base case provides a target for optimization and a road map to proceed to the solution.

19.2.4   Identifying and Prioritizing Key Decision Variables

Based on the Pareto ranking of effects on the objective function, the key decision variables are chosen. For the DME process, the methanol cost is high, but there are assumed to be no by-product reactions. The product DME is assumed to be at its purity specification. The only methanol loss is in the waste water; thus, the purity of the bottom product of T-202 could be a decision variable. Presently, approximately 0.27% of the methanol feed leaves in the waste water. Thus, the target savings attainable can be calculated as \$30,300/yr.

Medium-pressure steam is used in three units: E-201, E-204, E-206. The first of these is used to vaporize the reactor feed at 154°C, which is below the temperature of low-pressure steam. Since low-pressure steam is less expensive than medium-pressure steam, we should consider this change of heat source. Similarly, the duty of E-204 (a tower reboiler) does not necessarily require medium-pressure steam.

The base case single-pass conversion is 80%. If this can be increased, the costs (capital plus operating) of all units in the recycle loop will be decreased. This includes all the equipment in the flowsheet except P-201 and E-208. The conversion is set by the reactor inlet temperature (T_6) and pressure (P_6) and the volume of the reactor.

Thus, we have identified the single-pass conversion as an important dependent decision variable. With the techniques presented in Section 19.4, the optimum can be determined. The conversion is a secondary (decision) variable. There are an infinite number of temperature, pressure, and reactor-volume combinations for a given conversion. Prioritizing these three primary decision variables requires knowledge of the sensitivity of the objective function to changes in these variables. Although there are elegant mathematical techniques for estimating these sensitivities, the most efficient technique is often to evaluate the objective function at the limits of the variables. This is a standard experimental design technique (two-level factorial design) that is of great help in choosing key decision variables. If the cost changes little when the reactor pressure changes from its upper limit to its lower limit, another variable (such as temperature) should be chosen.

Another strategy to determine decision variables is to consider how the process is controlled. Any variable that must be controlled is a decision variable. There are alternate control strategies for equipment and for processes, but a well-designed control system reduces the degrees of freedom to zero without overconstraining the process. The controlled variable is a secondary decision variable; the variable manipulated by the final control element is the primary decision variable.

Considering the DME process in Figure B.1, the flowrate of methanol is a decision variable, as are the temperature of Stream 5 (controlled by steam pressure in E-201), the temperature of Stream 8 (controlled by the cooling water rate to E-203), the pressure of the reactor (controlled by the valve between Streams 8 and 9), and so on. The two columns have their own control systems, which show that the reflux to the column and the heat input to the reboiler are decision variables. Note that simple (two-product) distillation columns have two decision variables (besides the column pressure), but many different combinations are possible. Since the number of decision variables is equal to the number of controlled variables, which is equal to the number of operational specifications, one can look at any process simulator setup for a column and see the myriad choices.

The other type of decision variable is an equipment characteristic. The reactor volume and the number of stages are examples. Each specification that must be made before ordering a piece of equipment is a decision variable. Thus, the area of a heat exchanger and the aspect ratio of a packed-bed reactor are decision variables.

Once the decision variables have been identified and prioritized, the techniques of topological optimization (Section 19.3) and parametric optimization (Section 19.4) can be applied.

19.3   TOPOLOGICAL OPTIMIZATION

19.3.1   Introduction

As was discussed previously in Sections 19.1 and 19.2, there are essentially two types of optimization that a chemical engineer needs to consider. The first is termed topological optimization and deals with the topology or arrangement of process equipment. The second type is parametric optimization and it is concerned with the operating variables, such as temperature, pressure, and concentration of streams, for a given piece of equipment or process. In this section we will deal with topological optimization. Parametric optimization will be addressed in Section 19.4.

During the design of a new process unit or the upgrading of an existing unit, topological optimization should, in general, be considered first. The reasons for this are twofold. First, topological changes usually have a large impact on the overall profitability of the plant. Second, parametric optimization is easiest to interpret when the topology of the flowsheet is fixed. It should be noted that combinations of both types of optimization strategies may have to be employed simultaneously, but that the major topological changes are still best handled early on in the optimization process.

The questions that a process engineer needs to answer when considering the topology of a process include the following:

1. Can unwanted by-products be eliminated?
2. Can equipment be eliminated or rearranged?
3. Can alternative separation methods or reactor configurations be employed?
4. To what extent can heat integration be improved?

The ordering of the above questions corresponds approximately to the order in which they should be addressed when considering a new process. In this section, we will answer these questions and give examples of processes in which such topological rearrangements may be beneficial.

19.3.2   Elimination of Unwanted Non-Hazardous By-products or Hazardous Waste Streams

This is clearly a very important issue and should be addressed early on in the design process. The benefit of obtaining 100% conversion of reactants with a 100% selectivity to the desired product should be clear. Although this goal is never reached in practice, it can be approached through suitable choices of reaction mechanisms, reactor operation, and catalyst. As a chemical engineer, you may not be directly involved in the choice of reaction paths. However, you may be asked to evaluate and optimize designs for using alternative reactions in order to evaluate the optimum scheme.

In many cases, due to side reactions that are suppressed but not eliminated by suitable choice of catalyst and operating conditions, unwanted by-products or waste streams may be produced. The term unwanted by-product refers to a stream that cannot be sold for an overall profit. An example of such a by-product would be the production of a fuel stream. In this case, some partial economic credit is obtained from the by-product, but in virtually all cases, this represents an overall loss when compared to the price of the raw materials used to produce it.

Example 19.1

Consider the process given in Figure C.8, Appendix C, for the production of cumene. In this process benzene and propylene react in a gas-phase catalytic reaction to form cumene, and the cumene reacts further to form p-diisopropyl benzene (DIPB), a by-product that is sold for its fuel value.

$$C_6H_6 + C_3H_6 \rightarrow C_9H_{12}$$
cumene

$$C_9H_{12} + C_3H_6 \rightarrow C_{12}H_{18}$$
p-diisopropyl benzene

From the information given in Table C.14, estimate the yearly cost of producing this waste, Stream 14. Assume that the costs of benzene, propylene, and fuel credit are $0.27/kg, $0.28/kg, and $2.5/GJ, respectively.

From Table C.14, we have that the composition of the DIPB waste stream is 2.76 kmol/h DIPB and 0.92 kmol/h cumene. The standard heats of combustion for DIPB and cumene are given below:

cumene −5.00 GJ/kmol

DIPB −6.82 GJ/kmol

Revenue for sale as fuel = [(2.76) (6.82) + (0.92) (5.00)](8000)(2.5) = $468,000/yr
Cost of equivalent raw materials

$$\text{benzene} = (2.76 + 0.92)\,(78)\,(8000)\,(0.27) = \$620,000/\text{yr}$$

$$\text{propylene} = (2.76(2) + 0.92)\,(42)\,(8000)\,(0.28) = \$606,000/\text{yr}$$

$$\text{net cost of producing DIPB stream} = (620 + 606 - 468) \times 10^3 = \$758,000/\text{yr}$$

From the above example, it is clear that the production of DIPB is very costly and impacts negatively the overall profitability of the process. Improved separation of cumene would improve the economics, but even if all the cumene could be recovered, the DIPB would still cost an estimated $608,000/yr to produce.

Example 19.2

What process changes could be made to eliminate the production of the DIPB waste stream in Figure C.8?
There are several possible solutions to eliminate the DIPB stream:

1. Reduce the per pass conversion of propylene in the reactor. This has the effect of suppressing the DIPB reaction by reducing the cumene concentration in the reactor.
2. Increase the ratio of benzene to propylene in the feed to the reactor. This reduces the propylene concentration in the reactor and hence tends to suppress the DIPB reaction.
3. Obtain a new catalyst for which the DIPB reaction is not favored as much as with the current catalyst.

Each of these remedies has an effect on the process as a whole, and all these effects must be taken into account in order to evaluate correctly the best (most profitable) process. Some of these effects will be discussed in the following sections.

Alternatively, a hazardous waste stream may be produced. In such cases additional costly treatment steps (either on-site or off-site) would be required in order to render the material benign to the environment. The economic penalties for these treatment steps—incineration, neutralization, and so on—are often great and may severely impact the overall economic picture for the process. In addition to the economic penalties that the production of waste streams cause, there are political ramifications that may overshadow the economic considerations. For many companies, the production of hazardous wastes is no longer an acceptable process choice, and alternative reaction routes, which eliminate such waste streams, are aggressively pursued.

19.3.3   Elimination and Rearrangement of Equipment

Both the elimination and rearrangement of equipment can lead to significant improvements in the process economics. Algorithms for these topological changes are under development. However, the approach used here is one based on intuitive reasoning and illustrated by process examples. Both topics are considered separately below. The rearrangement of heat transfer equipment is considered as a separate topic in Section 19.3.5.

Elimination of Equipment.   It is assumed that the starting point for this discussion is a PFD in which all process equipment serves a valid function, that is, the process does not contain any redundant equipment that can be eliminated immediately.

The elimination of a piece of equipment is often the result of a change in operating conditions, and can thus be considered as the end product of a series of parametric changes. As an example, we consider the first alternative in Example 19.2.

Example 19.3

Evaluate the topological changes in the cumene PFD, Figure C.8, which can be made by reducing the per pass conversion of propylene in the reactor.

As the single-pass conversion of propylene (the limiting reactant) in the reactor is reduced, the DIPB production decreases due to the lowering of the cumene concentration in the reactor. At some point, the second distillation column and the associated equipment may be removed, since all the DIPB produced may leave the process in the cumene product, Stream 13.

Although the second distillation column, T-802, and associated equipment (E-805, E-806, V-804, and P-805 A/B), and utility costs may be eliminated, this may not be the most economical alternative. The increase in the size of other equipment in

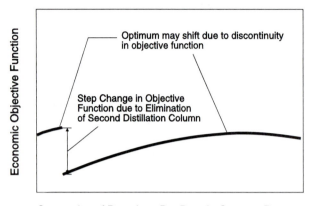

Conversion of Propylene Per Pass in Cumene Reactor

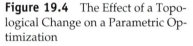

Figure 19.4   The Effect of a Topological Change on a Parametric Optimization

the front-end of the process, due to an increase in the recycle flows and the accompanying increase in utility costs, may overwhelm the savings from the elimination of the second tower and associated equipment. However, you should be aware of the consequences of reducing the DIPB production, since the elimination of the tower represents a step change in the profitability function that might otherwise be overlooked. This step change in the objective function is illustrated in Figure 19.4.

Rearrangement of Equipment.   There are certain guidelines that should be followed when the sequence of equipment is considered. Some are obvious. For example, one should try to pump a liquid rather than compress a gas; thus, it will always be better to place a pump before a vaporizer rather than a compressor after it. However, other topological changes are somewhat more subtle. The most common examples of equipment rearrangement are associated with the separation section of a process and the integration of heat transfer equipment. In this section we concentrate on the sequencing of separation equipment, and heat integration is covered in Section 19.3.5.

Example 19.4

Consider the dimethyl ether (DME) process shown in Figure B.1 (Unit 200) and Tables B.1 and B.2. The process is quite straightforward and consists of a gas-phase catalytic reaction in which methanol is dehydrated to give DME with no appreciable side reactions.

$$2CH_3OH \rightarrow (CH_3)_2O + H_2O$$

$$DME$$

The reactor effluent stream is cooled and sent to two distillation columns. The first column separates DME product from the water and unreacted methanol. The second column separates the methanol, which is recycled, from the water, which is then sent to a wastewater treatment facility to remove trace organic compounds.

Is there any economic advantage gained by changing the order of the distillation so that the water is removed first and the DME and methanol are separated in the second column?

There is no simple way to determine whether the separation sequence given in Example 19.4 should be changed. In order to evaluate which alternative is better, a rigorous parametric optimization for both topologies should be made and the configuration with the best economics should be chosen. Although it may not always be possible to determine which sequence is the best just by inspection, there are some guidelines that may help determine which sequences are worthy of further consideration. From Table 9.13 we have the following guidelines:

1. Perform the easiest separation first, that is, the one least demanding of trays and reflux, and leave the most difficult to the last.
2. When neither relative volatility nor feed composition vary widely, remove components one by one as overhead components.
3. When the adjacent ordered components in the feed vary widely in relative volatility, sequence the splits in order of decreasing volatility.
4. When the concentrations in the feed vary widely but the relative volatilities do not, remove the components in order of decreasing concentration.

In the next example we will apply these guidelines to our DME problem.

Example 19.5

Apply the guidelines for column sequencing to the DME process using the information given in Table B.2 and Figure B.1.

From Table B.2, the composition of the stream leaving the reactor and entering the separation section is as follows:

	Flowrate (kmol/h)	Mole Fraction	Relative Volatility ($P = 10.4$ bar)
DME	130.5	0.398	49.4
MeOH	64.9	0.197	2.2
Water	132.9	0.405	1.0

The relative volatilities for the components are taken from a simulation of the process, using CHEMCAD, with a UNIFAC estimation for the vapor liquid equilibrium (VLE). By applying the guidelines given above, it would appear that the easier separation is the removal of DME, and, according to Rule 1, this should be removed first, which is what is done currently. The other guidelines do not add any additional guidance, and we may conclude that the current sequence is probably the best. However, no mention is given in the guidelines of special considerations for water. Since water has a very high latent heat of vaporization, the duties of the condensers and reboilers in the columns will be higher than for similar flows of organic materials. By removing the DME first, water must be re-boiled in both columns. This suggests that there may be some advantage to removing the water first and then separating the DME and methanol. This case is considered in Problem 19.4 at the end of this chapter. Therefore, this example illustrates a new guideline that should be added to those in Chapter 9.

Special consideration is necessary when dealing with mixtures of polar compounds or other components that can form azeotropes or give rise to more than one liquid phase. As components are separated from a stream, the remaining mixture of components may fall into regions where two or more liquid phases are present or in which pairs of compounds form azeotropes. This can greatly increase the complexity of a separation and may strongly influence the sequence in which components or products are separated. This topic is considered in much greater detail by Douglas [2], and the reader is referred to this excellent reference for a more complete explanation and methods for evaluating azeotropic systems.

19.3.4   Alternative Separation Schemes and Reactor Configurations

Early in the design process, it is important to consider the use of alternative technologies to separate products from unused reactants and waste streams and alternative reactor configurations. The topic of alternative reactor configurations is considered in Chapter 14 and only alternative separation technologies are considered here. If one picks up any textbook on mass transfer or unit operations, it is immediately apparent that there exist a myriad of different technologies to separate chemical components. Despite the wide range of separation technologies available to the process engineer, when considering liquid-gas processes, the vast majority of separations are comprised of distillation, gas absorption and liquid stripping, and liquid-liquid extraction. According to Humphrey and Keller [3], 90 to 95% of all separations, product recoveries, and purifications in the chemical process industry consist of some form of distillation (including extractive and azeotropic distillation). The relative maturity of distillation technology coupled with its relatively inexpensive energy requirements makes distillation the default option for process separations involving liquids and vapors. Many notable exceptions exist, and alternatives to distillation technology may have to be used. Two examples are if the relative volatilities of two components are close to one (say less than 1.05), or if excessively high pressures or low temperatures are required to obtain a liquid-vapor mixture. We do not offer a comprehensive list of alternative separation techniques in this text. The approach taken here is to emphasize the importance of recognizing when a "nontraditional" technology might be employed and how it may benefit the process. Further details of the choice of alternative separation techniques are given in Table 17.1.

Example 19.6

In the toluene hydrodealkylation process shown in Figure E19.6 (taken from Figure 1.5), the fuel gas leaving the unit, Stream 16, contains a significant amount of hydrogen, a raw material for the process. Currently, there is no separation of Stream 8. The mixture of methane and hydrogen leaving V-101 is split into two, with one portion recycled and the other portion purged as fuel gas. What benefit would there be in separating the hydrogen from the methane in Stream 8 and recycling a hydrogen-rich stream to the reactor? What technology could be used to achieve this separation?

Significant benefits may be derived by sending a hydrogen-rich recycle stream back to the front end of the process. The methane that is currently recycled acts as a diluent in the reactor. By purifying the recycle stream all the equipment in the reactor loop, E-101, H-101, R-101, E-102, and C-101 could be made smaller (since the amount of methane in the feed to each equipment would be reduced), and the utility consumptions for the heat exchange equipment would also be reduced. In addition, the amount of hydrogen feed, Stream 3, would be reduced since far less hydrogen would leave in the fuel gas, Stream 16.

Several technologies exist to purify hydrogen from a stream of light hydrocarbons. It should be noted that distillation is not a viable option for this separation since the temperature at which methane begins to condense from Stream 8 is below −130°C. The two most likely candidates for this separation are membrane separation, in which hydrogen would preferentially diffuse through a polymer membrane, and pressure swing adsorption where methane would preferentially adsorb onto a bed of molecular sieve particles. In both of these cases, the potential gains outlined above are offset by the capital cost of the

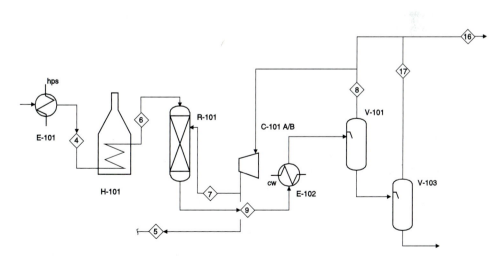

Figure E19.6   Toluene Hydrodealkylation Reactor and Hydrogen Recycle Loop

additional separation equipment and increased compression costs for the membrane case, where the permeate, high in hydrogen, would be obtained at a much lower pressure than the feed gas. In order to evaluate such process alternatives correctly, the economics associated with the new topology would have to be assessed, taking into account all the savings and costs outlined above. The case of implementing a membrane separator is considered in the problem report in Chapter 23.

19.3.5   Improvements in Heat Integration

This section is called improvements in heat integration since, even in a preliminary design, some form of heat integration is usually employed. Heat integration has been around in one form or another ever since thermal engineering came into being. Its early use in the process industries was most apparent in the crude preheat trains used in oil refining. In refineries, the thermal energy contained in the various product streams is used to preheat the crude prior to final heating in the fired heater, upstream of the atmospheric column.

The growing importance of heat integration can be traced to the large increase in the cost of fuel/energy starting in the early 1970s. The formalization of the theory of heat integration and pinch technology has been attributed to several researchers, Linhoff and Flower [4], Hohmann [5] and Umeda et al. [6]. The approach we use here follows that given by Douglas [2], and the interested student is encouraged to study this reference for additional insight into the broader concepts of energy integration.

A simple example is presented first to give the reader insight into the rationale for heat/energy integration.

Example 19.7

Figure E19.7 shows two configurations for the DME reactor feed and effluent heat exchange system. In both cases, the feed enters from the left at 154°C, it is heated to 250°C prior to being fed into the adiabatic catalytic reactor, R-201. The same amount of reaction takes place in both configurations, and the reactor effluent is then cooled to 100°C prior to entering the separation section of the process. The only difference between the two systems is the way in which the heat exchange takes place. In Figure E19.7a, the feed is heated with high-pressure steam and the effluent is cooled with cooling water. However, this does not make good economic sense. Since we generate heat in the reactor, it would make better sense to use this heat from the reaction to heat the reactor feed. This is what is done in Figure E19.7b. The reactor effluent is partially cooled by exchanging heat with the cool, incoming feed. Compared with the configuration of Figure E19.7a, the heat integration saves money in several ways: (1) The cooling water utility is reduced and the high-pressure steam is eliminated. (2) Heat exchanger E-203 is smaller since the duty is reduced, and E-202 is also smaller due to the fact that hps condenses at 254°C which means the ΔT driving force in the exchanger is very small and the area is large. The economics for the two exchangers and utilities for both cases are summarized below:

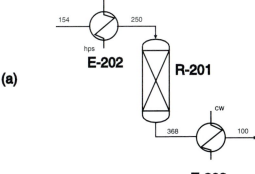

(a)

(b)

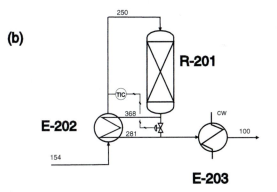

Figure E19.7   DME Reactor Feed and Effluent Heat Exchange System (a) Without Heat Integration, (b) With Heat Integration

	No Heat Integration	With Heat Integration
Fixed Capital Investment	$250,000	$129,000
Cost of Utilities	$105,400/yr	$16,670/yr
Net Present Value	−$1,225,000	−$373,000

The savings received over the life of the plant by using heat integration are (−373,000 + 1,225,000) = $852,000!

From the above example, it should be apparent that considerable savings may be achieved by integrating heat within the process. Rather than try to implement this heat integration on an ad hoc basis, such as in the above example, we present a formalized way of approaching these types of problems.

    We present the general algorithm to give the minimum number of exchangers requiring the minimum utility requirements for a given minimum approach temperature. The algorithm to solve the minimum utility, minimum number of exchangers (MUMNE) problem consists of the following steps:

1. Choose a minimum approach temperature. This is part of a parametric optimization since for every minimum approach temperature a different solution will be found.

2. Construct a temperature interval diagram.

3. Construct a cascade diagram and determine the minimum utility requirements and the pinch temperatures.

4. Calculate the minimum number of heat exchangers above and below the pinch.

5. Construct the heat exchanger network.

It is important to remember that the object of this exercise is to obtain a heat exchanger network that exchanges the minimum amount of energy between the process streams and the utilities and uses the minimum number of heat exchangers to accomplish this. This network is almost never the optimum economic design. However, it does represent a good starting point for further study and optimization.

Each of the five steps given above is considered in detail using an example problem that is outlined below.

Example 19.8

In a process, there are a total of six process streams that require heating or cooling. These are listed below along with their thermal and flow data. A stream is referred to as "hot" if it requires cooling and "cold" if it requires heating, the temperature of the stream is not used to define whether it is "hot" or "cold."

Stream #	Condition	Flowrate, $\dot{m}$ (kg/s)	C_p (kJ/kg°C)	$\dot{m} C_p$ (kW/°C)	T_{in} (°C)	T_{out} (°C)	$Q_{available}$ (kW)
1	Hot	10.00	0.8	8.0	300	150	1200
2	Hot	2.50	0.8	2.0	150	50	200
3	Hot	3.00	1.0	3.0	200	50	−450
4	Cold	6.25	0.8	5.0	190	290	−500
5	Cold	10.00	0.8	8.0	90	190	−800
6	Cold	4.00	1.0	4.0	40	190	−600
						Total	−50

Design the MUMNE network for this system.

It should be noted that the overall heat balance for these streams yields a net enthalpy change of −50 kW. This does not mean that a heat exchanger network can be designed to exchange heat between the hot and cold streams by receiving only 50 kW from a hot utility. This is because the above analysis takes into account only the first law of thermodynamics, that is, an enthalpy balance. In order to design a viable heat exchanger network, it is necessary to also consider the second law of thermodynamics, which requires that heat only flow from hot to cold bodies. As a consequence of this, the utility loads will in general be significantly higher than those predicted by a simple overall enthalpy balance.

Step 1: Choose a Minimum Approach Temperature.   This represents the smallest temperature difference that two streams leaving or entering a heat exchanger can have. Typical values are from 10°C to 20°C. We will choose 10°C for this problem, noting that different results will be obtained by using different temperature approaches. The range of 10°C–20°C given above is typical but not cast in concrete; indeed, any value greater than zero will yield a viable heat exchanger network.

Step 2: Construct a Temperature Interval Diagram.   In a temperature interval diagram all process streams are represented by a vertical line, using the convention that hot streams that require cooling are drawn on the left-hand side and cool streams requiring heating are drawn on the right. The left- and right-hand axes are shifted by the minimum temperature difference chosen for the problem, with the right-hand side being shifted down compared to the left. The temperature interval diagram for our problem is shown in Figure 19.5. Each process stream is represented by a vertical line with an arrow at the end indicating the direction of temperature change. Horizontal lines are then drawn through the ends of the lines, and the diagram divided up into temperature intervals. For

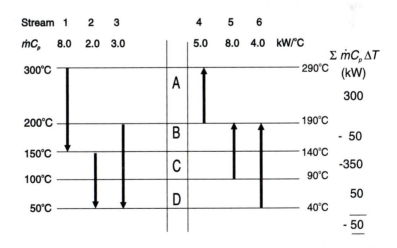

Figure 19.5   Temperature Interval Diagram for Example 19.8

our problem, there are four temperature intervals. The net amount of available energy from all the streams in a given temperature interval is given in the right-hand column. The convention of + for excess energy and − for energy deficit is used. Thus, if the right-hand column contains a positive number for a given temperature interval, this implies that there is more than enough energy in the hot streams to heat up the cold streams in that temperature interval. The sum of the numbers in the right-hand column is the net deficit or surplus enthalpy for all the streams, which for this example is −50 kW.

Step 3: Construct a Cascade Diagram.   The next step involves constructing a cascade diagram, which is illustrated in Figure 19.6. The cascade diagram simply shows the net amount of energy in each temperature interval. Since we can always transfer energy down a temperature gradient, if there is excess energy in a given temperature interval, we may "cascade" this energy down to the next temperature level. We continue to cascade energy in this manner, and the result is shown in Figure 19.6. From the cascade diagram, it is evident that there is a point in the diagram at which no more energy can be cascaded down, and that most often energy must be supplied from the hot utility to the process. This point is represented by line ab in the diagram. Below this line, we can continue the cascading process, but again, at some point, we must reject excess heat from the process to the cold utility. The line ab is termed the pinch zone or pinch temperature. If we try to transfer heat across the pinch zone, the net result will be that more heat will have to be added from the hot utility and rejected to the cold utility. The criterion of not transferring energy across the pinch zone and cascading energy whenever possible guarantees that the energy requirements to and from the utilities will be minimized. From Figure 19.6, we conclude that the minimum utility requirements for this problem are 100 kW from the hot utility and 50 kW to the cold utility, with the net difference equal to −50 kW, which is consistent

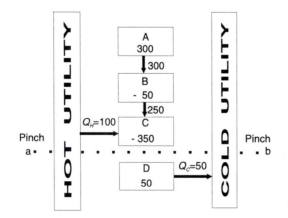

Figure 19.6   Cascade Diagram for Example 19.8

with the overall enthalpy balance. The pinch temperatures are 100°C and 90°C for the hot and cold streams, respectively.

Composite Temperature Enthalpy Diagram. Physically, the pinch zone, mentioned above, represents a point in the heat exchanger network at which at least one heat exchanger, or two streams, will have the minimum approach temperature set in Step 1. This point is perhaps more clearly shown in a composite enthalpy-temperature diagram. This diagram is essentially the same as the combination of all the T-Q diagrams for all the exchangers in the network. Such a diagram for this example is shown in Figure 19.7 and is constructed by plotting the enthalpy of all the hot streams and all the cold stream as a function of temperature, as shown below:

		Hot Streams	
Temperature Interval	*Temperature (°C)*	*Enthalpy of Hot streams in Temperature Interval (kW)*	*Cumulative Enthalpy of Hot Streams (kW)*
D	50	0	0
C	100	$(2+3)(100-50) =$ 250	250
B	150	$(2+3)(150-100) =$ 250	500
A	200	$(8+3)(200-150) =$ 550	1050
	300	$(8)(300-200) =$ 800	1850

		Cold Streams	
Temperature Interval	*Temperature (°C)*	*Enthalpy of Cold Streams in Temperature Interval (kW)*	*Cumulative Enthalpy of Cold Streams (kW)*
D	40	0	0
C	90	$(4)(90-40) =$ 200	200
B	140	$(8+4)(140-90) =$ 600	800
A	190	$(8+4)(190-140) =$ 600	1400
	240	$(5)(290-190) =$ 500	1900

The data above are plotted in Figure 19.7 as Curves 1 and 3, respectively. However, if no minimum temperature difference is used, then from Figure 19.7, it can be seen that the curves cross (Curves 1 and 2), which is physically impossible. We may, however, shift the curve for the cold streams to the right, and we can see that as we do this there exists a point at which the vertical distance between the hot and cold stream curves is at a minimum. This is the pinch zone. If we look at Curve 3, this corresponds to the case when the minimum approach temperature

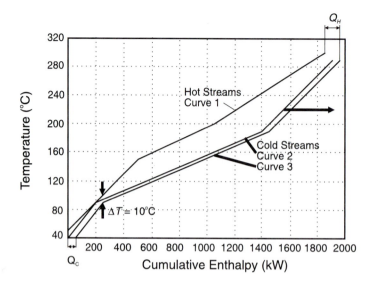

Figure 19.7 Composite Temperature-Enthalpy Diagram for Example 19.8

between the streams is 10°C, the value chosen in Step 1. Comparing this curve and the hot stream curve, we can see that the horizontal distance between the left- and right-hand ends of the curves represent the total duties from the cold and hot utilities, respectively, for the minimum utility case. Thus, the information shown in Figure 19.7 is consistent with the temperature interval diagram and the cascade diagram. In fact, it is exactly the same information just presented in a different way. From Figure 19.7, it should be clear that as the minimum approach temperature decreases, so do the minimum utility requirements. However, the temperature driving force within the heat exchanger network will also decrease and this will require larger, more expensive heat transfer equipment. Clearly, the optimum network must balance exchanger capital investment against utility, operating costs.

Step 4: Calculate the Minimum Number of Heat Exchangers.   Once the pinch temperatures have been found from Step 3, it is necessary to find the minimum number of heat exchangers required to carry out the heat transfer for the minimum utility design. From this point on, we will split the heat transfer problem into two and consider above and below the pinch as separate systems.

Above the Pinch.   The easiest way to evaluate the minimum number of heat exchangers required is to draw boxes representing the energy in the hot and cold process streams and the hot utility, as shown at the top of Figure 19.8. We now transfer energy from the hot steams and hot utility to the cold streams. These en-

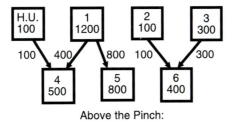

Above the Pinch:

Minimum Number of Exchangers = 5

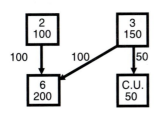

Below Pinch

Minimum Number of Exchangers = 3

Figure 19.8   Calculation of Minimum Number of Exchangers for Example 19.8

ergy transfers are indicated by lines with the amount of energy transferred shown to the side of the lines. Clearly, all the energy in the hot streams and utilities must be transferred to the cold streams. For each line that we draw, we require one heat exchanger. Thus, by minimizing the number of lines we minimize the number of heat exchangers. It should be pointed out that although the number of heat exchangers equals the number of connecting lines, the lines drawn at this stage may not represent actual heat exchangers. The actual design of exchanger network is covered in Step 5 below.

Below the Pinch.   We use the same method to calculate the minimum number of exchangers below the pinch. The diagrams for above and below the pinch are shown in Figure 19.8, and from this we see that five exchangers are required above the pinch and three below the pinch, or a total of eight heat exchangers for the entire network.

Step 5: Design the Heat Exchanger Network.   Again, we consider the systems above and below the pinch separately.

Design Above the Pinch.   To start, we draw the temperature interval diagram above the pinch (Figure 19.9). The algorithm we use to design the network starts by matching hot and cold streams at the pinch and then moving away from the pinch.

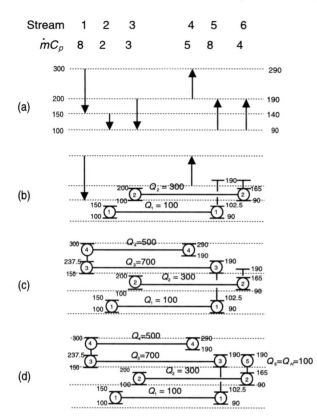

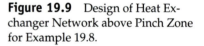

Figure 19.9   Design of Heat Exchanger Network above Pinch Zone for Example 19.8.

Design at the Pinch.   For streams at the pinch, we match streams such that $\dot{m}C_{p,hot} \le \dot{m}C_{p,cold}$. Using this criterion, we ensure that our ΔT_{min} from Step 1 is not violated. From Figure 19.9a, we see that we can match Stream 2 or 3 with Streams 5 or 6. Note, for this step, we consider only streams that are present at the pinch temperature. The next step is to transfer heat from the hot streams to the cold streams by placing heat exchangers in the temperature diagram. This step is shown in Figure 19.9b. There are two possibilities when matching streams at this point: exchange heat between Streams 2 and 5 and Streams 3 and 6, or exchange heat between Streams 2 and 6 and Streams 3 and 5. Only the first combination is shown in Figure 19.9b. A heat exchanger is represented by two circles connected by a solid line, each circle representing a side (shell or tube) of the exchanger. We try to exchange as much heat between streams as possible, and the temperatures of the streams entering and leaving the exchangers are calculated from an enthalpy balance. For example, consider the enthalpy change of Stream 5 as it passes through Exchanger 1. The total heat transferred is $Q_1 = 100$ kW and $\dot{m}C_{p,5} = 8$; therefore, the temperature change for Stream 5, $\Delta T_5 = 100/8 = 12.5°C$. Thus, the temperature change for Stream 5 through Exchanger 1 is 102.5°C to 90°C, as shown in Figure 19.9b.

Design Away from the Pinch.   The next step is to move away from the pinch and look at the remaining hot and cold streams. There are several ways in which we may exchange heat from Stream 1 (the only remaining hot stream) and the three cold Streams 4, 5, and 6. The criterion for matching streams at the pinch does not necessarily hold away from the pinch; however, we should make sure that when the network is designed the following assumptions are still true:

1. We do not violate the minimum approach temperature of 10°C set in Step 1.
2. We end up with five exchangers for the design above the pinch as calculated in Step 3.
3. We add heat from the coolest possible source (This is explained in more detail in the section on Problems with Multiple Utilities).

The matching of streams and the final design of the network are shown in Figures 19.9c and 19.9d. From these figures, it is clear that we have a design with five heat exchangers, the minimum approach temperature is nowhere less than 10°C, and that we add heat to the process at the lowest temperature consistent with this system, that is, 190°C.

Design Below the Pinch.   The approach for the design below the pinch is similar to that described above for above the pinch. We start at the pinch and match streams and then move away from the pinch and match the remaining streams. The temperature interval diagram below the pinch is shown in Figure 19.10a.

Design at the Pinch.   For streams at the pinch we match streams such that $\dot{m}C_{p,hot} \geq \dot{m}C_{p,cold}$. Using this criterion, we ensure that our ΔT_{min} from Step 1 is not violated. From Figure 19.10a, we see that for the three streams at the pinch this criterion cannot be met. This problem can be overcome by splitting the cold Stream 6 into two separate streams. However, before we consider this, let us see what happens if we try to match streams that violate the above condition. In Figure 19.10b, Stream 2 is matched with Stream 6. The net result is impossible, since it would cause a temperature cross in the exchanger, which violates the Second Law of Thermodynamics. In order to maintain the minimum temperature approach set in Step 1, we split Stream 6 into two equal streams, each having an $\dot{m}C_p = 2.0$. We can now match these split streams with hot Streams 2 and 3 without violating the criterion above. The net result is shown in Figure 19.10c, from which it can be seen that the minimum temperature difference is always greater than or equal to 10°C. It should be noted that Stream 6 could be split in an infinite number of ways to yield viable solutions—for example, it could be split into streams with $\dot{m}C_p$ values of 3 and 1.

Design Away from Pinch.   The final step is shown in Figure 19.10d, where the third exchanger is added to transfer the excess heat to the cold utility.

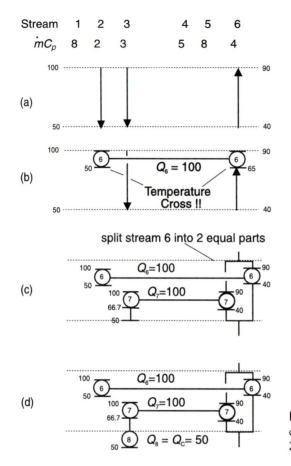

Figure 19.10   Design of Heat Exchanger Network below the Pinch Zone for Example 19.8

The final heat exchanger network is shown in Figure 19.11. The exchangers are now represented by single circles with fluid flowing through both sides. This network has the minimum number of heat exchangers, eight, for the minimum utility requirements, $Q_H = 100$ kW and $Q_C = 50$ kW, using a minimum approach temperature, $\Delta T = 10°C$. As mentioned earlier, different results will be obtained by using different minimum approach temperatures.

Up to this point, we have concentrated only on the topology of the exchanger network, that is, the interaction between the different streams required to give the minimum utility case. To complete the design, it is necessary to estimate the heat transfer area of the exchanger network, which in turn allows a capital cost estimate to be made. Heat transfer coefficients for the heat exchangers in Figure 19.11 can be estimated and the size of the exchangers can be calculated. By adjusting the minimum approach temperature, a series of case studies can be carried out to evaluate the effect of minimum approach temperature on a suitable

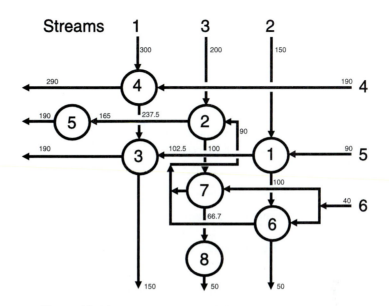

Figure 19.11   MUMNE Network for Example 19.8

economic objective function, for example, net present value. An example of this type of optimization is given in Problems 19.5 and 19.6 at the end of the chapter.

Problems with Multiple Utilities.   For more complicated problems, several pinch temperatures may occur, and energy may have to be added from more than one hot utility and rejected to more than one cold utility. In general, it makes good engineering sense to add heat from the coolest "hot" utility and reject heat to the hottest "cold" utility. For example, if heat must be rejected from the process at a temperature of 190°C, it could be rejected to the cooling water utility. However, it makes more sense—it is more profitable—if this excess energy can be used to make medium pressure steam, which can be used elsewhere in the plant. Likewise, if we need to add heat to the process at a temperature of 80°C, it makes little sense to use valuable high pressure steam (at 250°C); rather, low pressure steam should be utilized. The use of multiple utilities are required in Problems 19.9 and 19.10 at the end of this chapter.

Handling Streams with Phase Changes.   The area of heat integration and pinch technology, in general, is quite broad and many topics have not been covered here. The MUMNE approach outlined above will give a reasonable first approximation to the optimum heat integration scheme for a given process and is, therefore, useful in the preliminary design of a process. Although not stated explicitly, the analysis for the MUMNE design above assumes that the streams are single phase and that the specific heat capacities of the streams are constant over

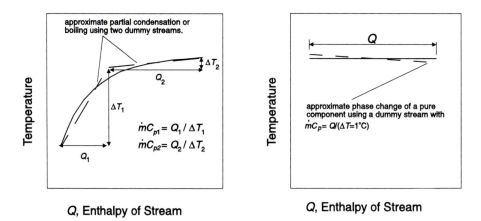

Figure 19.12   Use of Dummy Streams for Phase Changes

the temperature range in the process. For streams that undergo a phase change or for which the specific heat capacities are not constant, the analysis becomes more complicated. However, for such streams, we can use dummy streams that have constant heat capacities. This concept is illustrated in Figure 19.12. The left-hand diagram shows a situation where a partial phase change is occurring. In this case, we can approximate the enthalpy change by using two dummy streams with constant heat capacities as shown in the diagram. For a single component changing phase, the T-Q diagram is a horizontal line, and for this case we use an arbitrarily small ΔT of 1°C to get the dummy heat capacity. For large processes, the hand calculations described above become unwieldy and very time consuming. Fortunately there are several software packages on the market that can be used to optimize heat exchanger networks for complicated processes, for example, HERO v3.2©CHEPRO Ltd.

19.4   PARAMETRIC OPTIMIZATION

In optimizing a chemical process, it is necessary that the key decision variables be identified early on in the optimization procedure. This is necessary in order to reduce the computational effort and time and make the problem tractable. The choice of key decision variables is crucial to the efficiency of the optimization process. An exhaustive list of potential decision variables is not presented here. However, some important variables that should be considered for most processes are listed below:

1. Operating conditions for the reactor, for example, temperature, pressure, concentration of reactants. The temperature range may be restricted by cata-

lyst properties, that is, catalyst may sinter at high temperatures or be inactive at low temperatures.

2. Single-pass conversion in the reactor. The selectivity will be determined by the conditions mentioned in (1) above and the single-pass conversion.

3. Recovery of unused reactants.

4. Purge ratios for recycle streams containing inerts.

5. Purity of products (This is often set by external market forces?).

6. Reflux ratio and component recovery in columns, flow of mass separating agents to absorbers, strippers, extractors, and so on.

7. Operating pressure of separators.

Since most chemical processes utilize recycles to recover unused reactants, any changes in operating conditions that occur within a recycle loop will impact all the equipment in the loop. Consequently, the whole flowsheet may have to be resimulated and the economics reworked (capital investments and costs of manufacture) every time a new value of a variable is considered.

For variables that do not lie within a recycle loop, optimization may be simplified. An example is a distillation column in a separation sequence in which two products are purified and sent to storage. The operation of such a column does not impact any part of the process upstream and can, therefore, be considered independently, after the upstream process has been optimized. First, we will consider single- and two-variable optimizations for single pieces or small groups of equipment. Then we will look at overall process optimization strategies.

19.4.1 Single Variable Optimization: A Case Study on T-201, the DME Separation Column

When considering the optimization of a distillation column, the variables to be considered are: reflux ratio, operating pressure, percent recovery of key components, and purity of the products. For our initial case study, we shall consider that the column pressure and the feed to the column are fixed, from Figure B.1 and Table B.2 we have the following information:

feed, Stream 9	
temperature	89°C
pressure	10.4 bar
vapor fraction	0.148
molar flows (kmol/h)	
dimethyl ether	130.5 kmol/h
methanol	64.9 kmol/h
water	132.9 kmol/h
Total Flow	**328.3 kmol/h**

In addition, the product specification for the DME is that it be 99.5 wt% pure. We will assume that 98.9% of the DME in the feed must be recovered in the final product. We note that for this column, the pressure, reflux ratio, and % recovery of DME are all process variables that can be optimized. We first focus on the reflux ratio; later, we will relax the constraint on the operating pressure and carry out a two-variable optimization.

In order to proceed with the optimization, we must choose an objective function. For our example, considering the reflux ratio as our only decision or design variable, the only costs that will be affected by changes in reflux are the capital and operating costs associated with the column. In order to account correctly for both one-time costs and operating costs, we must use an objective function that takes into account the time value of money (see Chapter 5 for the different criteria used for assessing profitability). We will use the before tax NPV as our objective function:

$$
\begin{aligned}
OBJ &= FCI_{TM} + COM_d(A/P,i,n)(P/F,i,n_{startup}) \\
&= FCI_{TM} + (0.18FCI_{TM} + 1.23C_{UT})(A/P,i,n)(P/F,i,n_{startup})
\end{aligned}
\tag{19.1}
$$

It should be noted that the value of the cost of manufacturing term without depreciation, COM_d, in Eq. (3.2) includes a term with the fixed capital investment based on the total module cost (FCI_{TM}) and the cost of utilities. The other terms are not relevant to the optimization since they do not change with reflux ratio. Using a plant life of 10 years after start-up and a 10% internal rate of return and assuming a construction period of 1 year, Eq. (19.1) reduces to:

$$
\begin{aligned}
OBJ &= FCI_{TM} + (0.18FCI_{TM} + 1.23C_{UT})(6.145)(0.909) \\
&= 2.005FCI_{TM} + 6.871C_{UT}
\end{aligned}
\tag{19.2}
$$

The fixed capital investment term includes the total module costs for T-201, E-204, E-205, V-201, and P-202 A/B. The utility costs include the electricity for P-202 and the heating and cooling utilities for E-204 and E-205, respectively. A series of case studies were run (using the CHEMCAD process simulator) for different reflux ratios for this column and the equipment costs (evaluated from CAPCOST using mid-1996 prices) and utility costs (from Chapter 3) are presented in Table 19.4 along with the objective function from Eq. (19.2). A plot of the (R/R_{min}) vs. the objective function is shown in Figure 19.13. The optimum value of R/R_{min} is seen to be close to 1.12. Moreover, for values greater than 1.1, the objective function changes slowly with R/R_{min}. It should be noted that the results of this univariate search technique are presented as a continuous function of R/R_{min}. However, in reality, the objective function exists only at a set of points on Figure 19.13 (shown by the data symbols on the dotted curve), each point representing a column with an integer number of plates.

Table 19.4   Data for DME Column Optimization, R/R_{min} vs. OBJ

R/R_{min}	FCI (10^3)	Steam Cost (10^3/yr)	Cooling Water Cost (10^3/yr)	Electrical Cost (10^3/yr)	Total Utility Cost (10^3/yr)	OBJ from Eq. (19.2) ($103)
1.01	684	72.50	4.04	0.48	77.02	−1,911
1.02	509	72.72	4.05	0.48	77.24	−1,551
1.03	441	72.96	4.06	0.48	77.50	−1,417
1.04	411	73.15	4.07	0.48	77.70	−1,358
1.11	354	74.68	4.13	0.49	79.31	−1,255
1.27	342	78.16	4.28	0.51	82.95	−1,256
1.60	322	85.04	4.54	0.55	90.17	−1,265

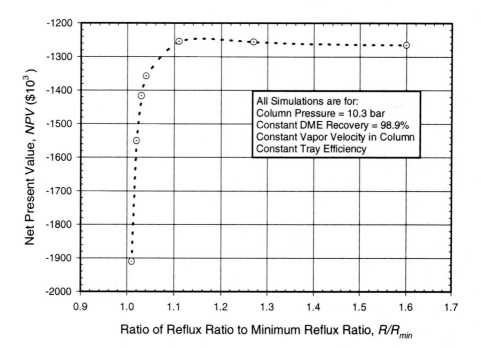

Figure 19.13   Single Variable Optimization for DME Column, T-201

19.4.2   Two-Variable Optimization: The Effect of Pressure and Reflux Ratio on T-201, the DME Separation Column

The effect of pressure on the operation of T-201 is considered next. It may be tempting to take the result from 19.4.1, and by holding the value of R/R_{min} constant at 1.12, carry out a univariate search on the pressure. However, the results from such a technique will not yield the optimum pressure or R/R_{min} values. This is because as pressure changes the optimum reflux ratio also changes. In order to optimize correctly this situation, we must vary both the pressure and R/R_{min} and determine the best combination. We can approach this problem in many ways. In the following example, we pick different pressures and repeat the procedure used in 19.4.1 for each pressure and then plot the results. This is not a particularly efficient procedure, but it yields plots that are easy to interpret. This is an example of a bivariate search technique. The data for this problem are presented in Table 19.5 and plotted in Figure 19.14. From the results shown in Figure 19.14, it is clear that the relationship between pressure and NPV at the optimum R/R_{min} is highly nonlinear. In addition, the optimum value of R/R_{min} does not remain constant with pressure as shown by the dotted line in Figure 19.14; although, for this example, $(R/R_{min})_{opt}$ does not change very much with pressure over the range considered here.

It was stated at the beginning of this section that the topology of the distillation column and associated equipment should remain fixed while carrying out the parametric optimization. This may unduly constrain the overall optimization of the process, and, for our case, limits the range of pressures over which the optimization may be considered. Moreover, when carrying out this optimization, it was necessary to change some of the utilities used in the process and this significantly impacted the results. Consider the information presented in Table 19.6 in

Table 19.5   Data for Two Variable Optimization of DME Column, T-201 (all costs in $1,000)

Pressure (bar)	13.5	11.5	10.3	9.0[1]	9.0[2]	7.5
R/R_{min}	OBJ	OBJ	OBJ	OBJ	OBJ	OBJ
1.01	−2,052	−1,975	−1,911	−1,890	−1,926	−5,203
1.02	−1,699	−1,613	−1,551	−1,511	−1,547	−4,847
1.03	−1,560	−1,474	−1,417	−1,373	−1,409	−4,719
1.04	−1,499	−1,411	−1,358	−1,310	−1,347	−4,665
1.11	−1,394	−1,312	−1,255	−1,213	−1,250	−4,616
1.27	−1,365	−1,288	−1,256	−1,204	−1,243	−4,715
1.60	−1,385	−1,309	−1,265	−1,216	−1,259	−4,955

[1]using lp steam [2]using mp steam

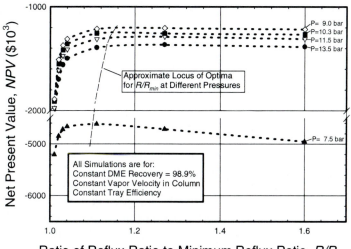

Figure 19.14 Optimization of DME Column, T-201, for Two Variables (Pressure and R/R_{min})

which the utility and fixed capital investment (*FCI*) breakdown is given for the case of R/R_{min} = 1.27 (which is close to the optimum value for all the cases). It is clear that several changes in utilities were implemented. Some were required in order to obtain viable processes, others were changed in order to improve the economics. First, consider the switch from medium-pressure steam to low-

Table 19.6   Breakdown of Costs and Process Information for R/R_{min} = 1.27 (all costs in $1,000 or $1,000/yr)

Pressure (bar)	13.5	11.5	10.3	9.0	9.0	7.5
mp steam	89.79	82.44	78.16	—	72.26	—
lp steam	—	—	—	62.80	—	56.67
cooling water	4.23[1]	4.26[1]	4.28[1]	8.61[2]	8.61[2]	—
refrig. water	—	—	—	—	—	539.35[3]
electricity	0.53	0.52	0.51	0.50	0.50	0.49
FCI	357	344	342	354	340	307
NPV	−1,339	−1,263	−1,230	−1,178	−1,217	−4,692
R_{min}	0.3528	0.3214	0.3024	0.2810	0.281	0.2562
$T_{condenser}$ (°C)	57	50	46	40	40	34
$T_{reboiler}$ (°C)	164	157	153	147	147	140

[1]$\Delta T = 10°C,$ [2] $\Delta T = 5°C,$ [3] $\Delta T = 10°C$

pressure steam for a process pressure of 9.0 bar. The results are shown in Tables 19.5 and 19.6 and are plotted in Figure 19.15. It is clear that there is a small but significant increase in the *NPV* when low-pressure steam is substituted for medium-pressure steam. The reason for this increase is that the reduction in utility costs (due to less expensive low pressure steam) outweigh the increase in equipment costs (due to a smaller temperature driving force and higher operating pressure in E-204, the column reboiler). A switch to low pressure steam is actually possible for all of the operating pressures shown with the exception of 13.5 bar, since the reboiler temperature for these cases is less than 160°C, the temperature at which low pressure steam condenses. The operating pressure at which the switch to low-pressure steam yields an increase in the objective function (*NPV*) is considered in Problem 19.11 at the back of this chapter.

Next, we consider a change in the cooling water utility. As the operating pressure of the column decreases, the bottom and top temperatures also decrease. One consequence is that low pressure steam may be used in the reboiler, as was considered above. In addition, as the top temperature in the column decreases, we reach a point where we can no longer use a temperature increase of 10°C for the cooling water. This occurs at an operating pressure of 9.0 bar. We can still use cooling water to condense the overhead vapor from the column, but we must reduce the temperature increase of the utility. The results shown are for a change in cooling water temperature of 30°C (in) to 35°C (out). As the operating pressure is reduced still further to 7.5 bar, the column top temperature decreases to 34°C. At this point, the use of cooling water becomes almost impossible, that is, the flow of

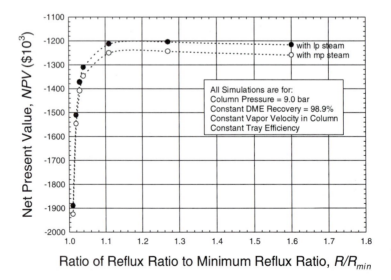

Figure 19.15   The Effect of Changing Utilities in the Reboiler for T-201 at a Column Pressure of 9.0 bar

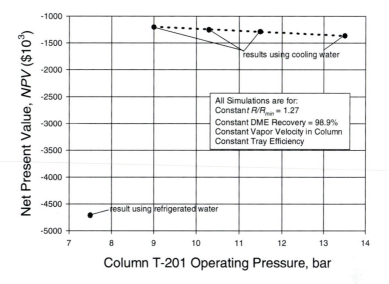

Figure 19.16   The Effect of the Operating Pressure of T-201 on the
NPV (all data for $R/R_{min} = 1.27$)

cooling water and area of the exchanger would become extremely large and any
slight changes in cooling water temperature would have a large impact on the
process. For this reason, refrigerated water is substituted as the cooling utility for
this case. The result is a very large increase in the cost of the utilities and a resulting decrease in the objective function (NPV) as shown in Figure 19.16.

The above example clearly illustrates the need to consider changes in utilities when process conditions change.

**During parametric optimization, changes in operating condi-
tions may require corresponding changes in utilities.**

19.4.3   Flowsheet Optimization Using Key Decision Variables

As stated previously, when the decision variables for an optimization lie within a
recycle loop, then evaluation of the objective function becomes more complicated. As an example, we will consider the DME flowsheet given in Appendix B,
Figure B.1. The number of possible decision variables, for even such a simple
flowsheet as this, are numerous. In this example, we will consider just three variables: single-pass conversion in the reactor (x), fractional recovery of the DME

product in Tower, T-201 (*f*), and reactor pressure (*p*). Even with just three variables, neither it is clear how many runs we should perform nor is it clear over what range of the decision variables runs should be performed.

The first step, therefore, in doing such optimizations, should be to choose the range over which the decision variables should be varied. The ranges chosen for this example are:

$$0.7 < x < 0.9, 0.983 < f < 0.995, 12.70 \text{ bar} < p < 16.70 \text{ bar}$$

The choice of these ranges is somewhat arbitrary, although it is useful to include maximum and/or minimum constraints as end points for these ranges. For example, it has been found from research in the laboratory, that, for the current catalyst, a maximum single-pass conversion of 90% and a maximum operating pressure of 16.70 bar should be used to avoid unwanted catalyst deterioration and undesirable side reactions. The upper limits for x and p represent upper constraints for these variables, which should not be exceeded. On the other hand, the lower limits for p and x are arbitrary and could be changed if the optimization results warrant looking at lower values. The choice of the range of f values is again arbitrary, although experience tells us that the fractional recovery of product (f) should be close to 1.

The next question to be answered is how many points should we choose for each variable. If we decided to use three values for each variable, then we would have a total of (3)(3)(3) = 27 simulations to run. A conservative estimate of the length of time to run a simulation, collect the results, evaluate the equipment parameters, price the equipment, estimate the utility costs, and finally evaluate the objective function is, say, 2 hours per run. This would yield a total time investment of (2)(27) = 54 hours of work! This is a large time investment, and would be much larger if, for example, we considered five variables, so it behooves us to minimize the number of simulations we carry out. With this in mind, we choose to use only the end points of the range for each variable, namely we will run (2)(2)(2) = 8 simulations. This still represents a large investment of time but is more reasonable than the previous estimate. Essentially what we are proposing to do is to carry out a 2^k factorial experiment (Neter and Wasserman [7]), where k is the number of decision variables (three for this case). In essence, each evaluation of the objective function (dependent variable) can be considered the result of an experimental run in which the independent variables (x, f, and p) are varied. The test matrix that we use is given in Table 19.7. The low range for each variable is designated 0, while the high range is designated 1. Thus, the run named DME011 represents the process simulation when the conversion is 0.7 (x is at its low value), fractional recovery is 0.995 (f is at its high value), and the pressure is at 16.7 bar (p is at its high value). For each of the eight cases shown in Table 19.7, a simulation was carried out and all equipment and operating costs determined. The results for these cases and the base case are also presented in Table 19.7.

Once the results have been obtained, we need to interpret them and determine the optimum operating conditions. The results in Table 19.7 are shown

Table 19.7   Test Matrix and Results for three Variable Optimization of DME Process

Run Number	variable 1 = x 0 = 0.7 1 = 0.9	variable 2 = f 0 = 0.983 1 = 0.995	variable 3 = p 0 = 12.7 bar 1 = 16.7 bar	FCI ($1,000)	Utilities ($1,000/yr)	OBJ (NPV) ($1,000)
Base Case	0.8	0.989	14.7	1,297	732	−7,630
DME000	0.7	0.983	12.7	1,378	782	−8,136
DME001	0.7	0.983	16.7	1,381	872	−8,760
DME010	0.7	0.995	12.7	1,393	780	−8,152
DME011	0.7	0.995	16.7	1,440	868	−8,851
DME100	0.9	0.983	12.7	1,210	539	−6,130
DME101	0.9	0.983	16.7	1,261	634	−6,885
DME110	0.9	0.995	12.7	1,232	537	−6,160
DME111	0.9	0.995	16.7	1,299	634	−6,961

graphically in Figure 19.17. In this figure, the region over which the optimization has been considered is represented by a cube or box with each corner representing one of the test runs. Thus, the bottom left-hand front corner represents DME000, while the top right-hand rear corner represents DME111, and so on. The values of the objective function are shown at the appropriate corners of the box. Also shown is the base case simulation (for the conditions in Table B.2), which lies at the middle of the box (this is due to choosing the ranges for each variable symmetrically about the base case value). We will first analyze these results in terms of general trends using a method known as the analysis of means. We will then fit a model for the objective function in terms of our decision variables and use this to estimate the optimum conditions, or more correctly, to give direction as to where the next test run should be performed.

Sensitivity of Objective Function to Changes in Decision Variables: Analysis of Means.   Using the data from Table 19.7 and Figure 19.17, we can estimate how a change in each variable affects the NPV. This is simply done by averaging all the results at a given value of one variable and comparing them with the average value at the other level of that variable. This yields the following results:

conversion, $x = 0.7$, $NPV_{avg,x=0} = (-8136 - 8760 - 8152 - 8851)/4 = -8475$
conversion, $x = 0.9$, $NPV_{avg,x=1} = (-6130 - 6885 - 6160 - 6961)/4 = -6534$

recovery, $f = 0.983$, $NPV_{avg,f=0} = (-8136 - 8760 - 6130 - 6885)/4 = -7478$
recovery, $f = 0.995$, $NPV_{avg,f=1} = (-8152 - 8851 - 6160 - 6961)/4 = -7531$

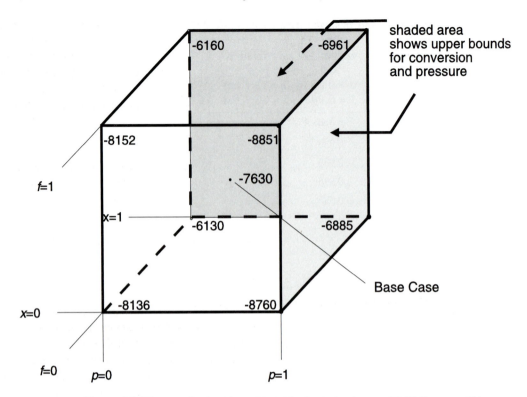

Figure 19.17   Results for Three-Variable Optimization on DME Process. Objective Function Values Are Shown at the Corners of the Test Cube

pressure, $p = 12.7$ bar, $NPV_{avg,p=0} = (-8136 - 8152 - 6130 - 6160)/4 = -7145$
pressure, $p = 16.7$ bar, $NPV_{avg,p=1} = (-8760 - 8851 - 6885 - 6961)/4 = -7864$

From these results, we conclude that the single-pass conversion has the greatest influence on the NPV, followed by the reactor pressure, with the recovery of DME having only a slight effect. Moreover, the results suggest that NPV will be maximized (least negative value) by using high conversion, low pressure, and low (within the range chosen) DME recovery. Although the maximum NPV that was obtained from these simulations was −$6,130,000 (DME100), it should not be assumed that this is the true maximum, or that the maximum lies within the range of decision variables considered thus far. In fact, the above analysis tells us that we should move to a new range for pressure and DME recovery.

   In order to estimate the maximum for the objective function, we will need to do further simulations at different conditions. We will use the results from Table 19.7 to choose the next values for the decision variables. From the results in Table 19.7 and Figure 19.17, we conclude that operating the reactor at the highest conversion allowable ($x = 0.9$) yields the optimum NPV and that reducing the reactor pressure also improves the *NPV*. Compared to these two effects, the recovery of DME is not very important and we will ignore it for now. It should be noted that for the given topology of the flowsheet, all the DME produced in the reactor leaves as product since the DME not recovered in T-201 simply gets recycled back to the reactor from the top of T-202. Thus, for this case, the DME recovery is not very important. However, this is usually not the case. By eliminating one decision variable from our search (f) and noting that another variable is constrained at its maximum value ($x = 0.9$) we have reduced the problem to a one dimensional search (p). The question is what should be the next value of p? Although we can decrease p by some arbitrary amount, we should note that we may run into another constraint as p is lowered. In fact, in Section 19.4.2, we identified a detrimental change in the *NPV* when the pressure in T-201 became so low that refrigerated water was required as the utility for the overhead condenser. Noting the pressure drops through the system and control valves, we estimate that the lowest pressure at which we can run the reactor and still operate T-201 at 9.0 bar, which is close to the lower limit for using cooling water, is $p = 10.7$ bar. Thus, our next simulation should be carried out at ($x = 0.9, f = 0.983, p = 10.7$).

Modeling the Objective Function in Terms of the Decision Variables.   It is useful to be able to estimate NPV values for new conditions before actually running the simulations. We can do this by using the results from Table 19.7 to model the NPV as a function of $x, f,$ and p. An infinite number of functional forms can be chosen for this model. However, we choose the following form:

$$NPV = a_0 + a_1 x + a_2 f + a_3 p + a_4 xf + a_5 xp + a_6 fp + a_7 xfp \tag{19.3}$$

where $a_0, a_1, \ldots a_7$ are constants that are fit using the data from Table 19.7. This model uses eight arbitrary constants, which will allow the function to predict the NPV exactly at each corner point of our experimental design. In addition, this form of model is a simple multivariable polynomial and the regression techniques to find $a_0, a_1, \ldots a_7$ are well established [7]. For our data, we get the following form for our objective function:

$$NPV = -63254 + 55556x + 49448f + 3573p - 44198xf - 3133xp \\ - 3677fp + 3021xfp \tag{19.4}$$

The accuracy of the model is tested by comparing the predictions of Eq. (19.4) with all the data in Table 19.7. The results are shown in Table 19.8.

From the results in Table 19.8, we can see that the model predicts the eight test runs exactly (differences in the last significant figure are due to round off errors in coefficients in Eq. (19.4)). The prediction for the base case is also good; although, the model tends to underpredict the *NPV* a little. We can also predict the expected value of *NPV* for our next simulation ($x = 0.9, f = 0.983, p = 10.7$) and from Eq. (19.4), we get *NPV* = −5,750. The actual value for the *NPV* for this new simulation is −5,947, which is higher than the predicted value by about 3%.

Table 19.8   Predictions of Model for NPV in Eq. (19.4) with Data from Table 19.7 (all figures are in $1,000)

Test Run	Result from Table 19.7	Prediction from Eq. 19.19
Base Case	−7,630	−7,502
DME000	−8,136	−8,134
DME001	−8,760	−8,758
DME010	−8,152	−8,150
DME011	−8,851	−8,849
DME100	−6,130	−6,127
DME101	−6,885	−6,881
DME110	−6,160	−6,157
DME111	−6,961	−6,957

At this point, clearly we are close to the optimum. Whether, further simulations are warranted depends on the accuracy of the estimate being used to obtain the costs and the extra effort that must be expended to analyze further simulations. If further refinement of the estimate of the optimum conditions is required, then the model, Eq. (19.4), can be modified to include terms in p^2 and used to refine further the grid of decision variables. Alternative approaches for more comprehensive models are considered by Ludlow et al. [8].

One final point should be made. This problem is fairly typical of real design optimizations in that the optimum conditions lie on several of the constraints for the variables. Even though we have used a simple polynomial model to describe the *NPV*, Eq. (19.4), which will tend to force the optimum to lie on the constraints, similar results would be obtained if we had used nonlinear models to describe the *NPV*.

19.5   LATTICE SEARCH TECHNIQUES VS. EXPERIMENTAL DESIGN

For problems involving many decision variables, there are three types of parametric optimization techniques:

1. *Analytical techniques based on finding the location where gradients of the objective function are zero.* These are effective if the objective function is continuous, smooth, and has only a few local extrema. Sometimes the objective function can be approximated as a smooth function, and techniques exist to condition an objective function to reduce the number of extrema. However, for the common, complex problems encountered in flowsheet optimization, analytical techniques are usually ineffective.

2. *Experimental design strategies such as those discussed in Sections 19.2 and 19.4.* These can be used when the conditions required for analytical techniques are not met. However, one must search the interior of the decision variable space extrema, which can be done by three-level (and higher-level) factorial designs [7,8]. Fractional factorial designs are essential for higher-level and higher-dimensionality problems. As the dimensionality (number of decision variables) and the number of constraints increase, so does the probability that the optimum lies on the boundary of the allowable search space. This makes experimental design techniques especially attractive.

3. *Pattern search techniques.* These are iterative techniques that are used to proceed from an initial guess toward the optimum, without evaluating derivatives or making assumptions about the shape of the objective function surface. From the initial guesses of the decision variables, a direction is chosen, a move is made, and the objective function is evaluated. If an improvement is detected, further moves are made in that direction. Otherwise, the search continues in a new direction. There are many strategies for choosing the search direction, deciding when to change direction, what direction to change to, and how far to move in each iteration. Crude methods of this type evaluate the objective function at each of the "nearest neighbor" points in each direction. The Simplex-Nelder-Mead [9] method uses $(n + 1)$ points when there are n decision variables to determine the next move. This is the smallest number of points that can give information about changes in the objective function in all directions. A related strategy is simulated annealing [10], in which all moves (good and bad) have non-zero probability of being accepted. Good moves are always accepted, and bad moves are accepted, but less frequently the larger the degradation in the objective function.

Both the pattern search and the experimental design techniques allow the use of discrete decision variables. Topological changes can always be represented by changes in discrete decision variables (such as those defining the flowsheet topology), so these techniques can accommodate both parametric and topological optimization.

We prefer the experimental design techniques during the early phases of design. They serve the multiple uses of prioritizing the decision variables, scoping the optimization problem to determine an estimate of the target improvement possible, and performing the parametric and topological optimizations.

19.6   PROCESS FLEXIBILITY AND THE SENSITIVITY OF THE OPTIMUM

Before any optimization result can be understood, one must consider the flexibility of the process and the (related) sensitivity of the result. A process must be able to operate with different feedstocks, under varying weather conditions (important for cooling utilities especially), over a range of catalyst activities, at different production rates, and so on. A process that is optimum for the "design" or "nameplate" conditions may be extremely inefficient for other operating conditions. The common analogy is a car that can operate at only one speed. Clearly, the design of the car could be optimized for this one condition but the usefulness of such a car is very limited.

The sensitivity of the optimum refers to the rate at which the objective function changes with changes in one of the decision variables. Here, there is an apparent contradiction. We choose decision variables because the objective function is most sensitive to them, but we therefore need to control the process very close to our optimum values of the decision variables to stay near the optimum. In fact, the objective function evaluations made during the optimization allow one to determine how precise the control must be and, most importantly, to determine the penalty for failure to control the decision variable to within prescribed limits. This is another of the interplays between optimization and process control.

The effects of process sensitivity and flexibility can best be shown with a family of curves such as shown in Figure 19.14. Regardless of the techniques used to *find* the optimum, it is only through these visual representations that the results of optimization can be effectively conveyed to those who will use them. They are essentially performance curves for the process, where the performance is measured as the value of the objective function.

19.7 SUMMARY

In this chapter, we have introduced several definitions commonly used in the area of optimization. In addition, the idea of applying a Pareto analysis to the base case was investigated. This type of analysis often reduces the number of decision variables that should be considered and the range over which the optimization should take place.

The differences between parametric and topological optimization were investigated and strategies were suggested for each type of optimization. The special case of pinch technology applied to heat integration was considered separately. The algorithm for analyzing the minimum utility, minimum number of exchangers problem (MUMNE) was outlined and was illustrated using an example.

Finally, the strategy for optimizing a process flowsheet using concepts from the statistical design of experiments was introduced. The method was then illustrated using the dimethyl ether (DME) flowsheet from Appendix B.

REFERENCES

1. Juran, J. M., "Pareto, Lorenz, Cournot, Bernoulli, Juran, and others," *Ind. Qual. Control*, **17**(4), 25 (1960)
2. Douglas, J. M., *Conceptual Design of Chemical Processes*, McGraw-Hill, New York, 1988.

3. Humphrey, J. L., and G. E. Keller, II, *Separation Process Technology*, McGraw Hill, New York, 1997, Chapter 2.

4. Linnhoff, B., and J. R. Flower, "Synthesis of Heat Exchanger Networks: 1. Systematic Generation of Energy Optimal Networks," *AIChE J.*, **24,** 633 (1978).

5. Hohmann, E. C., *Optimum Networks for Heat Exchange*, Ph.D. Thesis, University of Southern California, 1971.

6. Umeda, T., J. Itoh, and K. Shiroko, "Heat Exchanger System Synthesis," *Chem. Eng. Prog.*, **74** (7), 70 (1978).

7. Neter, J., and W. Wasserman, *Applied Linear Statistical Models*, Richard D. Irwin, Inc., Homewood, IL, 1974.

8. Ludlow, D. K., K. H. Schulz, and J. Erjavec, "Teaching Statistical Experimental Design using a Laboratory Experiment," *Journal of Engineering Education*, **84,** 351 (1995).

9. Nelder, J. A., and R. Mead, "A Simplex Method for Function Minimization," *Computer Journal*, **7,** 308 (1965).

10. Kirkpatrick, S., C. D. Gelatt, and M. P. Vecchi, "Optimization by Simulated Annealing," *Science*, **220,** 671 (1983).

PROBLEMS

19.1 Heuristics for process optimization are developed through experience. Based on your own experience in previous courses, employment, or other activities:
   a. Develop three new optimization heuristics.
   b. Describe your experiences that helped you formulate each heuristic.
   c. Give an example of how the heuristic would be applied to reduce an intractable optimization problem into a feasible optimization problem.

19.2 Apply the strategies from Section 19.2 to the Acrylic Acid process in Appendix B.
   a. Define an objective function.
   b. Identify the key decision variables.
   c. Prioritize these decision variables.
   d. Identify the constraints on the key decision variables.

19.3. Perform a parametric optimization for the second distillation column, T-202, in the DME process, Figure B.1, using reflux ratio and pressure as the independent variables and *NPV* as the objective function. You should follow the approach used in Section 19.4.2, which considered the first column, T-201, for this process.

19.4 For the DME process shown in Figure B.1, complete a parametric optimization for both columns when the order of separation is reversed. This means that the first column will remove water as a bottom product and the second column will separate DME from methanol. The approach used in Problem 19.3 should be followed here, with pressure and reflux ratio used as the independent variables and *NPV* as the objective function. Compare the results from Section 19.4.2 and Problem 19.3 with your results from this problem to determine which is the best separation sequence for the DME process.

19.5 For the heat integration problem developed in Section 19.3.5, construct the MUMNE heat exchanger network for the following cases

a. $\Delta T_{min} = 20$

b. $\Delta T_{min} = 5$

19.6 From the results of Problem 19.5 and the example in Section 19.3.5, estimate the before-tax *NPV* of each exchanger network assuming the following economic parameters: $i = 10\%$, $n = 7$ yr. You may estimate the heat transfer areas using the following film heat transfer coefficients:

> For streams 1, 2, and 5 use $h = 750 \text{ W}/\text{m}^2{}^\circ\text{C}$
>
> For streams 3, 4, and 6 use $h = 500 \text{ W}/\text{m}^2{}^\circ\text{C}$
>
> For cooling water use $h = 2000 \text{ W}/\text{m}^2{}^\circ\text{C}$ and for steam and molten salt use $h = 5000 \text{ W}/\text{m}^2{}^\circ\text{C}$

For temperatures greater than 254°C, assume that a molten salt stream is available at a cost of $7.0/GJ. This salt stream is available at 310°C and must be returned at 290°C. Also assume that the total module cost (TMC) of heat exchangers is given by TMC = $6000 (area)$^{0.47}$.

Which design do you recommend?

19.7 The temperature interval diagram for a process is shown in Figure P19.7. For this process, do the following:

a. Compute the missing values of $\dot{m}C_p$ for Streams 2 and 5 and the missing Q values for temperature intervals B and E.

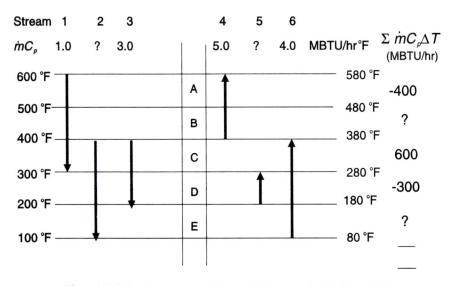

Figure P19.7   Temperature Interval Diagram for Problem 19.7

    b. Calculate the minimum hot and cold utility loads for this process subject to a minimum approach temperature of 20°F.

    c. Calculate the pinch temperatures for this process. You may assume that there is only one hot and one cold utility available.

    d. Calculate the minimum number of exchangers needed for the minimum energy case, for above and below the pinch.

    e. Design the MUMNE network for the process.

19.8 Four streams in a chemical process need to be heated or cooled. The thermal data for these streams is given below:

Stream No.	$\dot{m}C_p$ (10^3BTU/hr°F)	T_{in} (°F)	T_{out} (°F)
1	4.0	100	550
2	11.0	250	500
3	7.0	400	120
4	7.0	600	300

Assuming that only one hot and one cold utility are available, and using a minimum approach temperature of 20°F, do the following:

    a. Calculate the pinch temperatures.

    b. Calculate the minimum hot and cold utility duties for this process.

    c. Determine the minimum number of exchangers above and below the pinch required for Part (b) above.

    d. Design the MUMNE network for this process.

    e. Explain how your answers to the above questions would change given that the normal utilities (high-, medium-, and low-pressure steam, cooling water, and natural gas) are available?

19.9 Rework Problem 19.8 above assuming that the following utilities are available: high-, medium-, and low-pressure steam, cooling water, and for temperatures greater than 490°F, a molten salt stream is available at a cost of $7.0/GJ. Evaluate the operating cost and savings that this new network will give.

    Note: This problem may exhibit multiple utility pinches.

19.10 Design the MUMNE network for the DME process shown in Figure B.1 and Table B.2. Remember that the process streams flowing through the reboilers and condensers and any streams undergoing a partial change of phase will require dummy streams to simulate them. Draw the composite enthalpy-temperature diagram for this process. Assume a minimum approach temperature of 10°C. Estimate the savings in operating costs compared to the current system (Figure B.1).

19.11 For the DME column (T-201), find the operating pressure at which a switch from medium to low-pressure steam (for use in the reboiler, E-204) yields an improvement in the *NPV*. Refer to Section 19.4.2 for a discussion of what costs to consider for this problem.

SECTION

5

The Professional Engineer, the Environment, and Communications

Throughout this book, we have focused on the role that a chemical engineer plays in the analysis, synthesis, and design of chemical processes. In this section, that role is put in the context of the profession of chemical engineering, which is defined by the American Institute of Chemical Engineers as "the profession in which a knowledge of mathematics, chemistry, and other natural science gained by study, experience, and practice is applied with judgment to develop economic ways of using materials and energy for the benefit of mankind."

Far from being the "soft" side of engineering, the topics in this section are very much the crucial steps between a design in the mind of the engineer and the realization of a new or improved process or product in the service of humanity.

That translation from thought to reality depends very much on the engineer's ability to:

- Gain the necessary experience and knowledge.
- Earn the trust of society.
- Ensure that the process and products are safe.
- Explain the risks to the public.
- Protect the environment.
- Communicate effectively to a wide variety of audiences.

A chemical engineer is very much a member of a tiny technical elite of modern society. In the United States, less than 1% of the population is trained in engineering and the sciences. Chemical engineers account for only 0.05% of the population, yet design and manage the plants that produce such essential goods as pharmaceuticals, plastics, paper, fertilizers and pesticides, fuels, synthetic fabrics, and clean water. Chemical engineers use jargon and perform calculations that are mysterious at best to most people. Their success depends not only on performing the calculations but on convincing the public that they can add value to their quality of life.

The following chapters deal with these crucial issues.

Chapter 20: Ethics and Professionalism
Engineering problem solving within an ethical framework, legal responsibilities, and professional registration are developed. Ethics case studies and the content of the Fundamentals of Engineering (FE) examination are included.

Chapter 21: Health, Safety, and the Environment
Methods of analyzing risk are provided and the basic types of health, safety, and environmental regulations are explained, with references to government databases. Pollution prevention strategies are introduced through their relationships to HAZOP studies and to life-cycle analysis.

Chapter 22: Written and Oral Communications
Through a focus on audience analysis, strategies for improving the effectiveness of both written reports and oral presentations are explained. Commonly accepted (but frequently broken) rules for figures and tables are covered as are hints to effective use of communications software.

Chapter 23: A Report Writing Case Study
Following a sample student design report, this chapter offers models of both strong and weak written communications in several formats: memoranda, visual aides, and short design reports. A checklist of common errors is provided.

20

Ethics and Professionalism

Engineering has been described as "the strategy for causing the best change in a poorly understood or uncertain situation within the available resources" [1]. The realm of ethics and professionalism entails very real, poorly understood problems that are as challenging as any technical problems an engineer will face. This chapter presents heuristics, objective functions (i.e., ways of defining what is *best*), and constraint identification strategies that are especially useful in solving such problems.

As described in Chapter 9, all heuristics are, in the final analysis, fallible and incapable of justification. They are merely plausible aids or directions toward the solution of a problem [1]. Especially for the heuristics described in this chapter, we need to keep in mind the four characteristics of any heuristic:

1. A heuristic does not guarantee a solution.
2. It may contradict other heuristics.
3. It can reduce the time to solve a problem.
4. Its acceptance depends on the immediate context instead of on an absolute standard.

The fact that one cannot precisely follow all ethical heuristics all the time is to be expected, as it is with any set of technical heuristics.

The purpose of this chapter is to help you to develop strategies to make the best choice when faced with an ethical problem. The set of strategies developed will be different for each reader. We present below a general overview of engineering ethics and introduce a series of ethics scenarios. We have found that the best way to develop the facility to deal with ethical problems is, after reading the overview, to discuss these scenarios in small groups. Each group presents its solution to the class. Even with seemingly straightforward situations, the solutions can be diverse. A class discussion of the different solutions is followed by a reflection on what was discussed. Then we go on to the next problem. Many such scenarios are given at the end of this chapter and in books and articles [2,3,4,5,6,7].

20.1   ETHICS

Whenever chemical engineers develop products, design processes and equipment, manage process operations, communicate with other engineers and non-engineers, develop markets and sell products, lead other engineers, interact with clients, represent their firms to the government or to the public—in short when-ever chemical engineers do anything that impacts the lives of anyone—their choices of action are based on ethics. Even when faced with two different equations, we select one equation, based in part on our ethical values. Does the less precise equation include a safety factor that lowers the risk to our employer, employees, or the public? Do we spend more time to do more rigorous calculations, costing the firm more money but providing a better answer to the client? How do we decide?

In each of these circumstances, engineers apply their own moral standards, mindful of the legal requirements, using their personal code of ethics to make the decision. To help in the development of a personal code of ethics, which will provide a framework for making these decisions, we begin by identifying the three types of reasons for ethical behavior:

- Moral
- Legal
- Ethical

Although nearly all people share *some* fundamental moral ideals, each engineer has his or her own distinct set of moral principles. Typically, these principles are shaped by religion, conscience, and especially early childhood family experiences. The basic framework by which one decides what is right and what is wrong is generally very well developed by the time one reads an engineering text. Thus, this text will not take up the moral dimensions of engineering ethics other than to stress the importance of continually reminding oneself to be true to one's moral values as one works through ethical problems.

A few legal aspects of solving ethical problems are covered in Section 20.2.3, but the full legal consequences of engineering decisions are far beyond the scope of this book. The legal system (which includes government regulations) is a col-lection of rules of conduct for a society to assist orderly transactions between people. Chemical engineers should seek skilled legal advice whenever these rules, or the consequences for not following them, are unclear.

The aspect of ethical decision making we cover here is that commonly referred to as engineering ethics. There are generally accepted codes of conduct for engineers, although, as will become clear, they are too broad to be used alone as prescriptions for engineering choices in difficult situations. Engineering ethics is the system of principles and strategies that engineers use to solve complex problems involving other people's lives. It includes aspects of moral principles and legal responsibilities, as well as recognized codes of ethics and generally accepted norms of engineering and business behavior.

20.1.1   Moral Autonomy

In as much as engineers do not all share a single set of moral principles by which to make ethical decisions, we fully expect that different readers will make different decisions, especially in complex situations. The goal of this chapter is not *uniformity* of decisions by all engineers but *autonomy* of each engineer to make the right decision. In this context, the right decision can be identified by the use of a heuristic. The right decision is one that is

- Consistent with the engineer's moral principles
- Consistent with the generally accepted codes of engineering conduct
- Consistent with obligations that the engineer has accepted
- Consistent with the law
- Consistent with the applicable Code of Ethics

But, most importantly, **the right decision is one that the engineer can live with.** Of course, it is always possible that one person's decision would not be acceptable to another person.

> **The ability to make one's own ethical decisions is known as *moral autonomy*.**

Moral autonomy does not require that you be able to look back and always be confident that the choice made was the best of all possible choices. Although this is the goal, it is a moving target. Rather, you are exercising moral autonomy if you are in control of your decision, if you make the choice based on a reasonable analysis of the potential consequences consistent with your moral, legal, and ethical beliefs, rights, duties, and obligations. If you do not understand your moral principles, if you have no strategy for ethically analyzing a situation, or if you defer your own ethical responsibilities to others, you cannot claim to be exercising moral autonomy.

The goal of this chapter is to help you develop moral autonomy. Previewing the kinds of ethical problems that you are likely to have to resolve is the most powerful tool you can use to learn about and to develop your moral autonomy.

20.1.2   Rehearsal

When learning any new skill, one usually practices or rehearses. To learn to apply the ideal gas law in process calculations, one does end-of-chapter problems in thermodynamics. Understanding the theory behind the ideal gas law is no guarantee that one will be able to solve applied problems. Some people can and some people cannot. But few would argue that one can solve a problem as quickly, as easily, or as correctly the first time as the nth time one solves it. Such is the power of rehearsal.

Rehearsal becomes more important when decisions must be made quickly, extremely accurately, or under great stress. And great stress often accompanies ethical problems. Take a look at Example 20.1, for instance.

Example 20.1

The Falsified Data [2, Reprinted by special permission from *Chemical Engineering*, May 5, 1980 and September 22, 1980, Copyright © 1980 by McGraw-Hill, Inc., New York, NY 10020]

Jay's boss is an acknowledged expert in the field of catalysis. Jay is the leader of a group that has been charged with developing a new catalyst system, and the search has narrowed to two possibilities, Catalyst A and Catalyst B.

The boss is certain that the best choice is A, but directs that tests be run on both, "just for the record." Owing to inexperienced help, the tests take longer than expected, and the results show that B is the preferred material. The engineers question the validity of the tests, but, because of the project's timetable, there is no time to repeat the series. So the boss directs Jay to work the math backwards and come up with phony data to substantiate the choice of Catalyst A, a choice that all the engineers in the group, including Jay, fully agree with. Jay writes the report.

In this simple scenario, there is a great deal of stress. If Jay had never thought about what to do in such a situation, it is highly unlikely that he would make a decision that he could live with. It is much more likely that Jay would look back on the event and wish he could go back and make a different choice. Rehearsal gives us the opportunity to do just that. The first time we see the situation, we would like to be in a low-risk environment. If we make the wrong decision, it does not matter. The second time, we can make the better choice.

In early education and in skills training at any level, the goal of rehearsal is to work through a scenario that is likely to occur in the future and to develop the best response we can. In advanced professional education, and especially in engineering education, the goal of rehearsal is to work through a scenario representative of a broad range of situations that are likely to occur in the future and to develop a strategy for responding to the broad range of problems, many of which cannot be imagined. For ethical decision making, this strategy must be powerful, adaptive, and personal. Throughout your career, new ethical problems will arise. The key is to rehearse frequently, using example and homework problems and conceiving your own representative scenarios.

20.1.3   Reflection in Action

One of the characteristics of successful professionals in a variety of fields is frequent, *post-mortem* analysis. This self-imposed study of events that have occurred in one's professional life is called *reflection in action* [8]. After an engineering event has occurred in which ethical decisions were made, one sits down (individually or in a small group) and reviews the case, analyzing the facts, the missing information, the constraints, the unnecessary perceived constraints, the options considered, the options not considered, and the strategy used to arrive at the decision.

There are many reasons why reflection in action is so powerful, but we focus on two:

- It forces one to analyze the strengths and weaknesses of one's own strategy.
- It provides continual opportunities and encouragement for rehearsal.

Example 20.2

Reflection on the Falsified Data

Let us assume that Example 20.1 has occurred. We can use any standard problem-solving strategy [9] to reconstruct the scenario. Here we have used the McMaster 5-step strategy [10].

1. *Define.* Was the problem well defined by the participants? Was the real problem that the experimental results were unexpected? Was the real problem that now things were more uncertain than before? There are other possible definitions of the problem, but, if one defines the problem as how to deal with obviously flawed data, one may miss the entire point. The problem would be thereby unnecessarily overconstrained. What is the real problem? If the characters had defined the problem better, would they have reached a different conclusion? Would it have been a better conclusion?

2. *Explore.* What alternatives were explored? Are there other alternatives: requesting additional time to rerun the experiments, alerting the users of the data to their possible inaccuracy, writing a detailed analysis of the theoretical and experimental discrepancies, debriefing the technician who performed the experiments to determine if errors in technique could cause the discrepancy? Would a more careful exploration

of the alternatives have been helpful in this case? What brainstorming techniques might have been helpful?

3. *Plan.* Did the participants develop an adequate plan? What would (should) you have done differently?

4. *Do It.* Did the participants execute the plan well?

5. *Look Back.* Here is the play within the play. Would it have been helpful if the participants had reflected on past experiences? Would it have been helpful if the participants had projected what could happen as a consequence of their decision and had analyzed the expected effectiveness of their approach?

Because we do not know the outcome of the Falsified Data case, we cannot analyze fully the *Plan* and *Do It* stages. When the full case is reflected upon, we could, and would, do just that.

At the end of any rehearsal or reflection, one (or the group) should develop a list of heuristics to use in future ethical problem solving. These could be the heuristics that had been used effectively (in the case of a reflection) or new heuristics that can be used (in the case of a rehearsal). For Example 20.2, we might identify the following heuristics:

- Use a traditional problem-solving strategy for solving ethical problems.
- Consider the possibility that inexperienced people can be right.
- Debrief people fully before assuming facts about their actions.
- Consider what will happen if a specific decision is based on a false assumption.
- Be honest.
- Be concerned about the welfare of your company.
- Be concerned about the welfare of your employees.
- Do not let other people make ethical decisions for you.
-
-
-

20.1.4   Mobile Truth

It is quite natural for people to assume that groups to which they belong are right, and that other groups are wrong. This basic response gives rise to loyalty, strong familial pride and duty, willful obedience, and leadership. The strength of collective action depends on this response, which, in terms of ethical decision making in engineering, manifests itself in *mobile truth*.

Let us say that your AIChE student chapter and the IEEE student chapter are playing volleyball. If the referee makes a difficult call, you are apt to find the call right if it favors your team and wrong if it favors the other. You are not being unethical, but you are perhaps being unreasonable. Similarly, when you start to work for an organization, you begin to develop attitudes towards it that are similar to the bonding that occurs within families, nationalities, and schools. The faster these bonds develop, the faster you will be accepted. The stronger these bonds are, the more loyal you will become.

Example 20.3

If you work for Company A that produces polyethylene, you know that Type-A polyethylene is the best. Because you are offered a promotion, higher pay, and desirable relocation, you move to Company B. Almost immediately, Type-B polyethylene becomes the best. It could be because you are such a great chemical engineer, but it is probably because of mobile truth. The world did not change, but your frame of reference and, especially, your loyalty did.

> **When your affinity to a group clouds your ethical decision making, you are being affected by *mobile truth*.**

The point here is to learn to recognize mobile truth and to filter it out of ethical decision making. Obviously, a chemical engineer has ethical duties and obligations to an organization, but one must be ever-vigilant to identify the intrusion of mobile truth into the ethical decision-making process. The point is not just to make the process more "objective" or "fair," but rather to try to see the situation from the point of view of those outside the organization. Here are a few heuristics for avoiding being misled by mobile truth:

- Ask yourself if your decision would be different if you worked for another part of the company or for another company.
- Imagine that you live just outside the plant fence.
- Imagine that you work for the Environmental Protection Agency.
- Ask for the opinion of someone else in your organization. Explain the situation, suggesting that the facts pertain to a different organization. The response is likely to be less affected by mobile truth.

More heuristics can be developed through rehearsal and reflection.

Related to the concept of mobile truth is the concept of *post-rationalization*. Again, it is quite natural and normal for people to try to justify their actions (and the actions of their colleagues and organizations), whether they are defensible or not. Since complex ethical problems can be analyzed and evaluated in so many different ways, it is often easy to fool oneself into thinking that one has acted ethically. But one must keep in mind the frame of reference or point of view of others (outside your organization). If you want to be sure that what you have done or are about to do is ethical, imagine how someone outside the organization would view the decision. If there is a difference, chances are you were post-rationalizing (or pre-rationalizing).

20.1.5   Nonprofessional Responsibilities

Each chemical engineer has personal responsibilities: to family, to friends, to oneself. These responsibilities, like professional responsibilities, will change dramatically throughout one's life. The obvious example is one's family situation, which is likely to grow and change over a 40-year career. In general, it is easier to make ethical choices when they affect only oneself. Choices that affect one's nuclear family are especially difficult. Some choices might cost you your job, make you a social outcast, compel you to move from your home, even estrange you from your family.

Generally, one can mitigate consequences by

- Thinking early about the effects of your decisions on your family
- Taking into account likely changes in your family situation
- Most importantly, talking with your family about your decision

The choice of when to discuss a professional ethics problem with one's family (and with which members of the family) is certainly a difficult one. Considering what a dramatic effect your decisions can have on your family, it would be wise to consider a hypothetical situation before a real ethical problem arises. Think about various options of whom to tell what and when. Rehearsal can be done with one's family, with a peer group, or by oneself. Whichever you choose, it should make any potential conflict easier to resolve to everyone's satisfaction.

The crucial impact of ethical decisions on one's family is generally not fully appreciated until one is married or becomes a parent. In using rehearsal to prepare for difficult ethical situations, one should anticipate potential changes in family situation that might affect the decision. In a typical senior chemical engineering class, most students are single without children. However, there may be students with a significant range of circumstances. Some might be married without any children; several might be parents. Additionally, some may intend to be the only or the main potential wage earner, while others may expect to work or earn less than their husbands or wives. Whenever possible, it is better to form non-homogeneous groups for rehearsal of ethical scenarios. The addition of a married student to an otherwise single group usually changes the discussion substantially.

While discussing ethical scenarios, financial and other concerns also must be considered. Prospective chemical engineers often say, "I just wouldn't work for a company that would do that," or "I'd quit and get another job before I'd agree to that." But these comments are looked upon as naive by most who have made tough ethical choices, especially whistleblowers, as is discussed in Section 20.1.8. Often, chemical engineers who no longer have young children or who have saved well for retirement express more willingness to do what they feel is ethically correct. It is unclear whether this expressed willingness translates into action. Some choose to live more frugally, others choose jobs that require fewer or less difficult ethical choices. Those who become overcommitted—for example, financially—and then feel compelled to make the choice to perform an act that they consider ethically wrong, seldom would claim to be morally autonomous, let alone happy.

Example 20.4

A question of integrity [3, Reprinted by special permission from *Chemical Engineering*, March 2, 1987 and September 28, 1987, Copyright © 1987 by McGraw-Hill, Inc., New York, NY 10020]

Under the Toxic Substances Control Act (U.S. Public Law 94-469), all chemicals in general use, or that had been in use, were required to be placed on an inventory list in 1979. Some chemicals were omitted from the initial list through oversight. To produce a chemical not on the list, or a new chemical, a manufacturer must submit a pre-manufacturing marketing notification to the EPA.

(The purpose of notification is to allow EPA 90 days to review a chemical, to ensure that its production, distribution, and use will not be detrimental to human health and the environment. The Agency has the authority to place controls as necessary. Heavy penalties can be, and have been, assessed against violators of this regulation.)

Chris supervises a unit that has begun making a "new" chemical (one not on the list) and becomes aware that a pre-manufacturing notification has not been filed.

If Chris "blows the whistle," Chris's career with the company could be over, despite laws to protect whistleblowers. If Chris does nothing, workers may be exposed to the chemical without safeguards, and be harmed by it. And, if the company were caught, Chris's professional reputation could be stained, especially if it could be shown that the inaction resulted in harm to workers.

> Should Chris discuss this ethical problem with her family?
> What financial or other hardships might result if she reports the situation?
> What obligations does she have to her family in this regard?
> What obligations does she have to the community?
> What obligations does she have to her employer?

20.1.6   Duties and Obligations

Chemical engineers have certain duties by virtue of their positions, and they acquire other obligations in a number of ways: by accepting assignments, in joining professional organizations, and through their family choices, for example.

Throughout ethical problem solving, one needs to remind oneself of all of the duties and obligations to which one has agreed.

According to the National Science Foundation, less than 1% of the U.S. population are engineers. Of these, less than one in ten are chemical engineers. The American Institute of Chemical Engineers estimates that only 0.05% of U.S. citizens are chemical engineers. Clearly, the vast majority of people are not and do not think like chemical engineers. Few will understand what chemical engineers do, but all will be affected by what chemical engineers do. Keeping this fact in mind helps to put the awesome responsibilities of chemical engineers in perspective.

One's duties and obligations form the basis for some additional important heuristics in ethical problem solving:

- Remind yourself of relevant duties and obligations that you have accepted.
- Remind yourself of otherwise relevant duties and obligations that you have *not* accepted.
- When accepted duties and obligations are necessarily in conflict, rank these responsibilities.
- If you choose a solution that violates an obligation, discuss the decision with those to which the obligation was made and determine the consequences.

20.1.7   Codes of Ethics

Codes of ethics are formal obligations that persons accept when they join organizations or when they are allowed to enter a profession. In chemical engineering, there are three main types of codes of ethics: employer, technical society, and government. The employer-based codes of ethics are usually incorporated into the Codes of Business Conduct that one agrees to upon employment with a particular firm. These are covered in Section 20.4. The government-based codes are the Professional Engineer rules, regulations, and laws that exist in all states and territories of the United States and their counterparts throughout the world. These are covered in Section 20.2.

The most important technical society code for U.S. chemical engineers is the Code of Ethics of the American Institute of Chemical Engineers. Similar codes have been adopted by other engineering societies. When you sign your application for membership in AIChE, you agree to abide by this code (Figure 20.1).

The code is divided into three parts. The first part defines the profession of chemical engineering. The definition is quite broad, leading some to paraphrase it as chemical engineering is anything that a chemical engineer does. But the purpose of the definition is to define when an act is covered by the code. In this context, one might paraphrase it as: "Whenever anyone who is a member of AIChE performs any service that others might reasonably identify as chemical engineer-

---

AIChE Code of Ethics

Chemical engineering is the profession in which a knowledge of mathematics, chemistry and other natural sciences gained by study, experience, and practice is applied with judgment to develop economic ways of using materials and energy for the benefit of mankind.

Goals:

Uphold and advance the integrity, honor, and dignity of the engineering profession by:

- Being honest and impartial and serving with fidelity their employers, their clients, and the public
- Striving to increase the competence and prestige of the engineering profession
- Using their knowledge and skill for the enhancement of human welfare

Responsibilities:

1. Hold paramount the safety, health, and welfare of the public in performance of their professional duties.
2. Formally advise their employers or clients (and consider further disclosure, if warranted) if they perceive that a consequence of their duties will adversely affect the present or future health or safety of their colleagues or the public.
3. Accept responsibility for their actions and recognize the contributions of others; seek critical review of their work and offer objective criticism of the work of others.
4. Issue statements or present information only in an objective and truthful manner.
5. Act in professional matters for each employer or client as faithful agents or trustees, and avoid conflicts of interest.
6. Treat fairly all colleagues and co-workers, recognizing their unique contributions and capabilities.
7. Perform professional services only in their areas of competence.
8. Build their professional reputations on the merits of their services.
9. Continue their professional development throughout their careers, and provide opportunities for the professional development of those under their supervision.

Figure 20.1   AIChE Code of Ethics (Reprinted with permission of AIChE)

ing, that conduct is covered under the AIChE Code of Ethics." If you call yourself a chemical engineer, you will be held to the standards of chemical engineering, both technical and ethical.

The second part, called the *Goals,* describes four basic tenets of chemical engineering and places the profession in its proper perspective in relation to the society at large. This section serves at least two important purposes. First, as a chemical engineer reads the Goals, he or she realizes the very broad responsibilities to the society that may not be apparent in day-to-day work. A decision made by a chemical engineer might save thousands of lives by providing fertilizer to grow much needed food, or a decision might kill scores of people in a catastrophic release of toxic materials. These Goals are a powerful tool in helping chemical engineers to do the right thing. Secondly, the list of goals points out quite clearly that all goals cannot be met simultaneously all of the time. In fact, one frequently encounters ethical dilemmas, in which *no* choice is a perfect choice, completely satisfying all of our moral, legal, and ethical responsibilities.

The third part of the Code of Ethics concerns nine relatively specific responsibilities that chemical engineers have. Many ethical problems can be attacked by referring to these responsibilities. While reiterating and clarifying the responsibilities of chemical engineers for the safety and health of the society at large, they speak to the responsibilities to clients, employers, employees, and the profession itself. Of particular note are two principles that are sometimes overlooked by beginning engineers. Both refer to the need for continuing education. Principle 7 states that chemical engineers shall "perform professional services only in their areas of competence." It may be obvious that a chemical engineer should not be doing electrical engineering work without having had significant education in the relevant area of electrical engineering, but, even within chemical engineering, there are many areas that are not taught in even the most rigorous B.S.Ch.E. program. Therefore, throughout one's career, one is required by the Code to evaluate one's own chemical engineering capabilities continuously and to acquire any needed education (through reading, working with more experienced engineers, consulting experts, or taking courses) before accepting an assignment in any area of chemical engineering. This principle states that the chemical engineer, not some outside governmental agency, must take final responsibility for professional competence. A governmental agency may certify you as a Professional Engineer after you pass some day-long examinations and a multi-year internship, but you are still the responsible party for your own chemical engineering competence.

Principle 9 states that chemical engineers shall "continue their professional development throughout their careers, and provide opportunities for the professional development of those under their supervision." Many young engineers may think that one's chemical engineering education ends after the B.S., the M.S., or surely after the Ph.D., but the Code makes it clear that this is not the case. In the last 40 years, tremendous strides have been made in chemical engineering, and we certainly expect at least as great a change in the next 40 years. Thus, one is required to keep up with the latest advances through such activities as participating in in-house (i.e., within the company) training programs; taking continuing education courses offered by AIChE, universities, and other organizations; reading technical journals; consulting with experts; and so on. The facts that one learns during the B.S. experience are overshadowed by the strategies that one learns to attack problems. And some of the greatest of these problems are to evaluate one's knowledge, to decide what new material needs to be learned, and to develop and implement a plan to acquire that new knowledge.

But why are Principles 7 and 9 "ethical" considerations? When one claims to be a chemical engineer, society in general (and employers, clients, and employees in particular) puts trust in that individual. To earn that trust, one must be competent; however, most people do not have the background to judge the competence of a chemical engineer. Therefore, it is an ethical responsibility for the chemical engineer to practice within a scope of competency that can be defined completely only by that chemical engineer. And, as technology expands, a chemical engineer's capability must expand just to maintain a given scope of competency.

In addition to the AIChE Code of Ethics, two other codes are frequently used in chemical engineering. One is called the *Engineer's Creed*:

> *As a professional engineer, I dedicate my professional knowledge and skill to the advancement and betterment of human welfare. I pledge to give the utmost of performance, to participate in none but honest enterprise, to live and work according to the laws of man and the highest standards of professional conduct, to place service before profit, honor and understanding of the profession before personal advantage, and the public welfare above all other considerations. In humility and with need for divine guidance, I make this pledge.*

Some states incorporate this creed into the code of ethics for professional engineers, and some engineering colleges ask engineering graduates to recite this creed at the commencement ceremonies. The Engineer's Creed is a very much more general and "moral" (as opposed to ethical) obligation than is the AIChE Code of Ethics. The creed can serve as a bonding and an inspirational pledge.

The third common Code of Ethics is that of the National Society for Professional Engineers (Figure 20.2). This code is more detailed and more specific than is the AIChE Code of Ethics. It includes not only the canons and principles that are the total of the AIChE Code, but it also prescribes rather specific actions to take in specific circumstances. The NSPE Code applies to those who join the organization and to those licensed to practice engineering in states where this code is included in the Professional Engineers code. However, it is instructive for any chemical engineer to read the NSPE Code periodically, as a reminder of some of the ethical problems that arise in the profession.

20.1.8   Whistleblowing [12]

When a chemical engineer notices behavior that is possibly or potentially unethical, the question is, what action should he or she take? As noted in the AIChE Code of Ethics and in the NSPE Code of Ethics, there are specific avenues for action in the form of heuristics. For example, Responsibility 2 of the AIChE code requires members to "formally advise their employers or clients (*and consider further disclosure, if warranted*) if they perceive that a consequence of their duties will adversely affect the present or future health or safety of their colleagues or the public." Responsibility 3 requires chemical engineers to "offer objective criticism

NSPE Code of Ethics for Engineers

Preamble

Engineering is an important and learned profession. As members of this profession, engineers are expected to exhibit the highest standards of honesty and integrity. Engineering has a direct and vital impact on the quality of life for all people. Accordingly, the services provided by engineers require honesty, impartiality, fairness and equity, and must be dedicated to the protection of the public health, safety, and welfare. Engineers must perform under a standard of professional behavior that requires adherence to the highest principles of ethical conduct.

I. Fundamental Canons

Engineers, in the fulfillment of their professional duties, shall:

1. Hold paramount the safety, health, and welfare of the public.
2. Perform services only in areas of their competence.
3. Issue public statements only in an objective and truthful manner.
4. Act for each employer or client as faithful agents or trustees.
5. Avoid deceptive acts.
6. Conduct themselves honorably, responsibly, ethically, and lawfully so as to enhance the honor, reputation, and usefulness of the profession.

II. Rules of Practice

1. Engineers shall hold paramount the safety, health, and welfare of the public.
   a. If engineers' judgment is overruled under circumstances that endanger life or property, they shall notify their employer or client and such other authority as may be appropriate.
   b. Engineers shall approve only those engineering documents that are in conformity with applicable standards.
   c. Engineers shall not reveal facts, data, or information without the prior consent of the client or employer except as authorized or required by law or this Code.
   d. Engineers shall not permit the use of their name or associate in business ventures with any person or firm that they believe are engaged in fraudulent or dishonest enterprise.
   e. Engineers having knowledge of any alleged violation of this Code shall report thereon to appropriate professional bodies and, when relevant, also to public authorities, and cooperate with the proper authorities in furnishing such information or assistance as may be required.
2. Engineers shall perform services only in the areas of their competence.
   a. Engineers shall undertake assignments only when qualified by education or experience in the specific technical fields involved.
   b. Engineers shall not affix their signatures to any plans or documents dealing with subject matter in which they lack competence, nor to any plan or document not prepared under their direction and control.

Figure 20.2 NSPE Code of Ethics ([11] Reprinted with permission of NSPE)

     c. Engineers may accept assignments and assume responsibility for coordination of an entire project and sign and seal the engineering documents for the entire project, provided that each technical segment is signed and sealed only by the qualified engineers who prepared the segment.

3. Engineers shall issue public statements only in an objective and truthful manner.

     a. Engineers shall be objective and truthful in professional reports, statements, or testimony. They shall include all relevant and pertinent information in such reports, statements, or testimony, which should bear the date indicating when it was current.

     b. Engineers may express publicly technical opinions that are founded upon knowledge of the facts and competence in the subject matter.

     c. Engineers shall issue no statements, criticisms, or arguments on technical matters that are inspired or paid for by interested parties, unless they have prefaced their comments by explicitly identifying the interested parties on whose behalf they are speaking, and by revealing the existence of any interest the engineers may have in the matters.

4. Engineers shall act for each employer or client as faithful agents or trustees.

     a. Engineers shall disclose all known or potential conflicts of interest that could influence or appear to influence their judgment or the quality of their services.

     b. Engineers shall not accept compensation, financial or otherwise, from more than one party for services on the same project, or for services pertaining to the same project, unless the circumstances are fully disclosed and agreed to by all interested parties.

     c. Engineers shall not solicit or accept financial or other valuable consideration, directly or indirectly, from outside agents in connection with the work for which they are responsible.

     d. Engineers in public service as members, advisors, or employees of a governmental or quasi-governmental body or department shall not participate in decisions with respect to services solicited or provided by them or their organizations in private or public engineering practice.

     e. Engineers shall not solicit or accept a contract from a governmental body on which a principal or officer of their organization serves as a member.

5. Engineers shall avoid deceptive acts.

     a. Engineers shall not falsify their qualifications or permit misrepresentation of their or their associates' qualifications. They shall not misrepresent or exaggerate their responsibility in or for the subject matter of prior assignments. Brochures or other presentations incident to the solicitation of employment shall not misrepresent pertinent facts concerning employers, employees, associates, joint venturers, or past accomplishments.

     b. Engineers shall not offer, give, solicit or receive, either directly or indirectly, any contribution to influence the award of a contract by public authority, or which may be reasonably construed by the public as having the effect of intent to influencing the awarding of a contract. They shall not offer any gift or other valuable consideration in order to secure work. They shall not pay a commission, percentage, or brokerage fee in order to secure work, except to a bona fide employee or bona fide established commercial or marketing agencies retained by them.

Figure 20.2   (*Continued*)

III. Professional Obligations

1. Engineers shall be guided in all their relations by the highest standards of honesty and integrity.
   a. Engineers shall acknowledge their errors and shall not distort or alter the facts.
   b. Engineers shall advise their clients or employers when they believe a project will not be successful.
   c. Engineers shall not accept outside employment to the detriment of their regular work or interest. Before accepting any outside engineering employment they will notify their employers.
   d. Engineers shall not attempt to attract an engineer from another employer by false or misleading pretenses.
   e. Engineers shall not actively participate in strikes, picket lines, or other collective coercive action.
   f. Engineers shall not promote their own interest at the expense of the dignity and integrity of the profession.
2. Engineers shall at all times strive to serve the public interest.
   a. Engineers shall seek opportunities to participate in civic affairs; career guidance for youths; and work for the advancement of the safety, health, and well-being of their community.
   b. Engineers shall not complete, sign, or seal plans and/or specifications that are not in conformity with applicable engineering standards. If the client or employer insists on such unprofessional conduct, they shall notify the proper authorities and withdraw from further service on the project.
   c. Engineers shall endeavor to extend public knowledge and appreciation of engineering and its achievements.
3. Engineers shall avoid all conduct or practice that deceives the public.
   a. Engineers shall avoid the use of statements containing a material misrepresentation of fact or omitting a material fact.
   b. Consistent with the foregoing, Engineers may advertise for recruitment of personnel.
   c. Consistent with the foregoing, Engineers may prepare articles for the lay or technical press, but such articles shall not imply credit to the author for work performed by others.
4. Engineers shall not disclose, without consent, confidential information concerning the business affairs or technical processes of any present or former client or employer, or public body on which they serve.
   a. Engineers shall not, without the consent of all interested parties, promote or arrange for new employment or practice in connection with a specific project for which the Engineer has gained particular and specialized knowledge.
   b. Engineers shall not, without the consent of all interested parties, participate in or represent an adversary interest in connection with a specific project or proceeding in which the Engineer has gained particular specialized knowledge on behalf of a former client or employer.
5. Engineers shall not be influenced in their professional duties by conflicting interests.

Figure 20.2   (*Continued*)

    a. Engineers shall not accept financial or other considerations, including free engineering designs, from material or equipment suppliers for specifying their product.

    b. Engineers shall not accept commissions or allowances, directly or indirectly, from contractors or other parties dealing with clients or employers of the Engineer in connection with work for which the Engineer is responsible.

6. Engineers shall not attempt to obtain employment or advancement or professional engagements by untruthfully criticizing other engineers, or by other improper or questionable methods.

    a. Engineers shall not request, propose, or accept a commission on a contingent basis under circumstances in which their judgment may be compromised.

    b. Engineers in salaried positions shall accept part-time engineering work only to the extent consistent with policies of the employer and in accordance with ethical considerations.

    c. Engineers shall not, without consent, use equipment, supplies, laboratory, or office facilities of an employer to carry on outside private practice.

7. Engineers shall not attempt to injure, maliciously or falsely, directly or indirectly, the professional reputation, prospects, practice, or employment of other engineers. Engineers who believe others are guilty of unethical or illegal practice shall present such information to the proper authority for action.

    a. Engineers in private practice shall not review the work of another engineer for the same client, except with the knowledge of such engineer, or unless the connection of such engineer with the work has been terminated.

    b. Engineers in governmental, industrial, or educational employ are entitled to review and evaluate the work of other engineers when so required by their employment duties.

    c. Engineers in sales or industrial employ are entitled to make engineering comparisons of represented products with products of other suppliers.

8. Engineers shall accept personal responsibility for their professional activities, provided, however, that Engineers may seek indemnification for services arising out of their practice for other than gross negligence, where the Engineer's interests cannot otherwise be protected.

    a. Engineers shall conform with state registration laws in the practice of engineering.

    b. Engineers shall not use association with a nonengineer, a corporation, or partnership as a "cloak" for unethical acts.

9. Engineers shall give credit for engineering work to those to whom credit is due, and will recognize the proprietary interests of others.

    a. Engineers shall, whenever possible, name the person or persons who may be individually responsible for designs, inventions, writings, or other accomplishments.

    b. Engineers using designs supplied by a client recognize that the designs remain the property of the client and may not be duplicated by the Engineer for others without express permission.

    c. Engineers, before undertaking work for others in connection with which the Engineer may make improvements, plans, designs, inventions, or other records that may justify copyrights or patents, should enter into a positive agreement regarding ownership.

Figure 20.2   *(Continued)*

   d. Engineers' designs, data, records, and notes referring exclusively to an employer's work are the employer's property. Employer should indemnify the Engineer for use of the information for any purpose other than the original purpose.

As Revised July 1996

"By order of the United States District Court for the District of Columbia, former Section 11(c) of the NSPE Code of Ethics prohibiting competitive bidding, and all policy statements, opinions, rulings or other guidelines interpreting its scope, have been rescinded as unlawfully interfering with the legal right of engineers, protected under the antitrust laws, to provide price information to prospective clients; accordingly, nothing contained in the NSPE Code of Ethics, policy statements, opinions, rulings or other guidelines prohibits the submission of price quotations or competitive bids for engineering services at any time or in any amount."

Statement by NSPE Executive Committee

In order to correct misunderstandings which have been indicated in some instances since the issuance of the Supreme Court decision and the entry of the Final Judgment, it is noted that in its decision of April 25, 1978, the Supreme Court of the United States declared: "The Sherman Act does not require competitive bidding."
   It is further noted that as made clear in the Supreme Court decision:

1. Engineers and firms may individually refuse to bid for engineering services.
2. Clients are not required to seek bids for engineering services.
3. Federal, state, and local laws governing procedures to procure engineering services are not affected, and remain in full force and effect.
4. State societies and local chapters are free to actively and aggressively seek legislation for professional selection and negotiation procedures by public agencies.
5. State registration board rules of professional conduct, including rules prohibiting competitive bidding for engineering services, are not affected and remain in full force and effect. State registration boards with authority to adopt rules of professional conduct may adopt rules governing procedures to obtain engineering services.
6. As noted by the Supreme Court, "nothing in the judgment prevents NSPE and its members from attempting to influence governmental action . . ."

NOTE: In regard to the question of application of the Code to corporations vis-á-vis real persons, business form or type should not negate nor influence conformance of individuals to the Code. The Code deals with professional services, which services must be performed by real persons. Real persons in turn establish and implement policies within business structures. The Code is clearly written to apply to the Engineer and items incumbent on members of NSPE to endeavor to live up to its provisions. This applies to all pertinent sections of the Code.

Figure 20.2   *(Continued)*

of the work of others." Responsibility 4 states that chemical engineers shall "issue statements or present information only in an objective and truthful manner." It is clear that one has the responsibility to tell those who engage one's professional services when there is a problem or potential problem. It is clear that one should be truthful. What is not so clear is what a chemical engineer should do if, after such disclosure, the situation persists.

When is further disclosure "warranted"? What is further disclosure? Does a chemical engineer have a responsibility to disclosure further, if warranted? What are the likely consequences of further disclosure? These questions are likely to be the most difficult and most important of one's professional career.

Deciding that further disclosure is warranted and then making that disclosure is called whistleblowing. Most instances of whistleblowing share the following characteristics:

1. Whistleblowing rarely results in correction of the specific situation; however, it sometimes changes the prevailing strategy for decision making and thus reduces the chance of a further occurrence.
2. Whistleblowing often brings about severe personal and professional problems for the whistleblower.

Given the above characteristics, why would any chemical engineer be a whistleblower? All three codes of ethics mentioned in Section 20.1.7 require chemical engineers to dedicate their skills to the public welfare, and whistleblowing has led to improved automobile safety, safer nuclear and chemical plants, better control of toxic wastes, reduced government waste, and other laudable results. Whistleblowers themselves have stated that they could not stand the stress of nondisclosure, which they viewed as an abdication of their ethical responsibilities. In other words, they could be morally autonomous only by whistleblowing.

Some laws have been created that protect some whistleblowers in some circumstances. Federal government employees cannot be fired or otherwise punished for any disclosure of "a violation of any law, rule, or regulation, mismanagement, a gross waste of funds, an abuse of authority, or a substantial and specific danger to public health and safety"(Civil Service Reform Act of 1978). In some states, whistleblowing statutes have been passed. And one can make the argument that the Professional Engineers' Code of Ethics requires and, therefore, protects whistleblowing in some states. However, many states have "employment at will" laws, giving any employer the right to fire an employee for any (or no) reason. In other states, an employer might need to find some unrelated reason to fire the whistleblower.

One may go through one's entire chemical engineering career without whistleblowing, but it is unlikely that one would go through a 40-year career without having to face the question of whether to whistleblow. Thus, to be able to have the moral autonomy to make the decision, one must consider what one should do.

Before deciding to whistleblow, one needs to examine four key questions:

1. *What should one do to solve the problem without whistleblowing?* The assumption here is that a resolution of the problem through normal channels is likely to be more effective, more timely, and less stressful than whistleblowing would be. The AIChE and NSPE codes require that one attempt these avenues, if they are likely to be successful. Furthermore, most whistleblowers (successful or not) indicate that whistleblowing should be a last resort.

2. *Is whistleblowing likely to solve the problem?* One needs to be reasonably certain that there is a problem and that the disclosure outside normal channels will resolve it. Although merely exposing a problem may make you feel virtuous, such action would not be considered whistleblowing unless the problem were, in the view of the whistleblower, serious, and unless the disclosure were likely to effect a change for the better.

3. *What whistleblowing actions should one take?* Some whistleblowing occurs within an organization, the most common being "going over your boss's head." Other whistleblowing involves disclosures (anonymous or attributed) to the news media. Many levels of action between these two examples are possible. One must consider the action that is most likely to effect the change that one wants, consistent with the risks that one is willing to take. The goal is to change the situation for the better, not just to expose it.

4. *What are the potential consequences to one's personal and professional life?* The consequences of disclosure are often serious. Loss of one's job is a clear possibility. Identification as a whistleblower can derail or end one's career; finding another job may be very difficult. If the disclosure results in a plant closing, loss of jobs, or financial loss to the company or to other property owners, those affected may be very angry with the whistleblower. One's family may encounter cruel treatment from those who consider the whistleblower to be the problem. On the other hand, nondisclosure could result in danger to employees or to the general public, financial damage to shareholders or to taxpayers, or charges of a coverup. Most whistleblowers recommend early discussion of any potential whistleblowing action with one's family. Nearly all wish they had legal protection from retribution.

Example 20.4

Insider Information [3, Reprinted by special permission from *Chemical Engineering*, March 2, 1987 and September 28, 1987, Copyright © 1987 by McGraw-Hill, Inc., New York, NY 10020]

One day, Lee, a process engineer in an acrylonitrile plant, runs into a former classmate at a technical society luncheon. The friend has recently taken a job as a regional compliance officer for OSHA and reveals, after several drinks, that there will be an unannounced inspection of Lee's plant. In a telephone conversation a few days later, the friend mentions that the inspection will occur on the following Tuesday.

Lee believes that unsafe practices are too often tolerated in the plant, especially in the way that toxic chemical are handled. However, although there have been many small spills, no serious accidents have occurred in the plant during the past few years.

What should Lee do? Let us suggest a problem-solving strategy for this scenario, leaving the details to a class discussion:

1. *Define.* Determine what the problem is and what the ethically desired final outcome should be. Reread the relevant codes of ethics. What are Lee's obligations according to the AIChE Code of Ethics? According to the NSPE Code of Ethics? Are there conflicting obligations?

2. *Explore.* Brainstorm for solutions in a small group. Consider internal (within the company) actions, external actions, and non-action.

3. *Plan.* Rank the possible actions according to their likelihood to bring about the ethically desired result. Determine the most effective order for these actions. Which should be done first? Under what circumstances should you go to the next step? When would you involve your family and others? Remember that internal actions, if effective, are often the faster method to reach the desired solution.

4. *Do It.* Imagine that you followed through with your plan.

5. *Look Back.* Consider the consequences of the proposed actions. Do the likely consequences of the actions change the ranking or the ordering? Predict the final outcome of the action plan. Evaluate the outcome based on your moral, ethical, and legal responsibilities.

20.1.9   Ethical Dilemmas

Some ethical problems can be solved easily. In fact, one does this every day. The goal of this chapter is to help the reader to prepare for the difficult choices, the answers to the *ethical dilemmas*. Some problems (the examples in this chapter and the problems at the end, for instance) have no "perfect" solution. Whistleblowing, for example, may satisfy one ethical obligation while violating another. The same can be said of not whistleblowing. A simple example that many would judge an ethical dilemma is Example 20.5:

Example 20.5

Initial Employment [6]

Robin is a senior seeking employment. In January, Robin is offered a job by Company X for $3000/month and given 10 days to accept the offer. Robin accepts this offer. Two weeks later, Robin receives an offer of $3500/month and a more exciting position from Company Y. What should Robin do?

Many, at first, may think the answer is obvious. Some will wonder if this example is even an ethical dilemma. Yet if you ask five people what their responses would be, you are likely to get more than one answer. What are the student's obligations to Company X after accepting employment but before becoming an employee? Some companies have rescinded job offers after they have been accepted. Does this fact change Robin's obligations? Should Robin's family responsibilities affect the solution?

When faced with an ethical dilemma, *one must necessarily rank one's ethical obligations*. This is perhaps the important heuristic in solving ethical dilemmas, but here are some more:

- Rank order your ethical obligations again after brainstorming for solutions.
- Admit that you may not be able to satisfy all of your obligations, but then *try* to satisfy them all.
- Combine the individual actions identified in the brainstorming into action plans. Evaluate each entire plan (rather than individual actions) for its consequences and effectiveness.
- At some point in the decision-making process, involve those who are most affected by the consequences but who are not active participants in the solution (often family, friends, trusted colleagues).

20.1.10   Additional Ethics Heuristics

Many heuristics for the solution of ethical problems have been presented in Section 20.1. A crucial heuristic is **Always try to develop new heuristics.** In this vein, we offer more:

- Acquire all the information you can about the situation. The problem may be more serious than it first appears, or there may be no real problem at all.
- Be honest and open. This is especially important when dealing with those who are predisposed to distrust you.
- Acknowledge the concerns of others, whether you share their concerns or not.
- Remember that one is only as ethical as one can afford to be. One can enhance one's ability to afford to do the right thing by knowing the consequences and by balancing the responsibilities that one accepts with the rest of one's life.

For other heuristics and for cases to use for rehearsal, please see the references listed at the end of this chapter or consult the following web site

http://www.cwru.edu/affil/wwwethics/engcases.html

20.2 PROFESSIONAL REGISTRATION

To be recognized as a chemical engineer and to offer such services to the public, one must be a licensed (registered) Professional Engineer. Each state (and other U.S. jurisdictions) has its own Board of Professional Engineers and its own regulations on licensure and practice. However, engineers registered in one state can generally become registered in any other state: One needs only to certify the experience, testing, and the other requirements to the new state and pay a fee.

Although the Professional Engineer laws bar those not registered from offering their engineering services to the public, most states have a "corporate exemption." This exemption excludes from the licensure requirement engineers who offer their engineering services only to their employer and not to any outside firm or individual. If the firm offers engineering services, a licensed Professional Engineer must still sign, seal, and take responsibility for all such services. Stretched to the limit, this would mean that only the Chief Engineer of the firm would need to be a Professional Engineer. However, since the seal of the Professional Engineer on the engineering report or drawing certifies that the details of work have been checked by that engineer, in practice all principal engineers in the firm need to be licensed. In the language of the law, anyone in "responsible charge" of engineering work must be licensed.

There are many reasons to become a licensed Professional Engineer, even though many chemical engineers historically have not been:

1. One cannot be in "responsible charge" of engineering work that will be offered to other firms or to the public (or indeed call oneself an "engineer" in some states) without a P.E. license. Effectively, this means that chemical engineers who work for engineering design or engineering and construction firms need to be licensed early in their careers. Chemical engineers in many other kinds of firms will need to be licensed before they can rise above a certain technical level. And one cannot be a consultant or an expert witness without a license.

2. The "corporate exemption" described above is under review in many states and could be eliminated. This is a serious career concern, as most states have already eliminated the "grandfather clauses" that previously allowed someone with many years of significant experience in the field to be granted a P.E. license without the normal examinations. If the corporate exemption is rescinded, the status of engineers who are licensed will be enhanced, while other engineers will need to study for and pass the examinations and to document the appropriate years of engineering experience, just to retain their status.

3. Although states have generally exempted government employees from the P.E. requirements in the past, these exemptions are under review and are likely to be eliminated.

4. Professional Engineer registration is an indication to potential employers, as well as to the general public, of one's competence in the field.

There are two steps to becoming registered: Engineer-in-Training and Professional Engineer.

20.2.1 Engineer-in-Training

The first level of certification is as an Engineer-in-Training (EIT). As an Engineer-in-Training (also known as an Engineering Intern), you may begin to acquire engineering experience as an employee, but you may not offer your services as a chemical engineer directly to the public (or to firms other than your employer). To obtain EIT certification one must take and pass the Fundamentals of Engineering (FE) Exam, sometimes referred to as the EIT Exam. This is an 8-hour, closed-book, multiple-choice examination on a wide range of technical subjects typically covered in engineering programs. The exam is given twice a year (generally in October and April), on the same day in all U.S. jurisdictions.

To sit for the FE Exam, one must have completed or be about to complete a B.S. in an engineering field from a department that is accredited by the Accreditation Board for Engineering and Technology (ABET). Most U.S. chemical engineering departments are so accredited. Some states allow a person with a non-engineering degree (or a non-accredited degree) to take the examination after acquiring substantial experience. However, this alternative is being eliminated by many states. A fee is paid to the State Board, and references must be provided to certify that one is of good character. The address of the State Board for Professional Engineers can be obtained from any chemical engineering department.

The examination is best taken while one is still an undergraduate. Some states allow one to take the examination only in the last semester, and some allow it to be taken any time during the senior year. We recommend the earlier test date, when possible. The examination is written by the National Council of Examiners for Engineering and Surveying (NCEES, Clemson, SC 29633), from whom sample examinations and study materials can be obtained. A sample reference handbook is mailed directly to each applicant by NCEES. Although the examination is closed-book, a fresh copy of this handbook is supplied at the examination. Thus, you will know exactly what reference material will be provided so that you can study appropriately for the examination.

The morning session of the examination (4 hours) presently consists of 120 questions covering mathematics, engineering economics, chemistry, material science/structure of matter, statics, dynamics, mechanics of materials, fluid mechanics, thermodynamics, electrical circuits, ethics, and computers. The afternoon session (4 hours) consists of 60 questions specific to the engineering discipline of the applicant. For chemical engineering, the afternoon questions cover reaction engineering, thermodynamics, computer and numerical methods, heat transfer, mass transfer, material/energy balances, pollution prevention, process control, process design and economics evaluation, process equipment design, process safety, and transport phenomena. The afternoon questions are more involved than are the morning questions. Over the years, the definitions of these categories have changed; the material supplied after your application for the examination should be consulted. The best strategy for preparing for the examination is to obtain one or more of the many books written specifically for these examinations and to attend the refresher mini-courses that are offered through many engineering schools. Small study groups (4–6 students) are especially helpful in preparing for the FE examination. If you have not taken engineering courses outside of chemical engineering, substantial study of these areas is essential. After the examination, it generally takes up to four months before you are notified of the results. The national passing rate for chemical engineers is about 60–65%, which is above that for all engineers. Rules for retaking the examination vary from state to state.

Many engineers who take the FE exam after they graduate wish that they had taken it earlier. The examination is geared towards the material covered in B.S. engineering programs, and many engineers become more specialized during their careers. Those engineers would require more preparation for the FE examination. Also, the designation of Engineer-in-Training on one's credentials certainly makes a candidate more attractive to a potential employer. Finally, the next step in the registration process requires specific engineering experience, and many states only count experience that is gained after registration as an Engineering-in-Training. Thus, if you decide to become a P.E. at some later date, you may have to wait more than four years to become registered. Significant opportunities may be lost in the meantime.

20.2.2   Registered Professional Engineer

After becoming an Engineering-in-Training, one must acquire considerable engineering experience before one can take the Principles and Practice Exam (often referred to as the P.E. Exam). In most states, four years of responsible engineering work must be certified by engineers for whom you have worked or engineers who have reviewed your work in detail and can attest to its quality. In some states, only registered Professional Engineers may certify your experience. Character references are also required.

The Principles and Practice Exam is an 8-hour examination given once or twice a year (depending on the state). A separate exam is given for each field of engineering. Problems on the exam for chemical engineering currently involve fluids, heat transfer, kinetics, mass and energy balances, mass transfer, material science, plant design, process control, and thermodynamics. The examination is open-book. Some of the questions are multiple choice. One chooses four problems from a set of 10 problems in the morning and then chooses another four problems from a different set of 10 problems in the afternoon. As for the FE exam, sample Principles and Practice Exams are available through NCEES, who writes the examination.

You will probably want to take a refresher course or to do significant studying before taking this examination. Numerous books are available for self-study, and some universities and professional organizations (such as AIChE) offer courses.

20.3 LEGAL LIABILITY [13]

Chemical engineers often encounter legal liability for their work. We cannot in this book provide legal advice, but we do point out some of the situations in which a chemical engineer should seek legal counsel. Perhaps the best advice is to take a short course in legal issues for engineers early in your career and seek the advice of a licensed attorney if you have any doubts about your legal rights, obligations, or liabilities.

When you apply for registration as an Engineer-in-Training or a Professional Engineer, you receive a copy of the appropriate state laws and regulations governing the practice of engineering, and you are required to certify that you understand them. These should be studied in detail. But beyond these specific licensing requirements, a chemical engineer must deal with government regulations, contracts, and issues of civil and even criminal liability.

For example, throughout a chemical engineering career, one will deal with government agencies such as the Environmental Protection Agency (EPA), Occupational Safety and Health Administration (OSHA), Department of Transportation (DOT), and others. Federal and state laws give these agencies authority to regulate industry and commerce by enacting various regulations. For federal agencies, these regulations are first published in the *Federal Register* and then periodically catalogued in the *Code of Federal Regulations* (CFR). These documents are available in most large libraries and in all law libraries. It is worthwhile to look at some of these regulations while still in school, to become aware of the scope, detail, and format of these documents. The regulations cover, for example, maximum concentrations for wastewater discharges, approved process safety procedures, and requirements for transportation of hazardous materials. They are generally written by technically trained professionals (often engineers), although the wording will have been checked by the legal staff. Thus, it is expected that competent chemical engineers can and will read, understand, and follow these regulations.

Sometimes the engineer's responsibilities are greater than they might first appear. For example, OSHA sets maximum permissible employee exposure levels for many substances, but adherence to these may not be enough, because the "General Duty Clause" of the OSHA Act of 1970 has been interpreted as requiring an employer to avoid exposing employees to any hazards that are known or that *should be known* to the employer. If a substance is known to cause harmful effects (e.g., such a study has been published or the company has proprietary knowledge of such effects), the employer must control employee exposure even though OSHA has no standard for that substance.

Contracts are another form of legally binding document, in which two or more parties each agree to accept one or more obligations. Before signing any written contract or agreeing to any oral contract, one should understand one's obligations and what consideration (e.g., payment) is promised in return from the other party. Upon obtaining employment, one may be asked to sign an employment contract in which consideration (salary) is offered in return for certain chemical engineering services to the firm, and for adherence to specific codes of conduct (see Section 20.4). If any aspect of a contract is unclear, one should obtain legal counsel. If one is signing the contract on behalf of the firm, the firm's attorney is consulted. If one is signing as an individual, private attorney is consulted. Whenever there is any indication that either party may not live up to their obligations, appropriate counsel should be consulted.

Beyond the legal issues of contracts, civil legal actions known as torts sometimes involve engineers. These suits are brought when some action or lack of action by the defendant is alleged to have caused injury (physical, financial, or emotional) to the plaintiff. Both compensatory damages (to pay for correction of the injury) and punitive damages (to punish the party who did the injury) can be recovered by the plaintiff. Again, good advice here is to consult legal counsel whenever such circumstances arise or are anticipated to arise.

Finally, engineers can face criminal prosecution for such actions such as falsifying records submitted to federal regulatory agencies or willfully subjecting employees to hazardous environments. While such actions are rare, the penalties are severe.

20.4   BUSINESS CODES OF CONDUCT [14,15]

Most firms have formal codes of conduct that must be adhered to as a condition of employment. Often one is asked to sign these on the first day of work. As with any contract, it is important to read and to understand all of the details and to then fulfill all of the obligations undertaken. If an employee does not adhere to the code, he or she may be fired.

In consideration for employment, a chemical engineer (or any employee) accepts what are called fiduciary responsibilities. This means that trust has been placed in the engineer to act faithfully for the good of the firm. This is a general legal principle and is covered in Principle 5 of the AIChE Code of Ethics.

Related to fiduciary responsibilities is the avoidance of conflicts of interest. The focus here is to avoid circumstances where you have, or appear to have, contradictory obligations. For example, if you own stock in a valve manufacturer, it would be a conflict of interest if you were to order valves from that company for your present employer, if there were another supplier of equal quality valves at a lower price. In general, and especially in government service where the regulations are quite strict, one must avoid not only actual conflicts of interest but also apparent conflicts of interest. The test is that, if a reasonable person could reasonably assume there to be a conflict of interest, one has an apparent conflict of interest.

It is assumed that much information about employees will become known to the company. In a typical business code of conduct, one agrees to keep personnel information confidential. That is, one agrees not to release such information to anyone outside the firm and to release it only to those within the firm who have a clear need to know. Strict adherence to such an agreement is essential.

In the business code of conduct or in a proprietary secrecy or patent agreement, the employee agrees that certain knowledge gained and discoveries made through employment are the property of the firm and may not be divulged to others. It is crucial that engineers read and understand this agreement in detail before signing. Much of the value of a firm resides in proprietary knowledge, that is, knowledge that is not shared with others. Breaching a secrecy agreement can cost a firm, and the engineer could be in serious legal trouble. Similarly, a firm hires a chemical engineer to do work that often results in patentable discoveries. If such discoveries are made on company time, with company property, or with proprietary knowledge, the patent agreement will require disclosure to the firm and assignment of patent rights to the firm. If one does make discoveries completely outside the realm of employment, one should keep extremely careful and complete notes to avoid patent ownership difficulties.

If the firm has significant international operations, the code of conduct generally covers the conduct of employees representing the firm abroad. As business customs vary widely from country to country, this can be an especially important subject. Before leaving the country on assignment, it is crucial to obtain as complete knowledge as possible both about the customs and laws of the other country and about any changes in business procedure that are authorized by the firm. The U.S. government forbids representatives of U.S. corporations from engaging in certain business practices, even if they are customary in that country.

A final aspect of business conduct that must be considered, and that might be included in the code of conduct, is employee relations. Specific guidelines for hiring and firing of employees must be followed to avoid legal difficulties. Whenever you are in a position to hire or fire, you must be proactive to learn the appropriate company procedures. There are specific legal requirements about what information you cannot request from job applicants and what information you can use to make employment decisions. Also, most business codes of conduct now include requirements to avoid any discriminatory or harassing behavior. Good advice here, as mentioned above regarding conflicts of interest, is to avoid even the appearance of discrimination or sexual harassment.

REFERENCES

1. Koen, B. V., *Definition of the Engineering Method*, American Society for Engineering Education, Washington, DC, 1985.

2. Kohn, P. M., and R. V. Hughson, "Engineering Ethics," *Chemical Engineering*, May 5, 1980, 97–103. Responses in: Hughson, R. V., and P. M. Kohn, "Ethics," *Chemical Engineering*, September 22, 1980, 132–147.

3. Matley, J., and R. Greene, "Ethics of Health, Safety and Environment: What's 'right'," *Chemical Engineering*, March 2, 1987, 40–46. Responses in: Matley, J., R. Greene, and C. McCauley, "Ethics of Health, Safety, and Environment: CE readers say what's 'right'," *Chemical Engineering*, September 28, 1987, 108–120.

4. Mascone, C. F., A. G. Santaquilani, and C. Butcher, "Engineering Ethics: What are the right choices?" *Chemical Engineering Progress*, April 1991, 61–64. Responses in: Mascone, C. F., A. G. Santaquilani, and C. Butcher, "Engineering Ethics: How ChEs Respond," *Chemical Engineering Progress*, October 1991, 73–82.

5. Rosenzweig, M., and C. Butcher, "Can You Use That Knowledge?" *Chemical Engineering Progress*, April 1992, 76–80. Responses in: Rosenzweig, M., and C. Butcher, "Should You Use That Knowledge?" *Chemical Engineering Progress*, October 1992, pp. 7 and 85–92.

6. Woods, D. R., *Financial Decision Making in the Process Industries*, Prentice-Hall, Englewood Cliffs, NJ, 1975.

7. Woods, D. R., "Teaching Professional Ethics," *Chemical Engineering Education*, **18**(3), 106, 1984.

8. Schön, D. A., *The Reflective Practitioner: How Professionals Think in Action*, Basic Books, New York, 1983; Schön, D. A., *Educating the Reflective Practitioner*, Jossey-Bass, San Francisco, 1987.

9. Fogler, H. S., and S. E. LeBlanc, *Strategies for Creative Problem Solving*, Prentice Hall, Englewood Cliffs, NJ, 1995.

10. Woods, D. R., *A Strategy for Problem Solving*, 3rd edition, Department of Chemical Engineering, McMaster University, Hamilton, Ontario, 1985.

11. *The NSPE Ethics Reference Guide*, Publ. #1107, National Society of Professional Engineers, Alexandria, VA, 1993.

12. Martin, M. W., and R. Schinzinger, *Ethics in Engineering*, 2nd edition, McGraw-Hill, New York, 1989.

13. Heines, M. H., and K. B. Dow, "Proprietary Information: What are your rights and responsibilities?" *Chemical Engineering Progress*, July 1994, 78–84.

14. National Academy of Sciences, National Academy of Engineering, and Institute of Medicine, *On Being a Scientist: Responsible Conduct in Research*, 2nd edition, National Academy Press, Washington, DC, 1995.

15. *Guidelines to Professional Employment for Engineers and Scientists*, 3rd edition, American Institute of Chemical Engineers, New York, NY, 1989.

PROBLEMS

1. Proprietary information [6]. I am a senior. Last summer, I had an excellent job designing a new type of heat exchanger that the company was developing. When I returned to school, my professor asked me to use my knowledge on a design project, and to explain heat exchanger design to other seniors in the design group. I really feel that I have something exciting to share with my fellow classmates. Is it ethical for me to share my experiences with my colleagues?

2. Medical school [6]. I am a senior in chemical engineering, but I plan to go to medical school next year. I have not been accepted yet. I want a good summer job, yet, if my application for medical school is turned down, I would want the job to be permanent. Jobs are hard to get, and the interviewers with whom I have talked so far will offer me permanent employment but not summer employment. I have three interviews left. What do I tell these interviewers?

3. Sophomore [6]. I am a sophomore. None of my classmates has a summer job. Jobs are hard to get, and I need money to pay for school next year. I have been offered a good-paying job with lots of engineering experience to work on the production of propellants/explosives for anti-personnel mines. Personally, I strongly believe that this product should not be manufactured. Do I accept the job? Do I tell the company of my personal beliefs?

4. Not a hazard as defined [3, Reprinted by special permission from *Chemical Engineering*, March 2, 1987 and September 28, 1987, Copyright © 1987 by McGraw-Hill, Inc., New York, NY 10020]. In a unit where grain is steeped, sulfur dioxide is added directly to grain and water. Operators have long complained about sulfur dioxide fumes, citing runny noses, teary eyes, coughing, and headaches. The concentration has been checked many times and has always measured lower than OSHA specifications. Management's stance has always been, "Don't spend money to fix something if it isn't broken." A few employees have quit, citing allergies and other medical problems. Operators in the area have requested an engineering study to remedy the situation. Chris is given the job of investigating whether the exhaust fan should be replaced with an expensive ventilation system. What are Chris's obligations? Can they all be met? What creative strategies can you suggest to Chris to deal with this situation?

5. Improving a Reaction [5, Reprinted with permission of AIChE]. Look up this article and write a one-page response to the case. Then, look up the reader responses. Write a one-page reflection on the case.

6. The falsified data strike back [2, Reprinted by special permission from *Chemical Engineering*, May 5, 1980 and September 22, 1980, Copyright © 1980 by McGraw-Hill, Inc., New York, NY 10020]. In the case described in Section

20.1.2, Jay has written the report to suit the boss, and the company has gone ahead with an ambitious commercialization program for Catalyst A. Jay has been put in charge of the pilot plant where development work is being done on the project. To allay personal doubts, Jay runs some clandestine tests on the two catalysts. To Jay's astonishment and dismay, the tests determine that while Catalyst A works better under most conditions (as everyone had expected), at the operating conditions specified in the firm's process design, Catalyst B is indeed considerably superior. What should Jay do?

7. Contact the Board of Registration for Professional Engineers in your area. Determine what code of ethics is required and what are the specific requirements for registration.

8. In a small group (approximately four students), obtain examples of codes of conduct from different companies and compare their features.

9. Develop five heuristics (not mentioned in this chapter) for ethical problem solving.

10. Develop a scenario of an ethical dilemma and, in a small group, use a problem solving strategy to analyze it.

11. Find the Process Safety Management regulation in the *Code of Federal Regulations* (29CFR1900.119) and write a two-page synopsis of it.

CHAPTER

21

Health, Safety, and the Environment

One goal of chemical engineering is to produce goods and services that enhance the quality of life. Chemical engineers have been at the forefront of efforts to improve health (e.g., pharmaceuticals), safety (e.g., shatterproof polymer glasses), and the environment (e.g., catalytic converters). Moreover, throughout the product life cycle, chemical engineers are concerned with potential harm to the health and safety of people and damage to the environment. In this chapter, we focus on assessment of these potential dangers and on proactive strategies to avoid them.

Although much of the chemical processing industries regards activities in improvement of health, safety, and the environment (HSE) as one general function, the U.S. government has separated the field into distinct categories, based on the varying rights afforded to different classes and for historical purposes. The health and safety of employees, for example, are regulated by the Occupational Safety and Health Administration (OSHA), while the health of non-employees and the environment are regulated by the Environmental Protection Agency (EPA). Other activities are regulated by the Department of Transportation or the Department of Energy, among other agencies.

An exact description of applicable laws, rules, and regulations would be out of date before any textbook could be printed; rather, this chapter describes the types of regulations that are relevant to chemical engineering and provides guidance on where to find the current (and proposed) regulations. The focus is on the general concepts and strategies of risk assessment and reduction that transcend the details of regulations.

21.1   RISK ASSESSMENT

There are many ways to view risks to a person's health or safety. Some people will knowingly accept tremendous risk (such as in rock climbing or smoking), some expose themselves to risk perhaps without knowing (such as by sunbathing), and some refuse to be exposed at all to certain risks (such as extremely low concentrations of pesticides in food). Generally, people are much more accepting of risks that they choose (voluntary exposures) than to risks that are forced upon them (involuntary exposures). Also, people are much more accepting of risks that they understand than they are of risks that they do not understand. For example, the population at large tends to accept that human activity often degrades rivers and lakes with biological and chemical pollution, but they do not accept any measurable radioactive emissions from power plants.

21.1.1   Accident Statistics

Engineers quantify risks to provide a rational basis for deciding what activities should be undertaken and which risks are worth the benefits provided. These decisions are frequently made by non-engineers; thus, it is essential that the measures that engineers use are understandable to the public. For this reason, several measures have been established.

The *OSHA incidence rate* is the number of injuries and illnesses per 200,000 hours of exposure. Any injury or work-related illness that results in a "lost work day" is counted in the ratio. Thus, minor injuries that can be treated with first aid are not counted, but counted injuries range all the way to death. The 200,000 hours is roughly equivalent to 100 worker-years. The OSHA incidence rate illustrates two features of risk measures: The details of the accident are not included, and these measures can be used for non-work-related exposures.

The *fatal accident rate* (FAR) is the number of fatalities per 10^8 hours of exposure. Only fatalities are counted, and the 100,000,000 hours is roughly equivalent to 1000 worker lifetimes. Although only deaths are counted, in many fields there is a strong correlation between numbers of injuries and numbers of deaths. Thus, the reasonable presumption is that a decrease in the FAR will also decrease the OSHA incidence rate.

For some exposures, the available data are insufficient to determine the total time of exposure. For these cases, the *fatality rate* can be used. This measure is the number of fatalities per year divided by the total population at risk. For example, the exposure of individual smokers is extremely variable because of differences in type of cigarette, number of inhalations per cigarette, and so on. However, one can determine the fatality rate (~0.005), which gives a rough estimate of the chance that a smoker will die this year from smoking-related causes of 0.5%.

Table 21.1 shows some representative numbers of these three risk measurements. One of the possibly surprising observations is that the chemical process industries are relatively safe (for workers) compared to many other industries.

Table 21.1   Comparison of Three Risk Measurements [7]

Activity	OSHA Incident Rate (injuries & deaths per 200,000 h)	Fatal Accident Rate (deaths per 100,000,000 h)	Fatality Rate (deaths per person per year)
Working in chemical industry	0.49	4	
Staying at home		3	
Working in steel industry	1.54	8	
Working in construction	3.88	67	
Traveling by car		57	170×10^{-6}
Rock climbing		4000	40×10^{-6}
Smoking (1 pack per day)			5000×10^{-6}
Struck by lightning			0.1×10^{-6}

There are many reasons for this, ranging from the high level of remote sensing and operation that separates workers from the most dangerous parts of the plant to the historical concern for the hazards of industrial chemicals. Regardless of the reasons, though, chemical engineers continue to try to improve the safety of chemical plants.

Another observation is that there are differences in relative rankings if one uses the different risk measurements. Perhaps most striking is that the FAR for employees of the chemical industry is only slightly higher than the FAR for people staying in their homes. This is a potentially misleading comparison, though. The portion of the population that works is on average more healthy than the portion who do not. Also, most people will take risks at home that they would not be allowed to take at work. Finally, a worker is probably more tired and prone to accidents at home after a hard day at work.

The main use of these statistics is not to compare one activity to another (unless they can be substituted for one another) but to monitor improvements to the health and safety of workers and others achieved by process modifications.

21.1.2   Worst Case Scenarios

Another measure of risk is to imagine the worst possible consequence of an operation. Such a study is called a worst case scenario study, and they are required by some government agencies. These studies have some drawbacks, but they can be very useful in identifying ways to avoid serious accidents. A related strategy that is routinely used throughout the industry for identifying potential hazards is called HAZOP and is discussed in Section 21.4.

The development of worst case scenarios is certainly subjective, but government agencies and organizations develop guidelines for this task. For example, there is certainly a chance that an asteroid will impact Earth directly in the middle of the chemical plant. Should this be the worst case scenario? Most people would argue that this takes the worst case scenario study beyond reason, but there are no clear-cut rules. The subjective rules that have been developed say things like the worst accident that might reasonably be assumed possible over the life of the facility. Different people would define "possible" (or "probable") in different ways. Is sabotage by an employee possible? Is the simultaneous failure of three independent safety systems probable? Sometimes, probabilities of occurrence can be estimated, but often they cannot.

All of this is not to say that worst case scenarios are not useful. On the contrary, they have been extremely helpful, but they are difficult to perform and potentially difficult to understand. They are most useful when specific guidelines are followed and when they are used to enhance safety by developing safeguards against the accident scenarios developed.

These guidelines are constantly changing, so they are not described in detail here. However, the most current guidelines should be obtained from the EPA, OSHA, or other agencies.

21.1.3   The Role of the Chemical Engineer

> **As the professional with the best knowledge of the risks of a chemical processing operation, the chemical engineer has a responsibility to communicate those risks to employers, employees, clients, and the public.**

It can be very difficult to explain FARs in deaths per 10^8 hours to the general, nontechnical public. However, the consequences of failure to explain rationally and honestly to the public the risks and the steps taken to reduce those risks are tremendous. The ethical role of the chemical engineer in this communication is discussed in Chapter 20. Beyond those responsibilities, the damage from poor communication can destroy an industry. For example, the nuclear power industry in the United States has been destroyed in large part by the failure to communicate risks to the general public in a way that they could understand. When even relatively minor accidents occur, many feel that they were misled by engineers who seemed to have said that there were no risks, no chance of an accident.

21.2   REGULATIONS AND AGENCIES

Rules and regulations arise from both governmental agencies and from non-governmental organizations (such as AIChE). These rules and regulations change constantly, and the most up-to-date rules should always be determined. The following sections describe the kinds of rules and regulations promulgated (put into effect by formal public announcement) by various organizations. The actual agency should be contacted directly for the latest rules.

Federal government rules and regulations are published in the *Federal Register* (FR), which is issued daily. Notices of proposed or pending regulations are also given in the FR. Periodically, the regulations of a specific type are collated and published in the *Code of Federal Regulations* (CFR). Both the FR and the CFR are available in large libraries and in all law libraries, as well as on the Internet. In Table 21.2, we give the Internet address for the FR and CFR as of the date this book is written; however, one must search the Internet, as such addresses change constantly.

In addition to the direct government sources, numerous private firms collate federal regulations (more quickly than does the government) and they sell their compendia, often tailored to the needs of the customer.

State and local government regulations are also available in hard copy and increasingly on-line form. However, it is best to contact these agencies directly to be sure that one has all the relevant regulations. Nongovernmental organizations can be contacted directly for their rules.

However, sources of information on the Internet are notoriously inaccurate and out-of-date. Although this may be the best place to begin a search for regulations, it must not be the last.

21.2.1   OSHA and NIOSH

In general, the Occupational Safety and Health Administration promulgates regulations having to do with worker safety and health in industries other than mining (MSHA serves a similar role there). The National Institute for Occupational Safety and Health (NIOSH) is a federal research organization that provides infor-

Table 21.2 Internet Addresses for Federal Agencies

Code of Federal Regulations (CFR)	http://law.house.gov/cfr.htm
Federal Register (FR)	http://law.house.gov/7.htm
DOT Regulations (FR and 40 CFR)	http://www.dot.gov/legandreg.htm
EPA Regulations (FR and 40 CFR)	http://www.epa.gov/epahome/rules.html
MSHA Regulations	http://www.msha.gov/REGSINFO.HTM
NIOSH Data bases	http://www.cdc.gov/niosh/database.html
OHSA Regulations (FR and 29 CFR)	http://www.osha-slc.gov

mation to employees, employers, and OSHA to help assess health and safety risks. Regulations are not promulgated by NIOSH, although it does certify various analytical techniques and equipment (such as respirators).

Three major laws and regulations in this area are: the OSHA Act [1], the Hazard Communication Standard (29CFR1910.1200), and the Process Safety Management rule (29CFR1910.119).

OSHA Act.   The original act of Congress that set up OSHA, The Occupational Safety and Health Act of 1970 [1], specified certain regulations and standards and required OSHA to promulgate others. Perhaps of most importance is the so-called "general duty clause" of the OSHA Act that "each employer shall furnish to each of his employees employment and a place of employment which are free from recognized hazards that are causing or are likely to cause death or serious physical harm to his employees." [1, Section 5.a.1] This clause has been interpreted to mean that an employer must avoid exposing employees to hazards that *should have been known* to the employer, whether or not that hazard is specifically regulated by OSHA. Thus, the responsibility of researching the literature to see if anyone has identified a hazard is placed on the employer. Most chemical engineers are employees, yet they often represent the employer and, therefore, assume the responsibilities under the general duty clause.

To search for published data on chemical and physical hazards, a good place to start is the Registry of Toxic Effects of Chemical Substances (see http://www.cdc.gov/niosh/rtecs.html) and the NIOSHTIC bibliographic data base (see http://www.cdc.gov/niosh/nioshtic.html).

Some specific regulations ensued from the OSHA Act, notably for the chemical industry exposure limits for certain substances in breathing air for employees. These limits are called permissible exposure limits (PEL) and are often given in parts per million (by volume). Related limits are updated yearly by the American Conference of Governmental and Industrial Hygienists (ACGIH, a nongovernmental association) and are called threshold limit values (TLV®) [2]. In fact, the original OSHA regulations merely made the ACGIH suggested TLVs the official government limits. The NIOSH also publishes recommended exposure limits (REL), but these are not legally binding regulations. Each of these exposure limits is based on the assumption that a typical worker can be exposed to that concentration of the substance for 8 hours a day, 5 days a week, for a working lifetime, without ill effects. These assumptions are based often on sketchy data, extrapolated from animal studies. All of these limits (PEL, TLV, REL) are given in the convenient *NIOSH Pocket Guide to Chemical Hazards* [3], available from NIOSH at a very small charge. Respirator requirements and a description of the health effects of exposure are also given.

The exposure limits are based on a time-weighted average (TWA) over an 8-hour shift, which means that higher concentrations are allowable, as long as the average over the shift is not above the PEL. When even short-term exposure to higher levels is harmful, there is a separate short-term exposure limit (STEL), which is a 15-minute time-weighted average concentration that must never be exceeded, or an OSHA "ceiling" concentration, which is an instantaneous concentration that must never be exceeded. Also, maximum concentrations from which one could escape within 30 minutes without experiencing escape-impairing or irreversible health effects are identified as immediately dangerous to life and health (IDLH) concentrations. The IDLH limit is used under conditions in which a respirator is normally used. In the event of a respirator failure, the employee might not be able to escape if the concentration is above the IDLH. These limits are also given in the NIOSH handbook [3].

Hazard Communication Standard.   This regulation is also known as the Worker Right to Know or simply HazCom. The reference is 29CFR1910.1200. It requires that the employer train all employees so that the employees understand the hazards of the substances that they are handling, exposed to, or potentially exposed to. It is explicitly stated that the employer has not met this standard merely by giving the employee the hazard information orally or in written form. The employer must make sure that the employee *understands* the hazards; thus, much effort in training goes into satisfying this requirement. Two very obvious results and requirements of this regulation are proper labeling of containers and availability of Material Safety Data Sheets (MSDS). Table 21.3 lists typical major sections of an MSDS, but the precise format of an MSDS is not presently defined by regulation, although some of the minimum requirements are, and these are also listed in Table 21.3. The MSDS must list the substances, their known hazards,

Table 21.3   Typical Sections of a Material Safety Data Sheet (MSDS)

1. Material Identification
2. Ingredients and Hazards
3. Physical Data
4. Fire and Explosion Data
5. Reactivity Data
6. Health Hazard Information
7. Spill, Leak, and Disposal Procedures
8. Special Protection Information
9. Special Precautions and Comments

Some Minimum Requirements [from 29CFR1910.1200(g)]

1. An MSDS is required for each "hazardous chemical," including those specifically listed in 29CFR1910 (subpart Z), any material assigned a TLV®, or any material determined to be cancer causing, corrosive, toxic, an irritant, a sensitizer, or has damaging effects on specific body organs.
2. Written in English

3. Identity used on label

4. Chemical name and common name of all ingredients that are hazardous and that are present in ≥1% concentration or that could be released in harmful concentrations

5. Chemical name and common name of all ingredients that are carcinogens and that are present in ≥0.1% concentration or that could be released in harmful concentrations

6. Physical and chemical characteristics of the hazardous chemical (such as vapor pressure, flash point)

7. Physical hazards of the hazardous chemical, including the potential for fire, explosion, and reactivity

8. Health hazards of the hazardous chemical, including signs and symptoms of exposure, and any medical conditions which are generally recognized as being aggravated by exposure to the chemical

9. Primary route(s) of entry

10. OSHA permissible exposure limit, ACGIH Threshold Limit Value, and any other exposure limit used or recommended by the chemical manufacturer, importer, or employer

11. Whether the hazardous chemical is listed in the National Toxicology Program (NTP) *Annual Report on Carcinogens* (latest edition) or has been found to be a potential carcinogen in the International Agency for Research on Cancer (IARC) *Monographs* (latest editions), or by OSHA

12. Any generally applicable precautions for safe handling and use that are known to the chemical manufacturer, importer, or employer preparing the material safety data sheet, including appropriate hygienic practices, protective measures during repair and maintenance of contaminated equipment, and procedures for clean-up of spills and leaks

13. Any generally applicable control measures that are known to the chemical manufacturer, importer, or employer preparing the material safety data sheet, such as appropriate engineering controls, work practices, or personal protective equipment

14. Emergency and first aid procedures

15. Date of preparation of the material safety data sheet or the last change to it

16. Name, address, and telephone number of the chemical manufacturer, importer, employer, or other responsible party preparing or distributing the material safety data sheet, who can provide additional information on the hazardous chemical and appropriate emergency procedures, if necessary

and procedures for proper handling of the material and for proper actions in an emergency. Although only hazardous materials must have these MSDSs, they are available for practically all substances and mixtures in commerce. They are available directly from the supplier or manufacturer or on the Internet (http://www.cdc.gov/niosh/msds.html). We recommend using the Internet only as a preliminary source for MSDSs. One should get an MSDS directly from more than one manufacturer if possible, because errors in MSDS do occur and because different MSDSs will contain different data. Figure 21.1 shows an MSDS sheet for benzene.

1   PRODUCT IDENTIFICATION

PRODUCT NAME: BENZENE
FORMULA: C6H6
FORMULA WT: 78.10
CAS NO.:   71-43-2
NIOSH/RTECS NO.: CY1400000
COMMON SYNONYMS: BENZOL; PHENYL HYDRIDE; COAL NAPHTHA
PRODUCT CODES: 9156,9256,9153,9154,9155,B717,9149
   EFFECTIVE: 01/22/87
   REVISION #04

PRECAUTIONARY LABELLING

BAKER SAF-T-DATA(TM) SYSTEM

        HEALTH   — 4 EXTREME (CANCER CAUSING)
        FLAMMABILITY — 3 SEVERE (FLAMMABLE)
        REACTIVITY   — 0 NONE
        CONTACT   — 1 SLIGHT
HAZARD RATINGS ARE 0 TO 4 (0 = NO HAZARD; 4 = EXTREME HAZARD).

LABORATORY PROTECTIVE EQUIPMENT

GOGGLES & SHIELD; LAB COAT & APRON; VENT HOOD; PROPER GLOVES; CLASS B
EXTINGUISHER

PRECAUTIONARY LABEL STATEMENTS

POISON DANGER
EXTREMELY FLAMMABLE
CAUTION: CONTAINS BENZENE, CANCER HAZARD HARMFUL IF SWALLOWED, IN-
HALED, OR ABSORBED THROUGH SKIN EXCEPTIONAL HEALTH HAZARD—READ MA-
TERIAL SAFETY DATA SHEET KEEP AWAY FROM HEAT, SPARKS, FLAME. AVOID
CONTACT WITH EYES, SKIN, CLOTHING. AVOID BREATHING VAPOR. KEEP IN TIGHTLY
CLOSED CONTAINER. USE WITH ADEQUATE VENTILATION. WASH THOROUGHLY
AFTER HANDLING. IN CASE OF FIRE, USE ALCOHOL FOAM, DRY CHEMICAL, CARBON
DIOXIDE—WATER MAY BE INEFFECTIVE. FLUSH SPILL AREA WITH WATER SPRAY.

SAF-T-DATA(TM) STORAGE COLOR CODE: RED STRIPE (STORE SEPARATELY)

2   HAZARDOUS COMPONENTS

COMPONENT   %   CAS NO.
BENZENE   90-100   71-43-2

Figure 21.1   MSDS Sheet for Benzene

3   PHYSICAL DATA

BOILING POINT: 80 C (176 F)   VAPOR PRESSURE(MM HG): 74.6
MELTING POINT: 6 C   (43 F)   VAPOR DENSITY(AIR=1): 2.77
SPECIFIC GRAVITY: 0.88   EVAPORATION RATE: N/A
 (H2O=1)   (BUTYL ACETATE=1)
SOLUBILITY(H2O):   NEGLIGIBLE (LESS THAN 0.1 %) % VOLATILES BY VOLUME: 100
APPEARANCE & ODOR: CLEAR COLORLESS LIQUID HAVING CHARACTERISTIC ARO-
MATIC ODOR.

4   FIRE AND EXPLOSION HAZARD DATA

FLASH POINT (CLOSED CUP: −11 C (12 F)   NFPA 704M RATING: 2-3-0
FLAMMABLE LIMITS: UPPER   8.0 %   LOWER 1.3 %

FIRE EXTINGUISHING MEDIA

USE ALCOHOL FOAM, DRY CHEMICAL OR CARBON DIOXIDE. (WATER MAY BE INEF-
FECTIVE.)

SPECIAL FIRE-FIGHTING PROCEDURES

FIREFIGHTERS SHOULD WEAR PROPER PROTECTIVE EQUIPMENT AND SELF-CON-
TAINED   BREATHING APPARATUS WITH FULL FACEPIECE OPERATED IN POSITIVE
PRESSURE MODE. MOVE CONTAINERS FROM FIRE AREA IF IT CAN BE DONE WITH-
OUT RISK. USE WATER TO KEEP FIRE-EXPOSED CONTAINERS COOL.

UNUSUAL FIRE & EXPLOSION HAZARDS

VAPORS MAY FLOW ALONG SURFACES TO DISTANT IGNITION SOURCES AND
FLASH BACK. CLOSED CONTAINERS EXPOSED TO HEAT MAY EXPLODE. CONTACT
WITH STRONG OXIDIZERS MAY CAUSE FIRE.

TOXIC GASES PRODUCED

CARBON MONOXIDE, CARBON DIOXIDE

5   HEALTH HAZARD DATA

THIS SUBSTANCE IS LISTED AS ACGIH SUSPECT HUMAN CARCINOGEN, NTP HUMAN
CARCINOGEN, IARC HUMAN CARCINOGEN (GROUP 1). ACCEPTABLE MAXIMUM
PEAK ABOVE THE ACCEPTANCE CEILING CONCENTRATION FOR AN EIGHT-HOUR
SHIFT = 50 PPM FOR 10 MINUTES; (PEL) CEILING = 25 PPM.

Figure 21.1   (*Continued*)

THRESHOLD LIMIT VALUE (TLV/TWA):   30   MG/M3 (10 PPM)
SHORT-TERM EXPOSURE LIMIT (STEL):   75   MG/M3 (25 PPM)
PERMISSIBLE EXPOSURE LIMIT (PEL):   30   MG/M3 (10 PPM)

TOXICITY: LD50 (ORAL-RAT)(MG/KG)   −4894
            LD50 (ORAL-MOUSE)(MG/KG)   −4700
            LD50 (IPR-RAT)(MG/KG)   −2.9
            LC50 (INHL-MOUSE-7H) (PPM)   −9980

CARCINOGENICITY: NTP: YES   IARC: YES   Z LIST: NO   OSHA REG: NO

EFFECTS OF OVEREXPOSURE

INHALATION MAY CAUSE HEADACHE, NAUSEA, VOMITING, DIZZINESS, NARCOSIS, SUFFOCATION, LOWER BLOOD PRESSURE, CENTRAL NERVOUS SYSTEM DEPRESSION. INHALATION OF VAPORS MAY CAUSE SEVERE IRRITATION OR BURNS OF THE RESPIRATORY SYSTEM, PULMONARY EDEMA, OR LUNG INFLAMMATION. LIQUID MAY BE IRRITATING TO SKIN AND EYES. PROLONGED SKIN CONTACT MAY RESULT IN DERMATITIS. EYE CONTACT MAY RESULT IN TEMPORARY CORNEAL DAMAGE. INGESTION MAY CAUSE NAUSEA, VOMITING, HEADACHES, DIZZINESS, GASTROINTESTINAL IRRITATION, BLURRED VISION, LOWERING OF BLOOD PRESSURE. IRREVERSIBLE INJURY TO BLOOD FORMING TISSUE MAY RESULT FROM CHRONIC LOW LEVEL EXPOSURE.

TARGET ORGANS

BLOOD, CENTRAL NERVOUS SYSTEM, EYES, SKIN, BONE MARROW, RESPIRATORY SYSTEM

MEDICAL CONDITIONS GENERALLY AGGRAVATED BY EXPOSURE

NONE IDENTIFIED

ROUTES OF ENTRY

INGESTION, INHALATION, EYE CONTACT, SKIN CONTACT, ABSORPTION

EMERGENCY AND FIRST AID PROCEDURES

CALL A PHYSICIAN. IF SWALLOWED, DO NOT INDUCE VOMITING. IF INHALED, REMOVE TO FRESH AIR. IF NOT BREATHING, GIVE ARTIFICIAL RESPIRATION. IF BREATHING IS DIFFICULT, GIVE OXYGEN. IN CASE OF CONTACT, IMMEDIATELY FLUSH EYES OR SKIN WITH PLENTY OF WATER FOR AT LEAST 15 MINUTES.

Figure 21.1   (*Continued*)

6   REACTIVITY DATA

STABILITY: STABLE   HAZARDOUS POLYMERIZATION: WILL NOT OCCUR
CONDITIONS TO AVOID: HEAT, FLAME, OTHER SOURCES OF IGNITION
INCOMPATIBLES: STRONG OXIDIZING AGENTS, SULFURIC ACID,
                                           NITRIC ACID
DECOMPOSITION PRODUCTS: CARBON MONOXIDE, CARBON DIOXIDE

7   SPILL AND DISPOSAL PROCEDURES

STEPS TO BE TAKEN IN THE EVENT OF A SPILL OR DISCHARGE

WEAR SELF-CONTAINED BREATHING APPARATUS AND FULL PROTECTIVE CLOTH-
ING. SHUT OFF IGNITION SOURCES; NO FLARES, SMOKING OR FLAMES IN AREA.
STOP LEAK IF YOU CAN DO SO WITHOUT RISK. USE WATER SPRAY TO REDUCE VA-
PORS. TAKE UP WITH SAND OR OTHER NON-COMBUSTIBLE ABSORBENT MATERIAL
AND PLACE INTO CONTAINER FOR LATER DISPOSAL. FLUSH AREA WITH WATER.
J. T. BAKER SOLUSORB(R) SOLVENT ADSORBENT IS RECOMMENDED FOR SPILLS
OF THIS PRODUCT.

DISPOSAL PROCEDURE

DISPOSE IN ACCORDANCE WITH ALL APPLICABLE FEDERAL, STATE, AND LOCAL
ENVIRONMENTAL REGULATIONS.

EPA HAZARDOUS WASTE NUMBER: U019 (TOXIC WASTE)

8   PROTECTIVE EQUIPMENT

VENTILATION: USE GENERAL OR LOCAL EXHAUST VENTILATION TO MEET TLV RE-
    QUIREMENTS.
RESPIRATORY PROTECTION: RESPIRATORY PROTECTION REQUIRED IF AIR-
    BORNE CONCENTRATION EXCEEDS TLV. AT CONCENTRATIONS ABOVE 10 PPM,
    A SELF-CONTAINED BREATHING APPARATUS IS ADVISED.
EYE/SKIN PROTECTION: SAFETY GOGGLES AND FACE SHIELD, UNIFORM, PRO-
    TECTIVE SUIT, POLYVINYL ALCOHOL GLOVES ARE RECOMMENDED.

9   STORAGE AND HANDLING PRECAUTIONS

SAF-T-DATA(TM) STORAGE COLOR CODE: RED STRIPE (STORE SEPARATELY)

Figure 21.1   (*Continued*)

SPECIAL PRECAUTIONS

BOND AND GROUND CONTAINERS WHEN TRANSFERRING LIQUID. KEEP CONTAINER TIGHTLY CLOSED. STORE IN A COOL, DRY, WELL-VENTILATED, FLAMMABLE LIQUID STORAGE AREA.

10   TRANSPORTATION DATA AND ADDITIONAL INFORMATION

DOMESTIC (D.O.T.)

PROPER SHIPPING NAME	BENZENE (BENZOL)
HAZARD CLASS	FLAMMABLE LIQUID
UN/NA	UN1114
LABELS	FLAMMABLE LIQUID
REPORTABLE QUANTITY	1000 LBS.

INTERNATIONAL (I.M.O.)

PROPER SHIPPING NAME	BENZENE
HAZARD CLASS	3.2
UN/NA	UN1114
LABELS	FLAMMABLE LIQUID

Figure 21.1   *(Continued)*

Process Safety Management of Highly Hazardous Chemicals (29CFR1910.119).   This OSHA regulation applies to essentially all of the chemical processing industries and requires action in 13 different types of activities as shown in Table 21.4. When OSHA promulgated this broad regulation in the early 1990s (final rule effective May 26, 1992), it used the already existing rules of the American Institute of Chemical Engineers' *Guidelines for Technical Management of Chemical Process Safety* [4] and the American Petroleum Institute's *Recommended Practices 750* [5] as guides. In fact, as is often the case, OSHA essentially gave the force of law to these voluntary nongovernmental standards.

Process Safety Management (PSM) embraces nearly the entire safety enterprise of a chemical process organization. It requires training of employees, written operating procedures, specific quality in the engineering design of components and systems, very specific procedures for some activities, investigation and reporting of accidents that do occur, and an internal audit of the safety enterprise of the company. Below is a description of each of the thirteen components.

1. **Employee participation.** The employer must actively involve the employees in the development of and implementation of the safety program. Employees are more likely to understand the hazards and to follow the established safety procedures when they are involved early and continuously in

**Table 21.4 Process Safety Management
of Highly Hazardous Chemicals
(29CFR1910.119)**

1. Employee Participation
2. Process Safety Information
3. Process Hazards Analysis
4. Operating Procedures
5. Training
6. Contractors
7. Pre-Startup Safety Review
8. Mechanical Integrity
9. Hot Work Permit
10. Management of Change
11. Incident Investigation
12. Emergency Planning and Response
13. Compliance Safety Audit

the development of the safety program. This item was added to the earlier API and AIChE standards.

2. **Process Safety Information.** The employer must research the materials, process, and operation to determine the potential hazards and keep in an immediately accessible form all safety information. This includes all MSDSs, but it also includes information on the process itself, such as up-to-date process flow diagrams and P&IDs.

3. **Process Hazards Analysis.** Before a process is started up and periodically thereafter (typically every 3 to 5 years or whenever significant modifications are made), a detailed study must be made of the process to determine potential hazards and to correct them. There are several approved procedures, and an organization can opt to use an alternate procedure if it can be shown to be as effective. In fact, most of the chemical processing industries use the HAZOP technique, which is described in Section 21.4. This technique is a modified brainstorming process in which potential hazards are identified, their consequences are determined, and an action is identified to deal with the hazard.

4. **Operating Procedures.** Written operating procedures must be available to operators. Deviations in the plant operation from these procedures must be noted. These procedures must include startup, shutdown, and emergency response to process upset.

5. **Training.** The employer must train all employees in the hazards present and procedures for mitigating them.

6. **Contractors.** The employer is responsible for the safe conduct of any contractors. Although the contractors are responsible for the safe conduct of the contractor's employees, the owner or operator of the plant who enters into a contract with the contractor remains liable for the safe operation of the contractor. This is an OSHA addition to the earlier API and AIChE standards.

7. **Pre-Startup Safety Review.** The regulation specifically requires that there be a review of the safety aspects of the process before any processing occurs on the site. The review must be documented, and any deviations of the plant as built from the design specifications must be addressed.

8. **Mechanical Integrity.** Vessels and other equipment must meet existing codes and be inspected during manufacture and after installation. Appropriate procedures for maintenance must be developed and followed.

9. **Hot Work Permit.** This is a very specific procedure by which a wide range of people in the plant are notified before "hot work" such as welding can occur. Many chemical plants use flammable materials, and everyone in the area needs to be informed so that no flammable vapors are released during the operation.

10. **Management of Change.** During accident investigations in the chemical process industries, it has often been found that severe incidents (involving deaths and massive destruction) occurred because equipment, processes, or procedures were changed from the original design without careful study of the consequences. Thus, the OSHA regulation requires companies to have in place a system by which any modification is reviewed by all of the appropriate people. For example, any change in the reactor design must be reviewed not only by the design engineer but also by the process engineer who can evaluate how the overall process is affected. The maintenance leader must also make sure that the modification does not adversely affect the maintenance schedule or the ability of workers to get to or to maintain the equipment.

11. **Incident Investigation.** When there is a hazardous process upset, it must be investigated and a written report must be developed indicating the details of the incident, the probable cause, and the steps taken to avoid future incidents.

12. **Emergency Planning and Response.** There must be a written plan and employees must be trained to respond to possible emergency situations.

13. **Compliance Safety Audit.** Periodically, all of the elements of the safety system (including items 1–12 above) must be audited to make sure that the approved procedures are being followed and that they are effective.

21.2.2   Environmental Protection Agency (EPA)

The role of the Environmental Protection Agency is to protect the environment from the effects of human activity. Although this is a very broad role, in the context of the chemical processing industries, this role usually relates to emissions of harmful or potentially harmful materials from the plant site to the outside by air or by water. There are three classes of such emissions: (1) planned emissions, (2) fugitive emissions, and (3) emergency emissions. In this section, some of the present regulations for these classes of emissions are described. There are many more regulations that are not mentioned here. In any facility, one must keep constantly aware of new and modified regulations through research, use of an environmental compliance consulting firm, or communication with the local, state, and federal environmental protection agencies.

Planned Emissions.   Any process plant will have emissions. These may be harmful to the environment, benign, or, in rare cases, beneficial. In any case, a permit is usually required before construction and/or operation of the plant. Significant modifications to the plant (especially if they change the design emissions) will likely require a modification to the permit or a new permit.

These emissions permits are normally obtained through the state environmental protection agency, but federal regulations must be met. In some regions, states, or localities, the requirements for the permit may be significantly more stringent than the federal EPA regulations. One must contact the local agencies. However, as an overview, one can search the EPA databases described in Section 21.2.

Permits frequently require an extensive Environmental Impact Statement (EIS), which details the present environment and any potential disturbances that the planned activity could produce. For process plants, these EISs are typically written by a team of chemical engineers, biologists, and others, and they deal not only with planned emissions but also with potential process upsets and emergencies. In this regard, the worst case scenario mentioned in Section 21.1.2 is used.

Permitting is based on assessment of potential degradation of the environment and, thus, both the level of emissions from the plant and the present level of contamination in the local environment is considered. There are National Ambient Air Quality Standards (NAAQS) for a few materials and National Emissions Standards for Hazardous Air Pollutants (NESHAP) and New Source Performance Standards (NSPS) for these and others. Major sources (defined as those plants that emit more than some threshold annual quantity such as 25 tons of hazardous air pollutants) must meet the most stringent emissions criteria and require more permits. Similar standards are applied for water quality and for discharges into the water. Many of these regulations are promulgated based on the Clean Air Act and Clean Water Act, among others.

Beyond the effects on the environment after emissions are fully dispersed in the air or the water, there can be acute, short-term effects on nearby populations. Often chemical engineers perform dispersion studies to determine the range and longevity of the plume of harmful materials that emanates from the point of discharge into the air or water.

Fugitive Emissions.   Historically, we have been concerned with stack or routine discharge emissions. However, since the Clean Air Act Amendments of

1990, so-called fugitive emissions have been under close scrutiny. These emissions are not planned but rather occur because of leaks in valves, pumps, flanges, and so on. These emissions are especially prevalent in moving equipment (e.g., pumps and valves), where there must be a seal between the process side and the outside that allows movement of a shaft. Typically, these seals are made of a packing material that offers a tortuous, high-pressure-drop path between the inside and the outside. Some leakage occurs, and often, for liquid systems, the small leakage provides lubrication for the shaft. For such a packing seal, the lower the fugitive emissions, the higher the frictional losses from the shaft rotation.

Various designs are available to reduce fugitive emissions, including magnetic drives, double seals, canned pumps, bagged valves, and bellows. Discussions of these are not included here.

As plants are designed for lower stack emissions, the fugitive emissions are increasingly the dominant emission type. Thus, the emissions of thousands of valves, pumps, and flanges must be estimated and reported to the EPA. There are several procedures to make these estimates [6], from crude assumptions that all valves leak at the same rate to experiments in which a working valve is bagged in plastic and the fugitive emission rate is directly measured. One must consider all fugitive emissions when specifying components that could leak. Low-leakage or no-leakage pumps are more expensive than those of standard design, but they may mean the difference between obtaining a permit or not.

The focus of the Clean Air Act Amendments of 1990, Title I, is the release of volatile organic compounds (VOC), which are precursors to the photochemical production of ozone (smog), especially in areas that have not attained the NAAQS. The definition of a VOC is any organic compound with an appreciable vapor pressure at 25°C. Hazardous Air Pollutants (HAP) are also regulated through Title III of the Act.

Emergency Releases.   Process upsets can create catastrophic releases of hazardous materials, and regulations require that there be an effective plan to deal with these occurrences and that the consequences for affected populations not be too serious. As mentioned above, worst case scenarios and dispersion modeling are used to make this assessment.

One such regulation is the Emergency Planning and Community Right to Know Act (EPCRA) of 1986, also known as SARA, Title III. This regulation requires plants to provide the local community with information about potentially hazardous or toxic materials or processes. Further, the plant must work with the local community to develop effective emergency procedures that will be implemented automatically in the event of an accident. A Local Emergency Planning Committee is formed of members of the local government, emergency response organizations, and plant personnel. Also, releases of certain hazardous substances must be reported to the EPA, and a compilation of these releases made available to the public through the Toxic Release Inventory System (http://www.epa.gov/enviro/html/tris/tris_overview.html).

Through the Department of Transportation, regulations require all over-the-road transport vehicles to carry a manifest of hazardous materials that is immediately available to all emergency personnel, in the event of an accident. Also, the DOT and the U.S. Coast Guard regulate the conditions under which hazardous cargo can be transported. For example, these regulations frequently require stabilizing additives to prevent runaway polymerization.

Many other EPA regulations that are beyond the scope of this book impact the operation of chemical processing facilities, including the Resource Conservation and Recovery Act (RCRA), Comprehensive Environmental Response, Compensation, and Liability Act known as Superfund (CERCLA), Superfund Amendments and Reauthorization Act (SARA), and Toxic Substances Control Act (TSCA).

EPA Risk Management Program.   The Clean Air Act Amendments of 1990 also "require the owner or operator of stationary sources at which a regulated substance is present to prepare and implement a Risk Management Plan (RMP) and provide emergency response in order to protect human health and the environment" [40 CFR 68]. The final rule was effective August 19, 1996 [61 FR 31667]. Full implementation is required by June 21, 1999. The RMPs must be registered with the EPA, they must be made public, and they must be periodically updated. The Risk Management Program must include three elements:

- hazard assessment
- prevention
- emergency response

This program is coordinated with OSHA's Process Safety Management (PSM). In fact, compliance with the PSM standard is considered equivalent to the prevention part of the RMP. The following overview of the Risk Management Program pertains to all plants covered under the PSM standard, which includes most plants in the chemical process industries.

The *Hazard Assessment* must include a worst case analysis, an analysis of non-worst case accidental releases, and a five-year accident history. The *worst case release scenario* is defined by EPA [40 CFR 68] as

> *the release of the largest quantity of a regulated substance from a vessel or process line failure, including administrative controls and passive mitigation that limit the total quantity involved or the release rate. For most gases, the worst case release scenario assumes that the quantity is released in 10 minutes. For liquids, the scenario assumes an instantaneous spill; the release rate to the air is the volatilization rate from a pool 1 cm deep unless passive mitigation systems contain the substance in a smaller area. For flammables, the worst case assumes an instantaneous release and a vapor cloud explosion.*

The EPA rule specifies default values of wind speed, atmospheric stability class, and other parameters for the development of the offsite consequence analysis for the worst case scenarios. It also specifies the endpoint for the consequence analysis, based on the calculated concentration of toxic materials, the overpressure (1 psi) from vapor cloud explosions, and the radiant heat exposure for flammable releases ($5\ kW/m^2$ for 40 seconds).

The Prevention Program is identical to the PSM standard, except that the Emergency Planning and Response item is covered under a separate category in the RMP.

The Emergency Response Program portion of the Risk Management Plan is coordinated with other Federal regulations. For example, compliance with the OSHA Hazardous Waste and Emergency Operations (HAZWOPER) rule (29 CFR 1910.120), the Emergency Planning and Response portion of the PSM standard, and EPCRA will satisfy this requirement in the RMP regulation.

The Risk Management Plan must also designate a qualified person or position with overall responsibility for the program, and it must show the lines of authority or responsibility for implementation of the plan.

Overall, then, the only additional RMP requirement for plants already covered by the OSHA Process Safety Management regulation is the Hazard Assessment (including offsite consequence analyses of worst case and non-worst case accidental release scenarios). This Hazard Assessment must not be confused with the Process Hazard Analysis (PHA). The Hazard Assessment is a study of what will happen in the event of an accidental release and usually includes air dispersion simulations, for example. The PHA (e.g., HAZOP) studies the hazards present in the process and seeks to minimize them through redesign or modifications to operating procedures.

21.2.3 Nongovernmental Organizations

Many professional societies and industry associations develop voluntary standards, and these are often accepted by government agencies and, thereby, given the force of law. Examples of such organizations and examples of their standards are:

- American Petroleum Institute (API), Recommended Practices 750
- American Institute of Chemical Engineers (AIChE)
  - Center for Chemical Process Safety (CCPS)
  - Design Institute for Emergency Relief Systems (DIERS)
- American National Standards Institute (ANSI)
- American Society for Testing and Materials (ASTM)
- National Fire Protection Association (NFPA), fire diamond
- American Conference of Governmental Industrial Hygienists (ACGIH), TLV®s
- Chemical Manufacturers' Associations (CMA), Responsible Care® program
- Synthetic Organic Chemicals Manufacturers Association (SOCMA)

- American Society of Mechanical Engineers (ASME), boiler & pressure vessel code

21.3   FIRES AND EXPLOSIONS

The most common hazards on many chemical plant sites are from potential fires and explosions. Whenever a fuel, an oxidizer, and an ignition source are present, the hazard exists. The detailed analysis of these hazards and their consequences are covered in other books [7,8]. Here, we introduce the terminology of the field.

21.3.1   Terminology

Combustion is the very rapid oxidation of a fuel. Most fuels tend to oxidize slowly at room temperature. As the fuel is heated, it oxidizes more rapidly. If the heating source is removed, the fuel cools and returns to a slow oxidation rate.

If a certain temperature (*auto-ignition temperature*) is exceeded, the heat liberated by the oxidation is sufficient to sustain the temperature, even if the external heating source is removed. Thus, above the auto-ignition temperature, the reaction zone will expand into other areas with appropriate mixtures of fuel and oxygen. The energy required to heat a small region to the auto-ignition temperature is called the *ignition energy* and is often exceedingly small.

A gaseous mixture of fuel and air will only ignite if it is within certain concentration limits. The *lower flammability (or explosive) limit* (LFL or LEL) is the minimum concentration of fuel that will support combustion and is somewhat below stoichiometric. The maximum concentration of fuel that will support combustion is called the *upper flammability (or explosive) limit* (UFL or UEL). Above the UFL, the mixture is too "rich," that is, it does not contain enough oxygen. These two limits (UFL and LFL) straddle the stoichiometric concentration for complete combustion of the fuel. It is because of the convenient upper flammability limit of gasoline that this fuel is not more dangerous than it is (see Problem 21.8). Any mixture within the flammability limits should be avoided or very carefully controlled.

The *flash point* of a liquid is the temperature at which the vapor in equilibrium with the standard atmosphere above a pool of the liquid is at the LFL. Thus, a low flash point indicates a potential flammability problem if the liquid is spilled. Diesel fuel, for example, is much safer than is gasoline because diesel has a higher flash point. Regulations for transportation and use of gasoline are, therefore, much more stringent than they are for diesel. Flash point can be measured by the *open-cup* or the *closed-cup* method. In the open-cup method, an open container of the liquid is heated, while a flare-up of the vapor is intentionally attempted with an ignition source. The temperature of first flare-up is the flash point. In the closed-cup method, the liquid is placed in a closed container and allowed to come to equilibrium with air at standard pressure. Ignition is attempted at increasing temperature. Although the open-cup and closed-cup flash points for many materials are very close, materials that vaporize slowly and disperse in the atmosphere quickly can have much higher open-cup flash points than their closed-cup flash points. Although the MSDS will give the flash point, one must notice which type of flash point is being reported.

Explosions are very rapid combustions in which pressure waves are formed and are the mode of propagation of the combustion. The combustion creates a local pressure increase, which heats the flammable mixture to its auto-ignition temperature. This secondary combustion causes the pressure wave to propagate through the mixture. The traveling pressure pulse is called a *shock wave*. Often, a strong wind accompanies the shock wave. The combination of shock wave and wind is called a *blast wave*, and this causes much of the damage from explosions. When the shock wave speed is less than the speed of sound in the ambient atmosphere, the explosion is called a *deflagration*. When the speed is greater than the speed of sound, the explosion is called a *detonation*. Detonations can cause considerably more damage from the combination of blast wave, *overpressure*, and concussion. The damage from an overpressure of only 1 psi on structures can be extensive. Such a pressure differential on a typical door, for example, would result in considerably over one ton of pressure on the door—enough to break most locks.

Of special concern when flammable gaseous mixtures (or dispersions of combustible dusts in air) are present is the so-called *vapor cloud explosion (VCE)*. If there is a natural gas leak, for example, the cloud (mostly methane) will spread and mix with air. The cloud, parts of which are within the flammable limits, can be quite large. If it ignites, the deflagration will cause a shock wave perpendicular to the ground that can cause great damage, often flattening buildings. When a liquid stored above its ambient boiling point suddenly comes in contact with the atmosphere (through a rupture in the tank, for example), the rapid release and expansion of the vapor can cause a massive shock wave. This phenomenon is called a *boiling-liquid expanding-vapor explosion (BLEVE)*. The failure of a steam drum, for example, can cause a BLEVE. When the BLEVE is of a flammable substance, the resulting cloud can explode. This combination of BLEVE and VCE is one of the most destructive forces in chemical plant accidents. The classic example is a propane tank that ruptures when it becomes overheated in a normal fire. The propane is stored as a liquid under pressure. The tremendous quantity of propane that vaporizes rapidly mixes with the atmosphere and creates a massive VCE.

Runaway reactions are confined, exothermic reactions that go from their normal operating temperatures to above the ignition temperature, that is, they liberate more heat than can be dissipated. Thus, the temperature increases, increasing the reaction rate. Although there may be a steady state at a higher temperature (as there is in combustion), often the limits of mechanical integrity of the reaction vessel are reached before that point, and catastrophic failure of the vessel occurs. Such a failure can cause direct injuries, it can release toxic material, it may cause a fire, or it may lead to a BLEVE and/or a VCE.

To reduce the chance of a runaway condition during process upsets, the temperature difference between the reacting mixture and the cooling medium should be kept small. This may seem counterintuitive. However, consider a case where the temperature driving force is 1°C. If the temperature of the reacting mixture increases by 1°C during a process upset, the driving force for cooling has doubled! If the heat-exchange system had been designed for a 10°C driving force, that same upset would result in only a 10% increase in cooling. In systems with a chance for runaway, increased heat-transfer area is the cost of an inherently safe system.

A common scenario for an accident involving an exothermic reaction is the *loss of coolant accident (LOCA)*. Unless the cooling system is backed up to the extent that it is essentially 100% reliable, one must consider this scenario in designing the vessels and the pressure relief systems.

21.3.2   Pressure Relief Systems

During a severe process upset, the pressure and/or temperature limits of integrity for vessels can be approached. To avoid an uncontrolled, catastrophic release of the contents or the destruction of the vessel, pressure relief systems are installed. Usually, these are relief valves on vessels or process lines that open automatically at a certain pressure. Downstream, they are connected to flares (for flammable or toxic materials), scrubbers (for toxic materials), or a stack directed away from workers (for materials such as steam that present physical hazards). The design of the pressure relief system is especially important, as the worst case scenario must be considered, which is sometimes the simultaneous failure of multiple relief systems, as was the case for the Bhopal tragedy in 1984.

The design of such systems is complicated by several factors. The devices are designed to operate under unsteady conditions. Therefore, a dynamic simulation is required. Also, the flow through the relief system may be single-phase or two-phase flow. For two-phase flow, not only are the calculations more difficult, but more factors affect the pressure drop, such as whether the line is horizontal or vertical.

Besides the relief valves (which are called safety valves, relief valves, pressure relief valves, or pop valves depending on service), rupture disks are used to open the process to the discharge system. Rupture disks are specially manufactured disks that are installed in a line, similar to the metal blanks that are used between flanges to close a line permanently. However, the disks are designed to fail rapidly at a set pressure. Ideally, the rupture disk allows no flow when the pressure is below the set pressure and it immediately offers no resistance to flow when the pressure hits the set point.

21.4   HAZOP

Under the "Process Hazard Analysis" requirement of the Process Safety Management of Highly Hazardous Chemicals regulation (29 CFR 1910.119), employers must, by May 26, 1997, complete such an analysis of all covered processes using one or more of the following techniques:

- What-If
- Checklist
- What-If/Checklist
- Hazards and Operability Study (HAZOP)
- Failure Mode and Effects Analysis (FMEA)
- Fault Tree Analysis (FTA)

or "an appropriate equivalent methodology." The most widely used technique in the chemical process industries is HAZOP. The OSHA regulation specifically refers to the AIChE Center for Chemical Process Safety for details of process hazard analysis methods [9].

HAZOP is a modified brainstorming technique for identifying and resolving process hazards. It consists of asking questions about possible deviations that could occur in the process (or part of a process) under consideration. A HAZOP is always done in a group, and the regulation requires that the team have "expertise in engineering and process operations," have "experience and knowledge specific to the process being evaluated," and be "knowledgeable" about the HAZOP methodology. As with any brainstorming process, the ideas and suggestions can come very quickly, and there must be an identified "scribe" ready with appropriate software to capture them. Various software packages are available to speed this process and to offer additional "triggers" in the brainstorming process.

The first step in a HAZOP is to identify the normal operating condition or purpose of the process or unit. This is called the *intention*. Next, a *guide word* is used to identify a possible *deviation* in the process. For example, the intention may be to keep the temperature in a vessel constant. The guide words are:

- None, No, or Not
- More of
- Less of
- More than or As well as
- Part of
- Reverse
- Other than

In our example, there may be no coolant flow. Once such a possible deviation is identified, the team notes any possible *causes* of the deviation. If there are any safety *consequences* of the deviation, those are noted. Suppose the coolant flow ceased because of a pump failure. The consequence may be a runaway reaction. The *action* to be taken is assigned by the HAZOP team. In this case, the action might be an assignment for the process engineer to investigate a back-up pumping system. The team then goes on to the next possible deviation, until all reasonable deviations have been considered. The team does not solve the safety problem during the

Table 21.5. HAZOP for the Feed Heater of the Hydrodealkylation (HDA) Process (very incomplete)

Process Unit: H-101, Feed Heater, Figure 1.5

Intention: To provide feed to the Reactor (Stream 6) at 600°C.

Guide Word	Deviation	Cause	Consequence	Action
No	No flow in Stream 4	Blockage in line	Fluid in H-101 overheats	Consider an interlock on fuel gas flow
More of	Higher temperature Stream 6	Sudden reduction in Stream 6 flowrate	Reactor overheats	Consider an interlock on fuel gas flow
Less of	Lower temperature Stream 6	Flame out	Cold shot to R-101, quenching reaction	Include automatic flame detection with re-ignition cycle
As well as	Liquid drops in fuel gas	Failure of V-101 demister	Flame becomes erratic	Robust demister design
Part of	No toluene in Stream 6	P-101 not working	No reaction in R-101	Install low flow alarm on Stream 2
Reverse	Decrease in Stream 6 temperature through H-101	No probable cause		
Other than	Impurities in Stream 6	Impurities in feed or overheating in tubes	Impurities in product and/or catalyst deactivation	Monitor concentrations and H-101 temperatures

HAZOP; its job is to identify the problem and to assign its resolution to a specific person. An example of part of a HAZOP for the Fired Heater (H-101) of the Hydrodealkylation of Toluene process (Figure 1.5) is shown in Table 21.5.

The OSHA Process Safety Management regulation requires that the actions assigned be taken in a timely manner and that all process hazard analyses be updated at least every five years.

21.5   INHERENTLY SAFE DESIGN

Although safety controls can be added to existing processes, a more effective and efficient strategy is called *inherently safe design* [10]. This strategy is based on a hierarchy of six approaches to process plant safety. The idea is to streamline the process to eliminate hazards, even if there is a major process upset.

1. **Substitution.** One avoids using or producing hazardous materials on the plant site. If the hazardous material is an intermediate product, alternate chemical reaction pathways might be used, for example. In other words, the most inherently safe strategy is to avoid the use of hazardous materials.

2. **Intensification.** One attempts to use less of the hazardous materials. In terms of a hazardous intermediate, the two processes could be more closely coupled, reducing or eliminating the inventory of the intermediate. The inventories of hazardous feeds or products can be reduced by enhanced scheduling techniques such as Just-In-Time (JIT) manufacturing [11].

3. **Attenuation.** Reducing, or attenuating, the hazards of materials can often be effected by lowering the temperature or adding stabilizing additives. Any attempt to use materials under less hazardous conditions inherently reduces the potential consequences of a leak.

4. **Containment.** If the hazardous materials cannot be eliminated, they at least should be stored in vessels with mechanical integrity beyond any reasonably expected temperature or pressure excursion. This is an old but effective strategy—to avoid leaks. However, it is not as inherently safe as substitution, intensification, or attenuation.

5. **Control.** If a leak of hazardous material does occur, there should be safety systems that reduce the effects. For example, chemical facilities often have emergency isolation of the site from the normal storm sewers, and large tanks for flammable liquids are surrounded by dikes that control any leaks from spreading to other areas of the plant. Scrubbing systems and relief systems in general are in this category. They are essential, because they allow controlled, safe release of hazardous materials, rather than the uncontrolled, catastrophic release from vessel rupture.

6. **Survival.** If leaks of hazardous materials do occur, and they are not contained or controlled, the personnel (and the equipment) must be protected. This lowest level of the hierarchy includes fire fighting, gas masks, and so on. Although essential to the total safety of the plant, the greater the reliance on survival of leaks rather than elimination of leaks, the less inherently safe the facility.

21.6   POLLUTION PREVENTION

The Pollution Prevention Act of 1990 established a hierarchy of environmental management strategies roughly analogous in the environmental area to the inherently safe design hierarchy in the safety area. Rather than a specific regulation or set of regulations, pollution prevention has become a guiding philosophy for a broad range of voluntary programs and regulations. Pollution prevention is also known as source reduction (sometimes known as waste minimization) and is at the top of the hierarchy:

1. **Source Reduction.** The focus is to reduce the amount of any hazardous material, pollutants, and contaminants from entering the waste stream or being released to the environment by fugitive emission. The EPA includes the following in this category: toxics use reduction, raw material substitution, process or equipment modification, product redesign, training, improved inventory control, production planning and sequencing, and better management practices [12].

2. **Recycling.** When wastes *are* produced, efforts at recycling (including direct reuse) should be made. This strategy not only directly reduces the waste stream that must be treated, but, by reducing the production rate, it indirectly reduces the waste of the entire enterprise.

3. **Treatment.** This is the old way of dealing with environmental management. Wastes are created but then treated to make them benign. Such processes are called *end-of-pipe* practices, and they are now recognized as wasteful and less effective than source reduction or recycling.

4. **Disposal.** The least desirable waste management strategy is the disposal of the waste on land, in water, or into the atmosphere. This was the original strategy, before society was concerned about pollution. Remediation is the term for after-the-fact cleanup of these wastes. Although the focus of pollution prevention is on design changes to planned or currently operating plants, the challenges of remediation of our past pollution are monumental and will take generations to solve.

The effect of the Pollution Prevention Act of 1990 was to change the focus of EPA regulations from a concern merely for the output pollution of the enterprise to a focus on the ways pollution can be reduced and an incentive for companies to use source reduction. This focus on the upstream process is analogous to Total Quality Management (TQM), statistical process control (SPC) and other strategies that became widespread in industry in the early 1990s. These strategies for solving problems of quality of delivered products focus on the processes far upstream of the final manufacturing step.

A HAZOP-style guide-word technique for use in pollution prevention studies has been developed [13].

21.7 LIFE-CYCLE ANALYSIS

A life-cycle analysis or life-cycle assessment (LCA) is a detailed, technical study of the "environmental consequences of a product, production process, package, or activity [done] holistically, across its entire life cycle" [14]. The time frame of such an analysis is often referred to as "cradle-to-grave." The LCA consists of three stages:

1. **Inventory Analysis.** This is a quantitative study of the material and energy inputs and the air, water, and solid waste outputs of the entire life of the product or process.
2. **Impact Analysis.** The environmental consequences of both the inputs and the outputs of the inventory analysis are enumerated in this part of the LCA.
3. **Improvement Analysis.** The opportunities for improving the environmental consequences through modifications to the product or process are included in this section.

The first step in any life-cycle analysis is to define the boundaries of the analysis. If a vessel is used in the process, is the manufacture of the vessel within the boundaries? Is the production of the steel for the vessel included? Should the environmental effects of the mining of the iron ore be included? Or should the analysis begin with the on-site installation of the equipment? The choice of boundaries defines the scope of the analysis. A broader scope requires greater effort but leads to a more complete understanding of environmental effects.

The life cycle itself is divided into four stages:

1. Raw Materials Acquisition
2. Manufacturing
3. Use, Reuse, and Maintenance
4. Recycle and Waste Management

Simple life-cycle analyses are available [15]. They are most useful when parallel LCAs are done on competing products or processes, because they allow one to evaluate objectively the overall environmental impact of the choices. In this case, it is especially important that all parallel LCAs use the same boundaries.

21.8 SUMMARY

In this chapter, the overall framework of health, safety, and environmental activities in the chemical process industries is described. The specific regulations change constantly, and the cognizant agencies must be consulted for the current rules.

21.9 GLOSSARY

ACGIH: American Congress of Governmental Industrial Hygienists
AIChE: American Institute of Chemical Engineers
ANSI: American National Standards Institute
API: American Petroleum Institute

ASME: American Society of Mechanical Engineers
ASTM: American Society for Testing and Materials
BLEVE: Boiling-liquid expanding-vapor explosion
CAA: Clean Air Act
CCPS: Center for Chemical Process Safety
CERCLA: Comprehensive Environmental Response, Compensation, and Liability Act
CFR: Code of Federal Regulations
CMA: Chemical Manufacturers' Association
DIERS: Design Institute for Emergency Relief Systems
DOT: Department of Transportation
EIS: Environmental Impact Study (or Statement)
EPA: Environmental Protection Agency
EPCRA: Emergency Planning and Community Right to Know Act, also known as SARA, Title III
FAR: Fatal accident rate
FMEA: Failure modes and effects analysis
FR: Federal Register
FTA: Fault-tree analysis
HAP: Hazardous air pollutant
HAZOP: Hazard and operability study
IDLH: Immediately dangerous to life and health
LCA: Life-cycle analysis
LEL: Lower explosive limit
LFL: Lower flammability limit
LOCA: Loss of coolant accident
MSDS: Material Safety Data Sheet
MSHA: Mine Safety and Health Agency
NAAQS: National Ambient Air Quality Standards
NESHAP: National Emissions Standards for Hazardous Air Pollutants
NFPA: National Fire Protection Association
NIOSH: National Institute for Occupational Safety and Health
NIOSHTIC: NIOSH Technical Information Center
NSPS: New Source Performance Standards
OSHA: Occupational Safety and Health Administration
PEL: Permissible Exposure Limit
PHA: Process hazard analysis
PSM: Process Safety Management of Highly Hazardous Chemicals
RCRA: Resource Conservation and Recovery Act
REL: Recommended Exposure Limit
RMP: Risk Management Program (or Plan)
SARA: Superfund Amendments and Reauthorization Act
SOCMA: Synthetic Organic Chemical Manufacturers' Association
STEL: Short-term exposure limit

TLV®: Threshold limit value
TSCA: Toxic Substances Control Act
TWA: Time-weighted average
UEL: Upper explosive limit
UFL: Upper flammability limit
VCE: Vapor cloud explosion
VOC: Volatile organic compound

REFERENCES

1. "Occupational Safety and Health Act of 1970," Public Law 91-596, December 29, 1970.

2. *1996 TLVs® and BEIs®: Threshold Limit Values for Chemical Substances and Physical Agents and Biological Exposure Indices*, American Conference of Governmental Industrial Hygienists, Cincinnati, OH, 1996.

3. *NIOSH Pocket Guide to Chemical Hazards*, National Institute for Occupational Safety and Health, Cincinnati, OH, 1994. (Updated on the Internet at http://www.cdc.gov/niosh/npg.html)

4. *Guidelines for Technical Management of Chemical Process Safety*, American Institute of Chemical Engineers, New York, 1989.

5. *API Recommended Practices 750*, American Petroleum Institute, Washington, DC, 1990.

6. *Emission Factors for Equipment Leaks of VOC and HAP*, EPA-450/3-86-002, U.S. Environmental Protection Agency, Washington, DC, 1986; *Improving Air Quality: Guidance for Estimating Fugitive Emissions from Equipment*, Chemical Manufacturers Association, Washington, DC, 1989.

7. Crowl, D., and J. Louvar, *Chemical Process Safety*, Prentice-Hall, Englewood Cliffs, NJ, 1990.

8. Bodurtha, F. T., *Industrial Explosion Prevention and Protection*, McGraw-Hill, New York, 1980.

9. *Guidelines for Hazard Evaluation Procedures*, 2nd edition with worked examples, Center for Chemical Process Safety of the American Institute for Chemical Engineers, New York, 1992.

10. Kletz, T. A., *Cheaper, Safer Plants or Wealth and Safety at Work: Notes on Inherently Safer and Simpler Plants*, The Institution of Chemical Engineers, Rugby, England, 1984.

11. Hall, Robert W., *Attaining Manufacturing Excellence: Just-in-Time, Total Quality, Total People Involvement*, Business One Irwin, Homewood, IL, 1987.

12. *Pollution Prevention Directory*, EPA 742-B-94-005, U.S. Environmental Protection Agency, Washington, DC, 1994.

13. Doerr, W. W., "Guideword Technique for Identifying Pollution Prevention Opportunities," paper 244b, 1994 Annual Meeting of the American Institute of Chemical Engineers, San Francisco, November 1994.

14. *Life-Cycle Assessment: Inventory Guidelines and Principles*, EPA/600/SR-92/245, U.S. Environmental Protection Agency, Washington, DC, 1993.

15. Allen, D. T., N. Bakshani, and K. S. Rosselot, *Pollution Prevention: Homework & Design Problems for Engineering Curricula*, Center for Waste Reduction Technologies, American Institute of Chemical Engineers, New York, 1992.

PROBLEMS

1. You work for a chemical company with 30,000 employees. If your company has a typical safety record for the chemical industry, what is your best estimate of how many of your employees
   a. will succumb to a fatal accident while on the job this year?
   b. will be injured but not killed on the job?

2. Locate the nearest steam plant on campus.
   a. Develop two possible accident scenarios that should be considered when searching for worst case scenarios.
   b. List the safeguards that have been built into the system to mitigate some (or all) of these effects.

3. Find an MSDS for each of the components listed in the HDA process.

4. Find and summarize all regulations (safety, environmental, transportation) that you can find for benzene.

5. Assume that the Unit Operations Lab in your department must meet the Process Safety Management standard. Choose two of the thirteen components of PSM, and write a critical analysis of these aspects of lab operation.

6. Choose a unit operations laboratory experiment. Estimate the fugitive emissions using the techniques in Reference 6.

7. Some paints have a closed-cup flash point near room temperature, whereas they have no measurable open-cup flash point.
   a. Explain this apparent paradox.
   b. Which is the more useful flash point when using a paint?

8. Assume gasoline to be 87 vol% iso-octane (2,2,4 tri-methyl pentane) and 13 vol% n-heptane. At room temperature, the air above a pool of gasoline will become saturated.
   a. Is the air-gasoline mixture within its flammable limits?
   b. Will there be any location where the mixture will be within its flammable limits? Explain.

9. Perform a HAZOP on R-101 of the HDA process. Be sure to perform the analysis in a team.

10. Benzene in gasoline is now limited by regulation to a maximum of 0.25 wt%. Is the benzene concentration in the air above a pool of gasoline above or below the PEL? Show the effect of temperature.

22

Written and Oral Communications

Throughout your career as an engineer, chances are that you will draw on your skills in oral and written communication far more often than you ever imagined. Engineers report that they spend most of their time using some combination of the four communication skills: writing, speaking, reading, and listening [1]. The situation is the same for most engineering students, although their education is apt to have given them far more opportunity to read and to listen than to write and to speak.

Technical writing and oral presentation skills are particularly important to the design engineer. After all, the final product of a design project is generally an oral or written report (and sometimes both). Furthermore, throughout the design process, engineers use feasibility studies, environmental impact reports, meeting minutes, progress reports, status reports, appropriation requests, design notebooks, equipment specification sheets, and personnel evaluations, as well as various letters and memos, to document their work and to communicate their findings to colleagues, company management, clients or potential clients, and possibly the general public.

> Any evaluation of your success as an engineer, then, will be based in whole or in large part on the quality of your written and oral communications.

In this chapter, instruction in basic writing skills is not attempted. There are many good texts on technical communication [2,3,4]. However, we do provide guidance in adapting this skill to the area of engineering communication that focuses on process design. The formats and guidelines for written and oral design reports are explained, and strategies for improving one's communication are given. Other types of written reports common in engineering, such as laboratory reports and technical journal articles, are not covered. Many of the same strategies, however, apply to these genres, even though the specific guidelines may not.

A sample student design report and the visuals from the oral presentation of the project are given in Chapter 23 of this book. Also included there is a critique of these reports. Examples of the problems described in this chapter are given in the sample report.

22.1   AUDIENCE ANALYSIS

In all engineering communication, a crucial step is audience analysis. Written or oral reporting is, by definition, a transfer of information. One's success at this transfer hinges on presenting the type, detail, and scope of information that the audience expects with a clarity that avoids misunderstandings. It is important that you:

1. *Identify who your intended audience is.* Will your report or oral presentation be for fellow chemical engineers, other engineers, managers, executives, clients, or the general public? Both the appropriate level of detail and focus will vary with the background of the audience. Senior executives, even if they were trained years ago as chemical engineers, may not quickly understand the mathematical details of your design. More importantly, they will not be that interested. Rather, they need more of a big-picture focus. How might what you have done impact the bottom line? The public's perception of the company? The long-term viability of your company's technical advantage over its competitors? There may be a broad range of audience backgrounds, especially if the primary recipients of your report forward it on to others.

2. *Determine what response you want from your audience.* Do you want them to approve the project? Do you want them to abandon the project? Do you want them to leave the presentation or finish reading the report confident that you have presented the best alternatives in a way that will help them make the decision? Do you want them to order the correct equipment and materials? Do you want them to be confident that you have studied the most important environmental impacts of the project? Do you want to be

certain that your safety recommendations are immediately accepted and implemented?

3. *Imagine yourself to be one of your audience.* After writing the first draft, read the report as if you were one of the people identified in Step 1 above, with the indicated background and areas of concern. Read the report or hear the oral presentation as they would, and imagine what response you would have. What would you do after reading or hearing the report?

4. *Iterate.* Multiple drafts are essential. As is the case when solving any complex problem, one must attempt a solution, evaluate it, and then improve upon it. In succeeding drafts, reconsider your likely audience response. If it is not what you want, you must continue to revise.

22.2 WRITTEN COMMUNICATION

Some of the purposes of putting ideas into writing are:

1. To provide a permanent record
2. To get someone to do something (action writing)
3. To instruct someone (informative writing)

Often in engineering writing, all three purposes apply. What you write is a permanent record of your analysis, your conclusions, and your recommendations. Unlike oral communication, written documents must stand on their own, as the author is generally not present to answer questions about the content. Also, these documents will be referred to more than once, often much later than when they were written. Precise wording is crucial, since decisions will often be made by the reader without any further consultation with the writer. Ambiguous words or phrases, gaps in logical developments, and undocumented assumptions can lead the reader to make a wrong decision.

To help both the writer and the reader, various formats are used in written engineering communications. They serve both as a guide for the information flow of the document and as a rough checklist to be sure that important content is not omitted. The reader anticipates the location of specific content within the document; thus, the format helps the reader to understand the document as well as to skip to those sections of most importance when pressed for time. In the following sections, several formats common to chemical process design are described. The detailed guidelines for each format vary from organization to organization, and a sample format for design reports used at West Virginia University is given in section 22.2.8.

Although a typical set of guidelines is given, it is by no means the only (or necessarily the best) such set for design reports. In fact, the best guidelines in a particular organization are the ones agreed to by that organization. They are, in the best of circumstances, an agreement between both parties, whose purpose is to make the communication between the writer and the reader more complete, reliable, and efficient. The first step then is to determine what are your organization's guidelines.

22.2.1   Design Reports

In general, a design report presents the results of a design effort to those who will either implement the design or decide whether the design will be implemented. As with the design process itself (described in Chapter 2), design reports are of varying scope, detail, and complexity. Early on, the order-of-magnitude estimate leads to a relatively short report whose purpose is to persuade the reader either to commit to go forward to the study-estimate stage or to abandon the project in favor of other opportunities. The audience for such a report is typically upper management.

The goal of these early reports is to present the data required for decisions that are quite difficult to make. When the final project may require a capital expenditure of millions or even hundreds of millions of dollars, the stakes are high, and the audience needs enough information to be reasonably confident that their decision is the right one. The audience requires:

- A careful analysis of a base case with sufficient detail to provide a context
- Evidence that the most reasonable options have been considered
- A careful analysis of these options
- A statement of and justification for any significant assumptions made
- A finite number of options with a clear analysis of the advantages and disadvantages
- An estimate of the accuracy of the analysis

The audience for such a report is likely to be less interested in the details of proven technology than in the details and risks of new processes, catalysts, or equipment. The financial, environmental, health, and public perception aspects are likely to be of high concern.

Later design reports (based on preliminary design, definitive, and detailed estimates) are apt to have more technical audiences of process engineers, engineering managers, project engineers, purchasing agents, and so on. They will be more interested in the details that allow the design to proceed to a final, constructed, and operating plant. One outcome of any successful design report, however, is the informed decision of the readers to do something—to commit the company to more detailed design (or not), to order equipment, to start up a plant, to go after new markets, or to modify a process to increase yield or reduce environmental impact.

22.2.2 Transmittal Letters or Memos

The report stands on its own, but it is usually sent to the readers with a cover sheet that gives a broad overview of the document. This is a memo if the recipient is a fellow employee of your organization; it is a letter if the recipient is a client or an employee of a different organization. Some engineers may mistakenly treat this transmittal memo or letter as merely a tag formally indicating who is doing the transmitting, when, and to whom. However, the real purpose and distinct value of this sheet is to convince the potential reader to become a bona fide reader. To do this, the transmittal memo must contain a very brief statement of the scope and of the most important conclusion or recommendation of the report. It must catch the attention of the recipient. The reward of a good transmittal memo is that your report is read (at least in part).

22.2.3 Executive Summaries and Abstracts

Design reports can be long, and they often must contain very detailed information that will be of interest to only a small fraction of the readers of the document. All readers, however, benefit from a broad overview of the project being reported, with key assumptions, results, conclusions, and recommendations given. This overview is called an executive summary. Its length is governed by the guidelines of the organization. Some companies impose a one- or two-page maximum, while some organizations (especially government agencies) encourage longer executive summaries that are, in effect, short reports. Some guidelines suggest that the executive summary should be no more than 5% or 10% of the length of the report. The key is to remember that a well-written executive summary will be short enough that busy executives will read it all while communicating enough information to satisfy these executives, who are unlikely to read the rest of the report.

Some guidelines fail to make the distinction between an executive summary and an abstract. Others define the executive summary as longer than an abstract or more focused on conclusions and recommendations than is the abstract, which may also summarize the procedures used in the analysis. In any case, the executive summary must be focused on the needs of executives whose job is to make the proper decision based on the work you have done, without reading the details contained in the full report. On the other hand, an abstract is a very boiled-down version of the complete report and should have the same focus as does the full report. The abstract provides enough information for the reader to decide whether to read the report, and it provides an overview that will help the reader to assimilate the information while reading the report. Even though an abstract or executive summary appears at the beginning of the report, it should be written last. Care should be taken when writing the executive summary and/or abstract that all important results and conclusions are included since these are often the only sections of the report that are read in their entirety.

An executive summary (or abstract) should be a synopsis of the project being reported, not a description of the report itself. The former is called an *informative* executive summary and may include statements such as "The total capital investment is $100 million." The latter would be a *descriptive* executive summary and would include less helpful statements such as "The total capital investment is given in Section 4 of the report." If the goal of the summary is to get information to a reader whose time is too valuable to read the report (but who must make an important decision), it is poor strategy to write an executive summary (descriptive) to convince readers that they must then read the report to get any useful information. Rather, the executive summary (informative) should contain all the information necessary for the executive reader to make the decision, while the body of the report contains the details that will be required by others to implement the decision.

22.2.4 Other Types of Written Communication

Progress reports are often required for long projects. The goals of these (usually short) reports are:

1. To indicate the progress made to date
2. To present a schedule for the remainder of the project
3. To explain any unanticipated problems or opportunities encountered
4. To propose or to report on changes in direction from those originally projected
5. To give a synopsis of the important findings of the project

Performance evaluations are periodic (often annual or semiannual) reports of the quality of an employee's work. They generally include:

1. A synopsis of the work done
2. An evaluation of the quality of that work
3. A list or narrative of the observed strengths and weaknesses
4. Recommendations for improvement
5. Qualitative or quantitative rankings on specific work attributes

Minutes are often taken during design meetings, especially when important decisions are made or when the meetings involve clients. The minutes include a list of important topics discussed, important decisions made, and crucial information communicated to the parties present, as well as the list of attendees and the time and place of the meeting. Without minutes, different attendees will have memories (or even will have notes) of the decisions made that are in conflict. Minutes create the permanent record of the meeting that allows everyone involved to know where things are headed. To be effective, they must be signed by the minutes author and by representatives of each of the groups at the meeting. This typically involves a total of three signatures (client, manager, and author).

Increasingly, *e-mail* within an organization and across the internet is the medium of choice for rapid communication. Unfortunately, some engineers (and others) fail to apply the same criteria for precision, clarity, and accuracy to e-mail messages that they do to other written communication. These messages can be extremely important, and, although they can be less formal, they must convey the message accurately. Also, some e-mail users forget that their messages are at least as permanent as a paper document. In fact, many more copies may stay around for much longer. In some organizations now, there is a rule that one should not put anything in an e-mail message that they would not want circulated in hard copy throughout the organization (or even outside). Also, any document put on the World Wide Web is potentially available to anyone in the world.

22.2.5   Exhibits (Figures and Tables)

Many of your important results will be presented in figures and tables. When the numerical values are important to the audience, a table is more appropriate. When the trend of the data or the difference between sets of data is important, a graph is better. Figures (graphs and pictures) are numbered sequentially in the order that they are cited in the report. If they are not cited, they are not included. Similarly, tables are numbered sequentially in the order that they are cited. In most reports, these exhibits appear close to the point where they are cited, but care should be taken not to interrupt the flow of the text too much.

The normal kind of graph that engineers use (with two linear scales) is sometimes called a scatter plot or an x-y plot. Care should be taken to choose properly the dependent and independent variables, and the axes must be labeled with the variable name. Usually, an axis is not labeled with the symbol for the variable (because these are not standardized), and it is never labeled with just the units of the variable. The units should be included in the axis label after the variable name. For example, "Manufacturing Cost ($/yr)" would be appropriate, but "($/yr)" would not. The title of the graph appears at the bottom of the figure, immediately after the figure number. The title should be descriptive and should not merely repeat the axis labels. For example, "Figure 3. Determination of Optimum Pipe Diameter" might be appropriate, whereas "Figure 3. Annualized Cost vs. Pipe Diameter" would not be acceptable.

Other types of figures used in design reports are process (and block) flow diagrams, piping and instrumentation diagrams, pie charts, histograms (bar charts), and scheduling charts. Each is numbered, cited in the text, and titled. Specific format instructions for flow diagrams and P&IDs are given in Chapter 1. Pie charts and histograms are used to show relative sizes of quantities. For example, the total operating cost has many components, but only the steam cost may be a significant fraction compared to the rest. A pie chart would show this. (Also, the total quantity that the pie represents should always be given.) Relationships between numerical variables, however, are not shown on pie charts or histograms. A scheduling chart is often included in reports of ongoing projects, and these are numbered, titled, and located in the same way as other figures.

Tables contain a number and descriptive title at the top, and each column (except the first in some cases) has a label with appropriate units. Appropriate numbers of significant digits should be shown, and the columns of numbers should line up by their decimal points.

22.2.6   References

When one uses the ideas or words of another author, proper attribution is made in the report. A mark (superscript number, number in brackets, or author name and date in parentheses) is shown in the text at the point where the work is used. The complete citation is given either in a footnote at the bottom of the page or at the end of the body of the report in a separate section. The format for the text mark and for the citation are set in the guidelines of the organization and must be followed.

There are two main reasons for citations of other work. The citation acknowledges the ideas as someone else's, not yours. Such acknowledgment is essential; its absence is plagiarism. (Note that citation is required even when there is no direct quotation of another author.) Also, the citations give the reader guidance to other written sources whose level of detail or focus are different from your report.

With the internet and the World Wide Web, it is easy to download text into a file on your computer. To cite this material properly in your report, you must be sure to keep enough authorship data in the file to be able to write the citation. At a minimum, you should include the internet locator (URL) so that you can go back and obtain the author, date, and other information.

22.2.7   Strategies for Writing

The writing process is different for each writer. Some people write best longhand, others at a computer. Isolation is better for some, being in the middle of the activity is better for others. Some prefer to spend more time per draft on fewer drafts, others need to get something down on paper (screen), even if it will be modified drastically before the final draft. Here is a strategy, though, that can be quite effective:

1. Define your audience and your purpose.
2. Outline what you want in the report. This may be formal or just a list of topics, statements, or visuals that are key to your message.
3. Start writing early. Writing the introduction is a way of exploring what the problem is. Once you have identified the problem, the focus of the rest of the report will be clearer. Think of the writing of the report as you would any problem and use one of the problem-solving strategies from this book.

4. In the first draft, focus on the facts and on the logic. Save editing for later.
5. Do not turn in a first draft as a final draft.
6. Set aside the writing for a day before you pick up the hard copy to revise it.
7. Revise for organization, logic, technical content, clarity, and so on.
8. Set aside the hard copy for a day.
9. Revise for grammar, spelling, punctuation, conciseness, word choice, and other elements of editing.
10. Have someone read it aloud to you.
11. Do the final revision.
12. Proofread the **hard copy**.

The above is an example of an ideal process. Other models could be found. We must admit that we do not always follow all these steps, but, the more closely we follow them, the better our writing is.

An important issue is writer's block, which is a state of such high anxiety about writing that writing becomes impossible. Much like stage fright, writer's block is autocatalytic. The writing paralysis caused by the anxiety creates increased anxiety, which leads to heightened paralysis. Identifying the cause leads naturally to the cure. The three most common causes are:

1. *Not fully understanding what you are trying to explain.* If you are confused about the technical content of your work, it is very difficult to put it down in words.
2. *Having no clear sense of audience or purpose.* If you do not know who will read your report, how can you communicate with them?
3. *Trying to turn out a final version painful word by painful word.* If you want to write the perfect report in one draft, that draft will take far longer than writing multiple drafts.

When a group writes a single report, there are three basic strategies:

1. *Assigned sections.* This is especially helpful for long reports or when different members of the group have different strengths. A drawback is that the logical connections between the sections can be missed, and this must be corrected by having all members of the group critique all sections of the report between drafts. Finally, one person needs to go through and modify the writing of the sections so that they are written in one style.
2. *Writing by committee.* On shorter reports, the entire group can meet together, talk about what should be in each section, and decide as a group what the wording should be. This is a difficult strategy to implement, and, in fact, confusing, unfocused writing is often characterized derisively as "written by committee."

3. *Assigned functions.* Members of the group can be assigned tasks such as: preparing the outline, writing the drafts, critiquing the drafts for technical errors, critiquing the drafts for expected audience response, checking for consistency and logical connections, and so on. Again, coordination is the key. Everyone reads all drafts.

22.2.8   WVU Guidelines for Written Design Report

The following guidelines were written by the chemical engineering faculty at West Virginia University.

The format for a written design report differs from that of a laboratory report. A laboratory report tells a scientific story, starting with theory, proceeding through results, discussion, and conclusion. It is usually assumed that the readers will read the entire report, probably straight through from front to back. That is not the case for design reports, however. Often, management will read only the conclusions or the recommendations—or perhaps those plus the abstracts. Thus, in a design report, the most important conclusions should appear early in the report, with details presented later for the readers who delve further into the report. Such is the way of business, where you must effectively convey the bottom line to someone who may not have the time to read the entire report.

We suggest the following format for design reports.

Letter of Transmittal

This is a memorandum (if addressed to others at the author's company) or a letter (if addressed to persons at another company) that identifies the report. In order to get the readers' attention, several sentences summarizing the bottom line are essential. You should always sign any letter and initial any memo. Your signature indicates that you acknowledge responsibility for the letter or memo. Only a signed document can be considered official; an unsigned document may be a draft or other unofficial or unauthorized version.

Title Page

As well as giving the report title, the title page must include names of all contributors to the report, the date the report is finished, and the name and address of the author's organization.

Abstract or Executive Summary

An executive summary is an informative summary of the report that focuses on the concerns of management. The conclusions and recommendations are given, but the details of the other parts of the report are given only when they are unusual or unexpected and, therefore, impact the risk of decisions made by management. An abstract, on the other hand, is a synopsis of the report, and it has the same focus as the full report. Its purpose is to give an overview of the document, so that potential readers can decide whether to read the full document and to help them put the report details into perspective. For some of our reports, an abstract is appropriate; however, the senior design project report will require an executive summary. Be sure to ask which is required for a given assignment.

Abstracts and executive summaries should convey to the readers what you did, what you conclude, and what you recommend in a clear, concise, convincing manner. This section is for the readers who do not need any details or for the readers who are deciding whether to read further. It should be clear to readers who do not look at any other parts of the report. Summarize the bottom line; do not discuss computational details unless they are unique and applicable beyond the report at hand. In an executive summary, do not be afraid to use a few well chosen graphs, pie charts, or histograms to emphasize your important points, but choose these wisely in order to keep the length of the executive summary down.

An abstract or executive summary should start on a new page, and nothing else should appear on the same page.

Table of Contents

Only long reports need a table of contents. Regardless of whether you include one, all pages of your report should be numbered, preferably at the top right corner or top center (the latter permits easy two-sided copying).

Introduction

The introduction sets the stage for the remainder of the report. You should include a definition of the problem, the background, the key constraints on the solution, and how the problem was attacked. When possible, give some perspective on the specific problem in the context of the larger business picture.

Results

The following are essential components of a results section:

1. Labeled and dated flowsheet, with any oversized flowsheets bound so that they can be easily unfolded and read.
2. Material and energy balance tables, with temperature, pressure, phase, total mass and molar flowrate, and component mole flowrates.
3. Manufacturing cost summary, itemizing yearly revenue and expense (income from product sales, expenses for raw materials, utilities (itemized), equipment costs if calculated as an annual cost, personnel, etc.).
4. Investment summary, including costs to build and install plant now (if appropriate to goals of problem), itemized by piece of equipment.

5. Equipment summary, listing equipment to be purchased and installed, with specifications. This item can be combined with #4 above if the list is not too long.

All of the above should not simply appear without description. The results section is held together by prose that gives the readers a road map through the tables and figures, interpreting the data and drawing the readers' attention to the most significant results. Whether you use figures or tables is your choice. For each item, you must decide which conveys your results more efficiently. Do not use both a figure and a table to illustrate the same point. That would be redundant, wasting both your and your readers' time.

Whichever you choose, figures and tables have a specific format. They are numbered in the order in which they appear in the report. They should appear on the pages immediately following your first reference to them. Figures have a caption (title) at the bottom, which, if a graph, should not simply repeat the axes (bad: y vs. x; good: plot illustrating ...). Tables have a title at the top. All exhibits are called either a figure or a table. When you refer to one, Figure # or Table # should be considered a proper name and, therefore, be capitalized. Finally, avoid colors in figures and tables so that regular copiers can be used.

The following terminology is used (especially in software) to define the orientation of a table or figure on a page. Portrait refers to the way typed text appears, with the long dimension of the paper vertical. Landscape refers to text, figures, or tables appearing with the long dimension of the paper horizontal. Landscape figures and tables should always be oriented so that they can be viewed properly if one rotates the opened report 90° clockwise.

In more sophisticated design projects, where you evaluate several cases, graphical or tabular comparisons are appropriate. In this instance, provide the five components listed above for the case you recommend unless instructed to submit a base case calculation first.

If you use a pie chart, show the total quantity (i.e., the whole pie) in a legend. When the independent variable is numerical, do not use "line charts."

Discussion

In the discussion section, you go into more detail for readers who still want more information and are willing to read further. Here you discuss the reasons for making choices and the reasons for discarding alternatives. You might also discuss non-routine or unique computational aspects.

Conclusion

No new results are presented in this section. Rather, you should present your important conclusions based on your analysis. These conclusions are also summarized in the abstract or the executive summary. For design reports, conclusions usually include overall economic measures (NPV, annual profit, etc.), process modifications, and a statement concerning the ability of the design presented to satisfy the goals of the project. You may use an introductory paragraph and then list conclusions as short, one sentence statements rather than as lengthy paragraphs.

Recommendations

This section includes recommendations for further action and/or further study. If there are few conclusions and recommendations, these two sections can be combined.

References

This section can be presented in one of two ways. If you put your reference list before the appendix, it should contain only those references cited in the sections of the report preceding it. References may be listed by number, and cited in the text by this number, either in parentheses or in brackets (preferred) or as a superscript. Another method is to cite the reference by the author and year of publication. You should consult the reference list in any of your chemical engineering texts for the correct citation format. If you put the reference section before the appendix, citations to references used in the appendix should be made by superscripts and the full citation should be given in a footnote on the page.

The other alternative is to place the reference section at the very end of the report and to cite all references used either in the body of the report and in the appendix.

Do not reference material that is not specifically cited in the report. Software is not referenced unless you use it as a source of data, as might be the case with a process simulator such as ChemCAD™.

Other Sections

Sometimes, especially for longer reports, specialized sections, such as Safety, Assumptions, Environmental Concerns, Risks can be included.

Appendix

The appendix is the place for your detailed calculations, computer programs, and so on. Provide a Table of Contents for the appendix so that the readers can easily find specific calculations. Although calculations may be hand written, they must be legible and easy to follow.

Equations

All equations are centered, with right justified numbers in parentheses. As with figures and tables, equations are cited by number, for example, Equation 4. Unlike figures and tables, an equation is referred to only after it appears, as shown below.

incorrect:

The relationship for the heat capacity difference is given by equation 1.

$$C_p - C_v = \frac{\alpha^2 VT}{\kappa_T} \qquad \text{(equation 1)}$$

For an ideal gas, this reduces to Equation 2.

$$C_p - C_v = R \qquad \text{(Eq. 2)}$$

correct:

The relationship for the heat capacity difference is

$$C_p - C_v = \frac{\alpha^2 VT}{\kappa_T} \qquad (1)$$

For an ideal gas, Equation 1 reduces to

$$C_p - C_v = R \qquad (2)$$

How Engineering Reports Are Used

An engineering report (including appendices) is practically never read in its entirety by a single person. For most of the users of these documents, their time is too valuable to allow them to sit down and read every word, and some will have no interest in the details of your project. However, you must assume that each word will be read by someone, sometime, and that you will not be around to explain any ambiguous passages. Your report must be useful to the following types of readers:

1. **The person who has only a few minutes to read the report.** This is often an intelligent, non-technical person who controls millions of dollars. You must be sure that this person can pick up your report, immediately find the important answers, for example, the "bottom line," and make the right decision. If the answers are not prominently featured in the Executive Summary or the Abstract, this reader will find your report to be of little value. You cannot afford that judgment.

2. **The technical manager.** You may assume that this person is a chemical engineer, but you may not assume any specific technical knowledge about the details of your project. This person is busy but may have enough time to read most of the report (but not the appendices). Few engineers read a report from beginning to end! They look for the answers quickly. As soon as these answers are found, they make a decision and stop reading. Sections might be read in the following order, for example, until the answers are found: Executive Summary, Recommendations, Conclusions, Results, Discussion, Introduction. Different readers will read the sections in different orders. You must, therefore, take special care to put information in the correct sections.

3. **The engineer who is going to use your design.** This chemical engineer needs to find details of how you did your calculations and how you reached decisions. The appendices are of special interest to this reader. However, time is of the essence. This reader wants to be able to go immediately to the page or two

in your appendices that deals with a specific detail. Without good organization and a good table of contents for the appendices, this is impossible. If this reader cannot find the right information, your effort has been wasted.

4. **Others.** Many others may try to read your report: mechanical engineers, chemists, perhaps even environmental activists. Think about these people, too.

What can you do to see whether your report meets the needs of these readers? Ask someone who did not author the report to read it, pretending to be one of these readers. A friend, a roommate, or a fellow student might qualify. If they cannot understand what you are trying to say, you have a problem. Remember, if the readers do not find what they are looking for or cannot understand what you are trying to say, it really does not matter whether the information is in the report somewhere or whether the results are of high quality.

22.3   ORAL COMMUNICATIONS

Public speaking can be scary at first, but oral communication is crucial in engineering design activities. When done well, it communicates information faster than can the written word, and it allows for immediate communication from the audience to the speaker and between members of the audience. Thus, oral presentations (with questions) provide for a type of feedback control on the communication process, whereas written communications are closer to feedforward control.

The key to effective oral communication is, as it is for written communication, audience analysis. Include visuals that focus the audience attention on the areas of the project analysis that both you and the audience feel are important. Be sure that the audience receives the information required for making the right decision, and be sure not to bore the audience. If the readers of your report get bored, they may come back to it later. If the attention of your listeners starts to drift, the information is lost.

Before the actual presentation, practice is essential. (It may surprise you to learn that most faculty do not practice their lectures.) If possible, practice in front of a real audience who have agreed to give you feedback. Otherwise, try taping (audio or video) your presentation and critiquing it yourself. If none of these techniques will work, go to a quiet room and give your presentation out loud to yourself. In any case, it is best if you use the actual equipment (overhead projector, slide projector, and/or computer) that you will use in the actual presentation. If this is impossible, use hard copies of your visuals.

Successful oral presentation requires a degree of self-confidence. To avoid a last-minute problem in this regard, give your practice talk far enough in advance that you can make corrections to your talk and/or your visuals. If your audience says that you need to make major changes, you need to be able to implement these suggestions.

Even when speakers are well-prepared and self-confident, often they are nervous. A variety of techniques can be useful to alleviate tension. Here are a few:

- Prepare notes that you can refer to if you draw a blank on what to say next or to be sure that you talk about the key points. The notes should be on stiff note cards that will be easy to hold still and that will not distract you or the audience.
- Practice the presentation in front of a group. You are likely to be most nervous during the first time that you give a presentation.
- On your note cards, print in bold letters "RELAX" or something else that will be a trigger to help you calm down.
- If you tend to speak too quickly during presentations, print "SLOW DOWN" on a note card.
- Even if you never refer to them, note cards can be a kind of "security blanket" that could help you relax.
- Look at a point just above the heads of your audience. They will think you are making good eye contact, but you can avoid really seeing them.
- Before the start of the presentation, take a deep breath or two to calm down.
- During the presentation, keep focusing on the remainder of the presentation. If you have made a mistake (by omitting to say something that you wanted to include or by not remembering why you chose the reflux ratio that you did), ignore it until the question-and-answer period at the end. Reminding yourself and your audience that your presentation was less than perfect will make it more so.
- If questions during your talk make you nervous, start your talk by asking that the audience save their questions until the end.
- Before the talk begins, go to the room and check it out. Where will you stand during the talk? Would you feel more comfortable and in control if the projector were moved? (Often it can be.) Do you want the lights high or low? By getting used to the environment and making small changes to it, you begin to feel more in control. In a classroom or conference room, you may be able to move chairs or to keep the front row empty, if that would help.
- Avoid looking at one person in the audience throughout the talk. If that person is interested, he or she will show this by facial expressions and you will feel good. However, if suddenly that person loses interest, you may be devastated.

- Do not allow yourself to be in a rush immediately before your talk. Sit down, take a short walk, whatever will calm you, but do not get into a serious conversation.
- Do not drink too much coffee or other caffeinated beverages before a talk.
- Be sure your slides are loaded correctly, your transparencies are in order, and the pointer is where you want it before the last minute.

During the question-and-answer period, you relinquish part of your control over the situation. You do not know what will be asked, but you must be prepared to answer. Dealing with this uncertainty requires additional strategies:

- Anticipate what the questions will be. Often, especially if you practice the talk with an audience, many of the questions become obvious. Prepare answers to each of these.
- Admit when you do not know the answer to the question. However, if you have a partial answer, offer that. You should not leave the questioner with the impression that you know nothing about the topic.
- Thank the audience member for the question, even if you do not know the answer. If it is a question that you, too, would like the answer for, tell the audience so. If you will be investigating the issue further, let them know that.
- Do not try to answer the question until you are sure that you know what is being asked. Ask the questioner for clarification if necessary.
- If you need time to think about your answer, restate the question. This not only gives you a few seconds to think, but it also helps to make sure that you are answering the right question.

22.3.1   Formal Oral Presentations

When you are talking to a group (sometimes a large group), the presentation can be quite formal. You may be introduced, and there may be applause after you are finished. In this type of presentation especially, it is important that you have a strong opening and closing and that you focus on keeping the attention of the audience.

Prepare simple, uncluttered visuals that everyone in the room can see. Be sure to see the room before your talk. If the large crowd is going to make you nervous, plan to use the techniques mentioned above to make it feel like a smaller audience.

22.3.2   Briefings

Briefings are a very common form of oral presentation, in which a small group of people (as in a conference room) listen to your talk. The size of the audience will mean that they are more likely to try to pay attention to your talk than a very large audience would, but you must directly address their needs for information. Again, slides or transparencies are usually used, and you should prepare a hard copy set of these for each audience member. As people more often take notes at a briefing than at a formal oral presentation, the hard copy will help them to focus on your talk. Also, they may find it more helpful to follow your talk by looking at your hard copies than at the screen.

Because the audience tends to have a narrow range of background, the briefing tends to be more detailed and the question period is usually more extensive.

22.3.3 Visual Aids

Most technical presentations are developed around the visuals. Start with a title slide and then a slide that gives the big picture. As with written reports, it is important to obtain the guidelines for oral presentations from your organization. Some are very prescriptive, others not.

A rough guideline is that each visual will be used for one minute; however, this timing varies greatly. Sketch your visuals crudely as you run through what you would like to say. Imagine that you are a typical audience member. Would you have received the right information? Revise your planned visuals to address the gaps or lack of continuity.

A mixture of text visuals and graphics is usually most effective. Tables rarely make good visuals. As you explain the meaning of a trend on a graphic visual, your audience is absorbing the trend itself. When you present your conclusion, a text visual will probably be better. The best way to improve your visuals (in fact, your entire presentation) is to reflect on your own response to the visuals and presentations of others. What kept your attention? What was confusing? What communicated the most high-quality information? Again, doing a dry run with a friendly audience is a good way to get constructive criticism and feedback in a comfortable environment.

Be prepared to point at the screen to direct the audience attention. Before the presentation, the pointer should not only be located, it should be tested by the presenter. One must avoid moving the pointer either so fast or so erratically that the audience is distracted. If the presenter is too nervous, it will be difficult to use any pointer without erratic movement. If this is the case, both hands should be used to steady the pointer. If all else fails, the presenter should find a way to avoid using the pointer. In any case, avoid moving the pointer except when it is in use. The audience will always try to follow a moving pointer.

Right-handed people should always position themselves so that the screen is to their right as they face the audience. Left-handed people should always position themselves so that the screen is to their left as they face the audience. This orientation makes it more comfortable to maintain eye contact with the audience when one is pointing to the screen. It is also easier to change transparencies from this orientation.

Slides, transparencies, computer projections, and videos can be used for the visual aids. Each has its own advantages, and one should consider all options before deciding how best to present the material. However, there are two common traps to avoid. Using more than one of these media in a presentation can be impressive if done well; it is invariably disastrous if done without a great deal of rehearsal. Also, one must be prepared with a strategy for continuing if the equipment malfunctions; this occurs especially frequently with state-of-the-art computer projection hardware and software available at the time this book was written.

There are many guidelines concerning the size of lettering and the density of information to use on visuals. However, the best strategy is to imagine what the audience response would be.

No visual should ever be used in a presentation unless it has been previewed with similar equipment, in a similar room, from the most remote position that an audience member can be located.

22.3.4   WVU Oral Presentation Guidelines

The following guidelines were written by the chemical engineering faculty at West Virginia University.

When presenting an oral report, it is important to realize that the audience cannot digest material in the same way they can when reading a report. There will be no time for them to reread a sentence or paragraph, or to study a table or figure. Therefore, it is incumbent upon the speaker to emphasize the important points. The recommendations that follow, though applicable to many types of oral presentation, are written within the context of a design presentation.

Design oral presentations are often organized as follows:

1. Tell the audience what the important results and conclusions are, in the context of the big picture.
2. Present the details of the work in a logical manner to convince the audience of the validity of the results and conclusions.
3. Present the audience with a plan (or alternate plans) of action, based on your results and conclusions.

With this in mind, here is one way to organize an oral presentation.

Title Page

Identify the report and the presenters on a visual aid.

Overview of the Problem and Your Solution

Orient your audience with the big picture. Include only the details that are most important for the audience to remember as you fill in the gaps later.

Outline

Use an outline of the presentation to prepare the audience for what comes next. As the oral presentation is different from the written report, this outline is not merely the table of contents from the report. Also, consider including a few words abstracting the contents of each section. This outline slide is optional, but many presenters find it helpful in keeping themselves focused.

Results

This section of the oral presentation is not merely a synopsis of the results section of a written report. For example, a detailed material and energy balance table will not be easily seen or quickly understood by, and not of much interest to, your audience. Also, what is effectively communicated in a table in a report might be best communicated orally using a graph or pie chart. Avoid using complex tables and figures with small print. These can neither be seen in the back of the room nor digested by anyone. Remember also that the audience can later read your written report; however, the oral presentation must help them to understand the context, importance, and limitations of your results and conclusions.

Discussion

Once again, the content is focused on the direct needs of the audience. In preparing this part of the presentation, you can start with the discussion section of the written report, but serious modification is necessary.

Conclusions

Here you remind your audience what your most important conclusions are, often as a list or outline.

Recommendations

Here you focus on the actions that you suggest the audience to take. Be as specific as you can be, but do not step beyond the bounds of your own analysis, making unqualified recommendations of investments or modifications that your analysis cannot justify.

Other Important Points When Making an Oral Presentation

Transparencies vs. Slides

Transparencies are more flexible: you can modify them in real time, use them out of order, or use overlays; but slides seem more professional, more formal. You need to decide which format makes you more comfortable, more in control. In either case, make sure that the people in the back row can read them, and that you give the audience enough time to assimilate the information presented before you go on to the next visual.

Content of Slides or Transparencies

Do not put too much on a slide or transparency. A detailed table may not be readable in the back row. When making a slide or transparency, put yourself in the audience and ask yourself if you could learn anything from it if you only saw it for 30 seconds or a minute. Short, concise statements of a few words, with the speaker providing a more detailed explanation, are sufficient to convey your points. Use colors sparingly. Test all of your visual aids in a similar sized and shaped room in advance.

Presentation Mechanics

Always face the audience. If you have to look at the screen, take a quick glance and then turn back to the audience. Do not just read the slides. If using overheads, you can look at the transparency on the projector rather than turning toward the screen. Be careful not to block the view of a portion of the audience. If someone else is changing your transparencies, stay back next to the screen. If you are changing your own transparencies, step back away from the projector after making the change.

Try to avoid the following nervous habits: chewing gum, playing with the pointer or something in your pocket, rocking from side to side, or giggling. Approach the oral report with confidence and a firm belief in your abilities and your work.

Voice

Speak clearly, enunciate carefully, avoid audible pauses, and project your voice.

Do Whatever You Can to Be Comfortable

You are in control, not the audience. Beforehand, arrange the room in whatever pattern makes you feel most comfortable. Do you want to point with your left hand or your right? Do you want the shades open or the lights out? Then, do not make last minute changes in your presentation. Do not start a conversation with anyone who is not supportive before the presentation. Immediately before your presentation, take a few deep breaths and yawn. This is easier to do if you are not in the presentation room. If you do not want to be interrupted with questions during the presentation, tell the audience so. And, if they still interrupt, politely tell them that you will be answering that question later. Assume that everything is going to go well.

Notes

To use or not to use? Do whatever will make you most comfortable and in control. If you read from a script, no one will believe that you know what is going on. But no one can remember every detail without notes. When you practice your presentation, try it with and without notes, if you are not sure which is better. As a novice, you might find that notes bolster your confidence; however, with practice, you should wean yourself from using notes.

Audience Analysis

Just as with a written report, think about the different backgrounds and needs of your audience. Will they get the right message and make the right decisions?

Question and Answer

Admit it when you do not know the answer, but try to give any relevant partial answer that you have. Most people only ask a question because they do not know the answer, either. Try to be responsive, not evasive. And prepare for Q & A by imagining the questions that will be asked.

Post Mortem

After the presentation, go with your friends from the audience to a different room and get feedback immediately. Ask your colleagues what you could have done better. If they tell you that you were perfect, tell them that they are not being very helpful. Request constructive, supportive criticism! This is the best time to find out what you did right and what you did wrong. If you wait more than about an hour, the feedback will not be detailed enough to help you.

22.4   SOFTWARE AND AUTHOR RESPONSIBILITY

Word processing, spreadsheet, and graphics software can help an author produce a better final report, but each software package has limitations. It is the author's responsibility to understand these limitations and to overcome them. Some of these software problems are the results of programming errors, some arise from misunderstandings by the software developers of the rules of writing, and some represent differences between technical writing and other forms of written communication.

Above all, though, writers should save their files frequently and make backup copies.

22.4.1 Spell Checkers

Misspellings have always been unacceptable in formal written work. Readers generally have a lower opinion of the ideas expressed when the prose contains misspelled words. With spell checkers being so available, spelling errors have come even more to indicate sloppiness and lack of respect for the audience.

Clearly, one cannot submit a report that has not been subjected to a spell checker. However, no spell checker is perfect. More importantly, they can only check whether the "word" in question is *a* true word, not whether it is *the* word that the author intended. If one writes "to" when "too" is meant (or "preform" instead of "perform"), the spell checker will not correct it.

All spell checkers allow the author to add words to the dictionary. Many technical terms are not in the standard software dictionaries, so it is tempting to put in many words common to the author's subject. However, this is a risky business. If one puts a misspelled word in the user dictionary, nearly all of the author's reports will have that misspelling, without the author having a clue! Also, it is usually far more difficult to remove a user-supplied word from the dictionary than it is to add one.

Most spell checkers automatically suggest another word when they detect a misspelling. If one too quickly accepts the suggestion, rather strange prose can arise. In a technical report, we once somehow accepted the suggestion of "voices" for "VOCs" (volatile organic compounds)!

22.4.2 Thesaurus

In most writing courses, students are warned not to misuse a thesaurus. It is tempting to use a new word, one that sounds more sophisticated. Unfortunately, the probability that the new word can be substituted directly (without a modification of meaning) is nil. The thesaurus is designed to be used to *remind* one of a similar word. If the thesaurus suggests a word that you have no memory of, do not use it. Many satirical papers have been written about the blind use of a thesaurus.

22.4.3 Grammar Checkers

Grammar checkers continue to improve, but most are based on simple rules that apply imperfectly to technical communication. For example, the use of the passive voice is frowned upon in much nontechnical writing, but it can be the more appropriate voice in a technical report. The passive voice places emphasis on the object receiving the action rather than on the subject taking the action. Although the passive voice is often overused in technical writing, it does have its place.

Crude distinctions are often made by spell checkers. For example, some will flag each instance of "affect" and ask if the author meant "effect" (or vice versa). This can be tiresome, but one must be sure to understand the significant difference between these words.

Especially in long sentences, grammar checkers tend to become confused and flag correctly written possessive constructions and subject/verb agreement. It is important not to accept the suggested changes too quickly.

Various readability indices are included in grammar checkers. These are crude measures of the ease with which a reader can read and understand a document. Most predate the computer grammar checkers. A document with a good readability index could be totally incomprehensible, and one with a poor index might be quite good. One must not read too much into these indices. However, they do focus on elements of good writing, such as short paragraphs, short sentences, short words, and straightforward grammatical constructions. Thus, one should watch the readability numbers for sudden shifts to longer words, sentences, and paragraphs. If the same information can be conveyed in a simpler format, it is likely to be more helpful to the reader.

22.4.4   Graphs

For much charting software, the default chart type is not the common *x-y* plot, even though these are by far the most common in technical communication. If the independent variable (on the abscissa or horizontal axis) is numerical and continuous, one must use this type of graph. Bar charts and histograms are used only for non-numerical or (sometimes) discrete variables. Unfortunately, there is a type of graph that looks vaguely like an *x-y* plot, but is more like a bar graph. It is called a line chart. The line chart has evenly spaced tick marks on the abscissa, but the numerical values at the ticks are not even intervals! Such a graph is very misleading in technical work and must be avoided.

Most graphing software will automatically choose the scale for both axes. Unfortunately, this is usually not acceptable. For example, it is common practice to use even numbers when both odd and even are not shown and to use numbers ending in 5 or 0 in preference to others. These conventions make it much easier for the reader. Also, there are often constraints (such as 0–100%), and it is sometimes useful to extend axes far beyond the last data point. For example, in a series of graphs, one should use the same scale on each to allow comparisons. Typical graphing software will do almost none of the above automatically. The best strategy is to use user-specified scales on all graphs.

The lettering of the various parts of the graph should be clean and simple, and the more important parts should be in larger font size. This is rarely done automatically. The convention is to put the title of a figure (such as a graph) below the figure, although when the graph is presented by itself (as on a slide) the title may appear at the top. Titles should be properly capitalized: The initial letter of each word (except articles, prepositions, and conjunctions) should be capitalized. All other letters should be lower case.

Strangely, some graphical software automatically inserts a legend on graphs. Unless there are multiple variables plotted, this should not be done. In any case, a direct label near the curve is more effective. One must avoid default legends such as "Series 1" that are distracting and offer no information.

Default headers and footers on graphs are seldom appropriate. One must override the default to number the pages consecutively in the report and to avoid double titles, multiple pages labeled "Page 1," and so on.

Most software connects plotted points with line segments (or splines) by default. This is often not appropriate. For example, if the points represent experimental data, only the points should be shown. Also, the joining of points obtained from process simulation on such a graph can obscure the maximum, minimum, and the shifts from one topology to another. More about these details is given in Chapter 19.

22.4.5   Tables

In some word processing and spreadsheet software, the default for numbers in table cells is centered. If this is the case, one must override the default and line up the numbers in a column by their decimal points. Only then can the readers easily compare the numbers or sum them. Also, the default format may be to include many significant digits. One must be careful to include only those digits that are, in fact, significant. Otherwise, the readers will either be misled about the precision of your answers or they will discount your answers if they realize that you have exaggerated their precision.

When numbers have four or more digits to the left of the decimal point, they must have commas to help the reader to see quickly the thousands, millions, and so on, places. Only if the numbers are too cumbersome should powers-of-ten notation be used. When it is used, so-called engineering notation is preferred to scientific notation. In engineering notation, the power of ten is always divisible by three (analogous to the commas mentioned above). In a column of numbers that all refer to the same kind of property, the format should be the same for each entry.

22.4.6   Colors and Exotic Features

Most software will allow multiple colors, fonts, and other features. Although their use may be tempting, too many changes can be distracting to the readers and cause sensory overload. On slides, use only as many colors as will clarify your results. Most members of your audience will not be able to distinguish between subtle shades in the time available; distinctions should be made with bold contrasts. However, one must not rely on color alone—almost any report will be copied in black-and-white! If your distinctions are not clear when the graphics are reproduced this way, there is a problem.

Reports in which dozens of fonts are used are distracting. Whenever the readers have any trouble focusing on the content, they tend to focus on the variations in font size or type. Thus, one should use a few font changes to focus attention, but the report should not appear to use every available font. Also, the more exotic the font, the less likely the file will be compatible with other word processors.

A very serious problem is the improper use of "3-D" figures. If a graph is a plot of a dependent variable versus one independent variable, the graph is only a 2-dimensional representation. Any attempt to make it "faux 3-D" will make it less readable and more confusing. Features such as giving a pie chart a constant thickness or the bars of a bar chart constant depth make the chart less useful in conveying data. Such gimmicks are viewed as such in technical reporting. The same can be said of "shadowing" of 2-dimensional figures.

When a graph is truly 3-dimensional (two independent variables and one dependent variable), one must be careful to use a perspective that allows the reader to determine easily the height (the dependent axis) at various points. The most useful such perspective is the "isometric" perspective, which involves rotating the straight-on coordinate system by 45° and then tilting it up in back by 30°. In this perspective, the distances perpendicular to the axes are not distorted.

22.4.7   Raw Output from Process Simulators

Process simulation software typically will produce a flowsheet, but this is probably not a true process flow diagram and should not be used as one in a report. There are two reasons for this. First, the simulator-produced flowsheet is unlikely to follow all the conventions of PFDs given in Chapter 1 regarding equipment symbols, line crosses, labels, and so on. Second, the process simulated is not the true process. One "unit" in the process simulator (such as a distillation column) may be several pieces of equipment (tower, condenser, condensate tank, etc.) that need to be shown on a PFD. It is common to simulate a single unit (such as a process-process heat exchanger) as two units to decouple the recycle calculations. Some units, such as a storage tank, require no calculations in the process simulator and thus are not shown on its flowsheet. However, the simulator flowsheet is essential in the appendix of the report if the simulator "report" is included. Care should be taken to use the same stream numbers in the simulation flowsheet as in the PFD, whenever possible.

The "report" from a process simulator should be included in the appendix, but it serves neither as the main PFD nor as the main flow table. These reports are formatted for use by the user of the software, not by the reader of a report. Therefore, they should be considered as are the calculations in the appendix. They are referred to by the engineers who need to know the details, but they are not of interest to other readers of the report.

22.5 SUMMARY

Effective written and oral communication is crucial to the success of chemical engineering projects and to the success of chemical engineers. The keys to effective technical communication are: performing an audience analysis, following the format of the organization, and obtaining and acting on feedback from colleagues.

REFERENCES

1. Leesley, M.E., and M. L. Williams, "All a Chemical Engineer Does Is Write," *Chemical Engineering Education*, **12** (4), 188 (1978).
2. Munter, M., *Guide to Managerial Communication*, 3rd ed., Prentice-Hall, Englewood Cliffs, NJ, 1991.
3. *Effective Communication for Engineers*, McGraw-Hill, New York, 1974.
4. Ulman, J. N., and J. R. Gould, *Technical Reporting*, 3rd ed., Holt, Rinehart and Winston, New York, 1972.

PROBLEMS

22.1 Obtain the written report guidelines used in your department for laboratory reports. Compare them to those in Section 22.2.8 and explain the differences.

22.2 Exchange a report that you have written for one by a classmate. Spend three minutes skimming the report, and write down answers to the following three questions:
  • Why was the report written?
  • What is the author's key conclusion or recommendation?
  • What would you do, based on the report?
  Then discuss your answers to these questions with each other.

22.3 Obtain the written report guidelines from a firm that employs engineers. Compare them to those in Section 22.2.8 and explain the differences.

22.4 Ask a chemical engineering student who has not yet taken design to read a report you have written and to point out the difficult passages. Rewrite the report so that it satisfies those concerns.

22.5 After giving an oral presentation, perform a post mortem analysis with a couple of members of the audience.

22.6 Write a short (one paragraph) audience analysis for a design report in your design course.

22.7 Choose a graphing, drawing, or spreadsheeting software package. Investigate the features of the package and make a list of those features that can lead to poor visual aids.

22.8 Run a grammar checker and spell checker on one of your reports. Accept every change suggested.

23

A Report Writing Case Study

The purpose of this chapter is to illustrate some of the principles of report writing that were presented in Chapter 22. In order to accomplish this, we present a written report for a project that a junior engineer might be given in the first several years of employment. This chapter is split into several sections. Each section addresses some of the common errors that are made in report writing. Examples of a poor and improved cover memoranda, poor and improved graphics, and poor and improved writing styles are given. A checklist of common errors is also included. Finally, an example is presented of a improved written report that illustrates many of the principles outlined in this chapter and Chapter 22.

23.1 THE ASSIGNMENT MEMORANDUM

The assignment memorandum for the project considered in this chapter is shown in Figure 23.1.

This is a good assignment memorandum because it communicates to the junior engineer in a concise manner *what* to do, *why* to do it, *when* to have it completed, and *who* else is involved or interested in the project. Everything is stated clearly; nothing is left for interpretation.

By indicating *who* is being copied on the memo, the junior engineer knows all of those involved in the loop. The engineer also knows for whom the final report will be prepared. An essential step in preparing a report is knowing the audience.

By indicating that there are attachments to the memorandum, the recipient of the memo knows if the document is complete.

MEMORANDUM

TO: Lee Madera, Junior Process Engineer

FROM: Chris Stafford, Senior Process Engineer

RE: Benzene Production

DATE: January 12, 1998

COPIES: R. T. Hemrick, Principal Process Engineer
                                    M. R. Johnson, VP Engineering
                                    S. E. Kelley, VP Project Engineering
                                    W. C. Lin, VP Sales

ATTACHMENTS: Preliminary design of benzene process

Currently, the prices of benzene and toluene are such that the production of benzene from toluene via the catalytic hydrodealkylation of toluene is not profitable. However, the price of benzene over the past 15 years has fluctuated wildly (from a low of $0.21 per kg to nearly $0.67 per kg). Our company is interested in carrying out a feasibility study to determine the minimum price differential between benzene and toluene which will allow the toluene hydrodealkylation process to be profitable. It is recognized that currently the preferred method of producing benzene from toluene is via the disproportionation reaction to yield both benzene and xylenes. However, at present our company has no use for the xylene and would prefer to make just benzene.

     With this in mind, your assignment is to determine the process that will minimize the price differential between toluene and benzene required to yield an $NPV = 0$. This design represents a discounted break-even analysis of the process and will be used as an internal benchmark for comparing competing alternatives to produce benzene. In your analysis, you should use the following economic parameters:

(i)   Internal after-tax hurdle rate of 10%, and a taxation rate of 35%
(ii) MACRS depreciation over 6 years for all fixed capital investment
(iii) A project life of 15 years
(iv) A production rate of 68,000 tonne per year of 99.5 wt% pure benzene

The attached preliminary design for a toluene HDA process should be used as a starting point (base case) for your study.

     Submit your findings as a short report, not exceeding 8 pages of double spaced text, plus any tables and figures. Put the details of all your calculations, a PFD, flow table, and so on, in a clearly indexed appendix.

     This report is due on January 26, 1998 and will be read by several managers as well as other technical and sales executives. You will also present your major findings in a 15-minute oral presentation on January 28, 1998.

Figure 23.1   Example of a Good Assignment Memorandum

The first paragraph provides perspective on the problem (*why* it is assigned). This is essential for the junior engineer to make rational decisions during the assignment.

The second and third paragraphs clearly outline *what* is to be done, what the constraints are, and what the deliverables are. If this section were not clear, the junior engineer would either have to guess about the constraints or the specifics of the assignment or have to go back to the senior engineer and ask. If the senior engineer were unavailable to answer questions, this might delay the project. In any case, time would be wasted.

The final paragraph states *when* the assignment is due and when the presentation will be made.

23.2   RESPONSE MEMORANDUM

An example of a poor response memorandum is given in Figure 23.2. An example of a improved response memorandum is given in Figure 23.3.

In order to explain the difference between the "improved" and "poor" response memos, it is necessary to understand who reads different portions of reports. A secretary only reads the cover memorandum subject in order to determine where to file the report. The senior process engineer, the person who gave you the assignment, may read the entire report. The principal process engineer

---

M E M O R A N D U M

TO: Chris Stafford, Senior Process Engineer

FROM: Lee Madera, Junior Process Engineer

RE: Benzene Production

DATE: January 25, 1998

COPIES: R. T. Hemrick, Principal Process Engineer
                     M. R. Johnson, VP Engineering

ATTACHMENTS: The Benzene Report

In response to your memorandum, the attached report details the results of my study on the production of benzene via the catalytic hydrodealkylation of toluene. This process is based on the production of 68,000 tonnes per year of 99.5 wt% benzene. A summary of all the major equipment and operating costs along with other pertinent economic and process information is provided in the report.

Figure 23.2   Example of a Poor Response Memorandum

MEMORANDUM

TO: Chris Stafford, Senior Process Engineer

FROM: Lee Madera, Junior Process Engineer

RE: Benzene Production

DATE: January 26, 1998

COPIES: R. T. Hemrick, Principal Process Engineer
                             M. R. Johnson, VP Engineering
                             S. E. Kelly, VP Project Engineering
                             W. C. Lin, VP Sales

ATTACHMENTS: Report entitled: "Evaluation of the Minimum Breakeven Price Differ-
                             ential for Benzene and Toluene"

In response to your memorandum of January 12, 1998 regarding the benzene production process, the attached report details the results of my study on the production of 68,000 tonnes per year of 99.5 wt% benzene via the catalytic hydrodealkylation of toluene. This process yields a discounted break-even cost differential between benzene and toluene of $0.153 per kg, $0.034 per kg less than for the base case. At the current market price for benzene of $0.27 per kg, the price of toluene would have to drop nearly 50% (from the current value of $0.23 per kg to $0.117 per kg). The fixed capital investment for this project is $5.14 million, and the annual manufacturing costs are $25.42 million per year. A summary of all the major equipment and operating costs along with other pertinent economic and process information is provided in the report.

   If you have any questions regarding this report prior to my presentation on Wednesday, January 28, 1998, please feel free to contact me at extension 999.

Figure 23.3   Example of an Improved Response Memorandum

only reads the entire report if the results are interesting or controversial. The vice presidents will probably not read the entire report. For those who do not read the entire report or who need to decide whether to read the report, the information provided in the cover memorandum is essential.

   At any time, there will be many reports circulating within a company. Occasionally, a report may become detached from its cover memorandum. If the attachment is listed as *The Benzene Report*, the cover memorandum may never be matched with the correct report if they become separated. If your company makes benzene, there will probably be many "benzene reports" circulating at one time. Therefore, the complete title should be included on the cover memorandum.

   On the "poor" memorandum, report copies are only sent to two individuals. The assignment memorandum was copied to four individuals. Always provide copies of your final product to everyone "in the loop" based on the original memorandum.

The key problem with the "poor" memorandum is that it basically states "here it is" and nothing else. The "poor" memorandum provides no information to allow any of the people who receive the report to determine rapidly what the conclusions were or to decide whether they want to read the entire report. Suppose that the conclusion that, if the company invested $100,000 in a process modification, they could raise the break-even purchase price of toluene to $0.25/kg. It is essential that everyone "in the loop" knew that piece of information immediately. Therefore, a cover memorandum must summarize the key conclusions. What was found, how much it will cost up front (capital cost, if applicable), and what the profitability (NPV, DCFROR, raw material purchase price) is must be stated. In a short report, which is likely to be the rule in industry, the cover memorandum takes the place of an abstract. Therefore, it is imperative to include key results in the cover memorandum.

23.3 VISUAL AIDS

In Figures 23.4 through 23.9 examples of poor and improved pie charts, tables, and plots are shown. Major points of criticism are shown on the "poor" figures (Figures 23.4, 23.6, and 23.8) in script font, and these errors have been remedied in the corresponding "improved" figures (Figures 23.5, 23.7, and 23.9). Not all the common errors can be shown on these figures, and a comprehensive checklist for figures, tables, and written text is included in Section 23.5.

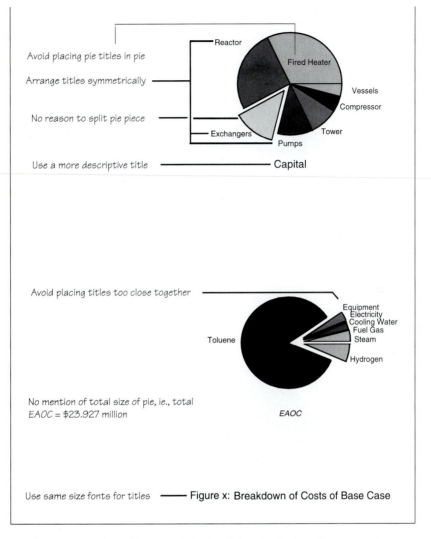

Figure 23.4   Some Common Mistakes Made in Pie Graph Presentation

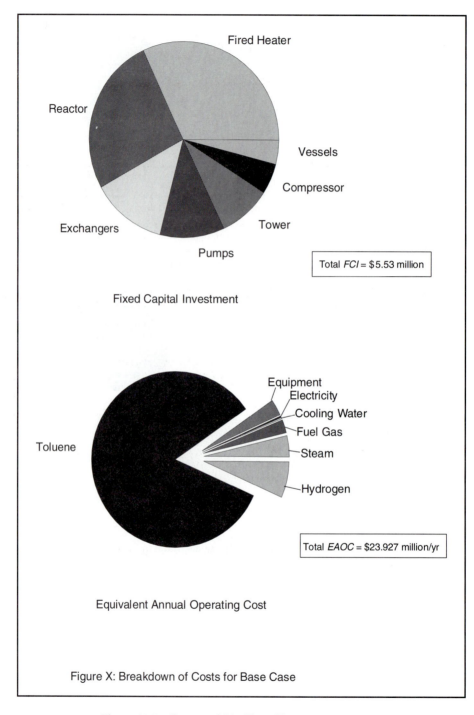

Fired Heater

Reactor

Vessels

Compressor

Exchangers

Tower

Pumps

Total *FCI* = $5.53 million

Fixed Capital Investment

Equipment
Electricity
Cooling Water
Fuel Gas
Toluene
Steam
Hydrogen

Total *EAOC* = $23.927 million/yr

Equivalent Annual Operating Cost

Figure X: Breakdown of Costs for Base Case

Figure 23.5   Corrected Pie Chart Shown in Figure 23.4

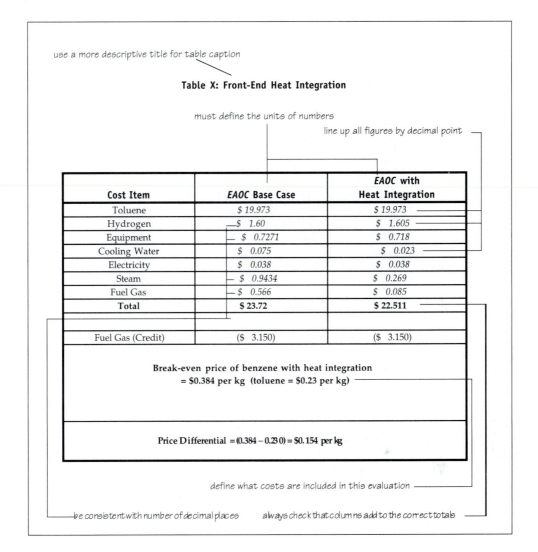

Figure 23.6   Some Common Mistakes Made in Table Presentations

Table X: Economic Impact of Front-End Heat Integration on Process Economics (all *EAOC* cost figures in millions)

Cost Item	*EAOC* Base Case	*EAOC* with Heat Integration
Equipment	$ 0.727	$ 0.718
Steam	$ 0.943	$ 0.269
Fuel Gas	$ 0.566	$ 0.085
Cooling Water	$ 0.075	$ 0.023
Electricity	$ 0.038	$ 0.038
Toluene	$ 19.973	$ 19.973
Hydrogen	$ 1.605	$ 1.605
Total	$ 23.927	$ 22.711
Fuel Gas (Credit)	($ 3.150)	($ 3.150)

Break-even price of benzene with heat integration = $0.384 per kg (toluene = $0.23 per kg)

Break-even costs include utilities, raw materials, maintenance, labor, fixed capital investment, etc.

Price Differential = (0.384 − 0.230) = $0.154 per kg

Figure 23.7   The Improved Table Shown in Figure 23.6

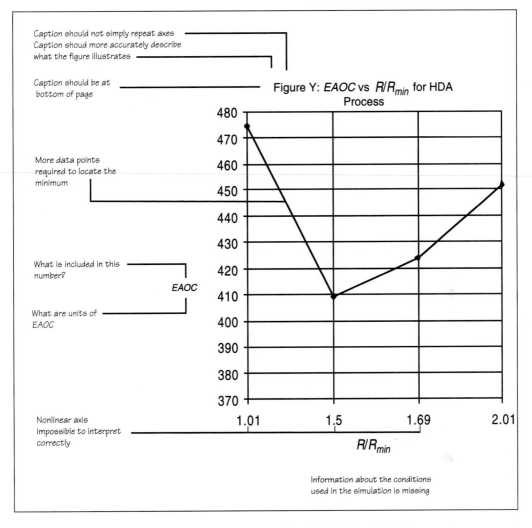

Caption should not simply repeat axes
Caption shoud more accurately describe
what the figure illustrates

Caption should be at
bottom of page

Figure Y: *EAOC* vs *R*/*R*_{min} for HDA Process

More data points
required to locate the
minimum

What is included in this
number?

EAOC

What are units of
EAOC

Nonlinear axis
impossible to interpret
correctly

R/*R*_{min}

Information about the conditions
used in the simulation is missing

Figure 23.8   Some Common Mistakes Made in Graphical Presentations

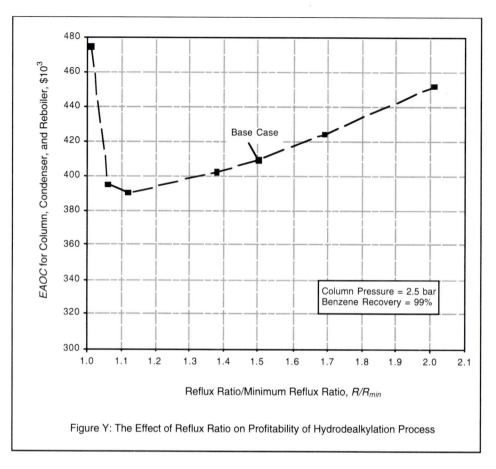

Figure Y: The Effect of Reflux Ratio on Profitability of Hydrodealkylation Process

Figure 23.9   Corrected Graph Shown in Figure 23.8

23.4 EXAMPLE REPORTS

Two examples of student reports follow. Section 23.4.1 contains an example of a portion of a student report with suggestions for improvement. Section 23.4.2 contains an example of an improved report.

23.4.1   An Example of a Portion of a Student Report

1. Introduction

The purpose of this report is to establish the minimum break-even price differential between benzene and toluene for the production of benzene using the catalytic hydrodealkylation of toluene. In this process, toluene is converted to benzene over a solid catalyst via the following reaction:

$$C_7H_8 + H_2 \rightarrow C_6H_6 + CH_4$$

This reaction is normally carried out at temperatures in the range of 580°C–660°C and at pressures of 35–70 bar. With the development of new catalysts, lower operating pressures, down to 25 bar, may be possible and I assumed this was feasible in this analysis. In the base-case process provided, the reactor consisted of a single-stage adiabatic packed bed of catalyst into which a small stream of recycle gas was fed for temperature control. Over the range of conditions considered here, there are essentially no side reactions.

> Didn't this info. come from Refs. 1 and 2? If so, you need to include the footnote here.

> Use words as well as symbols to identify the major substances in the reaction.

> Reference 3?

> Poor wording. Isn't this already the basis of the prelim. design provided with the assignment?

> Some readers are very uncomfortable with personal pronouns in technical reports. Are you sure that "I" is appropriate for your audience?

> Your readers want to know what process is best and when it should be used; they don't want a step-by-step history of what you did. Remember to focus on their needs.

2. Base-Case Evaluation

The first step was the analysis of the base case provided. The PFD for this base case is shown in Figure 1. According to the base-case report, the reactor inlet conditions of 600°C and 25 bar have been established to be close to the optimum. As a result, these parameters were not varied in the present study. A summary of the fixed capital investments, operating costs, and the break-even price differential for the base case is given in Table 1 and Figure 1. In order

> See your format instructions for ways to improve the organization and usefulness of your report.

> All the info. in the pie charts (Fig. 1) is given in Table 1. You don't need the figure.

> This is your 2nd Figure 1. Also "EAOC" is used in the figure but hasn't been defined yet.

to compare all the costs, I set up an EXCEL spreadsheet and the data for the base case was input into the program. These numbers are presented as equivalent annual operating costs by amortizing the one-time capital investments over the life of the project using a 10% discount rate. The break-even price of benzene for this base case is $0.417 per kg compared to the cost of toluene of $0.23 per kg. This yields a break-even cost differential of $0.187 per kg. The details of the break-even analysis are given in the appendix. The cost of manufacturing was estimated from the following equation:

$$COM_d = 0.180 * FCI + 2.73 \times C_{OL} + 1.23 * (C_{UT} + C_{RM})$$

where FCI = fixed capital investment, C_{OL} = cost of operating labor, C_{UT} = cost of utilities, and C_{RM} = cost of raw materials.

From Table 1, it is evident that the major costs will be associated with the purchase of toluene and hydrogen. The overall conversion of toluene in the base case will be 99.3%. Potential savings in toluene cost of approximately $140,000 per year may be realized. This savings would have a minor impact on the differential break-even price of benzene (approx. $ 0.003 per kg), therefore, the overall conversion of toluene is not considered a variable in the cases studied here. However, the hydrogen cost can be reduced significantly if a suitable separation technique can be found to purify the recycle gas, Streams 5

Margin annotations:

data were

Please use the guidelines. These details are not of interest to the readers.

with toluene costing

Don't include both * and × multiplication symbols.

Use present tense here.

;

and 7. Of the remaining costs, the steam, fuel gas, and equipment are the most significant.

> Wordy. Try "are the keys to the optimization."

The above items <u>provide a focus on where to concentrate the major optimization effort</u>. <u>In this regard</u>, a <u>two level</u> optimization strategy was

> Unnecessary and misleading.

> two-level

23.4.2   An Example of an Improved Student Written Report

1.   Introduction

The purpose of this report is to establish the minimum break-even price differential between benzene and toluene for the production of benzene using the catalytic hydrodealkylation of toluene. In this process, toluene is converted to benzene over a solid catalyst via the following reaction:

$$C_7H_8 + H_2 \rightarrow C_6H_6 + CH_4$$

$$\text{\textit{toluene}} \qquad\qquad \text{\textit{benzene}}$$

This reaction is normally carried out at temperatures of 580°C–660°C and at pressures of 35–70 bar [1,2]. With the development of new catalysts, operating pressures as low as 25 bar may be possible [3] and are assumed to be feasible in this analysis. In the base-case process provided, the reactor consists of a single-stage adiabatic packed bed of catalyst into which a small stream of recycle gas is fed for temperature control. Over the range of conditions considered here, there are essentially no side reactions.

2.   Base Case Evaluation

The PFD for this base case is shown in Figure 1. The previously reported optimum reactor inlet conditions of 600°C and 25 bar are used. A summary of the fixed capital investments, operating costs, and the break-even price differential for the base case is given in Table 1. They are presented as equivalent annual operating costs by amortizing the one-time capital investments over the life of the project using a 10% discount rate. The break-even sales price of benzene for this base case is $0.417 per kg for a toluene purchase price of $0.23 per kg (a break-even cost differential of $0.187 per kg). The details of the break-even analysis are given in the appendix. The cost of manufacturing is estimated using the following equation:

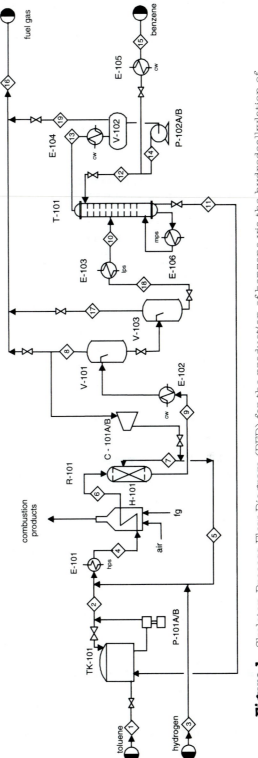

TK-101	P - 101 A/B	E - 101	H - 101	R - 101	C - 101 A/B	E - 102	V - 101	V - 103	E - 103	T - 101	E - 104	V - 102	P - 102 A/B	E - 105
Toluene	Toluene	Feed	Feed	Reactor	Recycle	Reactor	H.P	L.P	Benz.	Benz.	Benz.	Reflux	Reflux	Product
Storage	Pump	Preheat.	Heater		Gas	Effluent	Phase	Phase	Reboiler	Tower	Conden.	Drum	Pumps	Cooler
Tank					Compssor	Cooler	Separtor	Separtor						
								Tower						
								Feed						
								Heater						

Figure 1 Skeleton Process Flow Diagram (PFD) for the production of benzene via the hydrodealkylation of toluene.

665

$$COM_d = 0.180FCI + 2.73C_{OL} + 1.23(C_{UT} + C_{RM}) \qquad (1)$$

where FCI = fixed capital investment, C_{OL} = cost of operating labor, C_{UT} = cost of utilities, and C_{RM} = cost of raw materials.

Table 1 shows that the major cost is for toluene. The overall conversion of toluene in the base case is 99.3%, allowing a potential savings in

Table 1: Cost Summary for the Base Case Evaluation of Toluene HDA Process (All fixed capital and *EAOC* cost figures are given in millions.)

Equipment Type	Fixed Capital Investment	% of Total Fixed Costs
Fired Heater	$ 1.795	32
Reactor	$ 1.447	26
Exchangers	$ 0.695	13
Pumps	$ 0.579	11
Tower	$ 0.497	9
Compressor	$ 0.289	5
Vessels	$ 0.232	4
Total	**$ 5.534**	**100**

Cost Item	Equivalent Annual Operating Cost (*EAOC*)	% of Total *EAOC*
Equipment	$ 0.727	3.0
Steam	$ 0.943	3.9
Fuel Gas	$ 0.566	2.4
Cooling Water	$ 0.075	0.3
Electricity	$ 0.038	0.2
Toluene	$19.973	83.5
Hydrogen	$ 1.605	6.7
Total	**$23.927**	**100.0**
Fuel Gas (Credit)	($ 3.150)	

Break-even price of benzene = $0.417 per kg (for toluene at $0.23 per kg)
   Break-even costs include utilities, raw materials, maintenance, labor, fixed capital investment, and so on.

Price Differential = (0.417 − 0.230) = $0.187 per kg

toluene cost of approximately $140,000 per year. This savings would have only a minor impact (approx. $ 0.003 per kg) on the differential break-even price of benzene; therefore, the overall conversion of toluene is not considered a variable in the cases studied here. However, the hydrogen cost can be reduced significantly if a suitable separation technique can be found to purify the recycle gas, Streams 5 and 7. Of the remaining costs, the steam, fuel gas, and equipment are the most significant.

To concentrate the major optimization effort on costs of hydrogen, steam, fuel gas, and equipment, a two-level optimization strategy was employed. The first level focused on topological changes to the process and included the addition of a membrane separation unit to purify the recycle gas (Streams 5 and 7) and the implementation of a heat integration scheme. The second level of optimization focused on changes in operating parameters, particularly the column reflux ratio and the single-pass conversion in the reactor, because significant savings in utilities may be realized by changing these variables. The results of these optimizations are presented in the next sections.

3.   Topological Changes to Base Case PFD

3.1.   Membrane Separator

The first topological change attempted was the addition of a membrane separation unit to Stream 8 leaving the high-pressure phase separator. The membrane separation unit separates the recycle gas, sending a hydrogen-rich stream back through the compressor, C-101. This separation reduces the amount of methane in the recycle and the amount of hydrogen feed required is reduced. Several different cases were screened, and Table 2 shows the results of the best case, where significant reductions in steam, fuel gas, and hydrogen feed costs were obtained. However, these gains were more than offset by the increased

Table 2: Economic Impact of Membrane Separation Unit on Process Economics (Permeate available at 10 bar and 85% H$_2$ purity, all *EAOC* cost figures in millions)

Cost Item	*EAOC* for Base Case	*EAOC* with Membrane Separator
Equipment	$ 0.727	$ 1.370
Steam	$ 0.943	$ 0.780
Fuel Gas	$ 0.566	$ 0.456
Cooling Water	$ 0.075	$ 0.065
Electricity	$ 0.038	$ 0.320
Toluene	$19.973	$19.973
Hydrogen	$ 1.605	$ 1.338
Total	**$23.927**	**$24.302**
Fuel Gas (Credit)	($ 3.150)	($ 2.632)

Break-even price of benzene with membrane separator = $0.434 per kg   (toluene = $0.23 per kg)

Break-even costs include utilities, raw materials, maintenance, labor, fixed capital investment, etc.

Price Differential = (0.434 − 0.230) = $0.204 per kg

cost of electricity (for the compressor), the decrease in fuel gas credit, and the increase in equipment costs due to the larger compressor and the addition of the membrane separation unit. The net result was that the membrane separation unit provided no economic advantage. Consequently, this topological change is not recommended.

3.2. Heat Integration

The second topological change to the PFD was the addition of heat integration around the reactor, where the benefit would be greatest. Exchanging heat between the reactor effluent, Stream 9, and the high-pressure steam, Stream 4, can significantly reduce cooling water and fuel gas utilities. In addition, the cost of the front-end heat-exchange

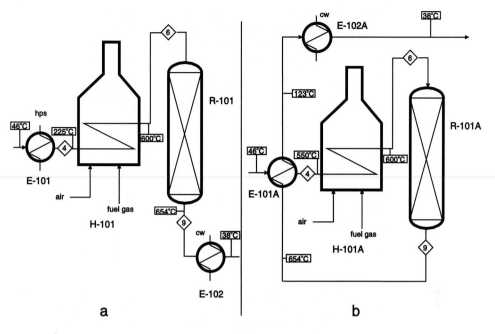

Figure 2 Front End of Toluene HDA Process (a) without Heat Integration and (b) with Heat Integration

equipment (E-101, E-102, and H-101) might also be reduced. Figure 2a shows the base-case configuration, and Figure 2b shows the optimized heat-exchange configuration around the reactor. Table 3 lists the savings in equipment and utility costs. The use of heat integration eliminates the high-pressure steam usage and reduces significantly the cost and utility demands of the fired heater, yielding a significant economic improvement by reducing the break-even price of benzene to \$0.384 per kg (break-even price differential of \$0.154 per kg).

4.   Parametric Changes to Base Case Operation

4.1.   Reflux Ratio

Table 1 shows that the two largest utility costs are steam and fuel gas. The heat-integration scheme outlined above can reduce these costs significantly. The next largest steam user after E-101 is E-106, the re-

Table 3: Economic Impact of Front-End Heat Integration on Process Economics (All *EAOC* cost figures in millions)

Cost Item	*EAOC* Base Case	*EAOC* with Heat Integration
Equipment	$ 0.727	$ 0.718
Steam	$ 0.943	$ 0.269
Fuel Gas	$ 0.566	$ 0.085
Cooling Water	$ 0.075	$ 0.023
Electricity	$ 0.038	$ 0.038
Toluene	$19.973	$19.973
Hydrogen	$ 1.605	$ 1.605
Total	**$23.927**	**$22.711**
Fuel Gas (Credit)	($ 3.150)	($ 3.150)

Break-even price of benzene with heat integration = $0.384 per kg (toluene = $0.23 per kg)

Break-even costs include utilities, raw materials, maintenance, labor, fixed capital investment, etc.

Price Differential = (0.384 − 0.230) = $0.154 per kg

boiler of T-101. The optimum reflux ratio for column T-101 is 1.12 times the minimum (Figure 3), significantly different from the 1.5 of the base case. For this calculation, it is assumed that costs other than the EAOC of the column, reboiler, and condenser are substantially unaffected by changes in column operation. The costs for the optimum reflux are compared to those of the base-case operation ($R/R_{min} = 1.5$) in Table 4. The overall effect is a small decrease in the break-even price differential of $ 0.0002 per kg.

4.2. Conversion

The final optimization attempted involved the single-pass conversion in the reactor (base case conditions, $T = 600°C$, $P = 25$ bar, conver-

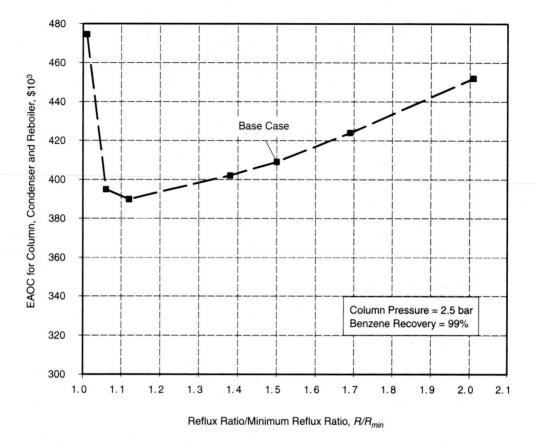

Figure 3 The Effect of Reflux Ratio on Profitability of Hydrodealkylation Process

sion = 0.75). The rationale for changing the conversion was that potential savings could be obtained by reducing the amount of toluene recycle; i.e., the size of equipment and utility usage in the recycle loop could
be reduced. The results of this optimization are shown in Figure 4,
where the break-even price for benzene is plotted as a function of single-
pass conversion. The optimum conversion is seen to occur at about 85%
with a break-even price for benzene of $0.383 per kg. The results for the
optimum conversion are compared to the base case in Table 5. The increase in break-even price for conversions greater than 85% is attributed to an increase in the amount of benzene leaving in the fuel gas.

Table 4: Economic Impact of Column Optimization on Process Economics (All _EAOC_ cost figures in thousands)

Cost Item	EAOC for Base Case R/R_{min} = 1.50	EAOC for Optimized Case R/R_{min} = 1.12
Column Equipment (T-101, E-102, E-106, V-102, P-102A/B)	$ 132.11	$ 148.03
Cooling Water	$ 11.04	$ 9.32
Steam	$ 275.90	$ 239.04
Total	**$ 418.05**	**$ 396.39**

Break-even price of benzene with column optimization = $0.4168 per kg   (toluene = $0.23 per kg)

Break-even costs include utilities, raw materials, maintenance, labor, fixed capital investment, etc.

Price Differential = (0.4168 − 0.230) = $0.1868 per kg

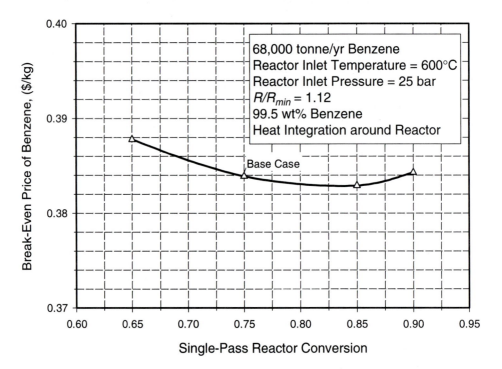

Figure 4 The Effect of Single-Pass Conversion on Break-Even Selling Price of Benzene

**Table 5: Economic Impact of Single-Pass Reactor
Conversion (plus heat integration) on Process
Economics (all _EAOC_ cost figures in millions)**

Cost Item	EAOC Base Case with Heat Integration Conversion = 0.75	EAOC Optimized Case with Heat Integration Conversion = 0.85
Equipment	$ 0.718	$ 0.675
Steam$ 0.269	$ 0.243	
Fuel Gas	$ 0.085	$ 0.068
Cooling Water	$ 0.023	$ 0.020
Electricity	$ 0.038	$ 0.035
Toluene	$ 19.973	$ 20.065
Hydrogen	$ 1.605	$ 1.605
Total	**$ 22.711**	**$ 22.711**
Fuel Gas (Credit)	($ 3.150)	($ 3.161)

**Break-even price of benzene with conversion = 0.85
= $0.383 per kg   (toluene = $0.23 per kg)**

Break-even costs include utilities, raw materials,
maintenance, labor, fixed capital investment, etc.

Price Differential = (0.383 − 0.230) = $0.153 per kg

The recovery of this "lost" benzene was not considered in the present analysis but is addressed below in the recommendations section.

5. Discussion

The results of the present study are summarized in Figure 5, where the results of the different case studies are compared. The minimum break-even price for benzene using the catalytic hydrodealkylation of toluene is $0.383 per kg. The recommended process uses significant heat integration around the reactor with a single-pass conversion in the reactor of 85%. For all cases considered, the ratio of hydrogen to toluene entering the reactor was maintained at 5.1:1 in order to suppress carbon formation. Finally, a reflux ratio of 1.12 times the minimum value was

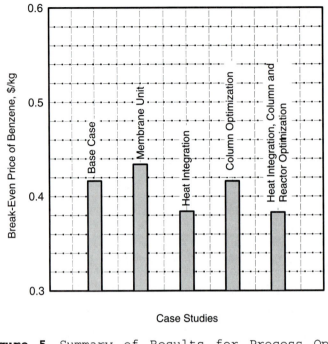

Figure 5 Summary of Results for Process Optimization of Toluene HDA Process

used in the benzene tower which was determined to be the optimum. The use of a membrane separation unit to purify the recycle hydrogen stream was not found to be economically attractive. The proposed process represents a significant improvement over the base case.

6. Conclusions

The optimum break-even price for benzene using this technology is estimated to be $0.383 per kg. With toluene priced at $0.23 per kg, this gives a break-even price differential of $ 0.153 per kg. Significant improvements from the base case were made, including heat integration in the front end, column optimization, and increasing the single-pass reactor conversion. The addition of a membrane separation unit to purify the recycle gas stream was found not to be profitable. In summary, the production of benzene from the hydrodealkylation of toluene is not profitable at current market conditions. A significant increase in the price differential between benzene and toluene (>$ 0.11 per kg) must occur before this process becomes economically feasible.

7.   Recommendations for Future Work

The loss of benzene to the fuel gas for the improved process presented above represents approximately $640,000 per year in extra raw material (toluene) costs. Although appropriate fuel credit was given in this study, it is recommended that a further study be carried out to investigate methods for recovering this lost benzene. An example of one such method is the use of a pre-fractionator prior to T-101, instead of the two flash separations, to obtain a sharper separation between the non-condensables and benzene and toluene. The maximum potential benefit of this recovery is a reduction in the break-even cost of benzene of $0.012 per kg; hence, this option should be considered.

The base-case operating conditions (temperature and pressure) of the reactor were used throughout this study. It is not expected that significant savings can be realized by altering these conditions since the reactor cost has very little impact on the overall break-even price of benzene.

8.   References

1. Weiss, A. H., and L. Friedman, "Development of Houdry Detol Process,", *Ind. Eng. Chem. Proc. Des. & Dev.*, 2, 163 (1963).
2. Tarhan, M. O., *Catalytic Reactor Design*, McGraw-Hill, New York, NY, 1983, p. 111.
3. Walas, S. M., *Chemical Process Equipment: Selection and Design*, Butterworth, Stoneham, MA, 1988, p. 29.

9. Appendix

23.5 CHECKLIST OF COMMON MISTAKES AND ERRORS

23.5.1 Common Mistakes for Visual Aids

1. When including columns of data in tables, the sum of the columns should be included and doubled-checked for correctness.
2. If numbers are included in a table but are not to be added, then care should be taken not to list these numbers in an unbroken vertical column. The natural tendency is for the reader to add the numbers, which may be inappropriate.
3. Either the number of decimal places or number of significant figures should be the same for all numbers appearing in a table (or report).
4. Place figure numbers and captions below the figure and table numbers and titles above the table.
5. A note regarding the units of the numbers appearing in a table should be included, for example, all numbers are in $ millions, or similar notation.
6. Pie diagrams should include the total value of the pie, for example, the total fixed capital investment is $500,000.

7. Avoid the use of redundant graphics. For example, a pie diagram would be redundant if all the same information were included in a table.

8. When presenting comparisons between different cases in the form of multiple tables, make sure that the order of items appearing in these tables is the same for all tables. If the order is changed, then comparisons are made very difficult.

9. When plotting data in the form of a figure, make sure that enough data are plotted. An example of insufficient data is given in Figure 23.8, where the optimum R/R_{min} value is almost certainly not at 1.50 as shown. Figure 23.9 shows a figure with an appropriate number of data points.

10. All figure and table numbers should be followed by a meaningful caption. Do not simply repeat the axis titles in the caption, for example, a plot of x vs. y. An additional caption describing the figure, separate from the one with the figure number, should not be included.

11. Never use line graphs. Line graphs are graphs that use an arbitrary x-axis scale having equal spacing between consecutive data points. Figure 23.8 illustrates this type of graph. These graphs are very difficult to interpret and often misleading.

12. Remember to place all landscape-oriented pages facing outwards—rotate 90° counter-clockwise.

13. Line up decimal points in columns of tables.

14. The same size font should be used for axis labels and axis titles. The font size should also be the same for all of the figure and table titles and be the same as that in the main text of the report.

15. If case studies are used in the report, the identification should be consistent throughout the report. For example, in a report with many different case studies, Case 2 should not be referred to as Case B in a table and Case II in a figure.

23.5.2   Common Mistakes for Written Text

Memoranda (Memos)

1. The list of people to copy in a reply memo should be the same as that used in the initiation memo.

2. Be careful to use the correct descriptive titles for attachments and memo subject. For example, for the cover memorandum in Figure 23.3, the subject should read "Evaluation of the Minimum Breakeven Price Differential for Benzene and Toluene" not "Benzene Production" as shown.

3. The significant results of the study or report should be briefly summarized in the memorandum. This enables the person reading it to quickly ascertain the major findings and prioritize the reading of the report (see Section 23.1).

Main Body of Written Report

1. An alternative to using first person narrative is the passive voice. Some authors claim (insist) that first person (In this report, *I* present *my* findings of a study on ...) is often clearer and more concise than the passive voice (In this report *the* findings of a study on ... *are* presented). However, for the novice, a report written in the first person often sounds (reads) unprofessional(ly), and it is safer to stick with the passive voice.

2. When writing chemical reactions, name all ambiguous chemicals in the reactions. For example, the reaction of toluene to yield benzene should be written as

$$C_7H_8 + H_2 \rightarrow C_6H_6 + CH_4$$
$$\quad\text{\textit{toluene}} \qquad\qquad \text{\textit{benzene}}$$

3. Details of calculation methods and software used should not appear in the main body of the report but rather in the appendix. Software may be cited if specific information is used, for example, if the second virial coefficient for methylene chloride was obtained from the CHEMCAD™ [5] databank.

4. Equation numbers should be included in () and be right justified, level with the equation. For example

$$COM_d = 0.180FCI + 2.73C_{OL} + 1.23(C_{UT} + C_{RM}) \tag{1}$$

In addition, the terms in the equation should always be defined either directly after the equation (which is preferred for written reports) or in a separate notation section at the back of the report (for books and technical papers).

5. The word data is plural: The optimization data *are* ...

6. New results should not be included in the Conclusions section of a report. Such information should have already been included in a separate Discussion section.

7. Try to make specific recommendations that can be quantified. Avoid stating the obvious, for example, find cheaper utilities or raw materials. These statements do not improve the report writer's credibility with the reader. Unless one has specific ideas in mind, this type of wishful thinking is detrimental to the credibility of the report.

8. In the References section, only references cited in the report should be included.

A

Cost Equations and Curves for the CAPCOST Program

The purpose of this appendix is to present the equations and figures that describe the relationships used in the capital equipment costing program CAPCOST© introduced in Chapter 2 and used throughout the text. The program uses a technique originally introduced by Guthrie [1] and modified by Ulrich [2]. The data from which the equations and the figures are generated are modified from Ulrich [2] (reproduced from Ulrich, G. D., *A Guide to Chemical Engineering Process Design and Economics*, Wiley, New York 1984. Copyright © 1984, by John Wiley & Sons, Inc., reprinted by permission of John Wiley & Sons, Inc). We start with a review in which we present the cost equations in a standardized way. The purchased cost of the equipment assuming ambient operating pressure and carbon steel construction, C_p, has the following form:

$$\log_{10} C_p = K_1 + K_2 \log_{10} A + K_3 (\log_{10} A)^2 \tag{A.1}$$

where A is the capacity or size parameter for the equipment. The pressure factor, F_p, for the equipment has the general form:

$$\log_{10} F_p = C_1 + C_2 \log_{10} P + C_3 (\log_{10} P)^2 \tag{A.2}$$

The units of pressure, P, are bar gauge or barg (1 bar = 0.0 barg). The pressure factors are always greater than unity, and should not be used outside the range of pressures given in the tables and figures.

The material factors, F_M, are dependent on the material of construction for a given piece of equipment and are listed in the appropriate section. Finally, the bare module cost, C_{BM}^o, and the bare module factor for the equipment, F_{BM}^o, are a function of the product of the material and pressure factor, thus:

$$C_{BM}^o = C_p F_{BM}^o = C_p(B_1 + B_2 F_M F_P) \tag{A.3}$$

Due to the diversity of the information, it is not always possible to correlate the data with the above equations. For some equipment, the bare module factor is presented in a different form, and some other minor deviations from the standard equations (A.1–A.3) are also used. When this is the case, the correlation to yield C_{BM}^o, or any other parameters, is given separately, directly following the table of factors for the piece of equipment.

The factors used in the above equations are given for nine different classes of equipment in the following sections. All the data are given for mid-1996, for which the CEPCI = 382. The data for all equipment classes are also plotted in Figures A.1–A.13 given at the end of this appendix.

A.1 HEAT EXCHANGERS

Capacity Parameter = Heat Transfer Area, A (m^2)

Table A.1 Correlation Coefficients for Heat Exchangers

Exchanger Type	K_1	K_2	K_3	C_1	C_2	C_3	B_1	B_2	A_{min} (m²)	A_{max} (m²)	P_{max} (barg)
Double Pipe	3.0238	0.0603	0	6.4945[1]	-6.6786	1.7442	0.74	1.21	0.2	10	300
Multiple Pipe	2.1138	0.9658	0	6.4945[1]	-6.6786	1.7442	0.74	1.21	10	75	300
Fixed Tube Sheet or U-Tube	3.2138	0.2688	0.07961	-0.06499[2]	0.05025	0.01474	1.80	1.50	4	900	140
Floating Head	3.4338	0.1445	0.10790	-0.06499[2]	0.05025	0.01474	1.80	1.50	10	900	140
Bayonet	3.5238	0.1916	0.09474	-0.06499[2]	0.05025	0.01474	1.80	1.50	10	900	140
Kettle Reboiler	3.5638	0.1906	0.11070	-0.06499[2]	0.05025	0.01474	1.80	1.50	10	100	140
Scraped Wall	3.7438	0.9270	0	6.4945[1]	-6.6786	1.7442	0.74	1.21	2	20	300
Teflon Tube	3.5738	0.4548	0	0	0	0	1.80	1.50	7	75	15
Air Cooler	3.6418	0.4053	0	-0.06154	0.0473	0	1.53	1.27	3.5	20,000	250
Spiral Tube	3.4088	0.6000	0.09944	-0.4045[3]	0.1859	0	0.74	1.21	.1	45	400
Spiral Plate	3.6788	0.4412	0	0	0	0	1.53	1.27	2	200	19
Flat Plate	3.8528	0.4242	0	0	0	0	1.53	1.27	15	1,500	19

[1] Pressure factors given are for $100 < P < 300$ barg, for $40 < P < 100$ use $C_1 = 0.6209$, $C_2 = -0.9274$, $C_3 = 0.3369$, for $P < 40$ $C_1 = C_2 = C_3 = 0$

[2] Pressure factors given are for when shell or both shell and tube are > 10 barg, when tubes only >10 barg use $C_1 = -0.04139$, $C_2 = 0.04139$, $C_3 = 0$

[3] Pressure factors given are for when shell or both shell and tube are > 10 barg, when tubes only >10 barg use $C_1 = -0.21150$, $C_2 = 0.09717$, $C_3 = 0$

Table A.2   Material Factors for Heat Exchangers

Exchanger Type	Shell—CS Tube—CS	CS Cu	Cu Cu	CS SS	SS SS	CS Ni	Ni Ni	CS Ti	Ti Ti
				Material Factor, F_M					
Double Pipe	1.00	1.25	1.60	1.70	3.00	2.80	3.80	7.20	12.00
Multiple Pipe	1.00	1.25	1.60	1.70	3.00	2.80	3.80	7.20	12.00
Fixed Tube Sheet or U-Tube	1.00	1.25	1.60	1.70	3.00	2.80	3.80	7.20	12.00
Floating Head	1.00	1.25	1.60	1.70	3.00	2.80	3.80	7.20	12.00
Bayonet	1.00	1.25	1.60	1.70	3.00	2.80	3.80	7.20	12.00
Kettle Reboiler	1.00	1.25	1.60	1.70	3.00	2.80	3.80	7.20	12.00
Scraped Wall	1.00	1.25	1.60	1.70	3.00	2.80	3.80	7.20	12.00
Spiral Tube	1.00	1.25	1.60	2.30	3.00	2.80	3.80	7.20	12.00

Teflon Tube Exchanger		Flat and Spiral Plate		Air Cooler	
Shell Material	F_M	Material in Contact with Process Fluid	F_M	Tube Material	F_M
CS	1.00	CS	1.00	CS	1.00
Cu	1.20	Cu	1.20	Al	1.50
SS	1.30	SS	2.30	SS	3.00
Ni	1.40	Ni	2.80	—	—
Ti	3.30	Ti	7.20	—	—

A.2   PROCESS VESSELS AND INTERNALS

A.2.1   Horizontal and Vertical Process Vessels

Capacity Parameter = Height or Length of Vessel, L (m)

The pressure factors for both horizontal and vertical vessels are given by the following correlation:

$$F_p = 0.5146 + 0.6838\log_{10}P + 0.2970(\log_{10}P)^2$$
$$+ 0.0235(\log_{10}P)^6 + 0.0020(\log_{10}P)^8 \qquad 3.7 < P < 400 \text{ barg}$$

$$F_p = 1.00 \qquad -0.5 < P < 3.7 \quad \text{barg}$$

$$F_p = 1.25 \qquad P < -0.5 \quad \text{barg}$$

Table A.3 Correlation Coefficients for Process Vessels

Diameter (m)	Vertical Orientation					Horizontal Orientation					Horizontal and Vertical
	K_1	K_2	K_3	L_{min} (m)	L_{max} (m)	K_1	K_2	K_3	L_{min} (m)	L_{max} (m)	P_{max} (barg)
0.3	3.3392	0.5538	0.2851	1.2	16	2.9202	0.5056	0.1261	1.0	20	400
0.5	3.4746	0.5893	0.2053	1.5	20	3.1032	0.5782	0.0632	1.5	25	400
1.0	3.6237	0.5262	0.2146	2.5	30	3.3592	0.5905	0.1106	2.2	30	400
1.5	3.7559	0.6361	0.1069	3.0	41	3.4204	0.8141	-0.0046	3.5	36	400
2.0	3.9484	0.4623	0.1717	4.0	45	3.7599	0.3683	0.1954	4.5	40	400
2.5	4.0547	0.4620	0.1558	5.0	50	3.6780	0.7120	0.0430	5.5	42	400
3.0	4.1110	0.6094	0.0490	6.0	50	3.7718	0.7159	0.0470	6.5	50	400
4.0	4.3919	0.2859	0.1842	7.0	50	4.1551	0.2238	0.2499	8.0	52	400

Table A.4   Material Factors and Coefficients for Eq. (A.3) for Process Vessels

	B_1	B_2
Horizontal	1.62	1.47
Vertical	2.50	1.72

Horizontal and Vertical	
Material of Construction	F_M
CS	1.0
SS clad	2.5
SS	4.0
Ni clad	4.5
Ni	9.8
Ti clad	4.9
Ti	10.6

A.2.2 Process Vessel Internals

Sieve Trays and Demister Pads.   Capacity Parameter = Diameter of Vessel, D (m)

$$C_P = 235 + 19.80D + 75.07D^2$$

$$C_{BM}^\circ = C_P N F_{BM} F_q$$

where
N = number of trays
F_q = quantity factor based on the number of trays in the vessel

Table A.5   Material and Quantity Factors for Sieve Trays

Number of Trays	F_q
1	3.0
4	2.5
7	2.0
10	1.5
>20	1.0

	F_{BM}	
Material of Construction	*Trays*	*Demister Pad*
CS	1.2	-
SS	2.0	1.2
Fluorocarbon	-	2.0
Ni-alloy	5.0	4.2

Tower Packings.   Capacity Parameter = Height of Packing, L (m)

Table A.6   Correlation Coefficients and Material Factors for Tower Packings

Diameter (m)	K_1	K_2	K_3	L_{min} (m)	L_{max} (m)
0.3	2.1630	0.9656	0	1.2	16
0.5	2.5210	0.9764	0	1.5	20
1.0	3.0169	1.0000	0	2.5	30
1.5	3.2160	0.9847	0	3.0	41
2.0	3.3848	0.9808	0	4.0	45
2.5	3.6023	0.9682	0	5.0	50
3.0	3.7921	0.9697	0	6.0	50
4.0	3.9986	0.9833	0	7.0	50

Material of Construction	F_{BM}
CS	1.2
Polyethylene	1.0
Porcelain	1.0
304 SS	2.2
316 SS	4.2

$$C_{BM}^o = C_p F_{BM}$$

A.3   PUMPS WITH ELECTRIC DRIVES

Capacity Parameter = Shaft Power, W_s (kW)

Table A.7   Correlation Coefficients for Pumps with Electric Drives

Pump Type	K_1	K_2	K_3	$W_{s,min}$ (kW)	$W_{s,max}$ (kW)	P_{max} (barg)
Reciprocating	3.9412	0.4170	0.09141	0.01	280	1200
Rotary Positive Displacement	3.6949	0.3590	0.05577	0.01	140	350
Centrifugal	3.5793	0.3208	0.02850	0.01	250	350

Pump Type	C_1	C_2	C_3	B_1	B_2
Reciprocating	0.3120	0.6320	0.0560	1.80	1.51
Rotary Positive Displacement	0.0231	0.7154	0.2615	1.80	1.51
Centrifugal	0.1682	0.3477	0.4841	1.80	1.51

$$F_p = C_1 + C_2 \log_{10} P + C_3 (\log_{10} P)^2$$

Table A.8   Material Factors for Pumps with Electric Drives

	Material Factors, F_M					
Pump Type	*Cast Iron*	*Cast Steel*	*Cu Alloy*	*SS*	*Ni Alloy*	*Ti*
Reciprocating	1.0	1.4	1.3	1.9	3.5	5.7
Rotary Positive Displacement	1.0	1.4	1.3	2.0	4.0	9.0
Centrifugal	1.0	1.8	—	2.4	5.0	—

A.4 FANS WITH ELECTRIC DRIVES

Capacity Parameter = Gas Flowrate, $\dot{v}$ (std m³/s)

Table A.9   Correlation Coefficients for Fans with Electric Drives

Type of Fan	K_1	K_2	K_3	$\dot{v}_{min}$ (std m³/s)	$\dot{v}_{max}$ (std m³/s)	P_{max} (barg)
Centrifugal Radial Fan	3.3391	0.2140	0.2297	1	250	1.0
Centrifugal Backward Curved Fan	3.1948	0.2280	0.2040	1	450	1.0
Axial Tube Fan	2.6806	0.4837	0.1333	1	250	1.0
Axial Vane Fan	2.9471	0.3302	0.1969	1	150	1.0

Table A.10   Pressure Factors for Fans with Electric Drives

Type of Fan	Pressure Factors, F_P [based on ΔP of fan (kPa)]				
	ΔP =1 kPa	2 kPa	4 kPa	8 kPa	16 kPa
Centrifugal Radial Fan	1.00	1.15	1.30	1.45	1.60
Centrifugal Backward Curved Fan	1.00	1.15	1.30	1.45	—
Axial Tube Fan	1.00	—	—	—	—
Axial Vane Fan	1.00	1.15	1.30	—	—

Figure A.11   Bare Module Factors for Fans with Electric Drives

	Bare Module Factor, F_{BM}			
Type of Fan	**CS**	**Fiberglass**	**SS**	**Ni alloy**
Centrifugal Radial Fan	2.2	4.0	5.5	11.0
Centrifugal Backward Curved Fan	2.2	4.0	5.5	11.0
Axial Tube Fan	2.2	4.0	5.5	11.0
Axial Vane Fan	2.2	4.0	5.5	11.0

$$C_{BM}^o = C_p F_p F_{BM}$$

A.5   COMPRESSORS AND BLOWERS WITHOUT DRIVES

Capacity Parameter = Fluid Power, W_f (kW)

Table A.12   Correlation Coefficients for Compressors and Blowers

Compressor Type	K_1	K_2	K_3	$W_{f,min}$ (kW)	$W_{f,max}$ (kW)
Centrifugal	2.9945	0.9542	0	50	8000
Axial	2.9945	0.9542	0	50	8000
Rotary	3.5116	0.6009	0	50	1100
Reciprocating	2.9945	0.9542	0	50	8000

Table A.13   Bare Module Factors for Compressors and Blowers

	Bare Module Factor, F_{BM}		
Compressor Type	**CS**	**SS**	**Ni alloy**
Centrifugal	2.5	6.3	13.0
Axial	3.5	8.8	18.0
Rotary	2.2	5.5	11.0
Reciprocating	2.9	7.3	15.0

$$C_{BM}^o = C_p F_{BM}$$

A.6   DRIVES FOR COMPRESSORS AND BLOWERS

Capacity Parameter = Shaft Power, W_s (kW)

Table A.14   Correlation Coefficients for Compressor and Blower Drives

Type of Drive	K_1	K_2	K_3	$W_{s,min}$ (kW)	$W_{s,max}$ (kW)	F_{BM}
Electric—Explosion Proof	2.3006	1.0947	−0.10160	3	6,000	1.5
Electric—Totally Enclosed	2.1774	1.0351	−0.08443	3	6,000	1.5
Electric—Open/Drip Proof	2.1206	0.9545	−0.06614	3	6,000	1.5
Gas Turbine	3.4171	0.6112	0	10	15,000	3.5
Steam Turbine	3.7222	0.4401	0	100	15,000	3.5
Internal Combustion Engine	2.6693	0.8074	0	7	15,000	2.0

$$C^o_{BM} = C_p F_{BM}$$

A.7   POWER RECOVERY EQUIPMENT

Capacity Parameter = Shaft Power, W_s (kW)

Table A.15   Correlation Coefficients for Power Recovery Equipment

Type of Machine	K_1	K_2	K_3	$W_{s,min}$ (kW)	$W_{s,max}$ (kW)	Bare Module Factor, F_{BM} CS	SS	Ni alloy
Axial Gas Turbine	3.5137	0.5888	0	100	4,000	3.5	6.0	8.0
Radial Gas or Liquid Expander	3.1143	0.6923	0	100	1,500	3.0	5.0	6.0

$$C^o_{BM} = C_p F_{BM}$$

A.8 FIRED HEATERS AND FURNACES

Capacity Parameter = Heat Duty, Q (kW)

Table A.16 Correlation Coefficients for Fired Heaters and Furnaces

Type of Heater	K_1	K_2	K_3	Q_{min} (kW)	Q_{max} (kW)	P_{max} (barg)
Reactive Process Heater— Reformer Furnace	2.6379	0.8179	0	3,000	150,000	200
Reactive Process Heater— Pyrolysis Furnace	2.5689	0.8067	0	3,000	150,000	200
Non-Reactive Process Heater	2.5526	0.7962	0	3,000	150,000	200
Thermal Fluid Heater— Hot Water	3.7038	0.1830	0.07876	100	20,000	—
Thermal Fluid Heater— Molten Salt, Mineral Oil and Silicon Fluids	3.8391	0.2201	0.06490	100	20,000	—
Thermal Fluid Heater— Diphenyl Based Oils	3.4549	0.5185	0.02081	100	20,000	—
Packaged Steam Boilers	2.1903	0.7644	0	120	20,000	70

Table A.17 Pressure and Bare Module Factors for Fired Heaters and Furnaces

Type of Heater	Pressure Factor, F_P				Bare Module Factor, F_{BM}		
	$P = 10$ barg	50 barg	100 barg	200 barg	Tube Material —CS	Tube Material —Alloy Steel	Tube Material —SS
Reactive Process Heater—Reformer Furnace	1.00	1.05	1.15	1.30	2.1	2.5	2.7
Reactive Process Heater—Pyrolysis Furnace	1.00	1.05	1.15	1.30	2.1	2.5	2.7
Non-Reactive Process Heater	1.00	1.10	1.25	1.40	2.1	2.5	2.7

$$C_{BM}^o = C_p F_{BM} F_P$$

Type of Heater	Pressure Factor, F_P			Steam Superheat Factor, F_{SH}			
	$P = 20$ barg	30 barg	40 barg	0°C (saturated)	50°C	100°C	150°C
Packaged Steam Boiler	1.00	1.25	1.70	1.00	1.10	1.15	1.20

$$C_{BM}^o = C_p F_P F_{SH} F_{BM} \quad (F_{BM} = 1.8)$$

Type of Heater	F_{BM}
Thermal Fluid Heater – Hot Water	2.2
Thermal Fluid Heater – Molten Salt, Mineral Oil and Silicon Fluids	2.2
Thermal Fluid Heater – Diphenyl Based Oils	2.2

$$C_{BM}^o = C_p F_{BM}$$

A.9 VAPORIZERS AND EVAPORATORS

Capacity Parameter = Total Vessel Volume, V (m^3) for Vaporizers
Capacity Parameter = Heat Transfer Area, A (m^2) for Evaporators

Table A.18   Correlation Coefficients for Vaporizers and Evaporators

Type of Equipment	K_1	K_2	K_3	A_{min} (m²)	A_{max} (m²)	V_{min} (m³)	V_{max} (m³)	P_{max} (barg)
Forced Circulation Evaporator	5.0775	0.69900	0	20	2,000	—	—	150
Falling Film Evaporator	4.5429	0.55440	0	30	300	—	—	150
Agitated (Scraped-Wall) Evaporator	4.5429	0.55440	0	2	20	—	—	150
Short Tube Evaporator	4.2470	0.58090	0	30	300	—	—	150
Long Tube Evaporator	4.3416	0.51610	0	100	10,000	—	—	150
Vaporizer— Jacketed Vessel	3.8620	0.20780	0.1901	—	—	1	100	320
Vaporizer— Internal Coil	4.0380	0.09142	0.2766	—	—	1	100	320
Vaporizer— Jacketed Vessel with Internal Coil	4.0380	0.09142	0.2766	—	—	1	100	320

Table A.19   Pressure Factors for Vaporizers and Evaporators

Type of Equipment	Pressure Factor, F_P													
	$P \leq 5$ barg	10 barg	15 barg	20 barg	40 barg	50 barg	80 barg	100 barg	150 barg	160 barg	320 barg
Forced Circulation Evaporator	1.00	1.00	1.05	1.10	1.17	1.20	1.26	1.30	1.5	—	—
Falling Film Evaporator	1.00	1.00	1.05	1.10	1.17	1.20	1.26	1.30	1.5	—	—
Agitated (Scraped-Wall) Evaporator	1.00	1.00	1.05	1.10	1.17	1.20	1.26	1.30	1.5	—	—
Short Tube Evaporator	1.00	1.00	1.05	1.10	1.17	1.20	1.26	1.30	1.5	—	—
Long Tube Evaporator	1.00	1.00	1.05	1.10	1.17	1.20	1.26	1.30	1.5	—	—
Vaporizer—Jacketed Vessel	1.00	1.40	1.60	2.00	3.00	—	4.30	—	—	6.50	13.00
Vaporizer—Internal Coil	1.00	1.40	1.60	2.00	3.00	—	4.30	—	—	6.50	13.00
Vaporizer—Jacketed Vessel with Internal Coil	1.00	1.40	1.60	2.00	3.00	—	4.30	—	—	6.50	13.00

Table A.20 Bare Module Factors for Vaporizers and Evaporators

Type of Equipment	Bare Module Factor, F_{BM}				
	CS	Cu Alloy	SS	Ni Alloy	Ti
Forced Circulation Evaporator	2.9	3.7	6.2	7.5	17.5
Falling Film Evaporator	2.3	3.0	5.6	6.9	17.0
Agitated (Scraped-Wall) Evaporator	2.3	3.0	5.6	6.9	17.0
Short Tube Evaporator	2.9	3.7	6.2	7.5	17.5
Long Tube Evaporator	2.9	3.7	6.2	7.5	17.5

Type of Equipment	Bare Module Factors, F_{BM}									
	CS	Cu	Glass Lined SS coils	Glass Lined Ni Coils	SS	SS Clad	Ni Alloy	Ni Alloy Clad	Ti	Ti Clad
Vaporizer— Jacketed Vessel	2.7	4.5	5.0[1]	5.0[1]	5.0	4.0	10.0	6.0	11.0	6.0
Vaporizer— Internal Coil	2.8	4.2	4.5	5.5	4.5	3.5	8.0	5.5	9.0	6.0
Vaporizer— Jacketed Vessel with Internal Coil	2.9	4.4	5.0	6.0	5.0	4.0	8.5	5.8	9.6	6.4

[1]Material factors given for glass lined vessel only

$$C^o_{BM} = C_p F_{BM} F_p$$

REFERENCES

1. Guthrie, K. M. (1969). "Data and Techniques for Preliminary Capital Cost Estimating," *Chem. Eng.*, 114–142, March 24.
2. Ulrich, G. D. (1984). *A Guide to Chemical Engineering Process Design and Economics*, Wiley, New York.

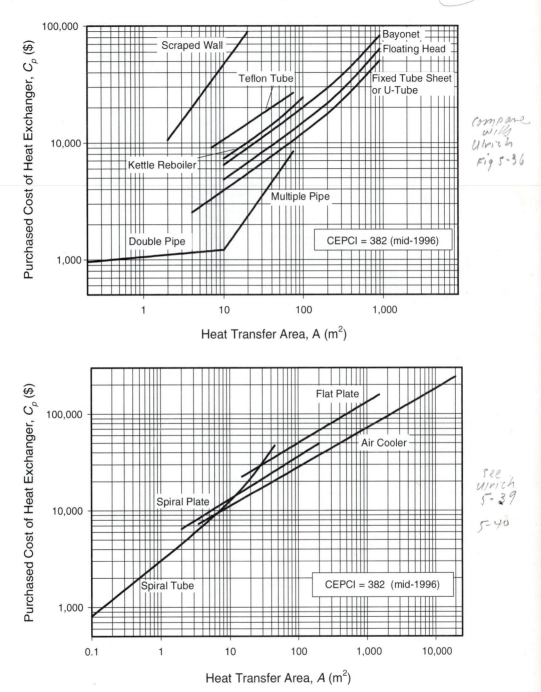

Figure A.1   Purchased Costs of Heat Exchangers

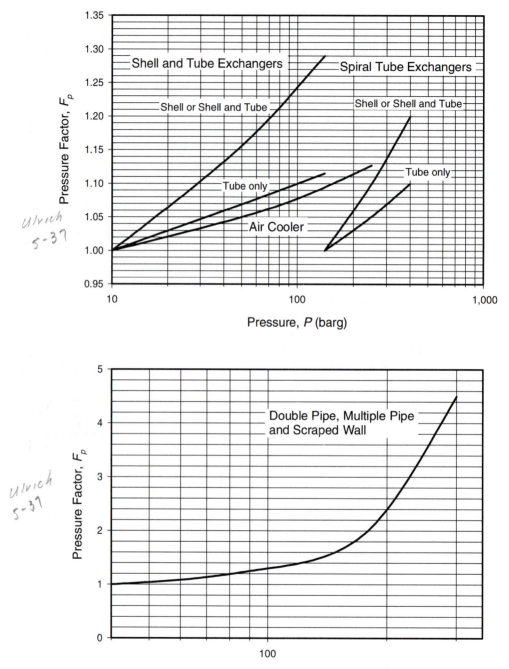

Figure A.2   Pressure Factors for Heat Exchangers

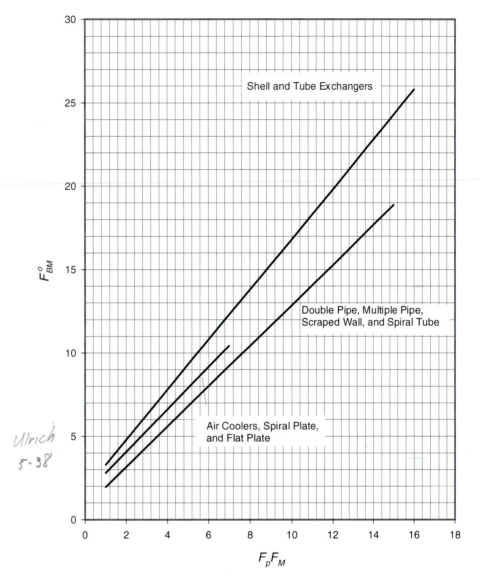

Figure A.3   Bare Module Factors for Heat Exchangers

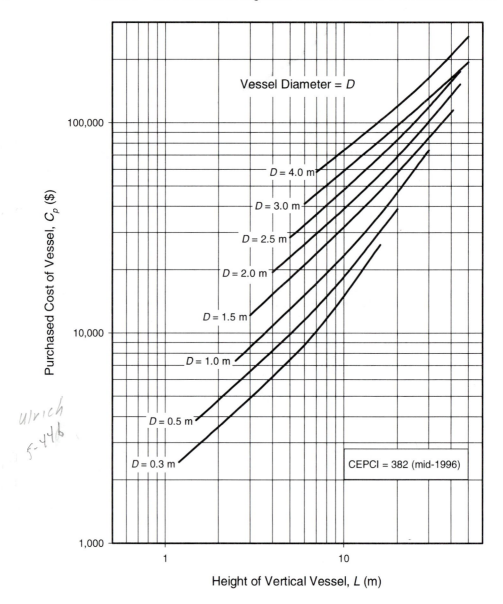

Figure A.4   Purchased Costs of Vertical Vessels

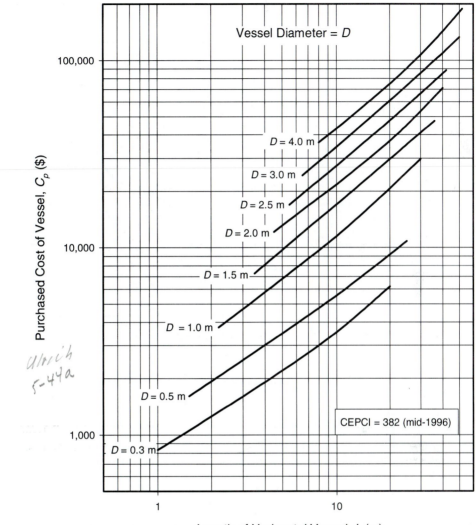

Figure A.5   Purchased Costs of Horizontal Vessels

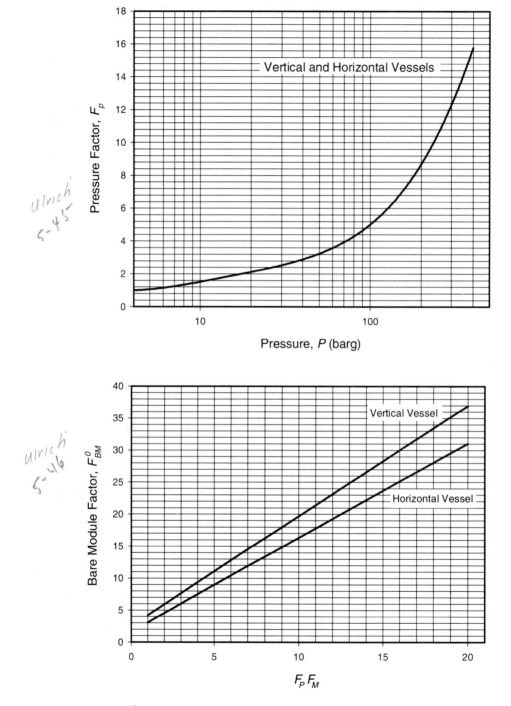

Figure A.6 Pressure Factors and Bare Module Factors for Process Vessels

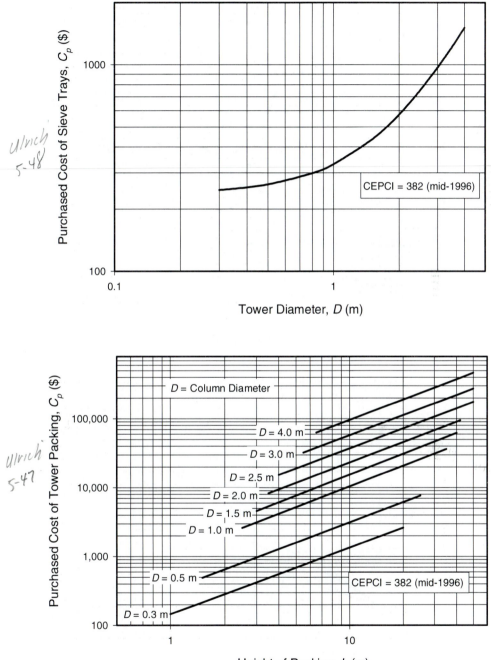

Ulrich 5-48

Ulrich 5-47

Figure A.7   Purchased Costs of Sieve Trays and Tower Packings

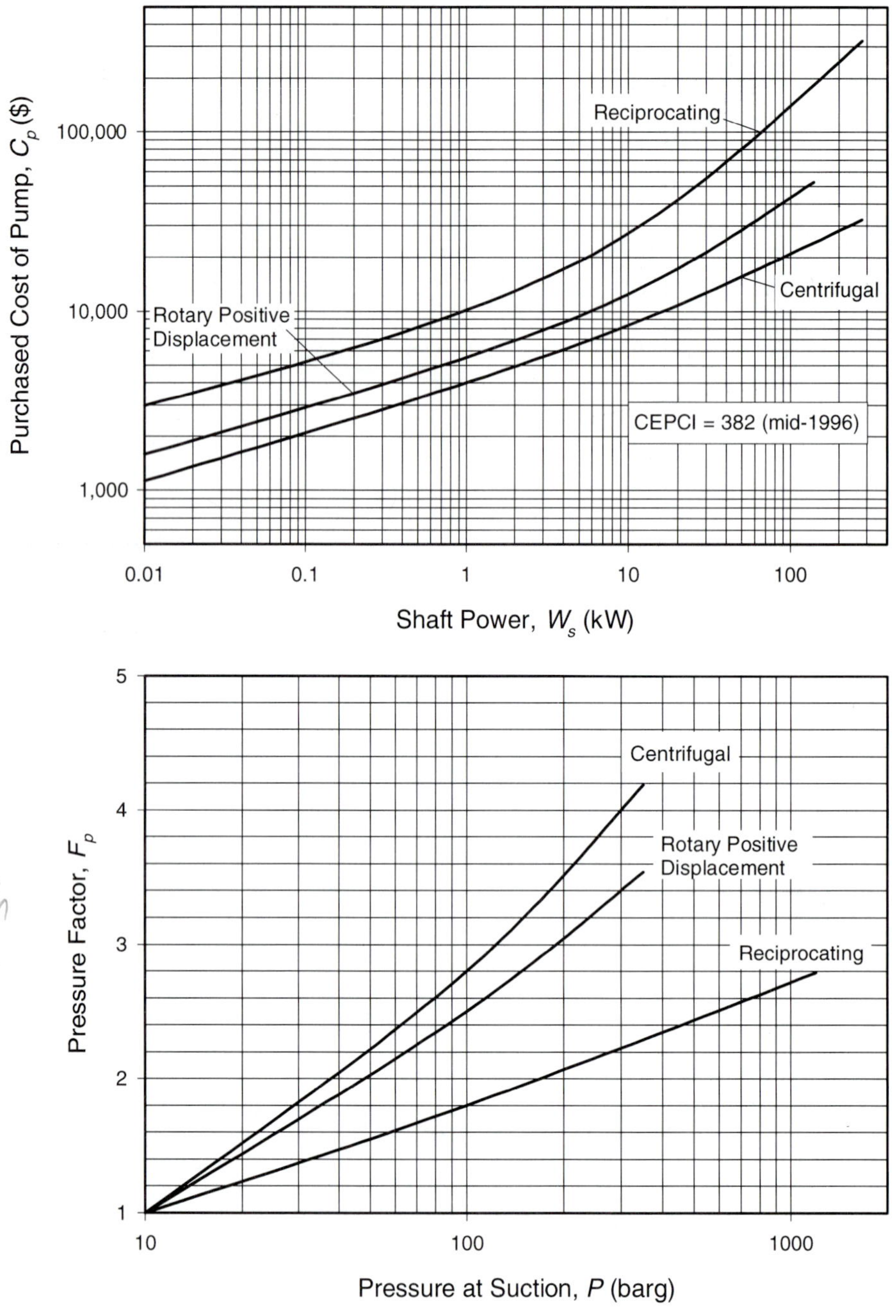

Figure A.8   Purchased Costs and Pressure Factors for Pumps with Electric Drives

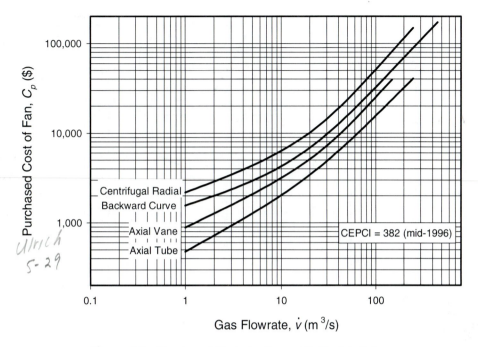

Figure A.9   Purchased Costs for Fans with Electric Drives

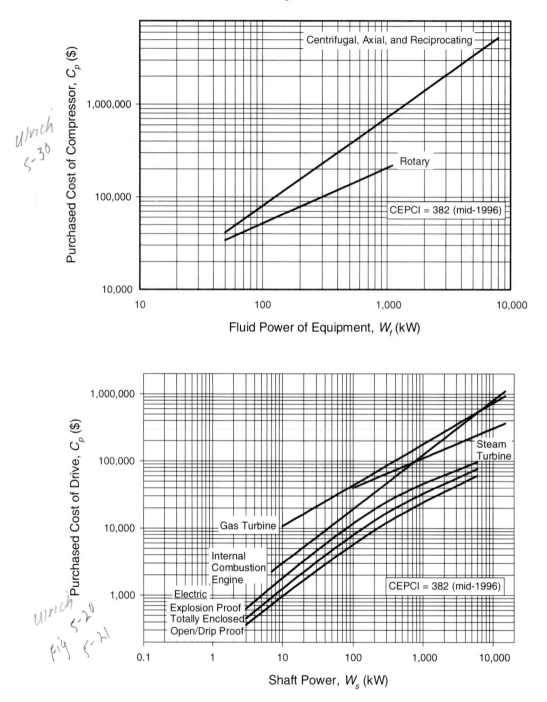

Figure A.10   Purchased Costs for Compressors and Drives

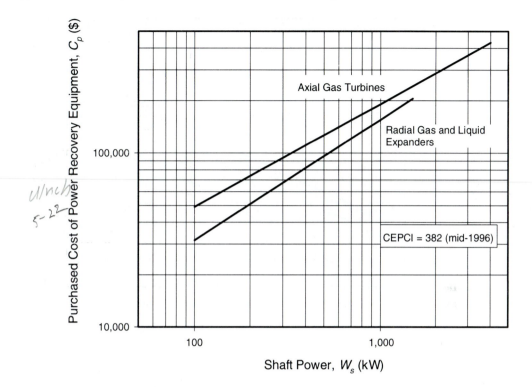

Figure A.11   Purchased Costs of Power Recovery Equipment

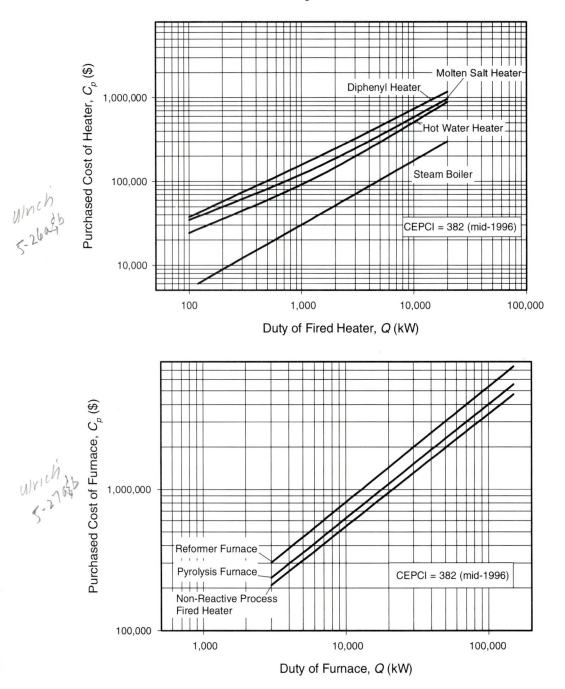

Figure A.12   Purchased Costs of Fired Heaters and Process Furnaces

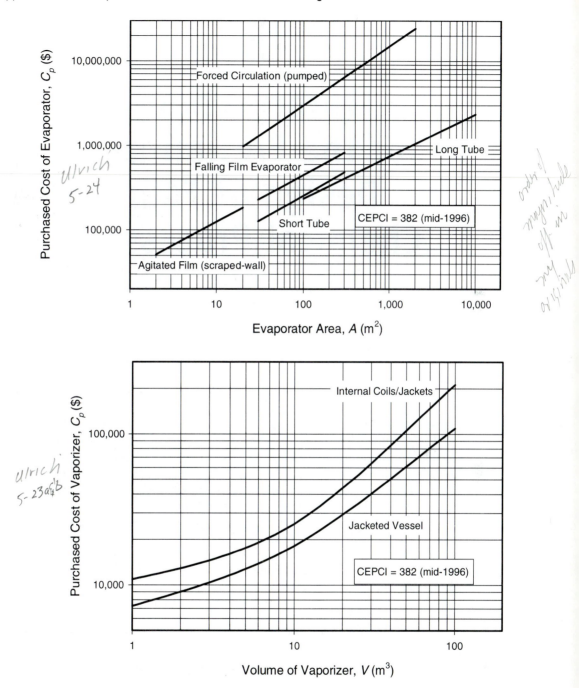

Figure A.13   Purchased Costs of Evaporators and Vaporizers

B

Information for the Preliminary Design of Four Chemical Processes

The processes listed in this appendix are used throughout the text to illustrate principles and to provide a body of data for examples and problems at the back of the chapters. The PFDs and corresponding heat and material balances are preliminary and are often obtained using thermodynamic packages available on a process simulator (note all flowsheets are simulated using CHEMCAD™ 3.1 or 3.2 software from ChemStations Inc., Houston, TX). In practice, this would allow reasonable economic evaluations of the processes, but would not be acceptable for the final design. In order for a process to be designed properly, it is imperative that the thermodynamic data used for the simulation be as accurate as possible. When the components are non-ideal (as would be the case when azeotropes or several liquid phases are present), then actual vapor-liquid or liquid-liquid equilibrium data, for the important binary pairs at the conditions used in the process, would be collected. From such data, liquid phase activity coefficients and/or binary interaction parameters would be calculated and used in the final simulations. In addition, the kinetic expressions given for the reactions are either approximate or fictitious. Thus, only approximate reaction rates and reactor sizes can be estimated from them.

It is also important to note that no optimization has been attempted on the processes in this section. Therefore, there is considerable room for improvement in all the processes. There are some "caricature" errors that have been included; these should stick out like a "sore thumb". One example is the use of a cooling water exchanger (E-301) to cool the circulating molten salt used for temperature control in the acrylic acid reactor in Figure B.2. This circulating stream is at a temperature of 250°C and must be cooled to 200°C, so this heat could be used to make medium pressure steam or to heat another process stream. Other aspects of the designs are more subtle and include sub-optimum reflux ratios in distillation columns and sub-optimal single-pass conversions in reactors. It is hoped that as you work through these flowsheets you will be able to improve significantly on the designs.

Finally, when making capital cost estimates for these processes, it may happen that the size of a given piece of equipment lies outside the range given in the costing program CAPCOST©. The program will automatically cost the equipment by splitting it up into the appropriate number of parallel and/or series units. This does not mean that there will be a need for more than one physical piece of equipment; rather, it is a conservative way to estimate the cost of large equipment.

B.1   PRODUCTION OF DIMETHYL ETHER (DME) VIA THE DEHYDRATION OF METHANOL

B.1.1   Process Notes

Dimethyl ether (DME) is used primarily as a propellant. DME is miscible with most organic solvents, it has a high solubility in water, and it is completely miscible in water and 6% ethanol [1]. Recently, the use of DME as a fuel additive for diesel engines has been investigated due to its high volatility (desirable for cold starting) and high cetane number. The production of DME is via the catalytic dehydration of methanol over an acid zeolite catalyst. The main reaction is as follows:

$$2\ CH_3OH \rightarrow (CH_3)_2O + H_2O$$
$$\textit{methanol} \qquad \textit{DME}$$

In the temperature range of normal operation, there are no significant side reactions.

A preliminary process flow diagram for a DME process is shown in Figure B.1, in which 50,000 metric tons per year of 99.5 wt% purity DME product is produced. Due to the simplicity of the process, a stream factor of 0.95 (8375 h/yr) is used. Preliminary equipment summaries and process stream information are given in Tables B.1 and B.2, respectively.

B.1.2   Process Description

Fresh methanol, Stream 1, is combined with recycled reactant, Stream 13, and vaporized prior to being sent to a fixed-bed reactor operating between 250°C and 368°C. The single-pass conversion of methanol in the reactor is 80%. The reactor effluent, Stream 7, is then cooled prior to being sent to the first of two distillation columns, T-201 and T-202. DME product is taken overhead from the first column. The second column separates the water from the unused methanol. The methanol is recycled back to the front end of the process, while the water is sent to waste water treatment to remove trace amounts of organic compounds.

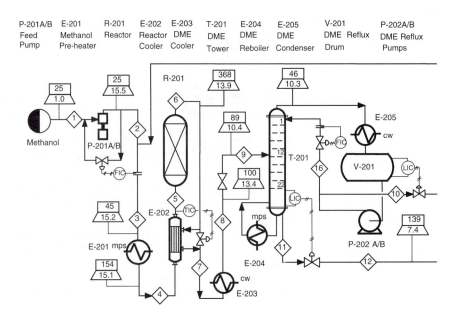

Figure B.1   Preliminary Process Flow Diagram (PFD) for the Production of Dimethyl Ether from Methanol (Unit 200)

B.1.3   Reaction Kinetics and Reactor Configuration

The reaction taking place is mildly exothermic with a standard heat of reaction, $\Delta H_{reac}(25°C) = -11,770$ kJ/kmol. The equilibrium constant for this reaction at three different temperatures is given below:

T	K_p
473 K (200°C)	92.6
573 K (300°C)	52.0
673 K (400°C)	34.7

The corresponding equilibrium conversions for pure methanol feed over the above temperature range is greater than 99%. Thus, this reaction is kinetically controlled at the conditions used in this process.

The reaction takes place on an amorphous alumina catalyst treated with 10.2% silica. There are no significant side reactions below 400°C. Above 250°C, the rate equation is given by Bondiera and Naccache [2] as:

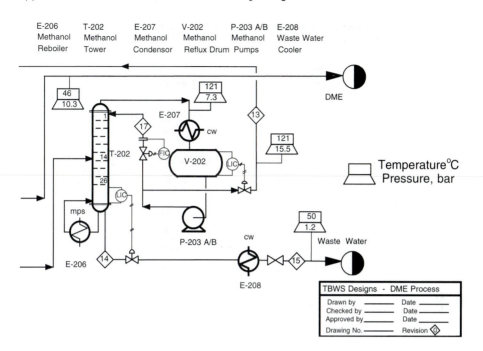

Figure B.1   (*Continued*)

$$- r_{methanol} = k_0 \exp\left[-\frac{E_a}{RT} \right] p_{methanol}$$

where $k_0 = 1.21 \times 10^6$ kmol/(m^3 reactor h kPa) , $E_a = 80.48$ kJ/mol, and $p_{methanol}$ = partial pressure of methanol (kPa).

Significant catalyst deactivation occurs at temperatures above 400°C, and the reactor should be designed so that this temperature is not exceeded anywhere in the reactor. The design given in Figure B.1 uses a single packed bed of catalyst which operates adiabatically. The temperature exotherm of 118°C, occurring in the reactor, is probably on the high side and gives an exit temperature of 368°C. However, the single-pass conversion is quite high (80%), and the low reactant concentration at the exit of the reactor tends to limit the possibility of a runaway. In practice, the catalyst bed might be split into two sections with an intercooler between the two beds. This has the overall effect of increasing the volume (and cost) of the reactor and should be investigated if catalyst damage is expected at temperatures lower than 400°C. In-reactor cooling (shell and tube design) and cold quenching by splitting the feed and feeding at different points in the reactor could also be investigated as viable alternative reactor configurations.

(*Text continues on p. 716*)

Table B.1 Preliminary Equipment Summary Table for DME Process

Equipment	P-201 A/B	P-202 A/B	P-203 A/B	V-201	V-202	T-201	T-202	R-201
MOC	Carbon Steel	Carbon Steel	Carbon Steel	Carbon Steel	Carbon Steel	Carbon Steel	Carbon Steel	Carbon Steel
Power (kW)	7.2	1.0	5.2	—	—	—	—	—
Efficiency	60%	40%	40%	—	—	—	—	—
Type/Drive	Reciprocating /Electric	Centrifugal/ Electric	Centrifugal/ Electric	—	—	—	—	—
Temperature (°C)	25	46	121	—	—	—	—	—
Pressure In (bar)	1.0	10.3	7.3	—	—	—	—	—
Pressure Out (bar)	15.5	11.4	16.0	—	—	—	—	—
Diameter (m)	—	—	—	0.96	0.85	0.79	0.87	0.72
Height/Length (m)	—	—	—	2.89	2.53	15.8	14.9	10.0
Orientation	—	—	—	Horizontal	Horizontal	Vertical	Vertical	Vertical
Internals	—	—	—	—	—	22 SS Trays 24 inch spacing	26 SS Trays 18 inch spacing	Packed bed section 7.2 m high filled with catalyst
Pressure (barg)	—	—	—	9.3	6.3	9.6	6.3	13.7

Table B.1 (*Continued*)

Equipment	E-201	E-202	E-203	E-204	E-205	E-206	E-207	E-208
Type	Float. Head Vaporizer	Float. Head	Float. Head Partial Cond.	Float. Head Reboiler	Fixed TS Condenser	Float. Head Reboiler	Float. Head Condenser	Float. Head
Duty (MJ/h)	14,400	2,030	12,420	2,490	3,140	5,790	5,960	1,200
Area (m²)	99.4	171.0	101.8	22.0	100.6	83.0	22.7	22.8
Shell Side								
Max Temp(°C)	154	250	280	153	46	167	121	167
Pressure (barg)	14.2	14.1	12.8	9.5	9.3	6.6	6.3	6.6
MOC	Carbon Steel	Carbon Steel	Carbon Steel	Carbon Steel	Carbon Steel	Carbon Steel	Carbon Steel	Carbon Steel
Phase	Boiling Liq.	V	Cond. Vapor	Boiling Liq.	Cond. Vapor	Boiling Liq.	Cond. Vapor	L
Tube Side								
Max Temp. (°C)	184	368	40	184	40	184	40	40
Pressure (barg)	10.0	12.9	4.0	10.0	4.0	10.0	4.0	4.0
MOC	Carbon Steel	Carbon Steel	Carbon Steel	Carbon Steel	Carbon Steel	Carbon Steel	Carbon Steel	Carbon Steel
Phase	Cond. Steam	V	L	Cond. Steam	L	Cond. Steam	L	L

Table B.2 Flow Summary Table for DME Process in Figure B.1

Stream No.	1	2	3	4	5	6	7	8
Temperature (°C)	25	25	45	154	250	368	281	100
Pressure (bar)	1.0	15.5	15.2	15.1	14.7	13.9	13.8	13.4
Vapor Fraction (molar)	0.0	0.0	0.0	1.0	1.0	1.0	1.0	0.0798
Mass Flow (tonne/h)	8.37	8.37	10.49	10.49	10.49	10.49	10.49	10.49
Mole Flow (kmol/h)	262.2	262.2	328.3	328.3	328.3	328.3	328.3	328.3
Component Mole Flow (kmol/h)								
Dimethyl ether	0.0	0.0	1.5	1.5	1.5	130.5	130.5	130.5
Methanol	259.7	259.7	323.0	323.0	323.0	64.9	64.9	64.9
Water	2.5	2.5	3.8	3.8	3.8	132.9	132.9	132.9

Table B.2   (Continued)

Stream No.	9	10	11	12	13	14	15	16	17
Temperature (°C)	89	46	153	139	121	167	50	46	121
Pressure (bar)	10.4	11.4	10.5	7.4	15.5	7.6	1.2	11.4	7.3
Vapor Fraction (molar)	0.148	0.0	0.0	0.04	0.0	0.0	0.0	0.0	0.0
Mass Flow (tonne/h)	10.49	5.97	4.52	4.52	2.13	2.39	2.39	2.17	3.62
Mole Flow (kmol/h)	328.3	129.7	198.6	198.6	66.3	132.3	132.3	47.1	113.0
Component Mole Flow (kmol/h)									
Dimethyl ether	130.5	129.1	1.4	1.4	1.4	0.0	0.0	46.9	2.4
Methanol	64.9	0.6	64.3	64.3	63.6	0.7	0.7	0.2	108.4
Water	132.9	0.0	132.9	132.9	1.3	131.6	131.6	0.0	2.2

Utility	mps	cw	mps	cw	cw	mps	cw	cw
Equipment	E-201	E-203	E-204	E-205	E-206	E-206	E-207	E-208
Temperature In (°C)	184	30	184	30	184	184	30	30
Temperature Out (°C)	184	40	184	40	184	184	40	40
Flow (tonne/h)	7.22	297.1	1.25	75.12	2.90	2.90	142.6	28.70

B.1.4 Vapor-Liquid Equilibrium (VLE)

The DME-water binary system exhibits two liquid phases when the DME concentration is in the range 34% to 93%. However, upon addition of 7% or more alcohol, the mixture becomes completely miscible over the complete range of DME concentration. In order to ensure that this non-ideal behavior is simulated correctly, it is recommended that binary VLE data for the three pairs of components be used in order to regress binary interaction parameters (BIPs) for a UNIQUAC thermodynamics model. If VLE data for the binary pairs are not used, then UNIFAC can be used to estimate BIPs. The results shown in Table B.2 were obtained using UNIFAC on the CHEMCAD™ Simulator (version 3.1) marketed by Chemstations, Inc. of Houston, TX.

REFERENCES

1. "DuPont Talks About its DME Propellant," *Aerosol Age*, May and June, 1982.
2. Bondiera, J., and C. Naccache, "Kinetics of Methanol Dehydration in Dealuminated H-Mordenite: Model with Acid and Basic Active Centres," *Applied Catalysis*, 69, 139–148, 1991.

B.2 ACRYLIC ACID PRODUCTION VIA THE CATALYTIC PARTIAL OXIDATION OF PROPYLENE [1,2,3,4,5]

B.2.1 Process Notes

Acrylic acid (AA) is used as a precursor for a wide variety of chemicals in the polymers and textile industries. There are several chemical pathways to produce AA, but the most common one is via the partial oxidation of propylene. The usual mechanism for producing AA utilizes a two-step process in which propylene is first oxidized to acrolein and then further oxidized to AA. Each reaction step usually takes place over a separate catalyst and at different operating conditions. The reaction stoichiometry is given below:

$$C_3H_6 + O_2 \rightarrow C_3H_4O + H_2O$$

Acrolein

$$C_3H_4O + \frac{1}{2}O_2 \rightarrow C_3H_4O_2$$

Acrylic Acid

Several side reactions may occur, most resulting in the oxidation of reactants and products. Some typical side reactions are given below:

$$C_3H_4O + \frac{7}{2}O_2 \rightarrow 3CO_2 + 2H_2O$$

$$C_3H_4O + \frac{3}{2}O_2 \rightarrow C_2H_4O_2 + CO_2$$

Acetic Acid

$$C_3H_6 + \frac{9}{2}O_2 \rightarrow 3CO_2 + 3H_2O$$

Therefore, the typical process set-up consists of a two-reactor system with each reactor containing a separate catalyst and operating at conditions so as to maximize the production of AA. The first reactor typically operates at a higher temperature than the second.

B.2.2   Process Description

The process shown in Figure B.2 produces 50,000 metric tons per year of 99.9% by mole AA product. The number of operating hours is taken to be 8000/yr, and the process is somewhat simplified since there is only one reactor [5]. It is assumed that both reactions take place on a single catalyst to yield AA and by-products. It is imperative to cool the products of reaction quickly to avoid further oxidation reactions, and this is achieved by rapidly quenching the reactor effluent with a cool recycle, Stream 8, of dilute aqueous AA in T-301. Additional recovery of AA and acetic acid (a by-product) is achieved in the absorber, T-302. The stream leaving the absorption section is a dilute aqueous acid, Stream 9. This is sent to a liquid-liquid extractor, T-303, to remove preferentially the acid fraction from the water prior to purification. There are many possible solvents which can be used as the organic phase in the separation; high solubility for AA and low solubility for water are desirable. Some examples include ethyl acrylate, ethyl acetate, xylene, diisobutyl ketone, methyl isobutyl ketone, and diisopropyl ether (DIPE) which is used here. The organic phase from T-303 is sent to a solvent recovery column, T-304, where the diisopropyl ether (and some water) is recovered overhead and returned to the extractor. The bottom stream from this column, Stream 14, contains virtually all the AA and acetic acid in Stream 9. This is sent to the acid purification column, T-305, where 95% by mole acetic acid by-product is produced overhead and 99.9 % by mole AA is produced as a bottom product and cooled prior to being sent to storage.

   The aqueous phase from the extractor, Stream 12, is sent to a waste water column, T-306, where a small amount of DIPE is recovered overhead and returned to the extractor. The bottom product, containing water and trace quantities of solvent and acid, is sent to wastewater treatment. Preliminary equipment summaries and process stream information are given in Tables B.3 and B.4, respectively.

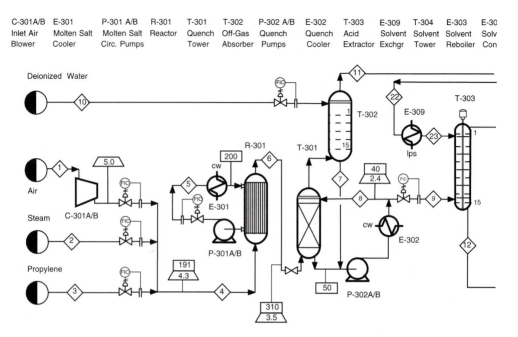

Figure B.2   Preliminary Process Flow Diagram (PFD) for the Production of Acrylic Acid from Propylene (Unit 300)

B.2.3   Reaction Kinetics and Reactor Configuration

The reactions taking place are kinetically controlled at the conditions used in the process, i.e., equilibrium lies far to the right. The reaction kinetics for the catalyst used in this process are given below:

$$C_3H_6 + \frac{3}{2}O_2 \rightarrow C_3H_4O_2 + H_2O \qquad \textit{Reaction 1}$$
$$\textit{Acrylic Acid}$$

$$C_3H_6 + \frac{5}{2}O_2 \rightarrow C_2H_4O_2 + CO_2 + H_2O \qquad \textit{Reaction 2}$$
$$\textit{Acetic Acid}$$

$$C_3H_6 + \frac{9}{2}O_2 \rightarrow 3CO_2 + 3H_2O \qquad \textit{Reaction 3}$$

$$\textit{where} \qquad -r_i = k_{o,i}\exp\left[-\frac{E_i}{RT}\right]p_{propylene}\,p_{oxygen}$$

Partial pressures are in kPa and the activation energies and pre-exponential terms for reactions 1–3 are given below:

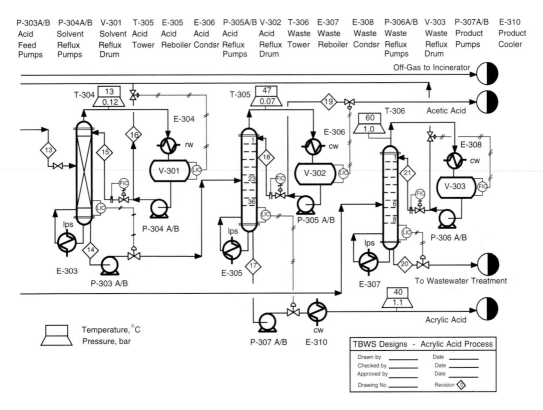

Figure B.2 (*Continued*)

i	E_i kcal/kmol	$K_{o,i}$ kmol/m^3 reactor h/(kPa)^2
1	15,000	1.59×10^5
2	25,000	1.81×10^8
3	20,000	8.83×10^5

The reactor configuration used for this process is a fluidized bed, and it is assumed that the bed of catalyst behaves as a well-mixed tank, i.e., it is isothermal at the temperature of the reaction (310°C). The gas flow is assumed to be plug flow through the bed, with 10% of the gas bypassing the catalyst. This latter assumption is made in order to simulate the gas channeling which occurs in real fluid bed reactors.

B.2.4 Safety Considerations

As with any reaction involving the partial oxidation of a fuel-like feed material (propylene), considerable attention must be paid to the composition of hydrocar-

(*Text continues on p. 727*)

Table B.3 Preliminary Equipment Summary Table for Acrylic Acid Process

Equipment	T-301	T-302	T-303	T-304	T-305	T-306	R-301*
MOC	Stainless Steel	Stainless Steel	Stainless Steel	Stainless Steel	Carbon Steel	Stainless Steel	Carbon Steel Fluidized Bed
Diameter (m)	5.3	3.5	2.2	7.5	2.4	2.3	3.6
Height/Length (m)	12	11	9.5	34	25	7.0	10
Orientation	Vertical	Vertical	Vertical	Vertical	Vertical	Vertical	Vertical
Internals	10m of high Efficiency Packing	15 Sieve Trays + Demister	15 Perforated Plates + Mixer	31 m of High Efficiency Structured Packing	36 Sieve Plates Stainless Steel	8 Sieve Plates Stainless Steel	1451 Heat Transfer Tubes (1420 m^2) Filled with molten salt
	Polyethylene	Stainless Steel	Stainless Steel	Stainless Steel			
Pressure (barg)	1.4	1.0	1.4	-1.0	-1.0	0.	3.0

*Installed cost of reactor (mid-1996) = 2×10^5 [Area (m^2)]$^{0.5}$

Table B.3 (*Continued*)

Equipment	P-301 A/B	P-302 A/B	P-303 A/B	P-304 A/B	P-305 A/B	P-306 A/B	P-307 A/B
MOC	Carbon Steel	Stainless Steel	Stainless Steel	Stainless Steel	Carbon Steel	Carbon Steel	Stainless Steel
Power (shaft) (kW)	32.3	106.2	0.9	51.3	1.2	9.0	1.0
Efficiency	75%	75%	40%	75%	40%	60%	40%
Type/Drive	Centrifugal/ Electric	Centrifugal/ Electric	Centrifugal/ Electric	Centrifugal/ Electric	Centrifugal/ Electric	Centrifugal/ Electric	Centrifugal/ Electric
Temperature (°C)	250	50	90	13	47	60	89
Pressure In (bar)	2.0	2.4	0.19	0.12	0.07	1.0	0.16
Pressure Out (bar)	3.6	3.9	2.05	4.62	3.31	4.62	2.46

Table B.3   (*Continued*)

Equipment	E-301	E-302	E-303	E-304	E-305	E-306	E-307
Type	Floating Head	Fixed TS	Floating Head Reboiler	Floating Head Condenser	Floating Head Reboiler	Fixed TS Condenser	Floating Head Reboiler
Duty (MJ/h)	83,400	70,300	101,000	108,300	2,230	2,280	21,200
Area (m²)	160	2550	891	7710	19.7	73.3	187
Shell							
Max Temp. (°C)	40	40	160	10	160	40	160
Pressure (barg)	4.0	4.0	5.0	4.0	5.0	4.0	5.0
Phase	L	L	Cond. Steam	L	Cond. Steam	L	Cond. Steam
MOC	Carbon Steel	Carbon Steel	Carbon Steel	Carbon Steel	Carbon Steel	Carbon Steel	Carbon Steel
Tube Side							
Max Temp(°C)	250	50	90	13	90	47	102
Pressure (barg)	2.0	2.0	-0.81	-0.88	-0.84	-0.93	0.1
MOC	Carbon Steel	Stainless Steel	Stainless Steel	Stainless Steel	Carbon Steel	Carbon Steel	Stainless Steel
Phase	L	L	Boiling Liq.	Cond. Vapor	Boiling Liq.	Cond. Vapor	Boiling Liq.

Table B.3   (*Continued*)

Equipment	E-308	E-309	E-310
Type	Fixed TS Condenser	Floating Head	Floating Head
Duty (MJ/h)	15,800	8,000	698
Area (m^2)	210	19.7	10.3
Shell Side			
Max Temp. (°C)	40	160	40
Pressure (barg)	4.0	5.0	4.0
Phase	L	Cond. Steam	L
MOC	Carbon Steel	Carbon Steel	Carbon Steel
Tube Side			
Max Temp. (°C)	60	40	89
Pressure (barg)	0.0	2.0	1.4
MOC	Carbon Steel	Carbon Steel	Stainless Steel
Phase	Cond. Vapor	L	L

Equipment	C-301 A/B	V-301	V-302	V-303
MOC	Carbon Steel	Stainless Steel	Carbon Steel	Carbon Steel
Power (shaft) (kW)	2260	—	—	—
Efficiency	77%	—	—	—
Type/Drive	Centrifugal Centrifugal –2 Stage/ Electric	—	—	—
Temperature (°C)	25	—	—	—
Pressure In (bar)	1.0	—	—	—
Pressure Out (bar)	5.0	—	—	—
Pressure (barg)	—	–0.88	–0.93	0.0
Diameter (m)	—	2.4	1.0	1.5
Height/Length (m)	—	7.2	2.5	4.5
Orientation	—	Horizontal	Horizontal	Horizontal
Internals	—	—	—	—

Table B.4 Flow Summary Table for Acrylic Acid Process in Figure B.2

Stream No.	1	2	3	4	5	6	7	8	9
Temperature (°C)	25	159	25	191	200	310	63	40	40
Pressure (bar)	1.0	6.0	11.5	4.3	3.0	3.5	2.0	2.4	2.4
Vapor Fraction	1.0	1.0	1.0	1.0	0.0	1.0	0.0	0.0	0.0
Mass Flow (tonne/h)	39.05	17.88	5.34	62.27	1075.0	62.27	3.08	1895.	27.46
Mole Flow (kmol/h)	1362.9	992.3	127.0	2482.2	—	2444.0	148.5	85200.0	1249.6
Component Mole Flow (kmol/h)					HiTec™ Molten Salt				
Propylene	—	—	127.0	127.0	—	14.7	—	—	—
Nitrogen	1056.7	—	—	1056.7	—	1056.7	—	—	—
Oxygen	280.9	—	—	280.9	—	51.9	—	—	—
Carbon Dioxide	—	—	—	—	—	60.5	—	—	—
Water	25.3	992.3	—	1017.6	—	1165.9	140.9	78870	1156.7
Acetic Acid	—	—	—	—	—	6.54	0.65	415	6.08
Acrylic Acid	—	—	—	—	—	87.79	6.99	5915	86.81
Solvent (Diisopropyl ether)	—	—	—	—	—	—	—	—	—

Table B.4 *(Continued)*

Stream No.	10	11	12	13	14	15	16	17	18
Temperature (°C)	25	48	40	40	90	13	13	89	47
Pressure (bar)	5.0	1.0	2.4	2.4	0.19	0.12	3.0	0.16	0.07
Vapor Fraction	0.0	1.0	0.0	0.0	0.0	0.0	0.0	0.0	0.0
Mass Flow (tonne/h)	2.54	37.35	20.87	143.0	6.63	155.3	136.4	6.26	5.28
Mole Flow (kmol/h)	141.0	1335.4	1156.9	1591.2	93.19	1705.7	1498.0	86.85	90.49
Component Mole Flow (kmol/h)									
Propylene	—	14.7	—	—	—	—	—	—	—
Nitrogen	—	1056.7	—	—	—	—	—	—	—
Oxygen	—	51.9	—	—	—	—	—	—	—
Carbon Dioxide	—	60.5	—	—	—	—	—	—	—
Water	141.0	150.2	1156.6	198.8	0.30	226.0	198.5	—	4.28
Acetic Acid	—	0.46	0.03	6.08	6.08	—	—	0.05	86.07
Acrylic Acid	—	0.98	—	86.81	86.81	—	—	86.80	0.14
Solvent (Diisopropyl ether)	—	—	0.30	1299.5	—	1479.7	1299.5	—	—

725

Table B.4   *(Continued)*

Stream No.	19	20	21	22	23
Temperature (°C)	47	102	60	13	40
Pressure (bar)	1.1	1.1	1.0	3.0	2.8
Vapor Fraction	0.0	0.0	0.0	0.0	0.0
Mass Flow (tonne/h)	0.37	20.84	37.37	136.4	136.4
Mole Flow (kmol/h)	6.34	1156.43	470.2	1498.5	1498.5
Component Mole Flow (kmol/h)					
Propylene	—	—	—	—	—
Nitrogen	—	—	—	—	—
Oxygen	—	—	—	—	—
Carbon Dioxide	—	—	—	—	—
Water	0.30	1156.4	126.8	198.7	198.7
Acetic Acid	6.03	0.03	—	—	—
Acrylic Acid	0.01	—	—	—	—
Solvent (Diisopropyl ether)	—	—	343.4	1299.8	1299.8

Utility	cw	cw	lps	rw	lps	cw	lps	cw	lps	cw
Equipment	E-301	E-302	E-303	E-304	E-305	E-306	E-307	E-308	E-309	E-310
Temperature In (°C)	30	30	160	5	160	30	160	30	160	30
Temperature Out (°C)	40	40	160	10	160	40	160	40	160	40
Flow (tonne/h)	1995.0	1682.0	48.5	5182.0	1.07	54.5	10.19	378.0	3.85	16.7

bons and oxygen in the feed stream. Operation outside the explosive limits is recommended for this reaction. However, the conditions used in this process lie within the explosive limits. The addition of a large amount of steam and the use of a fluidized bed reactor allow safe operation within the explosive limits. The second safety concern is that associated with the highly exothermic polymerization of AA which occurs in two ways. First, if this material is stored without appropriate additives, then free radical initiation of the polymerization can occur—this potentially disastrous situation is discussed by Kurland and Bryant [1]. Secondly, AA dimerizes when in high concentrations at temperatures greater than 90°C, thus much of the separation sequence must be operated under high vacuum in order to keep the bottom temperatures in the columns below this temperature.

B.2.5   Vapor-Liquid-Liquid Equilibrium

The use of a liquid-liquid extractor requires the use of a thermodynamic package (or physical property data) which reflects the fact that two phases are formed and that significant partitioning of the AA and acetic acid occurs with the majority going to the organic phase (in this case DIPE). Distribution coefficients for the organic acids in water and DIPE as well as mutual solubility data for water/DIPE are desirable. The process given in Figure B.2 was simulated using a UNIFAC thermodynamics package on CHEMCAD™ and should give reasonable results for preliminary process design. Much of the process background material and process configuration was taken from the 1986 AIChE student contest problem in Reference 5. The kinetics presented above are fictitious but should give reasonable preliminary estimates of reactor size.

REFERENCES

1. Kurland, J. J., and D. B. Bryant, "Shipboard Polymerization of Acrylic Acid," *Plant Operations Progress,* **6**(4) 203–207, 1987.
2. *Kirk-Othmer Encyclopedia of Chemical Technology,* 3rd Edition, vol. 1, 330–354, John Wiley and Son, New York, 1978.
3. *Encyclopedia of Chemical Processing and Design,* J. J. McKetta, and W. A. Cunningham, Ed. vol. 1, 402–428, 1976.
4. Sakuyama, S., T. Ohara, N. Shimizu, and K. Kubota, "A New Oxidation Process for Acrylic Acid from Propylene," *Chem. Technol.,* 350, June, 1973.
5. *The AIChE Student Annual 1986,* AIChE, B. Van Wie, and R. A. Wills editors, "1986 Student Contest Problem," 52–82, 1986.

B.3 PRODUCTION OF ACETONE VIA THE DEHYDROGENATION OF ISOPROPYL ALCOHOL (IPA) [1,2,3,4]

B.3.1 Process Notes

The prevalent process for the production of acetone is as a by-product of the manufacture of phenol. Benzene is alkylated to cumene which is further oxidized to cumene hydroperoxide and finally cleaved to yield phenol and acetone. However, the process shown in Figure B.3 and discussed here uses isopropyl alcohol (IPA) as the raw material. This is a viable commercial alternative and a few plants continue to operate using this process. The primary advantage of this process is that the acetone produced is free from trace aromatic compounds, particularly benzene. For this reason acetone produced from IPA may be favored by the pharmaceutical industry due to the very tight restrictions placed on solvents by the Food and Drug Administration (FDA). The reaction to produce acetone from IPA is given below:

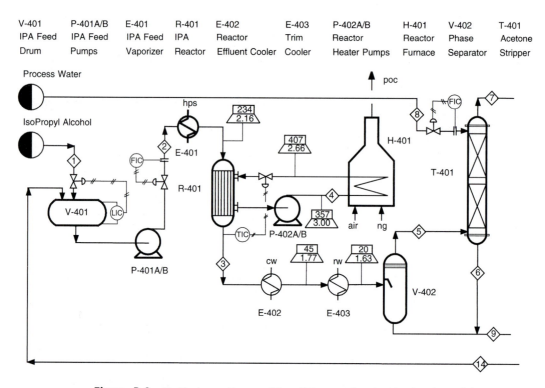

V-401	P-401A/B	E-401	R-401	E-402	E-403	P-402A/B	H-401	V-402	T-401
IPA Feed	IPA Feed	IPA Feed	IPA	Reactor	Trim	Reactor	Reactor	Phase	Acetone
Drum	Pumps	Vaporizer	Reactor	Effluent Cooler	Cooler	Heater Pumps	Furnace	Separator	Stripper

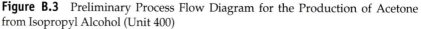

Figure B.3   Preliminary Process Flow Diagram for the Production of Acetone from Isopropyl Alcohol (Unit 400)

$$(CH_3)_2CHOH \rightarrow (CH_3)_2CO + H_2$$

Isopropyl Alcohol   Acetone

The reaction conditions are typically 2 bar and 350°C, giving single-pass conversions of 85–92 %.

B.3.2   Process Description

Referring to Figure B.3, an azeotropic mixture of isopropyl alcohol and water (88 wt% IPA) is fed into a surge vessel (V-401) where it is mixed with the recycled unreacted IPA/water mixture, Stream 14. This material is then pumped and vaporized prior to entering the reactor. Heat is provided for the endothermic reaction using a circulating stream of molten salt, Stream 4. The reactor effluent, containing acetone, hydrogen, water, and unreacted IPA, is cooled in two exchangers prior to entering the phase separator (V-402). The vapor leaving the separator is

(*Text continues on p. 735*)

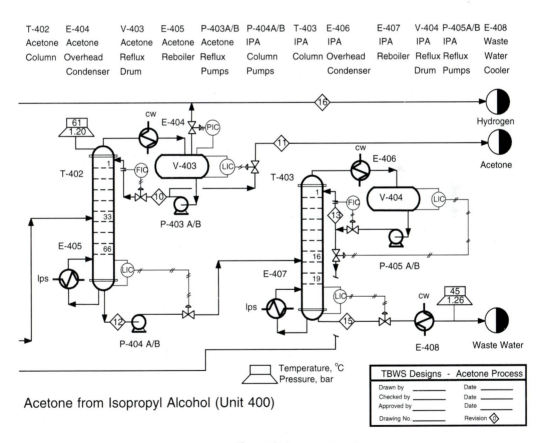

T-402	E-404	V-403	E-405	P-403A/B	P-404A/B	T-403	E-406	E-407	V-404	P-405A/B	E-408
Acetone	Acetone	Acetone	Acetone	Acetone	IPA	IPA	IPA	IPA	IPA	IPA	Waste
Column	Overhead	Reflux	Reboiler	Reflux	Column	Column	Overhead	Reboiler	Reflux	Reflux	Water
	Condenser	Drum		Pumps	Pumps		Condenser		Drum	Pumps	Cooler

Acetone from Isopropyl Alcohol (Unit 400)

TBWS Designs - Acetone Process

Figure B.3   (*Continued*)

Table B.5 Preliminary Equipment Summary Table for Acetone Process

Equipment	P-401 A/B	P-402 A/B	P-403 A/B	P-404 A/B	P-405 A/B	V-401	V-402
MOC	Carbon Steel	Carbon Steel	Carbon Steel	Carbon Steel	Carbon Steel	Carbon Steel	Carbon Steel
Power (shaft) (kW)	0.43	2.53	1.75	0.06	1.45	—	—
Efficiency	40%	50%	40%	40%	40%	—	—
Type/Drive	Centrifugal/ Electric	Centrifugal/ Electric	Centrifugal/ Electric	Centrifugal/ Electric	Centrifugal/ Electric	—	—
Temperature (°C)	25	400	61	90	83	—	—
Pressure In (bar)	1.13	1.83	1.41	1.93	1.42	—	—
Pressure Out (bar)	3.00	3.00	4.48	2.78	3.25	—	—
Diameter (m)	—	—	—	—	—	0.80	0.75
Height/Length (m)	—	—	—	—	—	2.40	2.25
Orientation	—	—	—	—	—	Horizontal	Vertical
Internals	—	—	—	—	—	—	SS Demister
Pressure (barg)	—	—	—	—	—	0.0	0.63

Table B.5 *(Continued)*

Equipment	V-403	V-404	T-401	T-402	T-403	H-401	R-401
MOC	Carbon Steel	Carbon Steel	Carbon Steel	Carbon Steel	Carbon Steel	Carbon Steel	Carbon Steel
Diameter (m)	0.83	0.93	0.33	1.25	1.36	—	1.85
Height/Length (m)	2.50	2.80	3.20	37.0	18.6	—	8.0
Orientation	Horizontal	Horizontal	Vertical	Vertical	Vertical	—	Vertical
Internals	—	—	2.5 m of Packing 1″ Ceramic Rashig Rings	66 SS Sieve Plates @ 18″ Spacing	19 SS Sieve Plates @ 24″ Spacing	—	448 2″ Diameter, 20′ Long Catalyst Filled Tubes
Pressure (barg)	0.2	0.2	1.0	0.4	0.4	—	2.0
Type	—	—	—	—	—	Fired Heater	—
Duty (MJ/h)	—	—	—	—	—	2,730	—
Area Radiant (m²)	—	—	—	—	—	10.1	—
Area Convective (m²)	—	—	—	—	—	30.4	—
Tube Pressure (barg)	—	—	—	—	—	2.0	—

Table B.5 *(Continued)*

Equipment	E-401	E-402	E-403	E-404	E-405	E-406	E-407	E-408
Type	Float. Head Vaporizer	Float. Head Partial Cond.	Float. Head Partial Cond.	Fixed TS Condenser	Float. Head Reboiler	Fixed TS Condenser	Float. Head Reboiler	Double Pipe
Duty (MJ/h)	3,550	3,260	563	3,095	3,500	7,340	7,390	174
Area (m²)	70.3	77.6	8.5	39.1	30.9	50.2	65.1	1.6
Shell Side								
Max. Temp (°C)	234	350	45	61	90	83	109	109
Pressure (barg)	1.0	1.0	1.0	0.2	0.4	0.2	0.4	0.4
Phase	Boiling Liq.	Cond. Vapor	Cond. Vapor	Cond. Vapor	Boiling Liq.	Cond. Vapor	Boiling Liq.	L
MOC	Carbon Steel	Carbon Steel	Carbon Steel	Carbon Steel	Carbon Steel	Carbon Steel	Carbon Steel	Carbon Steel
Tube Side								
Max. Temp (°C)	254	40	15	40	160	40	160	40
Pressure (barg)	41.0	3.0	3.0	3.0	5.0	3.0	5.0	3.0
Phase	Cond. Steam	L	L	L	Cond. Steam	L	Cond. Steam	L
MOC	Carbon Steel	Carbon Steel	Carbon Steel	Carbon Steel	Carbon Steel	Carbon Steel	Carbon Steel	Carbon Steel

Table B.6 Flow Table for Acetone Process in Figure B.3

Stream No.	1	2	3	4	5	6	7	8
Temperature (°C)	25	32	350	357	20	27	33	25
Pressure (bar)	1.01	2.30	1.91	3.0	1.63	1.63	1.50	2.0
Vapor Fraction	0.0	0.0	1.0	0.0	1.0	0.0	1.0	0.0
Mass Flow (tonne/h)	2.40	2.67	2.67	35.1	0.34	0.46	0.24	0.36
Mole Flow (kmol/h)	51.96	57.84	92.62	—	39.74	21.14	38.60	20.00
Component Mole Flow (kmol/h)				Molten Salt				
Hydrogen	—	—	34.78	—	34.78	0.00	34.78	—
Acetone	—	0.16	34.94	—	4.44	1.93	2.51	—
Isopropyl Alcohol	34.82	38.64	3.86	—	0.12	0.10	0.02	—
Water	17.14	19.04	19.04	—	0.40	19.11	1.29	20.00

Table B.6   *(Continued)*

Stream No.	9	10	11	12	13	14	15	16
Temperature (°C)	22	61	61	90	83	83	109	33
Pressure (bar)	1.63	1.5	1.5	1.4	1.2	1.2	1.4	1.2
Vapor Fraction	0.0	0.0	0.0	0.0	0.0	0.0	0.0	1.0
Mass Flow (tonne/h)	2.79	4.22	1.88	0.92	8.23	0.27	0.65	0.24
Mole Flow (kmol/h)	74.02	72.51	32.29	41.73	177.18	5.88	35.85	38.60
Component Mole Flow (kmol/h)								
Hydrogen	0.00	—	—	—	—	—	—	34.78
Acetone	32.43	72.46	32.27	0.16	4.82	0.16	—	2.51
Isopropyl Alcohol	3.84	0.05	0.02	3.82	115.10	3.82	—	0.02
Water	37.75	—	—	37.75	57.26	1.90	35.85	1.29

Utility	hps	cw	rw	cw	lps	cw	lps	cw
Equipment	E-401	E-402	E-403	E-404	E-405	E-406	E-407	E-408
Temperature In (°C)	254	30	5	30	160	30	160	30
Temperature Out (°C)	254	40	15	40	160	40	160	40
Flow (tonne/h)	2.09	77.90	13.50	74.00	1.68	176.00	3.55	4.16

scrubbed with water to recover additional acetone, and then this liquid is combined with the liquid from the separator and sent to the separations section. Two towers are used to separate the acetone product (99.9 mole %) and to remove the excess water from the unused IPA, which is then recycled back to the front end of the process as an azeotropic mixture. Preliminary equipment summaries and stream information are given in Tables B.5 and B.6, respectively.

B.3.3   Reaction Kinetics

The reaction to form acetone from isopropyl alcohol (isopropanol) is endothermic with a standard heat of reaction of 62.9 kJ/mol. The reaction is kinetically controlled and occurs in the vapor phase over a catalyst. The reaction kinetics for this reaction are first order with respect to the concentration of alcohol and can be estimated from the following equation [3,4]:

$$- r_{IPA} = k_0 \exp\left[-\frac{E_a}{RT} \right] C_{IPA} \quad \frac{kmol}{m^3 reactor\ s}$$

with $E_a = 72.38 MJ/kmol$, $k_0 = 3.51 \times 10^5 \frac{m^3 gas}{m^3 reactor\ s}$, $C_{IPA} = \frac{kmol}{m^3 gas}$

In practice, several side reactions can occur to a small extent. Thus, trace quantities of propylene, diisopropyl ether, acetaldehyde and other hydrocarbons and oxides of carbon can be formed [1]. The non-condensables are removed with the hydrogen while the aldehydes and ethers may be removed with acid washing or adsorption. These side reactions are not accounted for in this preliminary design.

For the design presented in Figure B.3, the reactor was simulated with catalyst in 2 inch (50.4 mm) diameter tubes each 20 feet (6.096 m) long and with a cocurrent flow of a heat transfer medium on the shell side of the shell and tube reactor. The resulting arrangement gives a 90% conversion of IPA per pass.

B.3.4   Vapor-Liquid Equilibrium

Isopropyl alcohol and water form a minimum boiling point azeotrope at 88 wt% isopropyl alcohol and 12 wt% water. Vapor-liquid equilibrium (VLE) data are available from several sources and can be used to back-calculate binary interaction parameters or liquid-phase activity coefficients. The process presented in Figure B.3 and Table B.6 was simulated using the UNIFAC thermodynamics package in the CHEMCAD™ simulator. This package correctly predicts the formation of the azeotrope at 88 wt% alcohol.

REFERENCES

1. *Kirk-Othmer:Encyclopedia of Chemical Technology,* 3rd Edition, vol. 1, 179–191, John Wiley and Son, 1976.
2. *Shreve's Chemical Process Industries,* G.T. Austin, ed., 5th Edition, 764, McGraw-Hill, 1984.
3. *Encyclopedia of Chemical Processing and Design,* Ed. J. J. McKetta, and W. A. Cunningham, vol. 1, 314–362, 1976.
4. C. Q. Sheely, *Kinetics of Catalytic Dehydrogenation of Isopropanol,* Ph.D. Thesis, University of Illinois, 1963.

B.4   PRODUCTION OF HEPTENES FROM PROPYLENE AND BUTENES [1]

B.4.1   Process Notes

This background information for this process is taken from Chauvel, et al. [1]. This example is an illustration of a preliminary estimate of a process to convert a mixture of C_3 and C_4 unsaturated hydrocarbons to 1-heptene and other unsat-

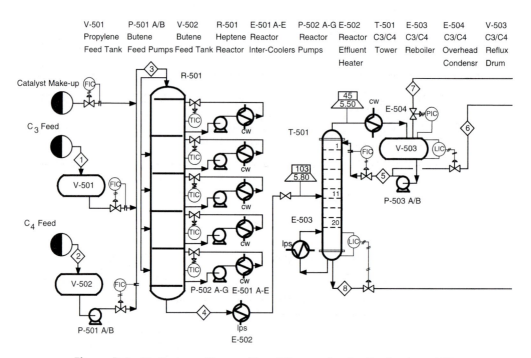

Figure B.4   Preliminary Process Flow Diagram for the Production of Heptenes from Propylene and Butenes (Unit 500)

urated products. The market for the 1-heptene product would be as a high-octane blending agent for gasoline or in the production of plasticizers. Based on preliminary market estimates, a production capacity of 20,000 metric tons per year of 1-heptene using 8000 operating hours/yr was set. This process differs from the other examples in Appendix B in several ways. First, the raw materials to the process contain a wide variety of chemicals. This is typical for oil refinery and petrochemical operations. Secondly, no specific kinetic equations are given for the reactions. Instead, the results of laboratory tests using the desired catalyst at different conditions and using different feed materials are used to guide the process engineer to an optimum, or close to an optimum, reactor configuration. The flowsheet in Figure B.4 and equipment summary and stream tables, Tables B.7 and B.8, have been developed using such information. It should be noted that a preliminary economic analysis, and hence the feasibility of the process, can be determined without this information, as long as yield and conversion data are available and the reactor configuration can be estimated.

(*Text continues on p. 745*)

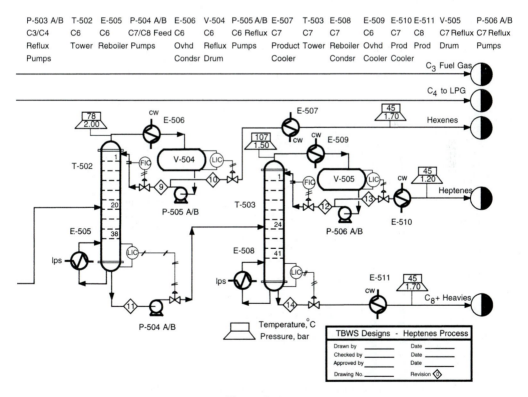

Figure B.4   (Continued)

Table B.7 Preliminary Equipment Summary Table for Heptene Process

Equipment	P-501 A/B	P-502 A-G[1]	P-503 A/B	P-504 A/B	P-505 A/B	P-506 A/B
MOC	Carbon Steel	Carbon Steel	Carbon Steel	Carbon Steel	Carbon Steel	Carbon Steel
Power (shaft) (kW)	6.75	5.13	2.75	0.66	2.15	1.93
Efficiency	40%	70%	40%	40%	40%	40%
Type/Drive	Centrifugal/ Electric	Centrifugal/ Electric	Centrifugal/ Electric	Centrifugal/ Electric	Centrifugal/ Electric	Centrifugal/ Electric
Temperature (°C)	25	45	45	151	78	107
Pressure In (bar)	2.97	8.00	5.50	2.50	2.00	1.50
Pressure Out (bar)	9.00	9.00	7.55	4.00	4.47	4.00

[1] 7 identical pumps – 5 operating + 2 spares

Table B.7   (*Continued*)

Equipment	V-501	V-502	V-503	V-504	V-505
MOC	Carbon Steel	Carbon Steel	Carbon Steel	Carbon Steel	Carbon Steel
Diameter (m)	1.40	1.90	1.10	0.95	0.92
Height/Length (m)	4.20	5.7	3.30	2.85	2.75
Orientation	Horizontal	Horizontal	Horizontal	Horizontal	Horizontal
Internals	—	—	—	—	—
Pressure (barg)	20.0	5.0	4.5	1.0	0.5

Equipment	R-501	T-501	T-502	T-503
MOC	Carbon Steel	Carbon Steel	Carbon Steel	Carbon Steel
Diameter (m)	3.00	1.05	1.10	0.90
Height/Length (m)	13.0	20.7	26.0	27.3
Orientation	Vertical	Vertical	Vertical	Vertical
Internals	Reactor Split into 5 Equal Sections	20 SS Sieve Plates @ 24" Spacing	38 SS Sieve Plates @ 18" Spacing	41 SS Sieve Plates @ 18" Spacing
Pressure (barg)	7.0	5.0	1.5	1.0

Table B.7 *(Continued)*

Equipment	E-501 A-E[1]	E-502	E-503	E-504	E-505	E-506
Type	Fixed TS	Float. Head Partial Vap.	Float. Head Reboiler	Fixed TS Condenser	Float. Head Reboiler	Fixed TS Condenser
Duty (MJ/h)	846	3,827	1,251	3,577	2,184	2,630
Area (m^2)	61.4	33.8	32.1	128.5	21.1	20.0
Shell Side						
Max. Temp (°C)	45	160	160	45	160	78
Pressure (barg)	8.0	5.0	5.0	4.0	5.0	1.0
Phase	L	Cond. Steam	Cond. Steam	Cond. Vapor	Cond. Steam	Cond. Vapor
MOC	Carbon Steel	Carbon Steel	Carbon Steel	Carbon Steel	Carbon Steel	Carbon Steel
Tube Side						
Max. Temp (°C)	40	103	151	40	135	40
Pressure (barg)	3.0	6.7	4.8	3.0	1.5	3.0
Phase	L	L+V	Boiling Liq.	L	Boiling Liq.	L
MOC	Carbon Steel	Carbon Steel	Carbon Steel	Carbon Steel	Carbon Steel	Carbon Steel

[1] Area and duty given for one exchanger, 5 identical exchangers are needed.

Table B.7 *(Continued)*

Equipment	E-507	E-508	E-509	E-510	E-511
Type	Double Pipe	Float. Head Reboiler	Fixed TS Condenser	Double Pipe	Double Pipe
Duty (MJ/h)	146	2,026	2,146	372	330
Area (m^2)	2.1	75.3	9.7	3.9	2.4
Shell Side					
Max. Temp (°C)	78	160	107	107	154
Pressure (barg)	1.0	5.0	0.5	0.3	0.7
Phase	L	Cond. Steam	Cond. Vapor	L	L
MOC	Carbon Steel	Carbon Steel	Carbon Steel	Carbon Steel	Carbon Steel
Tube Side					
Max. Temp (°C)	40	154	40	40	40
Pressure (barg)	3.0	1.0	3.0	3.0	3.0
Phase	L	Boiling Liq.	L	L	L
MOC	Carbon Steel	Carbon Steel	Carbon Steel	Carbon Steel	Carbon Steel

Table B.8   Flow Summary Table for Heptene Process in Figure B.4

Stream No.	1	2	3	4	5	6	7
Temperature (°C)	25	25	26	45	45	45	45
Pressure (bar)	11.62	2.97	8.00	7.7	7.50	6.50	5.00
Vapor Fraction	0.0	0.0	0.0	0.0	0.0	0.0	1.0
Mass Flow (tonne/h)	3.15	9.29	12.44	12.44	3.68	6.66	0.13
Mole Flow (kmol/h)	74.57	163.21	237.78	178.10	64.41	116.45	3.00
Component Mole Flow (kmol/h)							
Propane	3.56	—	3.56	3.56	0.31	0.56	3.00
Propylene	71.06	—	71.06	—	—	—	—
i-Butane	—	29.44	29.44	29.44	16.19	29.28	—
n-Butane	—	34.41	34.41	34.41	18.65	33.72	—
i-Butene	—	8.27	8.27	8.27	4.53	8.19	—
1-Butene	—	90.95	90.95	44.94	24.61	44.49	—
1-Hexene	—	0.14	0.14	21.21	0.12	0.21	—
1-Heptene	—	—	—-	26.53	—	—	—
1-Octene	—	—	—	7.41	—	—	—
1-Undecene	—	—	—	2.34	—	—	—

Table B.8   (*Continued*)

Stream No.	8	9	10	11	12	13	14
Temperature (°C)	151	78	78	135	107	107	154
Pressure (bar)	5.80	4.47	4.47	2.50	4.00	4.00	2.00
Vapor Fraction	0.0	0.0	0.0	0.0	0.0	0.0	0.0
Mass Flow (tonne/h)	5.64	5.79	1.86	3.79	4.30	2.53	1.26
Mole Flow (kmol/h)	58.65	69.84	22.44	36.22	43.78	25.76	10.46
Component Mole Flow (kmol/h)							
Propane	—	—	—	—	—	—	—
Propylene	—	—	—	—	—	—	—
i-Butane	0.16	0.50	0.16	—	—	—	—
n-Butane	0.69	2.15	0.69	—	—	—	—
i-Butene	0.08	0.25	0.08	—	—	—	—
1-Butene	0.45	1.40	0.45	—	—	—	—
1-Hexene	21.00	64.70	20.79	0.21	0.36	0.21	—
1-Heptene	26.52	0.84	0.27	26.26	43.28	25.47	0.79
1-Octene	7.41	—	—	7.41	0.14	0.08	7.33
1-Undecene	2.34	—	—	2.34	—	—	2.34

Table B.8 (*Continued*)

Utility	cw	lps	lps	lps	cw	lps	cw	cw	lps	cw	cw	cw	cw
Equipment	E-501 A-E	E-502	E-503	E-504	E-505	E-506	E-507	E-508	E-509	E-510	E-511
Temperature In (°C)	30	160	160	30	160	30	30	160	30	30	30
Temperature Out (°C)	40	160	160	40	160	40	40	160	40	40	40
Flow (tonne/h)	20.20[1]	1.84	0.60	85.50	1.05	62.90	3.49	0.97	51.30	8.90	7.89

[1] Flow of cooling water shown for 1 exchanger only

B.4.2   Process Description

Two liquid feed streams containing propylene and butene and a stream of catalyst slurried with 1-hexene are mixed at a pressure of approximately 8 bar prior to being sent to the reactor. The reactor consists of five essentially well-mixed sections, with similar concentrations in each section. Heat removal is achieved by using pump-arounds from each stage through external heat exchangers. The reactor effluent is partially vaporized before being fed to the first of three distillation columns. The first column (T-501) removes the unreacted C_3 and C_4 components, which are used subsequently as fuel (Stream 7) or sent to LPG storage (Stream 6). The next column (T-502) separates the 1-hexene product overhead (Stream 10) and sends the bottoms stream to the final column (T-503). In T-503, the main 1-heptene product (Stream 13) is taken overhead while the C_8 and heavier compounds are taken as the bottom product (Stream 14). The bottom product is processed off site to remove the heavy material and to recover spent catalyst.

B.4.3   Reaction Kinetics

The process given in Figure B.4 is based on the liquid-phase catalytic co-dimerization of C_3 and C_4 olefins using an organometalic catalyst. This catalyst is slurried with a small volume of the hexenes product and fed to the reactor with the feed streams. The volume of the catalyst stream is small compared to the other streams and is not included in the material balance given in Table B.8. In 1976, (CEPCI = 183) consumption of catalyst amounted to \$9.5/1000 kg of 1-heptene product, [1].

The primary reactions which take place are as follows:

$$C_3H_6 + C_3H_6 \rightarrow C_6H_{12}$$
1-Hexene

$$C_3H_6 + C_4H_8 \rightarrow C_7H_{14}$$
1-Heptene

$$C_4H_8 + C_4H_8 \rightarrow C_8H_{16}$$
1-Octene

$$C_3H_6 + 2C_4H_8 \rightarrow C_{11}H_{22}$$
1-Undecene

In order to maximize the selectivity of the heptene reaction, several reactor configurations were considered [1]. The reactor configuration which maximized the heptene production, in a minimum volume, was found to be a plug flow reactor in which the butene feed was introduced at one end and the propylene stream was injected along the side of the reactor. However, due to other considerations such as reactor complexity, it was finally decided to use a reactor with five equal stages in which the concentration in each stage is maintained approximately the same. Heat removal and mixing in each stage is accomplished by withdrawing a stream of material and pumping it through an external heat exchanger and back into the same stage of the reactor. The liquid cascades downwards from stage to stage by means of liquid downcomers. The inside of the reactor can thus be considered similar to a five-plate distillation column (without vapor flow). The distribution of the feeds into the different stages is not shown in Figure B.4 and the dimensions of the reactor are taken directly from Chauvel, et al. [1].

B.4.4   Vapor-Liquid Equilibrium

The hydrocarbon components used in the simulation can all be considered to be well behaved, i.e., no azeotrope formation. The simulations were carried out using the SRK VLE package using the CHEMCAD™ simulator.

REFERENCES

1. Chauvel, A., P. Leprince, Y. Barthel, C. Raimbault, and J-P Arlie, *Manual of Economic Analysis of Chemical Processes,* translated by R. Miller and E. B. Miller, McGraw-Hill, New York, 1976, 207–228.

C

Design Projects

INTRODUCTION

In this appendix, six design projects are presented. The projects are grouped in pairs with each pair of projects focused on a given process. The first project in each pair (Projects 1, 3, and 5) deals with an existing facility in which a problem has arisen or in which a change of operating conditions is desired. These types of problems are typically encountered in real operating plants, and would be the type of project a young engineer might encounter in his or her first year or two in the plant. These problems are often not completely defined and not all information is available. In addition, a solution to the problem is required in a fairly short period of time. This situation is not unheard of in the workplace and forces the engineer to make assumptions and analyze situations in which not all the operating variables are known. In analyzing these problems, the information in Section 3, Chapters 10–16, will be very useful. It is not recommended that these problems be given to students without some coverage of the material given in Section 3 of the text.

The second project in each pair (Projects 2, 4, and 6) is more like the typical "design" project that is covered in the senior capstone design course. Here, the student (or group of students) should be familiar with the information given in Sections 0, 1, and 4 and the guidelines for written reports given in Chapter 22.

In the authors' collective experience, the most worthwhile learning experience for the student is obtained during an oral presentation in which the student (or group of students) presents the results to a group of faculty and then defends his or her solutions in a 30–40 minute question and answer period. If this method of examination is chosen, the information in Chapter 22 on oral presentations will also be of use.

A note is made to any experienced engineer who may come upon these projects. It is recognized that some of the information given in the flow diagrams reflects poor (and sometimes extremely poor) engineering judgment. In fact, parts of some of the flow diagrams are almost caricatures of how the plant should be designed and operated. This is done intentionally, with the hope that the student will be able to find these gross errors and recommend improvements. The bottom line is that if, by doing these projects, the student can find errors and make improvements, then the transition from student to practicing engineer will have started, and we will have fulfilled our objective in writing this text.

Finally, background information for this appendix was obtained from the following sources.

1. *Kirk-Othmer Encyclopedia of Chemical Technology*, 3rd ed., Wiley, New York, 1978.
2. McKetta, J. J., and W. A. Cunningham (eds.), *Encyclopedia of Chemical Processing and Design*, Marcel Dekker, New York, 1976.
3. Rase, H. F., *Chemical Reactor Design for Process Plants*, Vol. 2, Wiley, New York, 1977.

1

Increasing the Production of 3-Chloro-1-Propene (Allyl Chloride) in Unit 600

BACKGROUND

You are currently employed by the TBWS Corp. at their Beaumont, Texas plant, and you have been assigned to the allyl chloride facility. A serious situation has developed at the plant, and you have been assigned to assist with troubleshooting the problems that have arisen.

Recently, your sister plant in Alabama was shut down by the EPA (Environmental Protection Agency) for violations concerning sulfur dioxide emissions from a furnace in their allyl chloride facility. Fortunately, the Beaumont facility had switched to natural gas as a fuel for their process in the early 1980s and, hence, is currently in compliance with the EPA and Texas regulations. However, the loss of the Alabama plant, albeit for a short time only, has put considerable pressure on the Beaumont plant to fulfill contractual obligations to our customers in Alabama for allyl chloride. Thus, part of your assignment is to advise management concerning the increase in production of allyl chloride that can be made at the Beaumont facility.

Another related issue that has been discussed by management is the long-term profitability of both allyl chloride facilities. Allyl chloride is used as a precursor in the production of allyl alcohol, glycerin, and a variety of other products used in the pharmaceutical industry. More efficient plants have been built recently by our competitors and we are being squeezed slowly out of the market by these rival companies. We still maintain a loyal customer base due to our excellent technical and customer service departments and our aggressive sales staff. However, we have been losing an ever-increasing share of the market since the mid–1980s. At present, the future looks bleak, and if the profitability and efficiency of our facilities do not increase in the near future, we may well be shut down in the next year or two, when some of our long-term contracts come up for renewal. A second part of your assignment is to look into the overall profitability of the Beaumont allyl chloride facility and determine whether any significant improvements in the overall economics can be made.

PROCESS DESCRIPTION OF THE BEAUMONT ALLYL CHLORIDE FACILITY

A process flow diagram (PFD) of the allyl chloride facility is provided in Figure C.1. This process (Unit 600) is the one to which you have been assigned.

Allyl chloride is produced by the thermal chlorination of propylene at elevated temperatures and relatively low pressures. Along with the main reaction, several side reactions also take place. These are shown below.

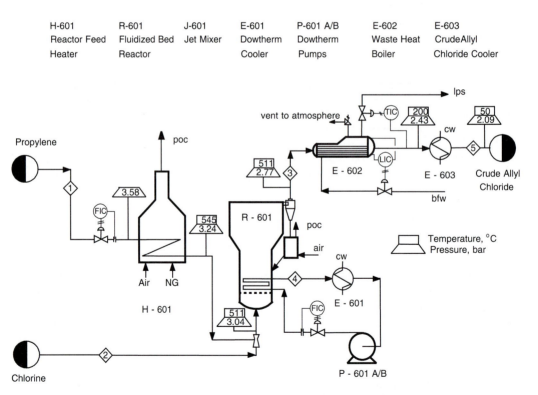

H-601	R-601	J-601	E-601	P-601 A/B	E-602	E-603
Reactor Feed Heater	Fluidized Bed Reactor	Jet Mixer	Dowtherm Cooler	Dowtherm Pumps	Waste Heat Boiler	CrudeAllyl Chloride Cooler

Figure C.1   Process Flow Diagram for the Production of Allyl Chloride (Reaction Section)

Main Reaction
$$C_3H_6 + Cl_2 \rightarrow C_3H_5Cl + HCl \qquad \Delta H_{reac,298K} = -112 \text{ kJ/mol}$$
allyl chloride

Side Reaction
$$C_3H_6 + Cl_2 \rightarrow C_3H_5Cl + HCl \qquad \Delta H_{reac,298K} = -121 \text{ kJ/mol}$$
2 chloro propene

Side Reaction
$$C_3H_6 + 2Cl_2 \rightarrow C_3H_4Cl_2 + 2HCl \qquad \Delta H_{reac,298K} = -222 \text{ kJ/mol}$$
di chloro propene

Side Reaction
$$C_3H_6 + 3Cl_2 \rightarrow 3C + 6HCl \qquad \Delta H_{reac,298K} = -306 \text{ kJ/mol}$$
carbon

The propylene feed is heated in a furnace, fired by natural gas, and brought up to reaction temperature (design conditions are given in Tables C.1 and C.2). The chlorine is mixed with the hot propylene in a mixing nozzle and then fed to the reactor.

During the thermal chlorination process, a significant amount of carbon can be produced, and it has a tendency to deposit on equipment that operates at temperatures above 400°C. For this reason, the reactor chosen for this process is a flu-

Table C.1 Flow Summary Table for Unit 600—Crude Allyl Chloride Production Facility

Stream No.	1	2	3	4	5
Temperature (°C)	25	25	511	400	50
Pressure (bar)	11.7	6.44	2.77	11.34	2.09
Vapor Fraction	1.0	1.0	1.0	0.0	1.0
Mass Flowrate (tonne/h)	3.19	1.40	4.59	16.63	4.59
Molar Flowrate (kmol/h)					
propylene	75.89	—	58.08	—	58.08
chlorine	—	19.70	—	—	—
allyl chloride	—	—	15.56	—	15.56
2-chloro propene	—	—	0.46	—	0.46
di chloro propene	—	—	1.81	—	1.81
hydrogen chloride	—	—	19.70	—	19.70
Carbon	—	—	—[1]	—	—
Dowtherm™ A	—	—	—	4.62 kg/s	—
Total Mole Flow (kmol/h)	75.89	19.70	95.61	4.62 kg/s	95.61

[1] Carbon is formed, but at a rate that does not affect the material balance.

Table C.1   **(*Continued*)**

Equipment	Fuel Gas (std m³/h)	Cooling Water (kg/s)	BFW (kg/s)	Steam (kg/s)
H-601	131²	—	—	—
E-601	—	14.5 $T_{in} = 30°C$ $T_{out} = 40°C$	—	—
E-602	—	—	0.333 (90°C)	0.333 (sat. at 6 bar)
E-603	—	6.82 $T_{in} = 30°C$ $T_{out} = 40°C$	—	—
²Actual gas flowrate shown, heater efficiency is 90%.				

idized bed with an inert solid, sand, on the reaction side. The sand provides a large surface area on which the carbon can deposit. It also acts as a scouring agent on the immersed heat transfer tubes in the reactor and prevents the build-up of carbon on the heat transfer surfaces. The carbon, which deposits preferentially on the sand, is removed by combustion in the solids regeneration unit shown in Figure C.1. The regenerated sand is sent back to the reactor, thus maintaining a constant inventory of solids in the reactor. The heat produced in the reactor, by the exothermic chlorination reactions, is removed via the heat transfer tubes through which is circulated a heat transfer medium. The heat transfer fluid is a commercially available coolant called Dowtherm A™. Physical properties of this fluid are included in Table C.3 of this assignment.

The gases leaving the reactor contain unreacted propylene along with the reaction products, as given in the flow table in Table C.1. These hot gases are cooled in a waste heat boiler and a trim cooler prior to being sent for further processing, including the refining of the allyl chloride and the separation and recycle of unused propylene.

SPECIFIC OBJECTIVES OF ASSIGNMENT

Your immediate supervisor, Ms. Jane Garcia, has taken you around the allyl chloride facility and told you some of the details of the plant operation. These are summarized below in the section on Additional Background Information. She has confirmed that the plant is currently operating at close to the design conditions given in Table C.1 and that the utility consumptions are within a few percent of those shown in Table C.1. In addition, she also provided you with a set of

Table C.2   Equipment Design Parameters (Unit 600)

J-601 Jet Mixer

Pressure Drop = 0.20 bar at design conditions
Operating Pressure = 3.24 bar (normal)
                               = 5.00 bar (maximum)

H-601 Reactor Feed Preheater

Process Side Conditions

Duty = 4000 MJ/h (normal)
           5400 MJ/h (maximum)
Operating Pressure   = 3.58 bar (normal)
                                   = 4.50 bar (maxi-
                                       mum)
Operating Temperature = 545°C (maximum)

R-601 Fluidized Bed Reactor

Operating Temperature = 511°C (normal)
                                    = 525°C (maximum)
Operating Pressure   = 3.04 bar (normal)
                                   = 4.50 bar (maximum)

Dimensions

Square cross section 3.1×3.1 m, Vessel height
    = 5 m
Fluidized bed height = 1.5 m
Heat transfer area = 23.0 m^2
Normal duty = 2188 MJ/h

P-601 A/B Dowtherm A Circulation Pumps

Operating Pressure   = 11.0 bar (normal)
                                   = 15.0 bar (maxi-
                                       mum)
Operating Temperature = 350°C (normal)
                                    = 400°C (maximum)
ΔP (normal) = 1.55 bar (22.6 psi)
ΔP (maximum) = 2.06 bar (30 psi)
Power (motor) = 2.5 kW
Flow Rate = 0.0068 m^3/s (normal)

E-601 Dowtherm A Cooler

Operating Pressure   = 11.0 bar (normal)
                                    = 15.3 bar (maxi-
                                        mum)
Operating Temperature = 400°C (maximum)
Duty = 2188 MJ/h

Heat Transfer Area = 2.6 m^2
Double pipe heat exchanger with Dowtherm
in inner tube

E-602 Waste Heat Boiler

Tube Side

Operating Pressure = 2.77 bar (normal)
                                = 3.50 bar (maximum)

Shell Side

Operating Pressure = 6.0 bar (normal)
                                = 8.0 bar (maximum)

Duty = 2850 MJ/h

Heat Transfer Area = 57.0 m^2

E-603 Crude Allyl Chloride Cooler

Tube Side

Operating Pressure = 2.43 bar (normal)
                                = 3.50 bar (maximum)

Shell Side

Operating Pressure = 4.0 bar (normal)
                                = 5.0 bar (maximum)

Duty = 1025 MJ/h

Heat Transfer Area = 52.0 m^2

battery limit conditions, Table C.4, for the utilities, feeds, and products which she has informed you are both current and accurate.

Your assignment is to provide a written report to Ms. Garcia by two weeks from now. This report, as a minimum, should contain the following items:

(i)   A cover letter to your supervisor.

(ii) An executive summary covering the following major points:

Table C.3   Properties of Dowtherm A—Heat Transfer Fluid

Properties of Dowtherm A are listed below:		
Temperature Use Range	Liquid	16°C – 400°C
	Gas	257°C – 400°C
Above 400°C Dowtherm A starts to decompose thermally.		
Liquid Properties for 350–400°C		
Thermal Conductivity	0.0943	W/m.K
Specific Heat Capacity	2630	J/kg K
Viscosity	1.4×10^{-4}	kg/m s
Density	680	kg/m^3
Vapor Pressure (400°C)	10.5	bar
Prandtl No. ($C_p\mu/k$)	3.9	

Findings on how much the throughput of Unit 600 can be increased in the short term (without the purchase of new equipment).

Findings of any potential improvements that will increase the profitability of Unit 600. You should provide an estimate of the impact of these changes (assume and internal discount rate for such improvements to be 15% p.a. before tax and all improvements should be calculated using a 5-year project life).

The impact that the proposed changes in operations might have on the environment and the health and safety of the plant personnel.

Recommendations for immediate changes in plant operations and an estimated time schedule in which these changes might be implemented.

(iii) A list of assumptions made in carrying out your study.

(iv) An appendix giving details of all important calculations made in your study.

The written report should follow the guidelines outlined in Chapter 22.

ADDITIONAL BACKGROUND INFORMATION

A process flow diagram is provided in Figure C.1, and flow summary and equipment summary tables are given in Table C.1 and C.2. This information is for the reaction section of Unit 600 only. The separation section is being studied by another group and you should not consider any changes for this section at this time. The separations section is shown in Figure C.3 and will be considered in Project 2.

Table C.4   Battery Limit Conditions for Feeds, Products, and Utilities (Unit 600)

Conditions at which feed and utility streams are available and at which products and utility streams must be returned to the boundary of the process are known as the battery limit conditions. For Unit 600 the battery limit conditions that exist are listed below. The limiting conditions are given at the equipment and take into account the pressure loss in the associated supply and return piping.

Utility	Condition at Equipment	
Cooling Water	5 bar,	30°C
Cooling Water Return	4 bar,	<45°C
Boiler Feed Water	6 bar,	90°C
High Pressure Steam[1]	41 bar,	saturated
Medium Pressure Steam[1]	11 bar,	saturated
Low Pressure Steam[1]	6 bar,	saturated
Natural Gas	4 bar,	25°C
Feeds and Products	**Condition at Process Boundary**	
Propylene	25°C, saturated vapor	
Chlorine	25°C, saturated vapor	
Crude Allyl Chloride	50°C, >2.09 bar (Stream 5)	

[1]Steam pressure at sources, such as waste heat boilers, may exceed these values in order to overcome pressure losses in header piping.

The data given in the tables and on the PFD reflect the current operating conditions and have been checked recently by your operations department. Some additional information regarding the allyl chloride facility has been provided by Ms. Garcia and is summarized below:

1. The temperature in the reactor should not exceed 525°C, since above this temperature, there is excessive coke production leading to operating problems in the downstream units. It is further recommended that the reactor temperature be maintained at close to 511°C during any changes in process operations.

2. All process exchangers using cooling water are designed to have a 5 psi (0.34 bar) pressure drop on the cooling water side for the design flow rate of cooling water. The velocity of cooling water at design conditions was set at 2 m/s and long-term operation at velocities above 3.5 m/s is not recommended due to increased erosion.

3. For the fluidized bed reactor, you may assume that the pressure drop across the bed of sand remains essentially constant regardless of the flowrate.

4. The cyclone and regenerator were designed by the vendor of the equipment to be considerably oversized and are capable of handling any additional loads that might be required during this temporary change in operations.

5. The heat transfer coefficient between the fluidized bed and the immersed heat transfer coils (the outside coefficient) is known not to vary much with fluidizing gas flowrate and may be assumed constant regardless of gas throughput. The heat exchanger in the fluidized bed may be reconfigured so that the three rows of tubes are piped in parallel.

6. Flow of all process and utility streams may be considered to be fully developed turbulent flow. Thus the pressure drop through the equipment will be proportional to the square of the velocity.

7. The conversion of propylene and chlorine in the fluidized bed will be virtually unaffected by changes in gas throughput. This is due to the long gas residence time in the reactor. In fact, the reactor's main purpose is to provide a large surface area for coke deposition and to provide good heat transfer. In addition, small changes in operating pressure of the reactor will not affect the selectivity of the reaction.

8. The crude allyl chloride (Stream 5 in Figure C.1) must be delivered to the separations section at a minimum pressure of 2.010 bar, and a maximum temperature of 50°C.

9. A manufacturer's pump curve for the circulating Dowtherm A™ pumps is provided for your use and is given in Figure C.2 of this assignment.

10. A set of original design calculations outlining the design of the units are provided for your information—some units were built slightly differently from these designs.

11. The level of bfw in E-602 is set so that all of the heat transfer tubes are covered.

You have also taken a tour of the plant recently, and in addition to confirming some of the points above with the operators, you make the following observations:

1. The Dowtherm A recirculation pump (P-601 A) is making a high-pitched whining noise.

2. Steam is leaking from the safety relief valve placed on top of E-602.

3. Some of the insulation on the pipe leading from the reactor R-601 to E-602 has come loose and is hanging from the pipe.

PROCESS DESIGN CALCULATIONS

Fluidized Bed Reactor, R-601

Heat generated in reactor = Q_R = 2188 MJ/h
Bed solids are 150 µm sand particles with density (ρ_s) of 2650 kg/m^3

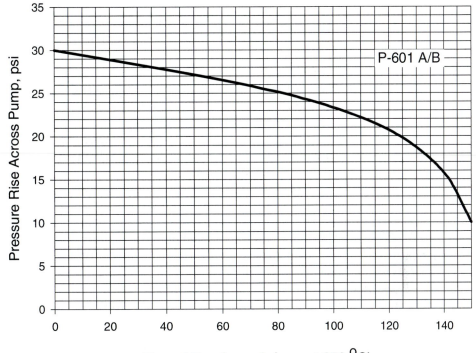

Figure C.2 axes: Pressure Rise Across Pump, psi vs. Flow of Dowtherm A (gpm at 350 °C), curve labeled P-601 A/B

Figure C.2   Pump Curve for P-601 A/B, Dowtherm A Circulation Pumps

At conditions in the reactor, the process gas has the following properties

$$\rho_g = 2.15 \text{ kg/m}^3 \text{ and } \mu_g = 2.25 \times 10^{-5} \text{ kg/m s}$$

Using the Correlation of Wen and Yu [1] we get:

$$Re_{p,mf} = [1135.7 - 0.0408 \, Ar]^{0.5} - 33.7$$

where $Ar = \dfrac{d_p^3(\rho_s - \rho_g)\rho_g g}{\mu_g^2} = \dfrac{(150 \times 10^{-6})^3 \, (9.81)(2650 - 2.15)(2.15)}{(2.25 \times 10^{-5})^2} = 372$

Therefore, $Re_{p,mf} = [1135.7 + 0.0408 \times 372]^{0.5} - 33.7 = 0.2246$

$$Re_{p,mf} = \frac{u_{mf} d_p \rho_g}{\mu_g} \quad \therefore \quad u_{mf} = \frac{(0.2246) \, (2.25 \times 10^{-5})}{(2.15) \, (150 \times 10^{-6})} = 0.0157 \text{ m/s}$$

Total volumetric flow of gas at inlet conditions to the bed = $V_{gas} = 0.5674 \text{ m}^3/\text{s}$

We need to get good heat transfer so we will operate the bed at 5 times u_{mf}, which puts us into the bubbling bed regime.

Free bed area (without h.t. tubes) = $V_{gas} / 5u_{mf} = 0.5674/(510.0157) = 7.2 \text{ m}^2$

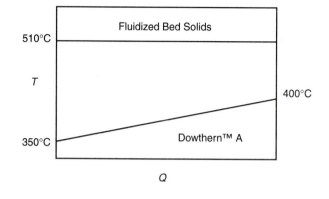

Look at heat transfer area for fluidized bed.

Heat transfer area required in bed = A_f
Overall heat transfer coefficient = U
Assume that the fluidized solids are well mixed and isothermal and assume that the cooling medium enters the bed at 350°C and leaves at 400°C, which is the maximum operating temperature for Dowtherm A.

$$\Delta T_{lm} = \frac{(160 - 110)}{\ln (160/110)} = 133.4°C$$

inside film heat transfer coefficient, h_i, is calculated from the Seider-Tate equation:

$$Nu = 0.023 \, Re^{0.8} \, Pr^{\,0.33} \, (\mu/\mu_w)^{0.14}$$

If we assume a velocity of 1.5 m/s inside a 3 inch sch 40 pipe (ID = 0.0779 m, OD = 0.0889 m), then

$$Re = (0.0779) \, (1.5) \, (680)/(1.4 \times 10^{-4}) = 567 \times 10^3$$

$$Nu = h_i d/k = 0.023 \, (567 \times 10^3)^{0.8} \, (3.9)^{0.33} = 1445$$

$$h_i = (1445) \, (0.0943)/(0.0779) = 1750 \, W/m^2K$$

outside film heat transfer coefficient, h_o = 267 W/m²K (from previous plant operating data) fouling coefficient on inside = 2500 W/m²K

$$d_o \, / \, d_i = 0.0889/0.0779 = 1.14$$

Ignoring the wall resistance, we get the overall transfer coefficient U_o from

$$U_o = \cfrac{1}{\cfrac{1}{267} + \cfrac{1.14}{2500} + \cfrac{1.14}{1750}} = 206 \, W/m^2K$$

$$A_0 = \frac{Q_R}{U_o \Delta T_{\text{lm}}} = \frac{2188 \times 10^6}{(3600)(206)(133)} = 22.2 \text{ m}^2$$

Now assuming tubes are 10 ft long and 3 inches in diameter (sch 40), the heat transfer area per tube is

$$\pi d_o L = (3.142)\,(0.0889)\,(10)\,(0.3048) = 0.8513 \text{ m}^2$$

$\therefore$ number of tubes required, $N_T = 22.2/0.8513 = 27$ tubes.

Using 3 layers of 9 tubes piped in series and placed in horizontal rows in the bed, each row occupies the following csa of bed

$$(9)\,(d_o)\,(L) = (9)\,(0.0889)\,(3.048) = 2.4 \text{ m}^2$$

$\therefore$ total csa for bed $= 2.4 + 7.2 = 9.6 \text{ m}^2$

Use a square bed with side dimensions $= (9.6)^{0.5} = 3.1$ m (10.2 ft)

Check Velocity in Tubes

csa for flow of Dowtherm A in tubes $= \pi\, d_i^2/4 = (3.142)(0.0779)^2/4 = 4.766 \times 10^{-3} \text{ m}^2$

flow of Dowtherm A $= 6.797 \times 10^{-3} \text{ m}^3/\text{s}$

$\therefore$ velocity of Dowtherm™ A in tubes $= 6.797 \times 10^{-3}/4.766 \times 10^{-3} = 1.43$ m/s $\Rightarrow$ assumption is **OK**

Pressure Drop in Tubes

$Re = 5.41 \times 10^5$ $\therefore$ friction factor, $f = 0.0045$ (with $e/d = 0.0006$)

$\Delta P = 2f L_{eq} \rho u^2/d = (2)\,(0.0045)\,(680)\,(1.43)^2\, L_{eq}/(0.0779) = 161\, L_{eq}$

Now L_{eq} = equivalent length of pipe in three rows of heat transfer pipes in fluidized bed

$= (27)\,(3.048)\,(1.5) = 123$ m (take this as 1.5 times length of pipe to account for fittings)

$\therefore \Delta P = (162)\,(123) = 0.20$ bar

Set bed height (height of sand above distributor plate) $= 1.5$ m

This gives a gas residence time in the bed of $(7.2)\,(1.5)\,(0.45) / 0.5674 = 8.6$ s This should be plenty of time since complete reaction should only take about 2–3 s

$$\Delta P_{bed} = h_{bed}\, \rho_{sand}\, (1 - \varepsilon)g = (1.5)\,(2650)\,(1 - 0.45)\,(9.81) = 0.214 \text{ bar}$$

Assume 0.04 bar for distributor loss and 0.064 bar for cyclones to give the overall equipment pressure drop

$$\Delta P_{reactor} = 0.214 + 0.04 + 0.016 = 0.27 \text{ bar}$$

Design of Fluidized Bed is given in sketch below.

side view of bed showing 3 rows of 9 tubes

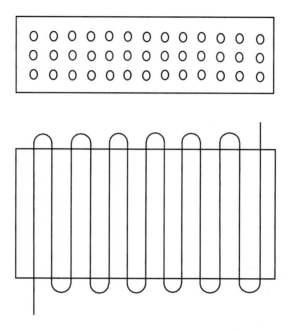

Overhead view of bed showing piping arrangement for one row; each row is piped in series with the row below.

For exchanger **E-601** we have $\Delta T_{lm} = (360 - 320)/\ln(360/320) = 340°C$ and $U = 850$ W/m^2 K (approximately equal resistances on both sides and small fouling resistances) heat transfer area A is given by

$$A = 2188 \times 10^6/(3600)(340)(850) = 2.1 \text{ m}^2$$

Use a double pipe heat exchanger since area is small

Cooling water flowrate $= 2188\times10^6 / (3600)(4180)(10) = 14.5$ kg/s

Dowtherm flowrate $= 2188 \times 10^6 / (3600)(2630)(50) = 4.62$ kg/s

Pressure drop across the exchanger $= 0.34$ bar for Dowtherm and cooling water

Velocity of cooling water through exchanger set at 2 m/s

Dowtherm A™ Cooling Loop

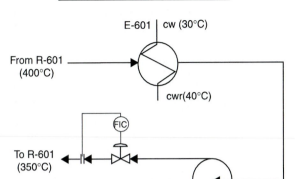

For pumps **P-601 A/B,** assume

> 0.34 bar (5 psi) pressure drop across exchanger on Dowtherm side
> 0.14 bar (2 psi) pressure drop for piping
> 0.85 bar (12.4 psi) pressure drop across the control valve
> 0.20 bar (2.9 psi) pressure drop across the reactor exchanger
> total loop pressure drop = 1.53 bar = 22.3 psi
> flow of Dowtherm = 4.62 / 680 = 6.797×10^{-3} m^3/s = 108 gpm

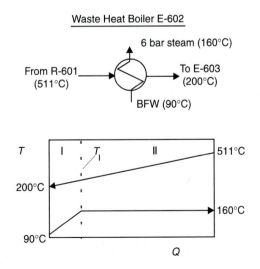

Power required for pumping liquid $= \dot{v}\Delta P = (6.797 \times 10^{-3})(1.53 \times 10^5) = 1.04$ kW

Assuming an efficiency of 45%, we get that the shaft power $= 1.04/0.45 = 2.31$ kW.

Use a 2.5 kW pump plus a spare.

Zone I $U = 90$ W/m^2 K (all resistance on gas side)
Zone II $U = 90$ W/m^2 K (all resistance on gas side)

Heat released as gas cools from 511 to 200°C is 2850 MJ/h = 792 kW

$$\Delta H_{bfw\text{-}steam} = 2380 \text{ kJ/kg}$$

steam flowrate $= 792/2380 = 0.333$ kg/s ($h_L = 376.9$, $h_{L,sat}$
$= 675.5$, $h_{V,sat} = 2756.9$ kJ/kg)

$$Q_{90-160°C,Liq} = (0.333)(675.5 - 376.9) = 99 \text{ kW}$$

$$Q_{160°C, Liq-Vap} = (0.333)(2756.9 - 675.5) = 693 \text{ kW}$$

For Zone I

$$(T_I - 200)/(510 - 200) = 99/792 \Rightarrow T_I = 238.8°C$$

$$\Delta T_{lm} = \frac{(239 - 160) - (200 - 90)}{\ln \dfrac{(239 - 160)}{(200 - 90)}} = 93.6°C$$

$$A_I Q_I/U_I \Delta T_{lm} F = 99 \times 10^3/(90)(93.6)(1.0) = 11.8 \text{ m}^2$$

For Zone II

$$\Delta T_{lm} = \frac{(239 - 160) - (511 - 160)}{\ln \dfrac{(239 - 160)}{(511 - 160)}} = 182.4°C$$

$$A_{II} Q_{II}/U_{II} \Delta T_{lm} = 693 \times 10^3/(90)(182.4) = 42.2 \text{ m}^2$$

Total Area $A = A_I + A_{II} = 11.8 + 42.2 = 54 \text{ m}^2$

$$C_{p,gas} = 1490 \text{ J/kg.K}$$

$$Q = m\, C_{p,gas} \Delta T = (4590)(1490)(200 - 50)/3600 = 0.285 \text{ MW}$$

Crude Allyl Chloride Cooler, E-603

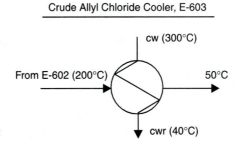

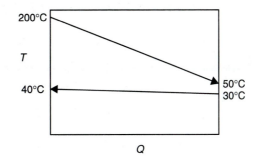

$$m_{cw} = (0.285 \times 10^6)/(4180)\,(10) = 6.82\ \text{kg/s}$$

$$\Delta T_{\text{lm}} = (160 - 20)/\ln(160/20) = 67.3°C$$

$$U = 90\ \text{W/m}^2\,\text{K}$$

$$A = Q/U\Delta T_{\text{lm}}F = (0.285 \times 10^6)/(90)\,(67.3)\,(0.95) = 49.5\ \text{m}^2$$

REFERENCE

1. Wen, C. Y., and Y. H. Yu, *AIChE J.*, **12,** 610 (1966).

2

Design and Optimization of a New 20,000 Metric Tons per Year Facility to Produce Allyl Chloride at La Nueva Cantina, Mexico

BACKGROUND

Recent developments at the Alabama plant have brought into question the stability of our allyl chloride production capacity. It has been noted that the North American market for allyl chloride consumption will probably grow extensively in the next decade and that additional production capacity may be needed. In order to maintain our current market share, and to remain competitive in this market, it may become necessary to build a new allyl chloride facility.

With this in mind, a new (grass roots) production facility to produce commercial grade allyl chloride is being investigated. This facility would most likely be situated in La Nueva Cantina, Mexico, close to our petrochemical facility, which would be able to supply propylene. The supply of chlorine from local manufacturers will also be plentiful. Your group has been given the task of doing a process estimate for this new facility.

ASSIGNMENT

Your assignment is to design and optimize a new grass roots 20,000 metric tons per year allyl chloride facility. This facility will produce commercial-grade allyl chloride from the synthesis reaction between propylene and chlorine. The front end process is to use fluidized bed technology similar to that currently in use in Unit 600 at Beaumont, Texas (see project 1). This assignment concerns the design of a new plant and, except for the fluidized bed reactor unit, the plant may be re-configured in any way that you feel is appropriate. For the fluidized bed reactor, kinetic equations are not available, therefore, your design should be similar to the Beaumont unit. With this in mind, the reactor temperature should be close to 511°C, and the ratio of propylene to chlorine into the reactor should remain the same. Although the selectivity of the reaction will change with changes in reactor pressure, you may assume for this preliminary design that the selectivity is unaf-fected by changes in pressure. Your final design should be the one that maxi-mizes the net present value (*NPV*) of the project under the following economic constraints:

> After tax internal hurdle rate = 10%
> Depreciation = MACRS (6 year schedule, see Prob. 4.18)
> Taxation rate = 15%
> Labor costs to be based on U.S. equivalent wage rate
> Construction period = 2 years
> Project plant life = 15 years after startup

Additional process information and some hints regarding problem-solving strat-egy are given below. You should plan to submit your written report in four weeks. The guidelines for written and oral reports given in Chapter 22 should be followed for this assignment.

PROBLEM-SOLVING METHODOLOGY

The optimization of a process as large and complicated as this one is not a trivial matter. To help you in this endeavor the following hints are given. (You may use all or none of them as you see fit.)

1. When initially running case studies using a process simulator, you should use any shortcut methods for simulating distillation columns that are avail-able. Avoid the rigorous methods for this type of preliminary work. You may want to use the rigorous (tray-to-tray) methods at the end of your work for the optimized case.
2. Establish a base case from the information provided. This should include operating costs and capital investments for the process shown in Figures C.1, C.3, and C.4, using the new production rate of 20,000 metric tons/year. You should make sure that you have all the equipment and costs included.
3. See where the major capital investment and major variable operating costs lie and use this information to focus on where your optimization should begin.

Figure C.3   Separations Section of Allyl Chloride Production Facility (Unit 600)

Note: This is a multi-variable optimization and within the time frame given for solving this problem, you can afford to concentrate on only a few variables. Identify the key decision variables to be manipulated, that is, the variables that have the greatest impact on the *NPV*.

4. Do not let the computer turn your mind to mush! Think about how changes in operating variables and other things will affect the economics, and try to predict the direction in which the *NPV* will change as each variable changes. You should be able to rationalize, at least qualitatively, why a certain change in a variable causes the observed change in the *NPV*.

PROCESS INFORMATION

To help you in this task the following information has been included with this assignment.

1. A flowsheet and a process description for the existing separations section of Unit 600 are included in Figures C.3, C.4 and Tables C.5 and C.6.
2. Product specifications are included in Table C.7.

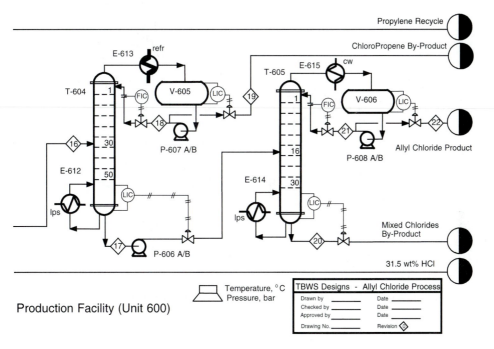

E-610 E-611 V-604 P-604A/B P-605A/B T-604 E-612 E-613 V-605 P-606A/B P-607A/B T-605 E-614 E-615 V-606 P-608A/E
C3   C3   C3   C3 Bottm C3 Reflux Chl-P Chl-P Chl-P Chl-P Chl-P Botm Chl-P   Allyl Allyl Allyl Allyl Allyl
Condsr Reblr Reflux Pumps   Pumps   Tower Reblr Condsr Reflux Pumps   Pumps   Tower Reblr Condsr Reflux Reflux
              Drum   Drum   Drum   Pumps

Figure C.3 (*Continued*)

3. Note that the current process (Unit 600) is very energy intensive, since very cold temperatures and expensive refrigeration are used throughout the process. Remember that the existing Beaumont plant was designed over thirty years ago when electricity and energy were relatively cheap. The optimal process configuration for today's conditions may be significantly different from that of the existing facility!

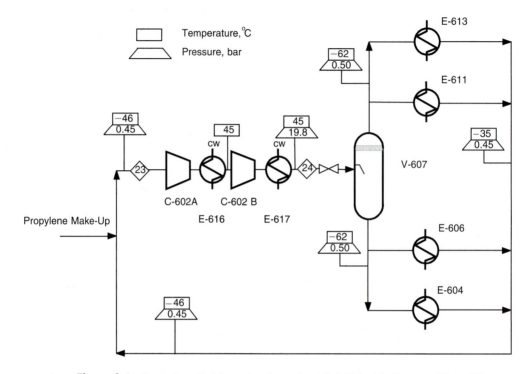

Figure C.4   Propylene Refrigeration Loop for Allyl Chloride Process (Unit 600)

Table C.5 Flow Summary Table for Allyl Production Process, Figures C.3 and C.4

Stream No.	5	6	7	8	9	10	11	12	13	14
Temperature (°C)	50	−50	−50	46	−57	−56	25	10	10	45
Pressure (bar)	2.09	1.50	1.50	1.50	1.40	1.5	3.0	1.4	1.3	19.9
Vapor Fraction	1.0	1.0	0.0	0.0	0.0	0.087	0.0	0.0	1.0	0.0
Mass Flowrate (tonne/h)	4.59	0.27	4.32	1.45	0.19	3.14	1.48	2.17	2.44	2.44
Molar Flowrate (kmol/h)										
propylene	58.08	2.72	55.36	0.55	3.65	57.53	—	—	57.53	58.08
allyl chloride	15.56	0.01	15.55	15.55	—	0.01	—	0.01	—	—
2-chloro propene	0.46	—	0.46	0.46	—	—	—	—	—	—
di-chloro propene	1.81	—	1.81	1.81	—	—	—	—	—	—
hydrogen chloride	19.70	4.19	15.51	0.01	1.03	19.69	—	19.11	0.58	—
water	—	—	—	—	—	—	82.00	81.88	0.12	—
Total Molar Flowrate (kmol/h)	95.61	6.92	88.69	18.38	4.68	77.23	82.00	101.00	58.23	58.08

Stream No.	15	16	17	18	19	20	21	22	23	24
Temperature (°C)	−40	59	60	30	30	106	55	55	−46	45
Pressure (bar)	1.5	2.0	1.5	1.4	1.4	1.5	1.4	1.4	0.45	19.9
Vapor Fraction	0.0	0.0	0.0	0.0	0.0	0.0	0.0	0.0	1.0	0.0
Mass Flowrate (tonne/h)	0.03	1.42	1.39	1.52	0.04	0.21	0.47	1.18	25.31	25.31
Molar Flowrate (kmol/h)										
propylene	0.81	—	—	—	—	—	—	—	602.5	602.5
allyl chloride	—	15.54	15.52	0.84	0.02	0.08	6.13	15.44	—	—
2-chloro propene	—	0.46	0.01	18.99	0.45	—	—	0.01	—	—
di-chloro propene	—	1.81	1.81	—	—	1.80	—	0.01	—	—
hydrogen chloride	0.01	—	—	—	—	—	—	—	—	—
water	—	—	—	—	—	—	—	—	—	—
Total Molar Flowrate (kmol/h)	0.82	17.81	17.38	19.83	0.47	1.88	6.13	15.46	602.5	602.5

Table C.6   Process Description of Unit 600

Refer to the process flow diagram for Unit 600, Allyl Chloride Purification Process, Beaumont, Texas (Figure C.3).

Crude allyl chloride, Stream 5, from the reaction section of Unit 600 (Figure C.1) enters the Allyl Product Cooler, E-604, at 50°C and 2.1 bar. This stream is cooled to –50°C using the circulating liquid propylene refrigerant. The two-phase mixture leaving E-604 is fed to V-601 where the liquid stream is taken off and is fed to the HCl column T-601. The HCl column removes essentially all the HCl and propylene from the cooled crude allyl chloride feed as overhead product at approximately –57°C. This stream is mixed with the vapor coming from V-601 and is fed to E-607 where it is heated with low pressure steam.

The bottom product from T-601 contains essentially all the chlorinated hydrocarbon derivatives and a small amount of propylene. This stream is fed to T-603 where the remaining propylene is removed as the overhead product at approximately –40°C. The bottoms product from T-603 is fed to column T-604 where 95 mole% chloro-propene is removed overhead at approximately 30°C. The bottom product from T-604 is fed to the Allyl Tower, T-605, where 99.9% by mole pure allyl chloride is removed as overhead product at 55°C and the sent to storage. The bottom product from T-605 contains 95 mole% 1,2, di-chloro-propene and this stream is sent to storage after being cooled in an offsite heat exchanger (not shown on Figure C.3).

The stream leaving E-607 is fed to T-602 where it mixes with water at 20°C. The flow of water is controlled to give an aqueous solution of hydrochloric acid with 31.5 wt% HCl. This concentration of acid is equivalent to a liquid density of 20° Baumé. The use of degrees Baumé is the common method by which hydrochloric acid is specified. The vapor stream leaving T-602 contains all the propylene and small amounts of water and HCl. This stream is sent to one of a pair of Acid Traps, V-603 A&B, where the water and HCl are removed (by adsorption onto activated carbon). The vapor stream leaving the absorbers is pure propylene. This propylene stream is sent to a two-stage compressor C-601 A&B with intercooler E-408 and condenser E-409. The stream leaving the condenser is a liquid at 45°C and is recycled to the propylene storage tank for Unit 600.

Four of the exchangers in Unit 600 (E-604, E-606, E-611, and E-613) require heat to be removed from the process stream at temperatures below 35°C. In order to do this, a refrigeration system is required. In Unit 600, this is achieved by circulating a stream of cold (–62°C) propylene through these exchangers. The refrigeration loop is shown in the second PFD for Unit 600, Figure C.4. The refrigeration is achieved by taking a high-pressure (20 bar) stream of liquid propylene (45°C) and flashing it down to low-pressure (0.5 bar). Cooled liquid and vapor propylene streams (–62°C) are sent to the four process exchangers where they provide the necessary cooling. The warmed propylene vapor (–46°C) is recycled back to the refrigeration loop compressors C-602 A&B, the intercooler E-616 and the condenser E-617. A makeup propylene stream is provided to account for minor system leaks.

Table C.7   Product Specifications and By-Product Costs (Unit 600)

Product Specifications		
Product	**Required Purity**	**Battery Limit Condition**
allyl chloride	> 99.9% by mole allyl chloride	Liq, $T < 55°C$, P >1.5 bar
By-Product	**Required Purity**	**Battery Limit Condition**
mixed chlorides	> 95% by mole 1,2 di-chloro-propene	Liq, $T < 50°C$, P >1.2 bar
chloro-propene	> 95% by mole chloro-propene	Liq, $T < 50°C$, P >1.2 bar
31.5 wt% hydrochloric acid	31.5 wt% hydrochloric acid ± 0.1 wt %	Liq, $T < 45°C$, P >1.2 bar

By-Product and Waste Stream Selling Prices/Costs	
By-Product	**Selling Price**
mixed chlorides	$ 0.10 /kg[+]
chloro-propene	$ 0.15 /kg[+]
31.5 wt% hydrochloric acid (20° Baumé)	From *Chemical Marketing Reporter*

Waste Stream	**Cost of Disposal**
Waste Acid Stream (cost of regenerating carbon)	$ 0.40 /kg of HCl + hydrocarbons collected on carbon

[+]These are credits that we will receive from our petrochemical complex for supplying these chemicals, which must meet the specifications given above. We may alternatively pay to dispose of these chemicals at a cost of $0.25 /kg. For this case, no specifications need to be met (i.e., these streams are now waste streams rather than by-products).

3

Scale-Down of Phthalic Anhydride Production at TBWS Unit 700

BACKGROUND

You have recently joined the TBWS Chemical Corporation. One of TBWS major businesses has always been production of phthalic anhydride from naphthalene. Phthalic anhydride production is integrated as part of a large chemical plant, in which naphthalene is produced and in which phthalic anhydride is immediately used to make polyester resins. In recent years, there have been some problems. Some end users have complained about the quality of the resins produced and have taken their business to other companies that produce phthalic anhydride from o-xylene. Therefore, our plant, which had been designed to produce 100,000 metric tons/year of phthalic anhydride from naphthalene, was scaled back to about 80,000 metric tons/year several years ago. We are now forced to scale down production once again, due to the loss of another large customer. Marketing informs us that we may lose additional customers. Research is working on development of catalysts for the o-xylene reaction, but their results are not expected for up to a year. There is an immediate need to determine how to scale down operation of our plant to 50% of current capacity (40,000 metric tons/year). We would like to accomplish this without a shutdown, since one is not scheduled for a few months. If 50% scale down cannot be achieved without a shutdown, we need to know how much scale down is possible immediately. Specifically, you are to determine the maximum possible scale down, up to 50%, under current operating conditions, and you are to define these operating conditions. Furthermore, you are to determine how the plant can be scaled down to 50% of current capacity, what operating conditions are required, and what capital expenditures, if any, are needed. Any suggestions for plant improvements that can be made during shutdown are encouraged. You must clearly define the consequences of any changes that you recommend for this process and the consequences of these changes on other processes that might be affected. It should be noted that one possible scenario is to operate the plant at design capacity for 6 months of the year and shutdown the plant for the remaining 6 months. Although this solution might work, we are reluctant to lay off our operators for half the year and also to purchase additional storage in order to store enough phthalic anhydride to supply our customers for the 6 months that the plant is down. You should not consider this option any further.

For your first assignment, you are to address the issues described above for the portion of the process before the switch condensers (see Figure C.5).

PHTHALIC ANHYDRIDE PRODUCTION

Unit 700 now produces about 80,000 metric tons/yr of phthalic anhydride. The feeds are essentially pure naphthalene and excess air. These are pressurized, heated and vaporized (naphthalene), and reacted in a fluidized bed with a vanadium oxide on silica gel catalyst. The reactions are:

$$C_{10}H_8 + \frac{9}{2}O_2 \rightarrow C_8H_4O_3 + 2H_2O + 2CO_2$$

naphthalene   *phthalic anhydride*

$$C_{10}H_8 + 6O_2 \rightarrow 2C_4H_2O_3 + 2H_2O + 2CO_2$$

maleic anhydride

$$C_{10}H_8 + \frac{3}{2}O_2 \rightarrow C_{10}H_6O_2 + H_2O$$

naphthoquinone

Additionally, the complete and incomplete combustion reactions of naphthalene also occur. The large exothermic heat of reaction is removed by molten salt circulated through coils in the reactor. The molten salt is used to produce high-pressure steam. Total conversion of naphthalene is very close to 100%. The reaction products proceed to a set of devices known as switch condensers. These are described in detail later. Design and operation of these devices is provided under contract by CONDENSEX. They guarantee us that their condensers can operate at any capacity and provide the same separation as in current operation, as long as the pressure and the composition of the condensable portion of Stream 10 remains constant. The net result of the switch condensers is that essentially all of the light gases and water leave as vapor, with small amounts of maleic and phthalic anhydrides, and that the remaining anhydrides and naphthoquinone leave as liquid. The liquid pressure is then reduced to vacuum for distillation.

P-701A/B	C-701	H-701	E-701	R-701	E-702	P-702A/B	E-703	P-703A/B
Naphthalene	Feed Air	Air	Naphthalene	Fluidized	Molten	Molten	BFW	BFW
Feed Pumps	Compressor	Preheater	Furnace	Bed Reactor	Salt Cooler	Salt Pumps	Preheater	Pumps

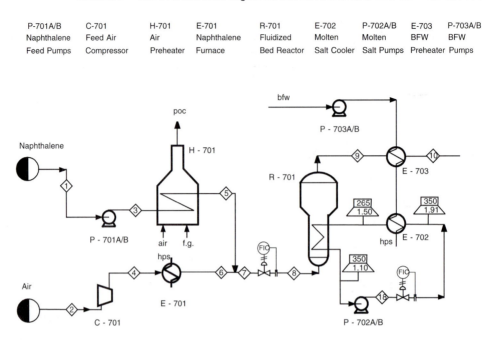

Figure C.5 Process Flow Diagram for the Production of Phthalic Anhydride from Naphthalene (Unit 700)

The first column removes maleic anhydride impurity overhead, and the second column removes the phthalic anhydride product overhead.

Organic waste is burned for its fuel value. The dirty air, Stream 11, must be treated. The anhydrides are scrubbed using water, which is then sent to the on-site waste water treatment facility.

OTHER INFORMATION

Other pertinent information is appended, including pump and compressor curves, Figure C.6 and Figure C.7, a flow summary table, Table C.8, and an equipment list, Table C.9.

ASSIGNMENT

Your assignment is to provide recommendations as to how much immediate scale-down is possible, and what, if any, modifications would be needed to scale down by 50%. For now, you are only to consider the portion of the process prior

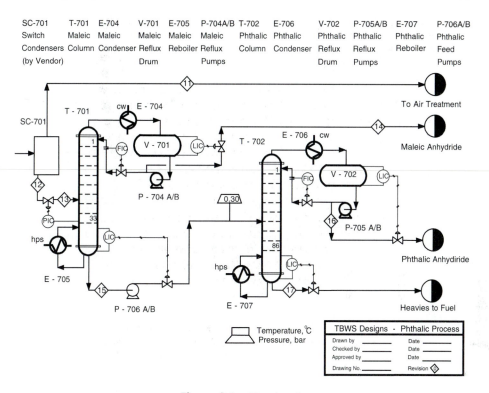

Figure C.5   (*Continued*)

to the switch condensers. You should also recommend any other changes that you feel should be made to improve performance in Unit 700. Since our plant is due for annual shutdown in a few months, we want specific recommendations as to what should be done at that time and the cost of these alterations and/or modifications.

Specifically, you are to prepare the following by . . . (two weeks from now)

1. A written report detailing the maximum scale-down possible, how to achieve 50% scale-down, recommendations, and costs associated with scaling-down production in Unit 700.
2. A list of new equipment to be purchased, including size, cost, and materials of construction.
3. An analysis of any change in the annual operating cost created by your recommended modifications.
4. A legible, organized set of calculations justifying your recommendations, including any assumptions made.

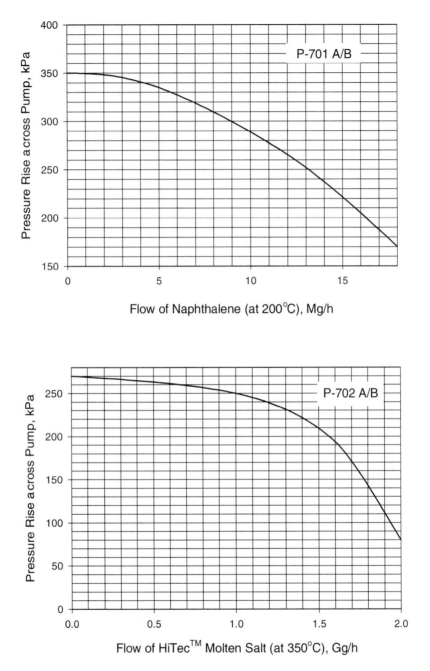

Figure C.6   Pump Curves for P-701 A/B and P-702 A/B

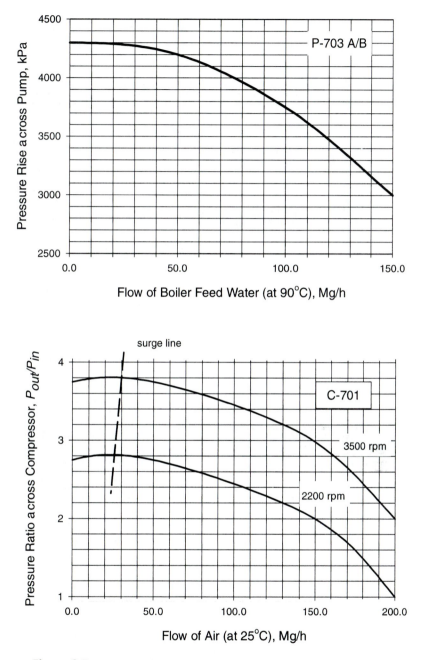

Figure C.7   Pump and Compressor Curves for P-703 A/B and C-701

Table C.8 Flow Summary Table for Current Operation of Phthalic Anhydride Production Facility, Unit 700 (see Figure C.5)

Stream No	1	2	3	4	5	6	7	8	9	10
Temperature (°C)	200	25	200	164	400	240	263	263	360	160
Pressure (bar)	0.80	1.01	3.35	3.10	2.85	2.85	2.75	2.25	2.00	1.70
Vapor mole fraction	0.0	1.0	0.0	1.0	1.0	1.0	1.0	1.0	1.0	1.0
Flowrate (tonne/h)	12.82	144.25	12.82	144.25	12.82	144.25	157.07	157.07	157.07	157.07
Flowrate (kmol/h)										
Naphthalene	100.0	—	100.0	—	100.0	—	100.0	100.0	—	—
Oxygen	—	1050.0	—	1050.0	—	1050.0	1050.0	1050. 0	469.0	469.0
Phthalic Anhydride	—	—	—	—	—	—	—	—	70.0	70.0
Maleic Anhydride	—	—	—	—	—	—	—	—	16.0	16.0
Naphthoquinone	—	—	—	—	—	—	—	—	2.0	2.0
Carbon Dioxide	-	—	—	—	—	—	—	—	306.0	306.0
Carbon Monoxide	—	—	—	—	—	—	—	—	50.0	50.0
Nitrogen	—	3950.0	—	3950.0	—	3950.0	3950.0	3950.0	3950.0	3950.0
Water	—	—	—	—	—	—	—	—	238.	238.
Total (kmol/h)	100.0	5000.0	100.0	5000.0	100.0	5000.0	5100.0	5100.0	5101.0	5101.0

Stream No	11	12	13	14	15	16	17	18
Temperature (°C)	131	131	131	141	241	190	241	350
Pressure (bar)	1.40	1.40	0.15	0.11	0.30	0.05	0.20	3.00
Vapor mole fraction	1.0	0.0	0.0	0.0	0.0	0.0	0.0	0.0
Flowrate (Tonne/h)	145.00	12.07	12.07	1.61	10.47	10.06	0.40	1624.72
Flowrates in kmol/h								HiTec™ Molten Salt
Naphthalene	—	—	—	—	—	—	—	—
Oxygen	469.0	—	—	—	—	—	—	—
Phthalic Anhydride	0.7	69.0	69.0	0.69	69.0	68.0	0.69	—
Maleic Anhydride	0.8	15.0	15.0	15.0	0.015	0.015	—	—
Naphthoquinone	—	2.0	2.0	0.08	1.9	0.002	1.91	—
Carbon Dioxide	306.0	—	—	—	—	—	—	—
Carbon Monoxide	50.0	—	—	—	—	—	—	—
Nitrogen	3950.0	—	—	—	—	—	—	—
Water	238.0	—	—	—	—	—	—	—
Total kmol/h	5014.0	86.0	86.0	16.0	71.0	68.0	2.60	17660

Table C.8 (*Continued*)

Stream No.	hps to E-701	conden. from E-701	hps to E-707	conden from E-707	hps to E-705	conden from E-705	cw to E-706	cwr from E-706	cw to E-704	cwr from E-704	bfw to E-703	bfw to E-702	hps from E-702
Temperature (°C)	254	254	254	254	254	254	30	45	30	45	91	173	254
Pressure (bar)	42.4	42.4	42.4	42.4	42.4	42.4	5.16	4.86	5.16	4.86	42.4	42.4	42.4
Vapor mole fraction	1.0	0.0	1.0	0.0	1.0	0.0	0.0	0.0	0.0	0.0	0.0	0.0	1.0
Total tonnes/h	6.73	6.73	9.36	9.36	1.89	1.89	267.84	267.84	19.62	19.62	104.98	104.98	104.98

Table C.9   Equipment Summaries (Unit 700)

Compressor (assume efficiency independent of flowrate)

C-701 centrifugal, 5670 kW, 80% efficient @ design flowrate and 3500 rpm can operate at two discrete rpm values, as shown on compressor curve; surge line also shown on compressor curve

Pumps (assume efficiency independent of flowrate)

P-701 A/B   centrifugal, 1.3 kW, 50% efficient
P-702 A/B   centrifugal, 54 kW, 70% efficient
P-703 A/B   centrifugal, 140 kW, 80% efficient
P-704 A/B   centrifugal, 0.5 kW, 40% efficient
P-705 A/B   centrifugal, 4.4 kW, 50% efficient
P-706 A/B   centrifugal, 0.8 kW, 40% efficient

Fired Heater (process fluid flows through a set of tubes with a natural gas or liquid fuel fired flame providing the radiant and convective heat transfer necessary to heat the fluid to the desired temperature)

H-701 fired heater, $Q = 9,350$ MJ/hr
consists of four identical banks of tubes—currently these are all in operation and are operating in parallel—piping and valving exist to run any or all tube banks in any configuration (i.e., series, parallel, etc.)—there is a control system that maintains the temperature of Stream 5 by measuring the temperature of Stream 5 and altering the natural gas and air feed rate.

Heat Exchangers (all one pass on each side, unless otherwise noted; h_i refers to tube side; tube wall resistance negligible, unless otherwise noted)

E-701   uses high-pressure steam, steam in shell, $Q = 11,370$ MJ/hr
$A = 695$ m^2, $U = 112$ W/m^2°C, $h_i = 114$ W/m^2°C

E-702   makes high pressure steam, steam in shell, Q = 215,407 MJ/hr
$A = 539$ m^2, $U = 2,840$ W/m^2°C, $h_i = 3,960$ W/m^2°C
hps supplies Unit 700 needs, excess steam used in Unit 300

E-703   preheats high-pressure bfw, bfw in shell, $Q = 36,900$ MJ/hr
$A = 1519$ m^2, $U = 57$ W/m^2°C, $h_i = 63$ W/m^2°C

E-704   total condenser for T-701, condensing fluid in shell
$A = 5.52$ m^2, $U = 600$ W/m^2°C, all resistance on water side

E-705   reboiler for T-701
$A = 50$ m^2, $U = 1,400$ W/m^2°C, approximately equal resistances

E-706   total condenser for T-702, condensing fluid in shell
$A = 51$ m^2, $U = 600$ W/m^2°C, all resistance on water side

E-707   reboiler for T-702
$A = 243$ m^2, $U = 1,400$ W/m^2°C, approximately equal resistances

Reactor

The reactor is a fluidized bed, which means that the bed temperature is essentially constant and equal to the exit temperature of the gas.

Table C.9   (*Continued*)

R-701   fluidized bed with vanadium oxide catalyst coated on silica gel
molten salt circulated in tubes to remove heat of reaction
heat exchange area = 15,850 m^2 (parallel tube banks within reactor)
U = 100 W/m^2°C, all resistance on reactor side
heat removal required = 2.154×10^5 MJ/hr
reactor pressure drop unaffected by flow rate

Molten Salt Loop

Molten salt is used to remove the heat generated in the reactor. It circulates in a closed loop and is thermally regenerated by making high pressure steam in E-702. The properties of this molten salt, known as HiTec™, may be found in the 6th edition of Perry's, page 9–77.

Switch Condenser

SC-701 There are three sets of condensers. Due to the low partial pressure of phthalic anhydride in the stream, it desublimates rather than condenses. Therefore, the process stream is cooled using a low temperature oil in tubes to promote desublimation. Then, after solid is loaded on the heat transfer surface, higher temperature oil is circulated in the tubes to melt the solid. There are three such devices, one operating in desublimation mode, one operating in melting mode, and one on standby. The net result is a liquid stream containing the condensables, and a vapor stream containing water and the non-condensables. These condensers are designed and maintained under contract by CONDENSEX. They indicate operation at any scale is possible as long as the pressure of Stream 10 remains within 10% of current operating conditions, and as long as the relative composition of the condensables remains approximately constant.

Distillation Columns (For both distillation columns, it may be assumed that weeping begins to occur at 35% of flooding. Both use high pressure steam and cooling water at the maximum allowable temperature rise.)

T-701   removes maleic anhydride impurity overhead
reflux ratio = 0.27
33 trays, 40% efficient, 12 in tray spacing, 2 in weirs
diameter = 0.84 m, active area = 75% of total area
Q_c = –1230 MJ/hr
Q_r = 3220 MJ/hr

T-702   removes phthalic anhydride product overhead
reflux ratio = 2.43
86 trays, 50% efficient, 18 in tray spacing, 0.75 in weirs
diameter = 4.2 m, active area = 75% of total area
Q_c = –16,810 MJ/hr
Q_r = 15,890 MJ/hr

Vessels

V-701   Diameter = 0.50 m, Length = 1.50 m

V-702   Diameter = 1.25 m, Length = 3.75 m

Air Treatment

The organics in Stream 11 are removed in a scrubber, with 10,000 kg of water needed per kg of organic, at the current operating conditions. If the organic content of Stream 11 becomes more concentrated, then the amount of water needed increases by 100 kg per 0.001 mass fraction of organic. The water is sent to the on-site waste water treatment facility.

REPORT FORMAT

This report should be brief. Most of the report should be an executive summary, not to exceed five double-spaced, typed pages, that summarizes your diagnosis, recommendations, and rationale. Figures and tables may be included (do not count against page limit) in the executive summary. An appendix should be attached that includes items such as the requested calculations. These calculations should be easy to follow. In general, the written report should follow the guidelines given in Chapter 22.

The Design of a New, 100,000 Metric Tons per Year, Phthalic Anhydride Production Facility

BACKGROUND

The operation of Unit 700, phthalic anhydride facility, has been successfully scaled down by 50%. Over the long term, we are still considering changing to o-xylene as the raw material. The catalysis and reaction engineering group has finished preliminary research and is very optimistic about its new catalyst. They promise that it will be superior to other versions of o-xylene to phthalic anhydride catalysts in that most side products are minimized. At this point we are uncertain as to whether Unit 700 will be retrofitted to accommodate the new catalyst or whether we will build a new, grass roots facility at another site, nearer to an o-xylene producer.

In order for us to have enough information to make an informed decision, we need a preliminary process design for a grass roots facility to produce phthalic anhydride from what may be assumed to be pure o-xylene. Your job is to prepare a preliminary design for the new 100,000 metric tons/yr phthalic anhydride from o-xylene plant, and it must be completed within the next month. You may assume that the o-xylene feed is available at 100°C and 1.1 bar and that the required purity for phthalic and maleic anhydride products are 99.9 wt% and 95.0 wt% respectively.

OTHER INFORMATION

Concentrated organic waste streams may be burned instead of natural gas only if a fired heater is included in the design. Dilute organic waste streams must be sent to a treatment facility, with the appropriate operating cost charged. The capital cost of this facility may be assumed to be included in the grass roots cost of the new facility. Other pertinent information is given below and in Tables C.10, C.11, and C.12.

Table C.10   Information on Reaction Kinetics (Unit 700)

The catalysis and reaction engineering group has obtained the following kinetic information regarding the o-xylene to phthalic anhydride reaction. The reactions are

$$C_8H_{10} + 3O_2 \rightarrow C_8H_4O_3 + 3H_2O$$

o-xylene   *phthalic anhydride*

$$C_8H_{10} + \frac{15}{2}O_2 \rightarrow C_4H_2O_3 + 4H_2O + 4CO_2$$

maleic anhydride

In addition, the complete combustion reaction for each organic component occurs. The reaction network is given below. As you can see, a major advantage of our catalyst is that CO formation is essentially eliminated and no heavy impurity is made.

$$\text{maleic anhydride} \xrightarrow{\ 5\ } CO_2$$

$$\text{o-xylene} \xrightarrow{\ 1\ } \text{phthalic anhydride} \xrightarrow{\ 2\ } CO_2$$

with branch $4 \nearrow$ from o-xylene to maleic anhydride and branch $3 \searrow$ from o-xylene to CO_2.

The catalyst must operate between 300°C and 400°C, and at pressures between 1 atm and 3 atm. In this range, the kinetics are as follows: (partial pressures in atmospheres, r_i in kmol hr^{-1} (kg catalyst)$^{-1}$, $k_0 = 1$ kmol hr^{-1}(kg catalyst)$^{-1}$atm^{-2}, $R = 1.987$ cal/K mole, and T is in K)

$$r_1 = k_1 p_{xy} p_{O_2} \qquad \ln\frac{k_1}{k_0} = -\frac{27{,}000}{RT} + 19.837$$

$$r_2 = k_2 p_{pa} p_{O_2} \qquad \ln\frac{k_2}{k_0} = -\frac{31{,}000}{RT} + 20.86$$

$$r_3 = k_3 p_{xy} p_{O_2} \qquad \ln\frac{k_3}{k_0} = -\frac{28{,}600}{RT} + 18.97$$

$$r_4 = k_4 p_{xy} p_{O_2} \qquad \ln\frac{k_4}{k_0} = -\frac{27{,}900}{RT} + 19.23$$

$$r_5 = k_5 p_{ma} p_{O_2} \qquad \ln\frac{k_5}{k_0} = -\frac{30{,}400}{RT} + 20.47$$

The lower flammability limit of o-xylene in air is 1 mole%, and the upper flammability limit is 6 mole%. For safety reasons, it is necessary that process conditions not be within these limits. It is also necessary that the o-xylene content of the reactor never exceed 10 mole%, because if that

Table C.10 (*Continued*)

limit is exceeded, the catalyst no longer operates at the desired selectivity, the reaction could become oxygen starved, forming significant amounts of CO and other undesired by-products.

At this time, we are unsure as to whether a packed bed reactor (shell and tube type, modeled as a plug flow reactor) or a fluidized bed reactor (modeled as an isothermal plug flow reactor with 10% feed gas bypass) is the better choice. Please address this in your preliminary design. For the shell-and-tube packed bed, the catalyst would be in the tubes. We do believe, however, that tube diameters exceeding one inch in a shell and tube configuration would not allow for rapid enough heat removal, causing significant hot spots, and subsequent catalyst damage.

For a fluidized bed, the following data may be assumed:

> spherical catalyst particle, diameter range d_p = 300–600μm
> catalyst particle density ρ_{cat} = 1600 kg/m^3
> void fraction at minimum fluidization ε_{mf} = 0.50
> heat transfer coefficient from fluidized bed to tube wall h = 300 W/m^2°C
> reactor should operate between $2u_{mf}$ (bubbling) and $50u_{mf}$ (turbulent)
> the reactor has a rectangular cross section (width of sides = w & v)
> range of acceptable side ratios 0.2 < v/w < 5
> maximum value of v or w = 8 m
> maximum volume of bed displaced by tube banks is 40%
> the cost of the fluidized bed should be estimated as 5 times the cost of a vessel of the same volume

for u_{mf}, use the correlation of Wen & Yu:

$$\text{Re}_{mf} = \frac{d_p u_{mf} \rho_g}{\mu} = \left[(33.7)^2 + \frac{0.0408 d_p^3 \rho_g (\rho_s - \rho_g) g}{\mu^2} \right]^{0.5} - 33.7$$

where ρ_g is the density of the gas in the fluidized bed (at average conditions) and ρ_s is the solid catalyst particle density (called ρ_{cat} above).

For a shell and tube packed bed, the following data may be assumed:

> catalyst particle diameter d_p = 3 mm
> catalyst particle density ρ_{cat} = 1600 kg/m^3
> void fraction ε = 0.50
> heat transfer coefficient from packed bed to tube wall h = 60 W/m^2°C
> use standard tube sheet layouts as for a heat exchanger
> shell diameter is a function of heat transfer characteristics and frictional losses

It is anticipated that a heat transfer fluid will be used in a closed loop to remove the highly exothermic heat of reaction from either type of reactor. We anticipate that all surplus high-pressure steam made can be sold elsewhere in the plant. Since we have had many years of successful operation using the HiTec™ molten salt, we anticipate using it again for the new process. However, you should feel free to recommend an alternative if you can justify superior performance at the same cost, or equal performance at a lower cost.

Table C.11   Design of the Switch Condensers and Air Treatment Costs (Unit 700)

These are a complex set of condensers. There are three. Because of the low partial pressure of phthalic anhydride in the stream, it desublimates rather than condenses. Therefore, the process stream is cooled using a low-temperature oil in tubes to promote desublimation. Then, after solid is loaded on the heat transfer surface, gas flow to this condenser is stopped, and higher temperature oil is circulated in the tubes to melt the solid. There are three such devices, one operating in desublimation mode, one operating in melting mode, and one on standby. The net result is a liquid stream containing the condensables and a vapor stream containing some maleic anhydride, some phthalic anhydride, and all of the non-condensables.

These condensers will once again be designed and maintained under contract by CONDENSEX. They indicate that operation at any scale is possible as long as the pressure of the feed to the condensers is between 1.70 and 2.00 bar. You may assume that all light gases are neither condensed nor dissolved, and that 99% of the organics are desublimated and melted. Based on past experience, Condensex suggests that we may estimate the capital cost of these condensers as 15% of all other capital costs for the new process, and that the annual operating cost is three times the cost of an equivalent amount of cooling water needed to satisfy the cooling duty from the energy balance on the condenser unit modeled as a component separator.

The following cost of treating the waste (dirty) air stream leaving the switch condensers should be used in your cost estimates.

$$\text{Air Treatment Cost} = \$10^{-4}V_{tot}(0.5 + 1000x_{or})$$

where V_{tot} = the total volume of "dirty air" to be treated, m^3 and x_{or} = mole fraction of organics in "dirty air" stream.

Table C.12   Simulator Hints (Unit 700)

The following hints were developed for students using the CHEMCAD ™ simulator. These should also provide help to people using other simulator packages.

Use SRK (Soave-Redlich-Kwong) for the VLE and enthalpy options in the thermodynamic package for all the units in this process.

For heat exchangers with multiple zones, it is recommended that you simulate each zone with a separate heat exchanger. For the switch condensers, use a component separator, and then calculate the heat duty from the inlet and outlet streams.

When simulating a process using "fake" streams and equipment, it is imperative that the process flow diagram that you present not include any "fake" streams and equipment. It must represent the actual process.

ASSIGNMENT

Your assignment is to provide:

1. An optimized preliminary design of a plant to make phthalic anhydride from o-xylene using the new catalyst.
2. An economic evaluation giving the *NPV* (net present value), after-tax, of the new project. For your evaluation you should use the following economic information:
   After tax internal hurdle rate = 9%
   Depreciation = MACRS (6 year schedule given in Prob. 4.18)
   Marginal taxation rate of 35%
   Construction period of 2 years
   Project plant life = 10 years after start-up

Specifically, you are to prepare the following by . . . (4 weeks from now):

1. A written report detailing your design and profitability evaluation of the new process.
2. A clear, complete, labeled process flow diagram of your optimized process.
3. A clear stream flow table giving the *T*, *P*, total flowrate in kg/hr and kmol/hr, component flowrate in kmol/hr, and phase for each important process stream.
4. A list of new equipment to be purchased, including size, cost, and materials of construction.
5. An evaluation of the after-tax *NPV*, and the discounted cash flow rate of return on investment (*DCFROR*) for your recommended (optimized) process.
6. A legible, organized set of calculations justifying your recommendations, including any assumptions made.

REPORT FORMAT

This report should be in the "standard" design report format. It should include an abstract, results, discussion, conclusions, recommendations, and an appendix with calculations. The report format rules given in Chapter 22 should be followed.

5

Problems at the Cumene Production Facility, Unit 800

BACKGROUND

Cumene (isopropyl benzene) is produced by reacting propylene with benzene. During World War II, cumene was used as an octane enhancer for piston engine aircraft fuel. Presently, most of the worldwide supply of cumene is used as a raw material for phenol production. Typically, cumene is produced at the same facility that manufactures phenol.

The plant at which you are employed currently manufactures cumene in Unit 800 by a vapor-phase alkylation process which uses a phosphoric acid catalyst supported on kieselguhr. Plant capacity is on the order of 90,000 metric tons per year of 99 wt% purity cumene. Benzene and propylene feeds are brought in by tanker trucks and stored in tanks as a liquid.

CUMENE PRODUCTION REACTIONS

The reactions for cumene production from benzene and propylene are as follows:

$$C_3H_6 + C_6H_6 \rightarrow C_9H_{12}$$

propylene   benzene   cumene

$$C_3H_6 + C_9H_{12} \rightarrow C_{12}H_{18}$$

propylene   cumene   p-diisopropyl benzene

PROCESS DESCRIPTION

The PFD for the cumene production process, Unit 800, is given in Figure C.8. The reactants are fed from their respective storage tanks. After being pumped up to the required pressure (dictated by catalyst operating conditions), the reactants are mixed, vaporized, and heated in the fired heater to the temperature required by the catalyst. The shell-and-tube reactor converts the reactants to desired and undesired products as per the above reactions. The exothermic heat of reaction is removed by producing high-pressure steam from boiler feed water in the reactor. The stream leaving the reactor enters the flash unit, which consists of a heat exchanger and a flash drum. The flash unit is used to separate the C_3 impurities, which are used as fuel for a furnace in another on-site process. The liquid stream from the flash drum is sent to the first distillation column, which separates benzene for recycle. The second distillation column purifies cumene from the p-diisopropyl benzene impurity. Currently, the waste p-DIPB is used as fuel for a furnace. The pressure of both distillation columns is determined by the pressure in the flash drum, that is, there are no pressure reduction valves downstream of the flash drum.

V - 801	P-801A/B	P-802A/B	E-801	H-801	R-801	E-802	V-802
Benzne	Benzene	Propylene	Feed	Reactor	Reactor	Reactor	Phase
Feed Drum	Feed Pumps	Feed Pumps	Vaporizer	Feed Heater		Effluent Cooler	Separator

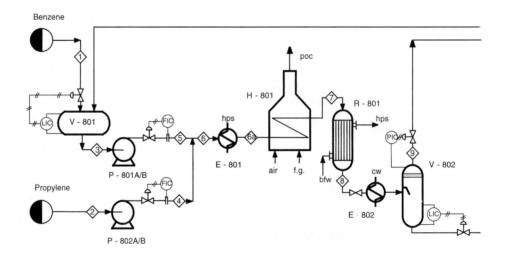

Figure C.8 Process Flow Diagram for the Production of Cumene Process (Unit 800)

RECENT PROBLEMS IN UNIT 800

Recently, Unit 800 has not been operating at standard conditions. We have recently switched suppliers of propylene; however, our contract guarantees that the new propylene feed be within specifications given in Table C.13.

Upon examining present operating conditions, we have made the following observations:

1. Production of cumene has dropped by about 8%, and the reflux in T-801 was increased by approximately 8% in order to maintain 99 wt% purity. The flows of benzene (Stream 1) and propylene (Stream 2) remained the same. Pressure in the storage tanks has not changed appreciably when measured at the same ambient temperature.

2. The amount of fuel gas being produced has increased significantly and is estimated to be 78% greater than before. Additionally, it has been observed that the pressure control valve on the fuel gas line (Stream 9) coming from V-801 is now fully open, while previously it was controlling the flow.

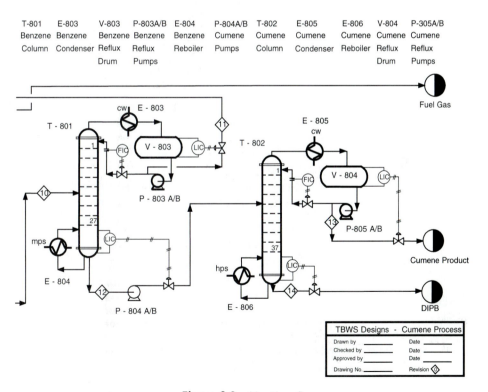

T-801	E-803	V-803	P-803A/B	E-804	P-804A/B	T-802	E-805	E-806	V-804	P-305A/B
Benzene	Benzene	Benzene	Benzene	Benzene	Cumene	Cumene	Cumene	Cumene	Cumene	Cumene
Column	Condenser	Reflux	Reflux	Reboiler	Pumps	Column	Condenser	Reboiler	Reflux	Reflux
		Drum	Pumps						Drum	Pumps

Figure C.8   (*Continued*)

Table C.13 Specifications of Products and Raw Materials

Raw Materials	
Benzene	>99.9 wt% purity
Propylene	≤ 5 wt% propane impurity
Product	
Cumene	>99 wt% purity

3. The benzene recycle Stream 11 has increased by about 5% and the temperature of Stream 3 into P-801 has increased by about 3°C.
4. Production of steam in the reactor has fallen by about 6%.
5. Catalyst in the reactor was changed 6 months ago, and previous operating history (over last 10 years) indicates that no significant drop in catalyst activity should have occurred over this time period.
6. DIPB production, Stream 14, has dropped by about 20%.

We are very concerned about this loss in production since we can currently sell all the material we produce.

Another problem that has arisen lately is the malfunction of the feed pumps. This problem arose during a very warm spell when the ambient temperature reached 110°F. A maintenance check showed that P-802 needed a new bearing, and this was taken care of, but P-801 seemed to be O.K. The ambient temperature has now returned to a mild 70°F, and both pumps seem to be working fine.

Currently, market conditions for cumene are very tight. We are in direct competition with some local companies that have recently built cumene plants. It appears that management is very concerned about our competitiveness since other producers in the area are beginning to undercut our prices. Management wants to find out if any significant savings in operating costs can be found for Unit 800.

OTHER INFORMATION

Other pertinent information is appended, including a flow table for the process streams at design conditions, that is, prior to the current operating problem, Table C.14; a utility summary table at design conditions, Table C.15; pump curves, Figures C.9 and C.10; a set of design calculations; and an equipment list, Table C.16.

Table C.14 Flow Summary Table for Cumene Production at Design Conditions, Unit 800 (Figure C.8)

Stream No.	1	2	3	4	5	6	6a	7
Temperature (°C)	25	25	41	28	44	41	214.0	350
Pressure (bar)	1.00	11.66	1.01	31.50	31.50	31.25	30.95	30.75
Vapor mole fraction	0	0	0	0	0.0	0.0	1.0	1.0
Flowrate (tonne/h)	8.19	4.64	16.37	4.64	16.37	21.01	21.01	21.01
Flowrates (kmol/h)								
Benzene	105.00	—	205.27	—	205.27	205.27	205.27	205.27
Propylene	—	105.00	2.89	105.00	2.89	107.89	107.89	107.89
Propane	—	5.27	2.79	5.27	2.79	8.06	8.06	8.06
Cumene	—	—	0.94	—	0.94	0.94	0.94	0.94
P-Diisopropyl Benzene	—	—	—	—	—	—	—	—
Total (kmol/h)	105.00	110.27	211.89	110.27	211.89	322.16	322.16	322.16

Stream No.	8	9	10	11	12	13	14
Temperature (°C)	350	90	90	57	179	178	222
Pressure (bar)	30.25	1.75	1.75	1.75	1.90	1.90	2.10
Vapor mole fraction	1.0	1.0	0.0	0.0	0.0	0.0	0.0
Flowrate (tonne/h)	21.01	1.19	19.82	8.18	11.64	11.08	0.56
Flowrates (kmol/h)							
Benzene	108.96	7.88	101.08	100.27	0.81	0.81	—
Propylene	8.86	5.97	2.89	2.89	—	—	—
Propane	8.06	5.27	2.79	2.79	—	—	—
Cumene	94.39	0.77	93.62	0.94	92.68	91.76	0.92
P-Diisopropyl Benzene	2.79	—	2.79	—	2.79	0.03	2.76
Total (kmol/h)	223.06	19.89	203.17	106.89	96.28	92.60	3.68

Table C.15   Flow Summary Table for Utility Streams in Unit 800

Stream Name	hps to E-801	condensate from E-801	mps to E-804	condensate from E-804	hps to E-806	condensate from E-806
Temperature (°C)	254	254	185.5	185.5	254	254
Pressure (bar)	42.37	42.37	11.35	11.35	42.37	42.37
Flowrate (tonne/h)	7.60	7.60	3.56	3.56	3.25	3.25

Stream Name	cw to E-802	cw from E-802	cw to E-803	cw from E-803	cw to E-805	cw from E-805
Temperature (°C)	30	45	30	45	30	45
Pressure (bar)	5.16	4.96	5.16	4.96	5.16	4.96
Flowrate (tonne/h)	261.30	261.30	85.88	85.88	87.50	87.50

Table C.16   Equipment Summary Table for Unit 800

Tanks (not shown on flowsheet)

TK-801   storage tank for benzene
There are two tanks, one feeding Stream 1 and one in a filling mode.
Each tank is 450 m^3.

TK-802   storage tank for propylene
There are two tanks, one feeding Stream 2 and one in a filling mode.
Each tank is 450 m^3.

Pumps (assume efficiency independent of flowrate)

P-801   centrifugal, 75% efficient, driver rated at 21.9 kW

P-802   centrifugal, 75% efficient, driver rated at 6.8 kW

P-803   centrifugal, 75% efficient, driver rated at 2.4 kW

P-804   centrifugal, 75% efficient, driver rated at 1.0 kW

P-805   centrifugal, 75% efficient, driver rated at 3.3 kW

Heat Exchangers (all one pass on each side, unless otherwise noted; h_i refers to tube side; tube wall resistance negligible, unless otherwise noted)

E-801   uses high-pressure steam, steam in shell, Q = 12,800 MJ/h
A = 20.8 m^2 in two zones
desubcooling zone: A = 13.5 m^2, U = 600 W/m^2°C, h_i = 667 W/m^2°C
vaporizing zone: A = 7.3 m^2, U = 1500 W/m^2°C, equal resistances on both sides

E-802   condenser for flash unit, process stream in shell, 1-2 configuration
Q = 16,400 MJ/h, A = 533 m^2

E-803   total condenser for T-201, condensing fluid in shell
A = 151 m^2, U = 450 W/m^2°C, all resistance on water side

E-804   reboiler for T-201
A = 405 m^2, U = 750 W/m^2°C, approximately equal resistances

E-805   total condenser for T-202, condensing fluid in shell
A = 24.0 m^2, U = 450 W/m^2°C, all resistance on water side

E-806   reboiler for T-202
A = 64.0 m^2, U = 750 W/m^2°C, approximately equal resistances

Fired Heater

H-801 Q = 6,380 MJ/h (heat actually added to fluid)
capacity 10,000 MJ/h of heat added to fluid
70% efficiency

Reactor

R-801   shell and tube packed bed with phosphoric acid catalyst supported on kieselguhr
boiler feed water in shell to produce high pressure steam
reactor volume = 6.50 m^3, heat exchange area = 342 m^2
234 tubes, 3.0 in (7.62 cm) ID, 6 m long
U = 65 W/m^2°C, all resistance on reactor side
heat removal required = 9,840 MJ/h

Table C.16   (*Continued*)

Distillation Columns

T-801 removes benzene impurity overhead for recycle
medium pressure steam used in reboiler
cooling water used in condenser, returned at maximum allowable temperature
reflux ratio = 0.44
27 trays, 50% efficient
24 in tray spacing, 3 in weirs
diameter = 1.13 m, active area = 75% of total area
Q_c = −5,390 MJ/h
Q_r = 7,100 MJ/h

T-802 removes cumene product overhead
high pressure steam used in reboiler
cooling water used in condenser, returned at maximum allowable temperature
reflux ratio = 0.63
37 trays, 50% efficient
24 in tray spacing, 3 in weirs
diameter = 1.26 m, active area = 75% of total area
Q_c = −5,490 MJ/h
Q_r = 5,520 MJ/h

Vessels

V-801 benzene feed drum   4.2 m length, 1.4 m diameter

V-802 flash drum   5.2 m height, 1 m diameter

V-803 T-801 reflux drum   4 m length, 1.6 m diameter

V-804 T-802 reflux drum   6.5 m length, 1.6 m diameter

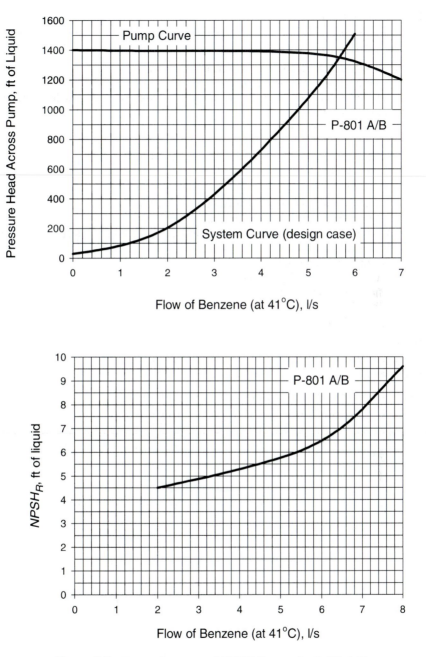

Figure C.9   Pump, System and NPSH Curves for P-801 A/B

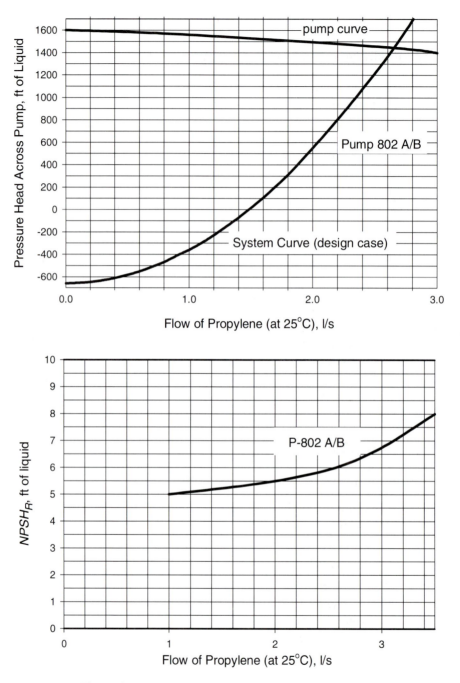

Figure C.10   Pump, System, and NSPH Curves for P-802 A/B

ASSIGNMENT

Specifically, you are to prepare the following by . . . (2 weeks from now):

1. A written report detailing your diagnosis of the operating problems with the plant along with your recommendations for solving these problems.
2. A list of new equipment to be purchased, if any, including size, cost, and materials of construction.
3. An analysis of any change in the annual operating cost created by your recommended modifications.
4. A legible, organized set of calculations justifying your recommendations, including any assumptions made.

REPORT FORMAT

This report should be brief. Most of the report should be an executive summary, not to exceed 5 double-spaced, typed pages, which summarizes your diagnosis, recommendations, and rationale. Figures and tables may be included (do not count against page limit) in the executive summary. An appendix should be attached which includes items such as the requested calculations. These calculations should be easy to follow. The guidelines given in Chapter 22 of this book should be followed.

PROCESS CALCULATIONS

Calculations for Fuel Gas Exit Line for V-802

Design flow of fuel gas = 1192 kg/h
Molecular weight of fuel gas = 59.9
Gas viscosity = 9.5×10^{-6} kg/m.s
Gas density = 1.18 (273) P / (293+90) (1.01) = 0.00876P kg/m^3 (P in bar)
Destination pressure (in burner in unit 900) = 1.25 bar
$\Delta P_{line} + \Delta P_{valve}$ = 1.75 –1.25 = 0.50 bar
ΔP_{valve} should be $\cong$ 0.30 bar and ΔP_{line} = 0.20 bar
length of line (Stream 9A) $\cong$ 125 m (equivalent length including fittings)
average pressure in line, P = (1.45 + 1.25) / 2 = 1.35 bar
density of gas in line = 0.00876 P = 1.18 kg/m^3
$\Delta P_{line} = 2f \rho u^2 L_e / d_{pipe}$

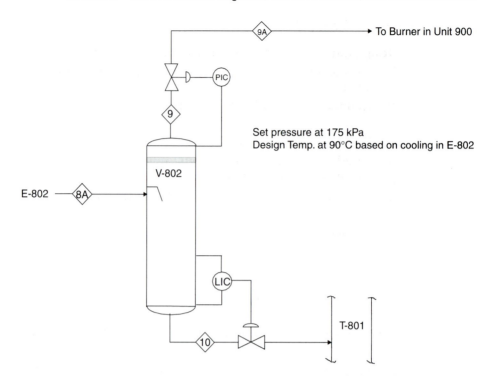

Set pressure at 175 kPa
Design Temp. at 90°C based on cooling in E-802

Look at 3″, 4″, and 6″ Sch 40 pipe:

d_{pipe} (nominal)	3″	4″	6″
d_{pipe}(inside) = d_i	0.0779 m	0.1022 m	0.1541 m
$u = 4Q/\pi d_i^2$	58.9 m/s	34.2 m/s	15.0 m/s
Re = $u\rho d_i/\mu$	5.69×10^5	4.34×10^5	2.88×10^5
e/d_i	0.00059	0.00045	0.0003
f (from friction factor diagram)	0.0046	0.0045	0.0042
$\Delta P_{line} = \dfrac{2f\rho u^2 L_{eq}}{d_i}$	0.603 bar	0.152 bar	0.018 bar

Choose 4″ sch 40 pipe.

$$\Delta P_{line} = 0.152 \text{ bar and } \Delta P_{valve} = 0.50 - 0.152 = 0.348 \text{ bar}$$

Calculations for P-801

Design Conditions (note that 1 kPa = 0.335 ft of water = 0.402 ft of benzene)

LAL (low alarm level) = 5 ft from ground and pump center line is 2 ft from ground

NOL (normal operating level) = 10 ft above ground level

NPSH Calculations (at LAL)

static head = 5 − 2 = 3 ft of benzene = h_{stat}

P_{supply} = 1.01 bar = 40.7 ft of benzene = h_{supply}

$\Delta P_{friction}$ (in supply line) = 1 psi = 2.8 ft of benzene = $h_{friction}$

Vapor Pressure of Stream 3

$$T = 30°C \quad P^* = 54 \text{ kPa}$$
$$T = 40°C \quad P^* = 72 \text{ kPa}$$
$$T = 50°C \quad P^* = 94 \text{ kPa}$$

Vapor Pressure of Stream 3 = 0.74 bar @ 41°C = 29.7 ft of benzene = h_{vp}

$NPSH_{available} = h_{supply} + h_{static} - h_{friction} - h_{vp} = 40.7 + 3 - 2.8 - 29.7 = 11.2$ ft

$NPSH_{required}$ (from pump curve) = 6.1 ft @ 5.5×10^{-3} m^3/s

∴ cavitation should not be a problem

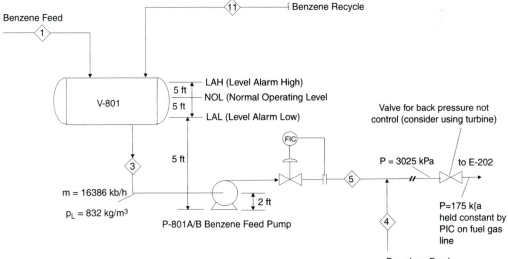

System Curve Calculations

$\Delta P_{friction}$ (discharge) = 31.50 − 1.75 = 29.75 bar = 1196 ft of benzene

$\Delta P_{friction}$ (suction) = 2.8 ft of benzene

$\Delta P_{discharge\text{-}tank}$ = 1.75 − 1.01 = 0.74 bar = 30 ft of benzene

ΔP_{static} = 0 (E-801 entrance @ 10 ft above ground level = NOL)

∴ Required head at design flow = 1199 + 30 + ΔP_{cv} = 1230 ft of benzene + ΔP_{cv}

From pump curve this gives us ΔP_{cv} = 135 ft = 3.36 bar (this is high but OK)

Calculations for P-802

Design Conditions (note that 1 kPa = 0.335 ft of water = 0.666 ft of propylene)

LAL (low alarm level) = 10 ft from ground and pump center line is 2 ft from ground

NOL (normal operating level) = 20 ft from ground

NPSH Calculations (at LAL)

static head = 10 − 2 = 8 ft of propylene = h_{stat}

$P_{supply} = P_{sat}$ (@25°C) = 11.66 bar = 777 ft of propylene = h_{supply}

$\Delta P_{friction}$ (in supply line) = 0.2 psi = 1 ft of propylene = $h_{friction}$ (3″ sch 40 pipe L_e = 20 ft)

Vapor Pressure of Stream 2 = 11.66 bar = 777 ft of propylene = h_{vp}

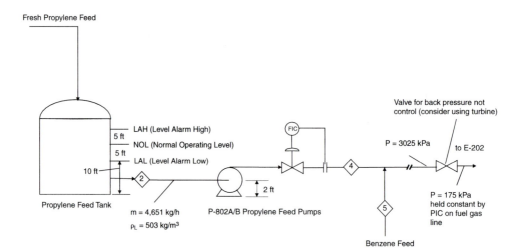

$NPSH_{available} = h_{supply} + h_{static} - h_{friction} - h_{vp} = 777 + 8 - 1 - 777 = 7$ ft of propylene
(@ propylene flowrate of 2.57×10^{-3} m^3/s)
$NPSH_{required}$ (from pump curve) = 6 ft

∴ cavitation should not be a problem (put note on P&ID to increase LAL to 12 ft to be safe)

System Curve Calculations

$\Delta P_{friction}$ (discharge) = 31.50 – 1.75 = 29.75 bar = 1981 ft of propylene
$\Delta P_{friction}$ (suction) = 1 ft of propylene
$\Delta P_{discharge\text{-}tank}$ = 1.75 – 11.66 = –9.91 bar = –660 ft of propylene
ΔP_{static} = –10 feet of propylene

∴ Required head at design flow = 1982 – 660 – 10 + ΔP_{cv} = 1312 ft of propylene + ΔP_{cv}
From pump curve this gives us ΔP_{cv} = 140 ft = 2.10 bar

6

Design of a New, 100,000 Metric Tons per Year Cumene Production Facility

BACKGROUND

In the opinion of our marketing research department, the demand for phenol-derived plasticizers is on the rise. Therefore, we are investigating the possibility of a new, grass roots phenol plant to handle the anticipated increase. Since phenol is made from cumene, a grass roots cumene plant would also be necessary. Given your experience in trouble-shooting our existing cumene process, we would like you to study the economics of a new cumene plant. Specifically, we would like a complete, preliminary design of a grass roots, 100,000 metric ton/yr cumene process using benzene and propylene.

We have a new, proprietary catalyst, and the kinetics are included in Table C.17. We would also like you to consider the economics of us continuing to use propylene with 5% propane impurity at $0.095/lb versus purer propylene feed. In preparing this preliminary design, you should assume that all steam made can be used elsewhere in the plant with the appropriate economic credit, that condensed steam can be returned as boiler feed water for the appropriate credit, and that fuel gas can be burned for credit at its LHV (lower heating value). Additional information is given in Table C.18.

ASSIGNMENT

Your assignment is to provide:

1. An optimized preliminary design of a plant to make cumene from benzene and propylene using the new catalyst.

Table C.17   Reaction Kinetics for Cumene Reactions (Unit 800)

The kinetics for the reactions are as follows:

$$C_3H_6 + C_6H_6 \xrightarrow{k_1} C_9H_{12}$$

propylene   benzene   cumene

$$r_1 = k_1 c_p c_b \quad \text{mole/g cat sec}$$

$$k_1 = 3.5 \times 10^4 \exp\left(\frac{-24.90}{RT}\right)$$

$$C_3H_6 + C_9H_{12} \xrightarrow{k_2} C_{12}H_{18}$$

propylene   cumene   p-diisopropyl benzene

$$r_2 = k_2 c_p c_c \quad \text{mole/g cat sec}$$

$$k_2 = 2.9 \times 10^6 \exp\left(\frac{-35.08}{RT}\right)$$

where the units of the activation energy are kcal/mol, the units of concentration are mol/l, and the temperature is in Kelvin.

For a shell-and-tube packed bed, the recommended configuration, the following data may be assumed:

catalyst particle diameter d_p = 3 mm

catalyst particle density ρ_{cat} = 1600 kg/m^3

void fraction ε = 0.50

heat transfer coefficient from packed bed to tube wall h = 60 W/m^2°C

use standard tube sheet layouts as for a heat exchanger

if tube diameter is larger than in tube sheet layouts, assume that tube cross sectional area is 1/3 of shell cross sectional area

2. An economic evaluation of your optimized process, using the following information:

After-tax internal hurdle rate = 9% p.a.

Depreciation = MACRS (6 year schedule see Prob. 4.18)

Marginal taxation rate = 35%

Construction period = 2 years

Project plant life = 10 years after start-up

Specifically, you are to prepare the following by . . . (4 weeks from now)

1. A written report detailing your design and profitability evaluation of the new process.

Table C.18 Additional Information (Unit 800)

Cost of Manufacture

In order to estimate the cost of manufacture (not including depreciation), COM_d, you should use the following equation:

$$COM_d = 0.180\,FCI + 2.73\,C_{OL} + 1.23\,(C_{UT} + C_{WT} + C_{RM}) \tag{3.2}$$

The current MACRS method for depreciation should be used in your calculations (see Problem 4.18).

Hints for Process Simulator

The CHEMCAD™ process simulator was used to generate the flow table given in Project 5. The hints given here are specifically directed to CHEMCAD™ users but should also be applicable for other process simulators.

Use SRK (Soave-Redlich-Kwong) thermodynamics package for VLE and Enthalpy calculations for all the equipment in the process.

For heat exchangers with multiple zones, it is recommended that you simulate each zone with a separate heat exchanger. Actual equipment may include several zones, so costing should be based on the actual equipment specifications.

For the reactor, you may use an isothermal reactor to estimate the volume of catalyst and heat exchange area. For more accurate results the temperature profile in the reactor should be modeled by completing a differential heat and material balance on the reactor.

For the distillation columns, you should use the shortcut method (SHOR) to get estimates for the rigorous distillation simulation (TOWR or SCDS). The shortcut method may be used until an optimum case is near. It is then expected that everyone will obtain a final design using rigorous simulation of the columns.

When simulating a process using "fake" streams and equipment, it is absolutely necessary that the process flow-diagram you present not include any "fake" streams and equipment. It must represent the actual process.

2. A clear, complete, labeled process flow diagram of your optimized process including all equipment and the location of all major control loops.

3. A clear stream flow table including T, P, total flowrate in kg/hr and kmol/hr, component flowrate in kmol/hr, and phase for each process stream.

4. A list of new equipment to be purchased, including size, cost, and materials of construction.

5. An evaluation of the annual operating cost for the plant.

6. An analysis of the after-tax NPV (10 years, 9%), and the discounted cash flow rate of return on investment ($DCFROR$) for your recommended process.

7. A legible, organized set of calculations justifying your recommendations, including any assumptions made.

REPORT FORMAT

This report should be in the "standard" design report format, consistent with the guidelines given in Chapter 22 of this text. It should include an abstract, results, discussion, conclusions, recommendations, and an appendix with calculations.

Index

LICENSE AGREEMENT AND LIMITED WARRANTY

READ THE FOLLOWING TERMS AND CONDITIONS CAREFULLY BEFORE OPENING THIS DISK PACKAGE. THIS LEGAL DOCUMENT IS AN AGREEMENT BETWEEN YOU AND PRENTICE-HALL, INC. (THE "COMPANY"). BY OPENING THIS SEALED DISK PACKAGE, YOU ARE AGREEING TO BE BOUND BY THESE TERMS AND CONDITIONS. IF YOU DO NOT AGREE WITH THESE TERMS AND CONDITIONS, DO NOT OPEN THE DISK PACKAGE. PROMPTLY RETURN THE UNOPENED DISK PACKAGE AND ALL ACCOMPANYING ITEMS TO THE PLACE YOU OBTAINED THEM FOR A FULL REFUND OF ANY SUMS YOU HAVE PAID.

1. **GRANT OF LICENSE:** In consideration of your payment of the license fee, which is part of the price you paid for this product, and your agreement to abide by the terms and conditions of this Agreement, the Company grants to you a nonexclusive right to use and display the copy of the enclosed software program (hereinafter the "SOFTWARE") on a single computer (i.e., with a single CPU) at a single location so long as you comply with the terms of this Agreement. The Company reserves all rights not expressly granted to you under this Agreement.

2. **OWNERSHIP OF SOFTWARE:** You own only the magnetic or physical media (the enclosed disks) on which the SOFTWARE is recorded or fixed, but the Company retains all the rights, title, and ownership to the SOFTWARE recorded on the original disk copy(ies) and all subsequent copies of the SOFTWARE, regardless of the form or media on which the original or other copies may exist. This license is not a sale of the original SOFTWARE or any copy to you.

3. **COPY RESTRICTIONS:** This SOFTWARE and the accompanying printed materials and user manual (the "Documentation") are the subject of copyright. You may not copy the Documentation or the SOFTWARE, except that you may make a single copy of the SOFTWARE for backup or archival purposes only. You may be held legally responsible for any copying or copyright infringement which is caused or encouraged by your failure to abide by the terms of this restriction.

4. **USE RESTRICTIONS:** You may not network the SOFTWARE or otherwise use it on more than one computer or computer terminal at the same time. You may physically transfer the SOFTWARE from one computer to another provided that the SOFTWARE is used on only one computer at a time. You may not distribute copies of the SOFTWARE or Documentation to others. You may not reverse engineer, disassemble, decompile, modify, adapt, translate, or create derivative works based on the SOFTWARE or the Documentation without the prior written consent of the Company.

5. **TRANSFER RESTRICTIONS:** The enclosed SOFTWARE is licensed only to you and may not be transferred to any one else without the prior written consent of the Company. Any unauthorized transfer of the SOFTWARE shall result in the immediate termination of this Agreement.

6. **TERMINATION:** This license is effective until terminated. This license will terminate automatically without notice from the Company and become null and void if you fail to comply with any provisions or limitations of this license. Upon termination, you shall destroy the Documentation and all copies of the SOFTWARE. All provisions of this Agreement as to warranties, limitation of liability, remedies or damages, and our ownership rights shall survive termination.

7. **MISCELLANEOUS:** This Agreement shall be construed in accordance with the laws of the United States of America and the State of New York and shall benefit the Company, its affiliates, and assignees.

8. **LIMITED WARRANTY AND DISCLAIMER OF WARRANTY:** The Company warrants that the SOFTWARE, when properly used in accordance with the Documentation, will operate in substantial conformity with the description of the SOFTWARE set forth in the Documentation. The Company does not warrant that the SOFTWARE will meet your requirements or that the operation of the SOFTWARE will be uninterrupted or error-free. The Company warrants that the media on which the

SOFTWARE is delivered shall be free from defects in materials and workmanship under normal use for a period of thirty (30) days from the date of your purchase. Your only remedy and the Company's only obligation under these limited warranties is, at the Company's option, return of the warranted item for a refund of any amounts paid by you or replacement of the item. Any replacement of SOFTWARE or media under the warranties shall not extend the original warranty period. The limited warranty set forth above shall not apply to any SOFTWARE which the Company determines in good faith has been subject to misuse, neglect, improper installation, repair, alteration, or damage by you. EXCEPT FOR THE EXPRESSED WARRANTIES SET FORTH ABOVE, THE COMPANY DISCLAIMS ALL WARRANTIES, EXPRESS OR IMPLIED, INCLUDING WITHOUT LIMITATION, THE IMPLIED WARRANTIES OF MERCHANTABILITY AND FITNESS FOR A PARTICULAR PURPOSE. EXCEPT FOR THE EXPRESS WARRANTY SET FORTH ABOVE, THE COMPANY DOES NOT WARRANT, GUARANTEE, OR MAKE ANY REPRESENTATION REGARDING THE USE OR THE RESULTS OF THE USE OF THE SOFTWARE IN TERMS OF ITS CORRECTNESS, ACCURACY, RELIABILITY, CURRENTNESS, OR OTHERWISE.

IN NO EVENT, SHALL THE COMPANY OR ITS EMPLOYEES, AGENTS, SUPPLIERS, OR CONTRACTORS BE LIABLE FOR ANY INCIDENTAL, INDIRECT, SPECIAL, OR CONSEQUENTIAL DAMAGES ARISING OUT OF OR IN CONNECTION WITH THE LICENSE GRANTED UNDER THIS AGREEMENT, OR FOR LOSS OF USE, LOSS OF DATA, LOSS OF INCOME OR PROFIT, OR OTHER LOSSES, SUSTAINED AS A RESULT OF INJURY TO ANY PERSON, OR LOSS OF OR DAMAGE TO PROPERTY, OR CLAIMS OF THIRD PARTIES, EVEN IF THE COMPANY OR AN AUTHORIZED REPRESENTATIVE OF THE COMPANY HAS BEEN ADVISED OF THE POSSIBILITY OF SUCH DAMAGES. IN NO EVENT SHALL LIABILITY OF THE COMPANY FOR DAMAGES WITH RESPECT TO THE SOFTWARE EXCEED THE AMOUNTS ACTUALLY PAID BY YOU, IF ANY, FOR THE SOFTWARE.

SOME JURISDICTIONS DO NOT ALLOW THE LIMITATION OF IMPLIED WARRANTIES OR LIABILITY FOR INCIDENTAL, INDIRECT, SPECIAL, OR CONSEQUENTIAL DAMAGES, SO THE ABOVE LIMITATIONS MAY NOT ALWAYS APPLY. THE WARRANTIES IN THIS AGREEMENT GIVE YOU SPECIFIC LEGAL RIGHTS AND YOU MAY ALSO HAVE OTHER RIGHTS WHICH VARY IN ACCORDANCE WITH LOCAL LAW.

ACKNOWLEDGMENT

YOU ACKNOWLEDGE THAT YOU HAVE READ THIS AGREEMENT, UNDERSTAND IT, AND AGREE TO BE BOUND BY ITS TERMS AND CONDITIONS. YOU ALSO AGREE THAT THIS AGREEMENT IS THE COMPLETE AND EXCLUSIVE STATEMENT OF THE AGREEMENT BETWEEN YOU AND THE COMPANY AND SUPERSEDES ALL PROPOSALS OR PRIOR AGREEMENTS, ORAL, OR WRITTEN, AND ANY OTHER COMMUNICATIONS BETWEEN YOU AND THE COMPANY OR ANY REPRESENTATIVE OF THE COMPANY RELATING TO THE SUBJECT MATTER OF THIS AGREEMENT.

Should you have any questions concerning this Agreement or if you wish to contact the Company for any reason, please contact in writing at the address below or call the at the telephone number provided.

PTR Customer Service
Prentice Hall PTR
One Lake Street
Upper Saddle River, New Jersey 07458

Telephone: 201-236-7105